THE COURSE OF EVOLUTION

McGRAW-HILL SERIES IN POPULATION BIOLOGY

CONSULTING EDITORS

PAUL R. EHRLICH, STANFORD UNIVERSITY
RICHARD W. HOLM, STANFORD UNIVERSITY

EHRLICH AND HOLM: THE PROCESS OF EVOLUTION
WATT: ECOLOGY AND RESOURCE MANAGEMENT
WELLER: THE COURSE OF EVOLUTION

THE COURSE OF EVOLUTION

J. MARVIN WELLER
PROFESSOR OF GEOPHYSICAL SCIENCES, EMERITUS
UNIVERSITY OF CHICAGO

RESEARCH ASSOCIATE OF THE
FIELD MUSEUM OF NATURAL HISTORY

ILLUSTRATED BY
HARRIET WELLER

McGRAW-HILL BOOK COMPANY
NEW YORK, ST. LOUIS, SAN FRANCISCO,
LONDON, SYDNEY, TORONTO,
MEXICO, PANAMA

THE COURSE OF EVOLUTION

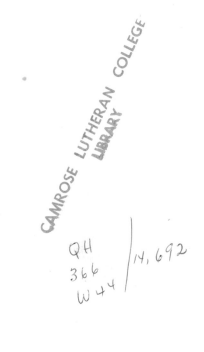

CONTENTS

PREFACE

Evolution is a subject that has inspired the writing of a multitude of books. They range in content all the way from the evidence for evolution, through its record as revealed by ancient or contemporary life and the history of evolutionary thought, to the philosophy growing out of evolutionary concepts, and so on. Authors have ranged from eminent specialists in several scientific fields to enthusiastic amateurs, facile perhaps in writing but lacking much essential knowledge. Many of these works are commendable and informative attempts to present certain facets of a very complex subject. Most, however, are imperfect or lacking in some way because evolution is too immense a subject to be handled well by any single person. No doubt this book is open to criticism of such a kind.

The historical aspects of evolution may be considered from two quite different points of view. The biologist looks into the past with the hope of explaining the modern world. The paleontologist, on the other hand, is interested primarily in the sequential development of life and regards the present as a moment in time no more important than any other moment. The biologist knows far more about modern organisms than the paleontologist can ever hope to learn about his fossils. The paleontologist, however, is continually discovering more about the order in which various types of life appeared and how they changed during the vast extent of geologic time. From the standpoint of established history, the importance of his knowledge needs no emphasis. Paleontologists, with their actual time perspective, certainly stand in an unrivaled position to view the record of life and interpret the course of evolution.

This book has been written from the paleontologist's point of view and in some respects it will not satisfy biologists principally interested in modern life. It is concerned mainly with organisms that are known to some reasonable extent as fossils. Considerable parts are devoted to animals and plants of kinds which are no longer living and which do not stand in an ancestral relationship to any modern life. Consideration of some other organisms, that may be very important in the world today, is omitted because paleontologists know little or nothing of them. Here and there much pure speculation is involved, but it is at least as well founded as much of the speculation made by biologists.

Books and articles on evolution written by paleontologists can mostly be assigned to one of two classes. First, there are those composed in a more or less popular vein that present a sort of moving-picture recounting in a rather superficial way how life has changed with passing time. Few of these are much concerned with the evolutionary influences that were at work or attempt to trace and explain the transformations that occurred in evolving lineages. In the second class are publications, generally devoted to some restricted group of organisms, which discuss in considerable detail the morphologic characteristics of members in presumably evolving lineages. Most of these publications are directed toward audiences of specialists and advanced students. They commonly include much technically worded descriptive matter, and the scientific names of many fossil organisms are presented as examples. Persons lacking a reasonably good prior understanding of the subject are more than likely to become seriously confused in their attempts to follow both facts and reasoning. Another common practice of specialists is to trace evolution without explanation by linking the names of plants or animals, evidently taking for granted that the reader is familiar with all the morphologic and other characters of these organisms. Few persons have this knowledge, and the uninformed reader is left totally without understanding of how evolution has been or may be traced.

This book is designed to occupy the middle ground by presenting information and ideas concerning what evolution has accomplished in the past in a reasonably comprehensive way and in a manner understandable to persons without a background of much specialized knowledge. There does not seem to be much point in attempting to draw up detailed genealogies because the fossil record is too incomplete and, commonly, too little understood to make this possible. Instead emphasis is given to the more or less obvious trends in evolution, showing the ways in which living things have slowly changed during more than half a billion years. These trends are not all clear, and different interpretations have been made concerning many of them. Nevertheless, most trends can be explained as resulting from continued adaptation to the ever-changing environments of the past. Finally some thoughts are presented with respect to evolution as seen in broad perspective, and a brief glance into the future is assayed.

Some elementary biologic knowledge by the reader is assumed, but he need not have had actual experience in either biologic or paleontologic studies. All descriptive details cannot be eliminated from the text, but an attempt has been made to present what is required in language shorn of unnecessary technicalities. Much morphologic information is provided in the illustrations, thus making its inclusion in the text superfluous. Some names, probably unfamiliar to the reader, must be used but these have been held to a reasonable minimum in the text. Anyone interested in greater morphologic detail and other names can find them by consulting the references listed in bibliographies appended to the chapters.

Little in this book is entirely original with me. Anyone with the time, inclination, access to a good library, and perseverance could locate most of the information for himself. As a paleontologist, I have gained my experience mainly from invertebrate fossils of late Paleozoic age. Much that I have written about such fossils and invertebrates of other ages has been drawn from lectures presented to my classes at the University of Chicago. For other knowledge, particularly for understanding of plants and vertebrates, I have been dependent on extensive reading and discussions with friends whose acquaintance with such organisms far exceeds my own. I owe much to them for both basic information and ideas.

Most sections of this book have been read by several friendly critics, whose comments and suggestions have been more than helpful. The fact that I do not list them here does not indicate failure to appreciate their interest and assistance. I have not, however, always heeded their probably excellent advice and, being nameless, they need not be embarrassed by my possibly faulty understanding or conclusions.

Many of the illustrations are copied with alterations from previously published figures, and I wish to express my gratitude for permission to adapt them for my use. All have been redrawn by my daughter, to whom I am greatly indebted for her skill and patience in carefully following my instructions. I have often been confused by figures in the literature whose orientations are not stated. To avoid similar confusion here the figures, with a few exceptions which are noted, are presented uniformly with the anterior directed either up or to the right. To facilitate comparison, the crown patterns of all teeth are shown with anterior to the right and the outside up. Illustrations representing evolutionary progress generally proceed from the bottom of the series upward, and those representing successive ontogenetic stages are arranged in the reverse order.

The illustrations are not numbered consecutively in the ordinary way. Instead each is identified by a number corresponding to the page where it appears. This makes finding them easy and removes the necessity of including page notations if figures are far distant from references to them in the text. Sources are given for all figures that are not original. If the source is not listed in the bibliography appended to the chapter, full reference is included in the caption. Tables are numbered similarly.

Certainly much more could be written about all the topics considered in this book. I hope that, in the space available, my presentation may be judged reasonably well balanced and logically arranged. I also hope that this book does modest justice to current knowledge and ideas about what I believe to be the most interesting and inspiring aspect of all nature.

J. MARVIN WELLER

LIST OF FIGURES

Figures are numbered to correspond with pages where they appear.

LIST OF TABLES

Tables are numbered to correspond with pages where they appear.

INTRODUCTION

THEORIES OF EVOLUTION – GENETICS
EVOLUTIONARY CONCEPTS – ORIGIN OF LIFE
BEGINNING OF FOSSIL RECORD – GEOLOGIC TIME

Life is the most interesting and intricate phenomenon of which mankind has any knowledge. All organisms in the modern world share unique chemical and physical attributes that are reflected in their construction, composition, and activities. Although organisms exhibit almost infinite diversity of form, an underlying similarity pervades them all. Every variety utilizes energy in the similar harmonious synthesis of innumerable complex but related interacting substances and in the reproduction of its kind. This is a strong suggestion that, if life can be explained, a single general explanation will account for it in every form.

THEORIES OF EVOLUTION
Any broad consideration of organisms living in the world today inevitably leads to speculation regarding their history and relations to one another. Both similarities and differences are so obvious that relationships graded in degree of closeness are indicated clearly. This has been reflected in systems of taxonomy ever since plants and animals first were classified. The concept of biologic evolution was required, however, to suggest that these similarities and differences are indications of various degrees of so-called blood relationship.

Early Ideas
The idea of some kind of evolution, or progressive change, in organisms is very old. Some of the ancient Greeks visualized nebulous transitions from the inorganic to the organic and from plants to animals in a series attaining more and more

perfection and leading up to man. Later in the Western world, complete faith in the scriptural story of creation effectively inhibited thinking about evolution. New and unorthodox views did not begin to form importantly until about the middle of the eighteenth century. In France the zoologist Buffon and in Sweden the botanist Linnaeus were among the first to suggest the vague possibility of some kind of evolution among organisms subsequent to the creation.

Lamarckism

The first completely integrated theory of biologic evolution was presented half a century later by the French zoologist Lamarck. This theory postulated that new characters were acquired by individual animals as the result of their needs or wants and that these characters, modified by use or disuse, were passed on by inheritance to their descendants. Lamarck visualized all animals as having diversified from a simple common beginning into very numerous increasingly more complex and highly organized lineages. His theory did not win much acceptance among contemporary naturalists, and it was ridiculed by some of the most prominent men of science at that time. Since then Lamarckism has had a history of alternate revivals and eclipses. In spite of the modern general opinion that the inheritance of acquired characters has been thoroughly discredited, distinct elements of this theory have persisted in the thinking of some biologists and paleontologists right up to the present day.

Darwinism

The announcement of Darwin's theory of evolution in the mid-nineteenth century marked the beginning of a new era in biologic science. Actually this theory embodied no fundamentally new idea, but it did combine older concepts in a fresh and convincing way and carried them to their logical conclusion. Darwin was particularly fortunate in his timing because the intellectual atmosphere in England was favorable for the consideration of a new materialistic theory of evolution, and he promptly gained the active support of several able and aggressive young biologists.

In its essentials, Darwin's theory had three bases: (1) The individuals making up a species vary more or less among themselves in form, physiology, and behavior. (2) The potential for reproduction of every species far exceeds that necessary to maintain its numbers. (3) Competition for food and living space results in the inexorable elimination of the weaker and less well-adapted individuals. This application of the principle of natural selection, or survival of the fittest, provided a fresh explanation for biologic evolution. It accounted more realistically for the gradual changes in the quality of a population that in time might result in its transformation into a new and different species. This theory did not, however, account for maintenance of the necessary variability upon which natural selection acts. Because of their complete ignorance of genetics, Darwin and most of his followers were constrained to accept the Lamarckian concept of the inheritance of acquired characters.

Modern Theories

The development of genetics at the beginning of the present century was an outcome of the rediscovery of Mendel's work, which had been published obscurely about 35 years before. Mendel had analyzed the results of breeding experiments with garden peas and demonstrated that their heredity is controlled by paired determinants which separate in the pollen and ovules and are recombined by fertilization. This was correlated with the work of cytologists who had observed the structure and changes in the nuclei of cells as these divide in living tissues. The rapid advancement of genetics soon provided a seemingly satisfactory explanation, which had been lacking previously, for the origin and persistence of variability within a species population. The evolutionary theory based on the combined principles of natural selection and genetics became known as Neo-Darwinism.

An understanding of the genetic relations of parents and offspring opened the way for the mathematical exploration of genetic action within whole populations. The results are almost entirely theoretical, but they indicate the possible or probable changes that may be effected in differently constituted populations under different circumstances. Insight into this aspect of genetics is particularly interesting and significant and aids in understanding the historical record of evolution as revealed by fossils. The modern refined theory that has emerged is commonly termed the synthetic theory of biologic evolution.

EVOLUTION AND PALEONTOLOGY

The statement commonly is made that evolution is a fact. Enough evidence has accumulated to convince almost all natural scientists that evolution has occurred. Such uniformity of opinion, however, does not constitute a proof, and evolution, like any other scientific theory, never can be absolutely proved. Evolution, therefore, is a theoretical process that seems to provide a uniquely satisfactory integration of a vast body of diverse biologic information which has not been explained convincingly in any other way. It is concerned mainly with what has happened in the past, and the contributions made by paleontology are particularly important. Presumably evolution is progressing at the present time and will continue in the future. The rate at which it proceeds, however, is so slow that in human experience very few of its results can be recognized with certainty. Also because of its slowness, the theory cannot be adequately tested experimentally. Observations and experiments with modern organisms, demonstrating genetic change and biologic transformation, are commonly cited as evidence of evolution, but they only establish that a mechanism exists which can explain what has been observed in nature.

Evolution is accepted with so few reservations by practically all biologists that it generally is discussed as though it surely has occurred. That is true of this book as well as of almost all modern scientific literature. Although biologists are in remarkable agreement concerning the reality of evolution, there are many different shades of opinion as to exactly how it has operated to accomplish what it has.

The older concepts of evolution were based on the observation of modern life alone. At Lamarck's time, paleontologic knowledge was so scanty that no appeal was made to it for support. Darwin marshaled the facts available to the geologists and paleontologists of his day. He argued that what was known of fossils was in harmony with his theory, but he felt the necessity to explain why fossils did not demonstrate evolution much more clearly. This situation now has changed, and paleontology provides unsurpassed evidence in favor of evolution. Although the fossil record is very incomplete, many examples are known that seem to illustrate long and continuous progressive evolutionary changes. Comparisons of modern organisms may indicate many evolutionary relationships, but only the historical record provided by fossils can reveal any of the details of the actual trends of evolution. The main part of this book is devoted to a presentation of what is known and thought about this evolutionary record and the conclusions that are drawn from it in an attempt to trace in a broad way the course of biologic evolution.

GENETIC PRINCIPLES

Genetics is the science of heredity. It is important in a consideration of evolution because it explains both (1) the uniformity and stability which are observed to characterize populations of varied individual organisms, and (2) the origin of changes which alter populations and may result in the emergence of new species and higher taxonomic groups. Although modern research has revealed the great complexities of many genetic processes, the basic principles of genetics are fairly simple.

Chromosomes
Chromosomes are tiny bodies that become most clearly visible in the nuclei of cells shortly before their division in living tissues. The number of chromosomes in a nucleus varies greatly between different organisms, but generally it is constant in

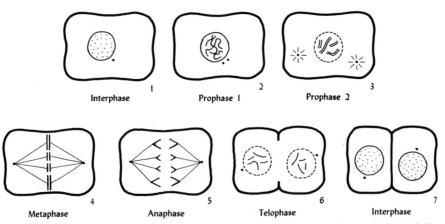

| Interphase | Prophase I | Prophase 2 |
| 1 | 2 | 3 |

| Metaphase | Anaphase | Telophase | Interphase |
| 4 | 5 | 6 | 7 |

Fig. 4 Diagram showing the successive stages in mitosis, or the division of ordinary cells, in which the chromosomes double and identical lots are segregated in each of the daughter cells.

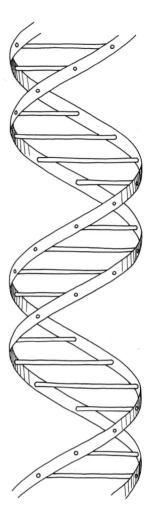

Fig. 5 Diagrammatic model of part of a long twisted DNA (deoxyribonucleic acid) molecule in a chromosome. It consists of two cross-bonded chains. (*Modified after Beadle and Beadle, 1966, "The Language of Life," fig. 92, p. 171.*)

most of the cells of any species. Ordinary body cells possess two almost identical sets of chromosomes whose corresponding members commonly can be identified by their similarity in size and shape. When cells divide, the chromosomes duplicate themselves by splitting lengthwise so that a complete lot is provided for each new cell (Fig. 4). Thus almost every cell in an organism possesses two sets of chromosomes identical to those present in every other cell.

The principal genetic controls are contained in chromosomes, but others occur outside the nucleus in the cell's cytoplasm or in some of the organelles it contains. Possibly these controls exist in structures comparable to chromosomes, but they are not so evident and their action is not completely understood.

Genes

A chromosome is a protein complex containing one or more double-stranded and twisted chain-like molecules of deoxyribonucleic acid, commonly termed DNA (Fig. 5). These molecules consist of successive segments known as genes which differ from one another by the order of only four related chemical structures arranged in the cross bonds between the strands (Fig. 6). A single chromosome may contain a thousand or more genes.

The genes possessed by any organism were derived from its parents and they control its heredity. They operate by synthesizing substances that pass out of the nucleus into the surrounding part of a cell. There these substances determine the cell's development and activity. Most physical and physiologic characters of organisms are governed by several to many genes, and a single gene is likely to affect several different characters.

The action of genes is influenced by their environments, which are (1) intrachromosomal, (2) intracellular, and (3) extracellular. The mutual relations of genes within a chromosome are important because they modify one another's actions and

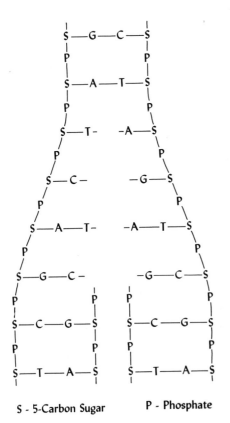

S - 5-Carbon Sugar   P - Phosphate

Purines   Pyrimidines

A - Adenine   T - Thymine

G - Guanine   C - Cytosine

Fig. 6 Diagram showing the essential chemical structure of DNA and probable method of reproduction. Structural requirements restrict the cross bonds to two types: A is always joined to T, and C is always joined to G. Both kinds of bonds can be oriented in two different ways. The cross bonds between the sugar-phosphate chains are weak; they break in the center (middle part of diagram), and each half of the molecule synthesizes its complement (lower part of diagram). Proteins may be attached to the phosphates. The hereditary code is provided by the order in which the four kinds of cross bonds occur. Part of the molecule including three or more cross bonds probably acts as a gene. Gene mutations result from accidental changes in the order of the cross bonds.

identical genes arranged in different orders may have different effects. The environment provided by a cell alters as tissues differentiate and genetic influences change. Finally, the external environment acts upon an organism as a whole and may result in modifying its form and physiologic traits. Therefore, organisms that are identical genetically may develop somewhat differently. The differences ordinarily are slight, but in some organisms and under certain circumstances they may be conspicuous and important.

Corresponding genes in corresponding chromosomes of the two sets possessed by cells may be identical or homozygous and act harmoniously together. On the other hand, they may be slightly different or heterozygous, in which case they are more or less in conflict. In a common situation of this kind, a gene in one chromosome seems to suppress completely the influence of the corresponding but different gene of the other chromosome. It is termed a dominant and the ineffective gene is a recessive. The homozygous or heterozygous condition of an organism ordinarily cannot be determined without analyzing the nature of its offspring.

The Function of Sex

Sex is not necessary for reproduction. Many organisms, both plants and animals, reproduce asexually, and evidently sexual combination serves another function. When the processes are considered by which sperms and eggs are formed and eggs are fertilized, the sexual method of reproduction is seen to preserve and increase within a population the variability without which natural selection could not operate.

Sex cells ordinarily are formed by two consecutive divisions in a process termed meiosis (Fig. 8). Cells of the sex organs have two sets of chromosomes like other body cells, a condition known as being diploid. When these cells mature, their chromosomes pair with the similar chromosomes of the two sets, which are likely to be heterozygous, lying closely beside each other. The chromosomes split, but the members of the newly formed pairs do not separate, as they do in ordinary cell division. Instead, the new pairs are drawn apart in opposite directions and are enclosed in the nuclei of two new cells. These new cells possess two sets of chromosomes, but these were derived from only half the chromosomes present in the former cell. In this process the original chromosome sets provided by the parents do not act as units and the direction in which any new pair of chromosomes may move depends on chance alone. Consequently the number of different combinations in which chromosomes derived from the two original sets can be segregated in the new cells increases exponentially with the number of chromosomes in a set.

This shuffling of chromosomes in many different ways is only half the story. While the doubled chromosomes are paired, they seem to break at corresponding places and exchange their severed parts in a process termed crossing-over (Fig. 9). Thus the chromosomes of the cells resulting from the first division of meiosis are not necessarily constituted similarly to the chromosomes of the former cell. Crossing-over is a common process and may take place in every pair of chromosomes. Some chromosomes may cross over at two or more positions. Crossing-over does not change the gross structure of a chromosome in any way because it still consists of the same number and sequence of genes as it did before. Genes, however, have changed places between the chromosomes. If the original chromosomes were heterozygous for some genes, the combinations of genes in the derived chromosomes may be different.

At the second cell division of meiosis the chromosomes do not double but those of each new pair are drawn apart in opposite directions, the direction in every case depending on chance alone, and two new nuclei appear in two new cells. These are the sex cells. Each of them has only one set of chromosomes, half the number of the original cell. This is a condition known as being haploid.

Fertilization is accomplished by the chromosomes of a sperm joining those in the nucleus of an egg. Thus the diploid condition is reestablished. These complicated processes characteristic of sexual reproduction mix up the chromosomes and genes so thoroughly that every individual is almost certain to be different from every other.

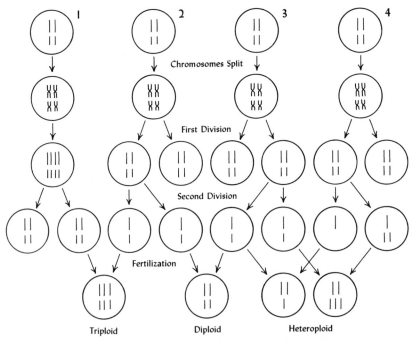

Fig. 8 Diagram illustrating the double cell division of meiosis, which produces sex cells, and subsequent fertilization. Meiosis proceeds normally in 2 and 3 and produces haploid sex cells. In 1, the first division of meiosis fails and the sex cells are diploid. In 4, the second division is irregular and the sex cells do not have complete sets of chromosomes. The union of two normal haploid sex cells produces a normal diploid individual. The union of a haploid and a diploid sex cell produces a triploid (polyploid) individual. The union of a haploid sex cell and one having an incomplete set of chromosomes produces a heteroploid individual.

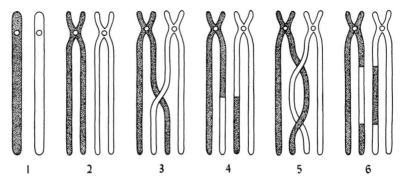

| l | 2 | 3 | 4 | 5 | 6 |

Fig. 9 Crossing-over may occur just before the first meiotic division. Two corresponding chromosomes (1) pair and split lengthwise (2), forming a tetrad. The two newly formed pairs are attached
at their centromeres and do not separate. The chromosomes cross over (3) and exchange parts
(4). Double or multiple crossing-over (5 and 6) also is possible. The two parts of the re-formed
chromosomes separate to complete the first meiotic division.

Mutations

Any change in genetic structure not resulting from the shuffling of chromosomes
or crossing-over is a mutation. Two kinds require distinction: (1) changes affecting
chromosomes without the alteration of any genes, and (2) changes in the genes
themselves. Mutations are genetic accidents, and they are relatively rare. The
frequency of any particular mutation may be in the neighborhood of 1 in 100,000
or even more, but so many possibilities exist that perhaps one individual organism
in ten is a mutant in some respect.

Chromosome mutations result from either (1) the breaking of chromosomes
and the rejoining of their parts in irregular ways different from the orderly exchange
of similar parts as in crossing-over (Figs. 10; 10a), or (2) the irregular segregation
of chromosomes in sex cells so that these cells are provided with more or less than
one complete set (Fig. 8-4). Mutations of these types are likely to result in irregular
pairing of chromosomes in the process of sperm and egg production. If sex cells
are formed they are likely to be so imperfect in their genetic equipment that reproduction is restricted or prevented. If reproduction does occur, however, new races
may be developed or genetic barriers may be set up that give rise to new species
or perhaps more rarely to new genera. Chromosome mutations introduce nothing
that is really new into the genetic system, and they cannot be an important factor
in sustained evolution.

Gene mutations are believed to be the consequences of local chemical changes
in the linking complexes of chromosomes. A variety of different mutations is possible with respect to any gene, and subsequent reverse mutations may reestablish
previous conditions. A gene mutation may have conspicuous results, but it is more
likely to effect only minor changes in form or physiology without impairing fertility
or setting up genetic barriers immediately. A succession of gene mutations indefi-

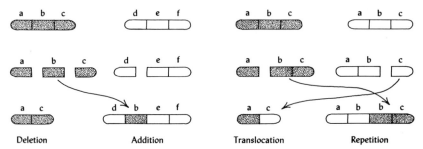

| Deletion | Addition | Translocation | Repetition |

Fig. 10 Diagrams illustrating mutation by the irregular breaking of chromosomes and the rejoining of their fragments. On the left, two dissimilar chromosomes break and a fragment of one becomes inserted in the other. This results in deletion in one chromosome and addition in the other. On the right, two similar chromosomes break at different places and exchange parts. This results in translocation (and deletion) in one chromosome and repetition (and translocation) in the other. Chromosomes seem to break and rejoin in still more complex ways.

nitely continued, however, can completely alter the genetic constitution of a lineage. Gene mutations introduce entirely new genetic structures and provide the mechanism that makes sustained evolution possible. (See Table 11.)

Many mutations are recessive, and therefore their effects may not be immediately apparent. They are likely to accumulate in heterozygous individuals in considerable numbers and great variety. The changes resulting from mutations vary greatly, but generally they are disadvantageous to individuals. This is not surprising, because all plants and animals are highly organized complex mechanisms closely adapted to particular environments and ways of life, and random changes are not likely to improve them. Some mutations are disastrous and prevent the normal development of an organism or result in its early death. A very few, however, may prove to be advantageous.

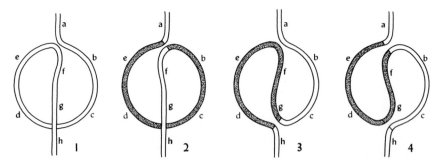

Fig. 10a Diagrams illustrating how a chromosome (1) may break and re-form (2 and 3) so that a segment is reversed or inverted. In 2, the normal order of parts (or genes) has been changed to a-e-d-c-b-f-g-h, and in 3 the order becomes a-b-c-g-f-e-d-h. Double inversion also is possible, as in 4, where the order becomes a-e-d-g-f-b-c-h.

Table 11   Sources of genetic variability

Genes:
  Biochemical mutations
  Translocation of genes
  Deletion of genes
  Addition or duplication of genes

Gene patterns in chromosomes:
  Altered by crossing-over
  Altered by inversion

Changed chromosome systems:
  By duplication, etc., of chromosome sets: polyploidy
  By gain or loss of individual chromosomes: hetero-
  ploidy

Segregation of chromosomes in meiosis:
  Changed combinations of heterozygous chromosomes

**Any genetic change to be viable must be internally com-
patible and its expression must meet external environmental requirements.**

POPULATION GENETICS

Individual organisms do not evolve; each is fixed by its inheritance. Evolution is made possible by mutations and rearrangements in the chromosomes that change the inheritance passed on by one generation to the next. Sustained evolution is accomplished by the accumulation of mutations that spread throughout a population whose character is thereby altered. Populations are the units that evolve. Therefore, a consideration of the genetics of whole populations is essential for an understanding of evolution.

Factors determining the genetic constitution of a population and the changes that may direct its evolution are complex. The most important are (1) mutation and gene recombination, (2) natural selection, which is the process of differential survival and reproduction of individuals of diverse genetic types, (3) population size, (4) population structure, and (5) breeding patterns. Altogether these provide so many variables that the relations are difficult to comprehend. Separately and in different simple combinations they can be considered mathematically, and the effects in an infinite population can be calculated. Real populations, however, may be very large, but they are not infinite. Probabilities with respect to them can be determined by the methods of statistics.

Mutation Pressure

Mutations presumably occur with a certain statistical regularity. They may be considered to set up a kind of pressure, forcing a population in the direction of genetic change. The intensity of this pressure with respect to any mutation is, of course, proportional to how often a mutation appears within a population.

The pressure exerted by a particular mutation if unopposed will push a population toward a completely homozygous new condition (Fig. 12, upper curve). In the early stages of such a process, the pressure is relatively great and genetic change is relatively rapid because the mutation can appear in many individuals (Fig. 13). As the mutation accumulates within a population, however, the number of non-mutant individuals becomes progressively smaller, new mutant individuals appear less frequently, and the pressure diminishes accordingly. If the pressure were unopposed, a completely homozygous condition would be approached but never attained in an infinite population. Any finite population, however, would sooner or later become completely homozygous.

Every mutation probably is opposed by a reverse mutation which exerts a contrary pressure and this, acting against the other, tends to prevent the attainment of a homozygous state (Fig. 12, lower curves). The rates of the two mutations are likely to be different, and genetic equilibrium would be reached when these rates, one declining and the other increasing, become exactly equal.

Mutation rates are not known with desirable accuracy, but by assuming different values, the results of their effects upon a population can be calculated. Thus

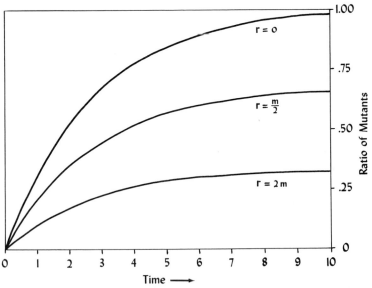

Fig. 12 Graph showing progressive increase in a particular mutation in a very large population. The lower scale is relative because it depends on mutation rate and length of generations. The upper curve shows increase without reverse mutations. It approaches 100 percent. The middle curve shows the result if a reverse mutation rate r is half the direct mutation rate m. It approaches 66⅔ percent. If the rates were equal, the number of mutants would approach 50 percent. Reverse mutation rates greater than the direct rates do not completely suppress increase in mutant individuals but result in approach to equilibrium at less than 50 percent. Thus the lowest curve shows approach to 33⅓ percent if the reverse mutation rate is twice the direct rate.

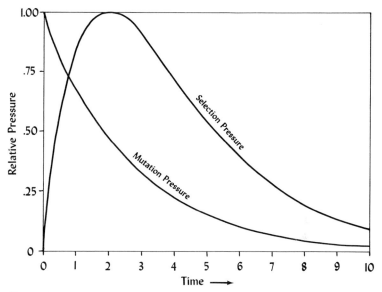

Fig. 13   Graph showing progressive changes in mutation and selection pressures in a very large population. The amplitudes of these curves bear no necessary relations to each other.

it is possible to reach an understanding of the influence of mutations in genetic change and the attainment of possible equilibrium.

Selection Pressure

Every mutation presumably is advantageous or disadvantageous to individuals in some degree, even though its influence may be very slight. Natural selection acts upon the differences induced by mutations and gene recombinations and favors or disfavors either or both the survival and reproduction of differently constituted individuals. Therefore, selection also can be considered to generate a pressure that directs genetic change by the preservation or elimination of mutant individuals.

If conditions are stable, selection pressure works in only one direction with respect to any mutation and its reverse mutation. Its strength is related to its advantage and the proportion of mutant individuals in a population (Fig. 14). The effectiveness of selection in changing the quality of a population is greatest at intermediate proportions of mutant to nonmutant individuals but becomes practically negligible if the proportion is very low or very high (Fig. 13).

Selection pressure cannot be adequately quantified, but again calculations involving assumed values indicate the relative influence of natural selection under different circumstances.

Complications

Mutations and natural selection exert pressures that act simultaneously on a popu-

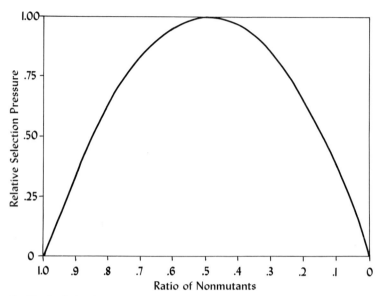

Fig. 14   Graph showing relations of selection pressure to the ratio of nonmutant individuals in a very large population.

lation either in the same or in opposite directions, and consequently they require consideration in combination with each other. Furthermore, selection may favor either dominant or recessive characters or both in different degrees, one or the other homozygous condition, or the heterozygous state. Finally certain mutations known as lethals result in the death of individuals. Some are deadly when heterozygous; they cannot become established in any population. Others have this effect only when homozygous; they can accumulate and persist indefinitely in a population in heterozygous individuals.

Mathematical formulas have been devised which take account of these and some other possibilities. They show, for infinite populations, the changes in genetic constitution that would occur between one generation and the next and also whether a homozygous condition or an equilibrium would eventually be approached. Most of these equations, however, are concerned only with a particular mutation and its reverse mutation. In nature many mutations appear and accumulate within the same population and natural selection acts upon them all at once. Two or more mutations may variously support or nullify each other as far as selection is concerned. It is even possible that two mutations, each individually disadvantageous, may be advantageous if present in the same individual. Complications of these kinds introduce so many variables that calculations have not been devised to accommodate them all.

Population Size

If there were no mutation or selection pressures, the proportions of a pair of corresponding but different genes within a population would fluctuate randomly on both sides of an average ratio. In an infinitely large population neither gene would become fixed, i.e., it would never completely replace the other, so that all individuals never would become homozygous. In any finite population, however, one gene sooner or later would become fixed. Calculations demonstrate that the average time required for fixation would decrease progressively in smaller and smaller populations.

Selection pressure opposes most random variation. In small populations the effect of selection pressure ordinarily is very slight, in intermediate-sized populations it becomes important, and in very large ones random variation may be almost entirely suppressed. Mutation pressure likewise commonly is an opposing force, and its influence on random variation is much the same. These relations are very important with respect to evolution. They indicate that (1) large populations are likely to evolve only slowly and adaptively, (2) moderate-sized populations can evolve considerably more rapidly, also probably in an adaptive manner, and (3) chance plays a much greater part in small populations, which can change very rapidly in either adaptive or nonadaptive ways. The last is likely to lead to quick extinction, but, rarely, new advantageous qualities may become fixed.

In population genetics, population size generally is considered to be the number of breeding individuals, which may be only a fraction of the total population. If, however, the sexes are disproportionally represented, the effective size is much nearer to the number of the less-abundant sex than to the total of breeding individuals. Likewise if the population size fluctuates considerably in a cyclic manner, the effective size is only slightly larger than the number of breeding individuals when the actual population is at a minimum.

Population Structure

Populations may be distributed nearly uniformly throughout extensive regions, but they are much more likely to be spread in varying concentrations from place to place, and variously sized subpopulations may be more or less separated from one another. Small isolated subpopulations might be expected to drift apart genetically rather rapidly. Calculations show, however, that the exchange of only one breeding individual in every other generation probably would prevent the fixation of genes in different ways in neighboring subpopulations. This holds for subpopulations of any size.

Breeding Patterns

Most studies of population genetics have been made with reference to randomly breeding, diploid, bisexual organisms. Many other breeding patterns are known, however, each of which introduces some variations or complications of results. Mathematical investigations could be made if the breeding and genetic systems of all organisms were understood, but much of the necessary knowledge has not been gained or the calculations have not been made. One special situation is provided by self-fertilization, which reduces heterozygosity one-half in every generation and leads rapidly to the development of pure homogygous races.

Other important variations in breeding patterns involve several types of asexual reproduction and selective mating. There is little reason to believe, however, that their study would lead to significant modifications in the more fundamental aspects of population genetics.

EVOLUTIONARY AND RELATED CONCEPTS

All evolution, since the first spark of life appeared, possibly has had a genetic basis, although the genetics of the earliest organisms must have been much simpler than any known today. Most of it can be explained by (1) the variability among individuals which make up populations, (2) natural selection, and (3) the adaptations which resulted. Nevertheless, a variety of popular evolutionary concepts have been formulated, some of them without reference to genetic principles. Several of these and also some others deserve brief notice.

Recapitulation

The so-called biogenetic law formulated in the familiar expression "ontogeny recapitulates phylogeny" has played a very important part in evolutionary thinking. The idea that each individual animal in its growth, and particularly in its embryonic development, passes through an orderly succession of morphologic stages similar to the adult evolutionary stages of its ancestral lineage, and that evolutionary advancement simply added new final stages to what had been accomplished previously, was widely current in the late nineteenth and early twentieth centuries. This concept of recapitulation has been vigorously challenged, however, and the contention has been made that embryonic stages resemble only ancestral embryonic or youthful stages.

Although it is now recognized that the biogenetic "law" is by no means of universal application, the complete rejection of this principle is not justified. Several seemingly well-authenticated examples are known to paleontologists in which post-embryonic growth is believed to recapitulate mature ancestral stages (see Figs. 101a; 229). Embryonic similarities, whether or not they ever characterized adults, are still relied upon by zoologists to establish the probable evolutionary relationships among very diverse groups of modern animals. The recapitulation principle, however, seems to have no clear application among plants.

Several other concepts are related to recapitulation. Accelerated development is considered to have pushed ancestral stages further and further back in the youthful development of individuals until the earliest stages were eliminated from the record of their growth. In a somewhat similar way, intermediate stages may have been telescoped until they disappeared. These modifications of recapitulation seem to be substantiated in some instances.

In a contrary way, arrested development has been considered to have had an opposite result. The persistence of a youthful stage or the precocious attainment of sexual maturity brought continued morphologic development to a halt so that adults resemble immature stages of their ancestors. Some possible examples in the fossil record have been recognized. Appeal, however, to such a process to explain some theoretical evolutionary transformations, such as the supposed persistence of an embryonic stage as an adult, is most highly speculative, to say the least.

Another concept is the exact reverse of recapitulation. This is that new evolutionary characters, introduced at a very early developmental stage, persisted longer and longer in descendant generations until at last they became characteristic of adults. Some rather convincing examples have been pointed out among the fossils (Figs. 222; 223).

Recapitulation, its modifications, and its opposite all seem to have some validity. Tests of these concepts can be made and decisions reached concerning the actual course of evolution in different groups of organisms only by the detailed study of the fossil record. Many opportunities exist for misinterpretation.

Higher and Lower Organisms

The concept of higher and lower organisms, or evolutionarily more advanced or more primitive plants and animals, is very common. Some useful comparisons can be made in this way, on a relative evolutionary basis, mainly among organisms that are not too distantly related to one another. Furthermore, certain groups of animals such as the sponges and coelenterates are validly considered to be lower in an evolutionary sense than molluscs, arthropods, or chordates because of their simpler organization. Plants, many of which can be graded according to the details of their reproduction, are better subjects for this kind of classification than are animals as a whole.

It is not possible to arrange all organisms in an evolutionary hierarchy. When this is attempted, the self-centered human viewpoint places mankind at the summit of all creation and makes comparisons on this basis. Human beings actually are no more highly evolved than most other groups of modern mammals. Furthermore, mammals as a whole cannot realistically be considered more evolutionarily advanced than, for example, cephalopods or insects. Probably the greatest inconsistency of all is reflected by the view that all protozoans should be classed as the lowest and most primitive of animals. Many of them almost certainly are as highly evolved in their own way as any other organisms.

The Tree of Life

Exclusively asexually reproducing organisms must receive their inheritance unchanged from a single parent unless this continuity is broken by mutation. Each individual is a simple branch in a potentially ever-expanding but unchanging sequence of populations. A mutation, however, may initiate a new type of organism. If the mutant and parental types remain in contact and in competition with each other, one is likely to be eliminated promptly by natural selection. Ordinarily it is the mutant which disappears, but when this is not so, an evolutionary step has been accomplished. If the two types are separated in some way or cease to be competitors, both may persist, and evolutionary branching will have resulted.

The earliest and most primitive organisms possibly reproduced only in this way. Their evolution, if it were known, could be represented as a simple branching "tree of life." Most evolution is customarily thought of in this way (Fig. 636), but for any other kind of reproduction such a representation is a generalization that ignores details.

Reproduction by sexual combination or any other method that accomplishes exchange or mixing of the genetic heritage introduces complications of this pattern. Inheritance passes to an individual from two immediate predecessors and from many more of former generations. This is true even if genetic mixing happens only occasionally among organisms whose reproduction is predominantly asexual. After every mixing, natural selection favors certain individuals or strains over the others, and the first step of the evolutionary process has been accomplished.

Because of local environmental conditions, natural selection is not likely to act identically in all parts of an extensive population composed of varied individuals. Subpopulations are changed slightly in different ways, and they interact with their neighbors. The subpopulations also, in a manner, are competitors, and natural selection again plays its part at a slightly higher level, permitting one subpopulation to expand and causing another to contract and possibly disappear. All this can take place without mutation. When mutations do appear, they may accelerate these processes. The tree of life does not accurately reflect relations within such populations as they evolve because the more evident evolutionary transformations arise from a base which is a tangle of individuals, strains, and subpopulations.

Such a pattern is characteristic, for example, of the vertebrates and many other animals that do not reproduce asexually. The situation with respect to plants, particularly angiosperms, is considerably different. Many plants may be compared to colonial animals like the corals, with each flower corresponding to a polyp produced by budding. A great variety of plants can and do spread asexually by the extension of runners, by the growth of new shoots which rise from roots, or in other ways. This permits the preservation of genetic structures which are so aberrant that the plants could not ordinarily reproduce by seed. Some of these plants are hybrids between related species or even genera (Fig. 638). Rarely by further hybridization or mutation they may suddenly produce new fertile species.

Complex evolution of this type is believed to have been fairly common among plants. New species arise from a base of anastomosing relationships, and these complications persist to and perhaps well beyond the species level. Therefore, the tree-of-life pattern is an accurate representation of plant evolution only above these levels.

The genetics of colonial animals like corals and bryozoans is not well known. Possibly their evolutionary patterns are comparable to those of plants.

Evolutionary Diagrams
Many evolutionary diagrams are structured like portions of the tree of life. They are

of two types: (1) Those intended to show only the relations of modern organisms and possibly others that are extinct terminal products of evolution. These have the recognized groups named only at the ends of branches. (2) Those intended also to show the relations of terminal groups to antecedent groups. These have branches diverging from named groups that may or may not include some modern organisms. If modern forms are not included, the ancient groups are considered to be extinct although they are survived by direct descendants differing from them in some important way. If one modern group is shown as being descended from another group that still exists, questions are always likely to arise. This is because the modern forms belonging to one group cannot be ancestral to another group. Therefore, an assumption necessarily has been made that if the actual ancestors of a group were known, they would be recognized as members of another group still represented by living organisms. All divisions between recognized ancestral and descendant groups are hazy and more or less subjective. Consequently opinions are sure to differ as to where the boundaries should be drawn.

Irreversibility of Evolution

Evolution is commonly stated to be irreversible. This is not strictly true; reverse mutations do occur, and some relatively simple lost structures have been observed to reappear in laboratory animals. Nothing in the genetic mechanism of evolution precludes reversion in the most detailed way. Divergence that serves to differentiate organisms from their ancestors, however, generally increases slowly as the result of many successive small mutations. By the time distinct species have developed, the number of mutations ordinarily has become so great that even nearly complete reversion is almost a statistical impossibility. Irreversibility, therefore, can be accepted as a valid evolutionary principle at the species and higher levels.

Competition

Natural selection results from differential reactions to environments and the active and continuous competition that begins with genes and progresses upward to perhaps the highest taxonomic levels. Every mutation, in a sense, struggles for survival. The conflict actually involves mutant and nonmutant individuals, and the outcome is determined by the material expression of the individual's entire genetic system.

Natural selection commonly is considered with respect to the individuals of a species population. Certainly, one individual's most effective competitors are likely to be other similar and associated individuals. Competition, however, does not stop here. It continues between strains, subpopulations, races, and whole species populations, and so on perhaps even up to phyla. As the biologic unit rises in the taxonomic scale, however, the process becomes progressively more diffuse. It is less and less a matter of simple competition between comparable individuals or groups and increasingly a matter of competition with a combination of other different organisms. Thus at higher levels, competition becomes more obviously related to general environmental adaptation which, of course, is its essence at any level.

Adaptation and Preadaptation

Adaptation has two aspects that are important in different ways: (1) Progressive adaptation contributes greatly to the success of organisms within the environments to which they are accustomed and to relatively minor environmental changes or fluctuations. This is a short-range process. (2) Preadaptation is the accumulation of real or potential traits that permit organisms to adapt rapidly to major changes in their environments. It is a chance process because the changes that a species may be called upon to face cannot be predicted.

Most organisms inhabit environments that vary somewhat from place to place and from time to time in both physical and biologic aspects. The extremes that any species can successfully endure are determined by the flexibility of its adaptations. This, in turn, depends on its genetic structure. Natural selection tends to narrow adaptability, and this may be immediately advantageous. In the long run, however, it is likely to be undesirable, because a narrowly adapted species may not be able to readapt to changed conditions. If not, it fails in its competition with other species and becomes extinct. Undoubtedly many species in the past have disappeared for this reason.

Both adaptation and preadaptation are accomplished mainly by selection. Directly adaptive mutations quickly achieve material expression and become dominant in a population. Potentially preadaptive mutations, on the other hand, are likely to accumulate within a population as recessives. Mutations of this kind cannot be recognized as preadaptive until organisms which possess them demonstrate new adaptive capabilities. Thus preadaptation is an undirected preparation of organisms that is followed by more or less rapid shifting from one kind of an environment into another somewhat different one. The shift may be temporal and accompany local environmental change, or it may be spatial movement of organisms into a different contemporary environment.

Progress versus Degeneration

Progress is an essential element in all evolution; it implies improvement with respect to success in life. Any change may be considered progressive if it contributes to either better adaptation or more efficient competition. The most progressive species are those whose populations are expanding and whose dominance within their organic communities is increasing in comparison with other related species or those following a similar way of life. Progress involves structural and functional improvements; it commonly is reflected by increased structural and behavioral complexity. Although it may seem contradictory, even species that are not expanding cannot be considered nonprogressive unless their adaptive evolution has come to a complete halt. Failure to expand may indicate only that such species are less progressive than their competitors.

The loss of structures or functions that have been useful to organisms without the development of new ones to take their places commonly is considered to be regressive evolution and, therefore, characteristic of degeneration. Such losses restrict species to narrower environments, reduce their opportunities for adaptation to other ways of life, and are likely to lead to their eventual extinction. Species of this kind, however, actually are becoming specialized. They are adapting in positive ways that increase their efficiency and contribute to their immediate success. Evolution of this type surely reflects a kind of progress right up to the time when an environmental change occurs to which the species is unable to respond successfully.

Individuals may exhibit truly degenerative traits that are disadvantageous. Such individuals generally are eliminated quickly from a population by natural selection. For this reason, major degenerative traits cannot accumulate and become characteristic of a species except possibly in very small populations and for very brief intervals of time. The identification of a race or species as degenerate, therefore, no matter how simplified or specialized it may be, misrepresents its true nature as far as the dynamics of evolution is concerned.

Orthogenesis

The idea that evolution is actuated by some purposive force or directed toward some predetermined goal is very old. Many paleontologists formerly were impressed by what they interpreted as undeviating evolution proceeding in certain definite directions for long periods of time. Some of them generalized that this apparent trend, termed orthogenesis, was characteristic of many evolutionary lineages. Such a view, however, has little modern following. Understanding of genetics necessitates the conclusion that long-term consistently oriented evolution is not probable. Furthermore, increasing paleontologic knowledge has demonstrated that certain supposed examples of straight-line evolution were misinterpreted. Although general directional trends are evident, these lineages now are believed to have fluctuated more or less erratically and even at times to have reversed direction. This is exactly what might be expected to result from adaptation to a succession of varying environments.

Nevertheless, some examples of what seem to be orthogenetic trends do remain (Figs. 552; 553). The explanation probably is that the opening of a totally new environment was followed by slow but persistent adaptation. This progressed in the only direction that adequately fitted organisms for the occupation of that environment. Close observation generally reveals that straight-line evolution with respect to one conspicuous character has been accompanied in the same organisms by the evolution of other characters that do not show a similar regularity. All evidence points to the conclusion that evolution has been opportunistic, in the sense that it has followed only those paths which clearly were open to it.

Racial Senescence

The notion has been popular in the past that the life history of a species, a genus, or any other taxonomic group is directly comparable to the life history of an individual. An early period of youthful vigor was supposed to pass into one of mature stability to be followed by increasingly decrepit old age that ends finally in extinction. Noteworthy evolutionary radiation, at least temporarily successful, was viewed as an indication of racial youth. An idea associated with orthogenesis was that a period of successful evolution developed a momentum that eventually carried lineages to inadaptive lengths. A second radiation of resulting short-lived supposedly inadaptive and abnormal forms was considered to identify the onset of senescence. Abnormality was seen in bizarre ornamentation, distorted growth, gigantism, etc., which were interpreted as the frantic efforts of a weakened and declining stock to escape extinction by discovering a way to attain rejuvenescence.

Any parallel drawn between racial and individual history appears to be totally unrealistic and unwarranted. It necessitates the assumption of decline in an unknown vital force. It denies the effectiveness of adaptation and natural selection. Moreover it is not supported by the paleontologic record. The whole concept of racial senescence seems to have been erected on false premises and the misinterpretation of the imagined similarities of two quite different processes.

Deceleration of Evolution

The idea has been expressed that evolution has slowed from some former relatively rapid rate and perhaps has nearly ceased or eventually will altogether fail. Also involved to some extent is the supposition that, in its early stages, evolution was little influenced by natural selection which has become increasingly important in later time. The basis for these views is provided by the fact that primary evolutionary diversification into phyla and other high taxonomic categories dates back to very ancient time and comparable differentiation has not been accomplished more recently. This fact probably also has encouraged the general concept of racial senescence.

The differences that divide the phyla and other high taxonomic categories certainly are more important than those which separate low categories such as genera and species. That early evolution was more effective in this way because it was more rapid or of a different type has seemed a natural conclusion. This does not harmonize, however, with known genetic principles, nor does the fossil record favor it over other theories that are believed to be more rational.

At every stage in the evolutionary process mutations may be expected to produce variations limited only by the potentialities of the parent stock and the viability of the variants in available environments. These variants probably produce at first a nearly continuous spectrum connecting extreme forms. Natural selection rapidly or gradually eliminates many of the variants, mostly intermediates, and at the same time continuing mutations accentuate differences in the lineages that survive. Thus as time progresses, diversity increases and the surviving lineages become more isolated from one another both genetically and morphologically.

If this pattern is correct, the degree of differentiation among organisms, measured conveniently in terms of the scale of taxonomic categories, is relative to the amount of elapsed time, and appeal to fundamentally different rates or types of evolution is unnecessary.

The contrary idea, that evolution has accelerated, has been expressed more rarely. To some extent, however, this concept may have more reality than the other, although it is unlikely that acceleration has been progressively more or less continuous. The point here is that the initiation of sexual reproduction and the resulting mixing of genetic inheritances so increased the potentiality for variation upon which natural selection operates that in this single step evolution probably was speeded up remarkably.

Parallel and Convergent Evolution

Organisms which resemble each other in conspicuous or apparently important ways are not necessarily related closely. Two processes that are known as parallel and convergent evolution have preserved or produced likenesses responsible for uncertainty and much possible confusion in the classification of organisms and the recognition of their evolutionary histories.

Some lineages derived from a common ancestral stock did not become increasingly different as they evolved, but followed more or less remarkably parallel evolutionary paths. Probably their genetic structures were not greatly altered and this predisposed them to similar chance mutations. If they continued to occupy comparable environments and did not change their ways of life importantly, they would be likely to react similarly to their surroundings and adapt in almost indistinguishable ways.

Other lineages, possibly of much more remote relationship, have evolved in such a manner that they converged morphologically and became increasingly more similar to each other in some respects. This may have resulted from the independent alteration or origin of features that (1) were similarly advantageous to different kinds of organisms, or (2) were adapted by different organisms to similar ways of life. Most convergent evolution seems to have been related more to similar functional adaptation than to similarity of environment.

Homology and Analogy

The recognition of genetic and evolutionary relations requires the correct interpretation of morphologic features. Confusion is possible because (1) similarities are not necessarily reliable evidence of close relationships, and (2) conspicuous differences may not indicate relations that are remote. This introduces the concepts of homology and analogy.

Homology signifies that particular anatomical parts of different organisms correspond exactly within the plan of a single morphologic system. The parts may be so similar that there is no problem and they can be recognized as variants which have resulted from evolutionary diversification. Examples are the legs of dogs and horses which evolved from the corresponding parts of some kind of ancestral animal. On the other hand, the parts may be so different that similarity of origin may not be immediately apparent, as with the leg of a dog, the flipper of a seal, and the wing of a bat. Divergence ordinarily resulted from gradual functional adjustment of organisms to different environments or different ways of life. Thus the digging foot of a pelecypod, the creeping foot of a gastropod, and the grasping foot of a cephalopod are all homologous.

Analogy signifies similarity in form or function in the parts of different orga-nisms that do not correspond within the plan of a particular system. In other words, the similarity is not related to evolution from the corresponding part of an ancestral organism. Similarities of this kind ordinarily developed by evolution in response to the adaptive adjustments of different organisms to similar environments or similar ways of life. Although morphologic similarities may be obvious, the fundamental underlying structures of analogous parts are likely to be quite different. This is clear, for example, with respect to the legs of arthropods and vertebrates or the wings of insects and birds. Equally important but less obvious are the differences between the eyes of some cephalopods and of vertebrates.

Many mistakes in the interpretation of homology and analogy are possible. Homologous relations may not be recognized, and analogy may be confused with homology.

Rates of Evolution

Evolution has two dynamic aspects for which estimates have been made in quanti-tative ways. These are (1) cumulative and progressive changes in evolutionary lineages or taxonomic groups of organisms, and (2) diversification or increase in the variety of organisms derived from common ancestors or included within particu-lar taxonomic groups. The first may be looked upon as progressing in a single general direction paralleling time, and the second as contemporaneous evolution radiating laterally in many different directions. Both can be estimated with respect to intervals of time, and both can be compared relatively between different groups of organisms.

Evolutionary rates can be assessed in terms of relative changes in homologous characters. This is possible, however, only within closely related groups of fossils, and it has not been common. Another more generally employed method compares the duration of fossil genera in millions of years or the averages for groups of related genera. The results are interesting and not without significance, but the opportunities for misinterpretation are very great. Two facts are important and need to be kept constantly in mind: (1) not all genera or other taxonomic groups of similar rank are equally significant in terms of evolutionary progress; and (2) the time represented by different segments of the geologic past is not accurately known. Thus the conclusion does not necessarily follow that Tertiary mammals evolved ten times as fast as Paleozoic pelecypods simply because figures show that mammalian genera persisted about one-tenth as long as the genera of pelecypods. Nevertheless, the rate of evolution does seem to have varied considerably between different groups of organisms and in the same group at different times.

Evolutionary Radiation

Relative evolutionary radiation is estimated by comparing counts of genera or other taxonomic categories in different groups of related organisms which lived during or made their first appearances in approximately equal intervals of time. Species rarely have been used, because the data with respect to them are uncertain and difficult to obtain. Estimates of this kind encounter the same two uncertainties mentioned in the last section—unequal significance of genera or other taxonomic groups and inaccurate calibration of geologic time. In addition a third important factor is the incompleteness of the fossil record. Thus comparisons between well-known floras or faunas from widely exposed and carefully studied strata with those from areally restricted or neglected beds are likely to suggest erroneous conclusions.

Counts of genera or other groups for the early, middle, and late parts of geologic periods commonly show rather regular fluctuations, with the highest counts in the middle intervals (Fig. 25). This has been interpreted as demonstration of the cyclic nature of evolutionary radiation. More probably these distributions indicate that the organisms of the early and late intervals are less completely known. With equal relative representation of all fossils, the irregularities almost certainly would be much reduced or would disappear. In their place longer-term and more significant fluctuations might be expected to become apparent.

Most estimates of evolutionary rates and radiation are likely to reflect the

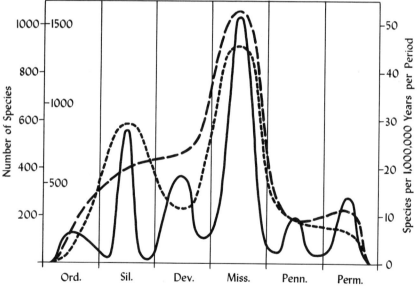

Fig. 25 Curves showing numbers of described Paleozoic crinoids of various ages. Solid curve (calibrated outside to left) shows number of species known from early, middle, and late parts of the periods. It suggests that the rate of radiating evolution fluctuated cyclically with the periods. Long dashes (calibrated inside to left) show numbers plotted for whole periods and suggest a longer-range pattern of evolution. Short dashes (calibrated to right) show numbers of species per 1,000,000 years as calculated for whole periods. The form of this curve depends on estimates of the relative durations of the periods. If the duration of the Silurian has been underestimated, the peak in the curve here is too high. If estimates of time are reasonably accurate, a curve of this last kind probably represents evolutionary radiation better than the others. (Data estimated from Moore, 1952, J. Paleontol., vol. 26, fig. 1, p. 342.)

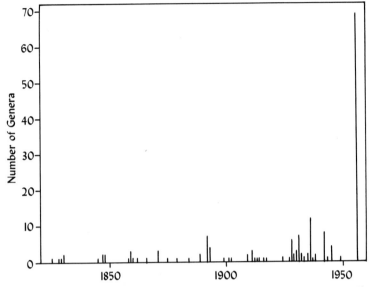

Fig. 26   An example of a "monographic burst" of genera. Shown here by years are the numbers of lower Middle Ordovician brachiopod genera described as new. Total number, 174; described in one work in 1956, 69 genera, or 40 percent. Obviously, any numerical comparisons of these brachiopods with those of other ages made before or after this date would seem to have very different implications. (*Data from Cooper, 1956, Smithsonian Inst. Misc. Collections, vol. 127.*)

unequal taxonomic treatment of the fossils of different ages and different groups. The publication of monographs naming ever more minutely discriminated fossils is sure to be a distorting factor (Fig. 26). Some distributions influenced in this way suggest brief intervals of so-called explosive evolution that are much exaggerated, to say the least. Obviously interpretations should be made with the greatest caution because, to a very large extent, names rather than real organisms are the basis for such comparisons.

Speciation

Genetics provides the mechanism which, by mutation and natural selection, accounts for evolution. Species ordinarily originate by the accumulation of many relatively small differentiating characters, although more rarely, especially among plants, the same results may be produced suddenly by hybridization. The small evolutionary steps are rarely evident in the fossil record. Paleontologists see mainly the abrupt appearances of already differentiated groups of organisms whose evolution was accomplished in unknown areas. Whether or not these larger steps were produced in the same way by small mutations has been much debated.

Some evolutionists have concluded that two processes, differing in the scale of their results, have been involved in evolution. This view drew some support from the idea that the features distinguishing genera and higher taxonomic groups, being superior to those of species, have had a different origin. Although this possibility cannot be ruled out with certainty in every instance, there seems to be no reason to believe that the small changes responsible for the divergence of species, if continued long enough, are not capable of accomplishing all evolutionary differentiation.

Extinction

Death is the inevitable consequence of life. Some individual organisms that reproduce vegetatively are potentially immortal, but sooner or later death overtakes them all. Species and evolving lineages have constantly replaced each other, but only a small fraction of them is known from their fossil remains.

The many millions or even billions of species which have inhabited the earth during its long geologic history have disappeared from the organic scene in one or the other of two ways: (1) by evolving into other species, or (2) by dying out completely. Evolving species make up the lineages which paleontologists try to trace. Strictly speaking, these species did not become extinct, because they were survived by offspring which, if known, are called by other names. Very many species and most of the older lineages, however, left no descendants and did become extinct.

No real difference can be recognized between the extinction of species and lineages. A lineage consists of species; it dies out when its final species becomes extinct. Physical and biologic environments have been in a state of constant change. Species and lineages survived as long as they succeeded in adapting to these changes. When they failed, extinction quickly followed.

TAXONOMY AND EVOLUTION

The organic world is so complex that if plants and animals were not classified in some way according to their similarities and differences, generalizations which aid in understanding them would be most difficult. Before Darwin's time classification was entirely practical and reflected only morphologic relationships. Since the widespread acceptance of evolution as a dominating process, however, taxonomy, or the theory and philosophy of classification, has changed. Many biologists and paleontologists now believe that classification, to be natural and proper, must reflect evolutionary patterns. Although viewpoints have altered in this way, the methods and practice of classification have shown remarkably little change.

Evolutionary patterns cannot be observed directly, but it is hoped that they can be reconstructed on the basis of characters that are believed to be primitive, inadaptive, or homologous. Characters certainly of these kinds, however, are not obviously different from any others, and ordinarily their identification depends on the more or less intuitive recognition of patterns that are presumed to be evolutionary. Reasoning in which patterns are inferred from selected characters and significant characters are identified on the basis of patterns surely is circular. If this is recognized, the results are likely to be lacking in conviction. Otherwise serious errors may be made which persist indefinitely.

Furthermore, classification ordinarily does not and perhaps cannot wait until the course of evolution is understood with some assurance. Many organisms, especially fossils, are classified before much is known about them. Even their morphology may not have been adequately investigated. Nevertheless if taxonomy ideally is considered to reflect evolutionary relations, the presumption may be made that currently accepted classification does actually reflect these relations. This of course is likely to involve faulty reasoning. All thoughtful taxonomists recognize the dangers of such presumptions. Many persons, however, overlook the possibility that conclusions concerning evolution may be seriously distorted because they are based on classifications not in accord with true evolutionary patterns. Particularly if the evidence of fossils is ignored or imperfectly evaluated, an outmoded system of classification may remain in vogue and influence some evolutionary thinking long after its inadequacy is apparent.

Paleontologists surely are more conscious of the time element connected with their studies than are many biologists who deal only with modern organisms. This naturally directs their attention to evolution as a process of great importance. Consequently, most paleontologists, especially those concerned with vertebrates, are likely to accept the view that classification not only should but also can reflect phylogenetic relations. Moreover many of them seem to believe that attention directed to classifying fossils is equivalent to the study of evolution.

This view, however, is not held by all. Some of the older and more experienced paleontologists who deal primarily with invertebrates are less idealistic and more practical in recognizing the difficulties and uncertainties of phylogenetic taxonomy. They, as well as a considerable number of neontologists, contend that classification based directly on the observable characteristics of organisms is desirable because it is more likely to reveal the true relations of organisms than classification based on inferences drawn from the observation of these characteristics.

ORIGIN OF LIFE

Life could have appeared on earth in one of only three ways: (1) it was created supernaturally; (2) it reached the earth from some external source; or (3) it evolved here in an orderly manner from nonliving matter. Almost all scientists dismiss the first as an event wholly incompatible with either experience or reason. The second solves no problem and only creates an extra one. If "astroplankton" colonized the earth, life originated somewhere else. Interplanetary and interstellar distances are so great and destructive radiation in outer space is so intense that the immigration of any living thing seems to be a happening of the utmost improbability. Therefore, life is believed to have originated on the earth very long ago in a much-simplified state which, in turn, resulted from a long sequence of chemical evolution.

Chemical Evolution

Chemical evolution leading up to life probably began on the youthful earth soon
after it acquired an atmosphere comparable physically to the modern one but
different from it chemically. This atmosphere is believed to have consisted mainly
of water vapor, hydrogen, ammonia, and methane (Fig. 29). No free oxygen was
present, and in the course of time most of the hydrogen escaped from the gravity
field of the earth. Energy derived from ultraviolet light and lightning decomposed
some of the atmospheric gases and their parts recombined more or less at random
to produce numerous more complex substances. Laboratory experiments with electric discharges demonstrate that this is possible and a variety of relatively complex
so-called organic compounds have been formed in such a manner.

    After water condensed and accumulated in an ocean, these compounds were
carried down by rain and gradually concentrated in solution. Additional random
chemical reactions took place, and larger "organic" molecules were built up. Every
chemical reaction that occurs in living cells can also occur inorganically. Only time

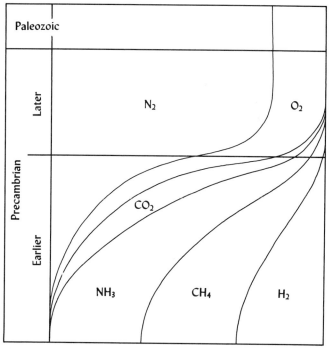

Fig. 29   Diagram showing presumed transformations in the earth's atmosphere.
Water vapor is omitted because of uncertainties concerning distribution of water
in atmosphere and hydrosphere. (*Modified from Kuznetsov, Ivanov, and Lyalikova,
1963, "Introduction to Geological Microbiology," fig. 1, p. xiv.*

is needed for these reactions to be completed, and many millions of years were available for chemical evolution in the primordial ocean.

All the postulated chemical reactions are reversible, but some of the larger molecules escaped that fate. Certain molecules became more stable, less soluble, and acquired the characters of colloids. These aggregated in clumps, and the constituents of some of them began to be arranged in structural patterns that influenced their adsorptive properties. The more perfect ones accepted only such material as was compatible with their structures. Some grew more or less steadily until they reached unstable size and then broke up into smaller parts, which in their turn grew.

Nonliving organic matter cannot persist for long in the modern world unless it is protected against consumption by animals as food, decomposition by bacteria, and atmospheric oxidation. The earth in its early stages, however, was lifeless and without free oxygen. Therefore, the colloidal aggregates persisted. Perhaps some of them competed for the same preformed compounds. Because of structural advantages, one kind may have multiplied more rapidly than others. Thus a very primitive type of natural selection may have operated.

Primitive Organisms

Achievement of the ability to synthesize material needed for their growth converted nonliving colloidal aggregates into what may be considered the most primitive of organisms. Instead of accepting only preformed molecules, they absorbed other substances and used chemical energy derived from a simple fermentation process to build the required compounds. This implies the presence of a catalyst or enzyme to direct and expedite such action. The ability to synthesize, even in the most simple way, conferred such an enormous advantage upon its first possessor that none of the other almost-living aggregates could compete with it. From this point on, successive mutations and the process of natural selection can account for the ramifying of all life from an ancestor of this kind.

Photosynthesis

When hydrogen of the original atmosphere was essentially depleted, two important things happened: (1) Oxygen produced by the photochemical decomposition of water vapor began to accumulate. This was acted upon in the upper atmosphere by ultraviolet light to form ozone, which has the property of absorbing high-energy radiation. (2) The production of "organic" compounds in the atmosphere almost ceased because the raw materials were much depleted and the energy needed for their formation was no longer received in large amounts. As the energy-producing compounds already present in the ocean were consumed by primitive organisms, life processes must have slowed and finally stopped had not another source of energy been exploited.

The energy that came to play the most important biologic role is that of ordinary sunlight. It is trapped by a colored substance, chlorophyll, and serves to decompose water, to free oxygen, and to produce carbohydrates from carbon dioxide.

Before the development of this process of photosynthesis, all organisms must have been essentially animal-like because they were wholly dependent on the capture of preexisting "organic" matter for the raw materials of life. The advent, however, of photosynthesis and typical plants which were able to manufacture food from carbon dioxide and water put an end to this dependence. From this time onward organisms were presented with opportunities for further development that were almost limitless.

Respiration

In an oxygenless environment, metabolic energy is derived from fermentation processes whose end products still possess much unexpended chemical energy. When oxygen, mainly produced by plants, began to accumulate importantly in the ocean and the atmosphere, a new source of energy became available. The complete oxidation of organic matter produces much more energy than fermentation. The use of oxygen, or respiration, as a primary source of energy marked the development of the first true animals. Thus the organic world was revolutionized by plants which synthesized food from a nearly inexhaustible supply of inorganic raw materials and then by animals which discovered a rapid and efficient way to obtain chemical energy for growth and movement.

BEGINNING OF FOSSIL RECORD

For all practical purposes, the fossil record begins in the rocks of the Cambrian Period. The most ancient fossils that are abundant and well preserved occur in sedimentary rocks of early Cambrian age, dating back to a time some 600 million years ago. A few much less satisfactory fossils have been found in somewhat older sediments, and traces believed to record the existence of organisms, probably algae, are present in some rocks of very much greater ages.

Cambrian fossil animals, all marine invertebrates, certainly are not the remains of primitive organisms. The commonest ones are trilobites which were highly organized complex arthropods and represent one of the most advanced great animal groups, or phyla. Other fossils indicate that all the principal animal phyla inhabiting the world today probably were already in existence before the beginning of the Cambrian Period. None of these creatures stood anywhere near the base of the evolutionary ladder, and they provide no evidence concerning the nature of the first living things. Therefore, the foregoing sketch of the origin of life and the progressive development of organisms is entirely theoretical. This sequence, however, seems reasonable and is consistent with much that is known about physics, chemistry, biology, and geology. If it is far from the mark in some respects, modifications probably can be made relatively easily in this theory.

Without much doubt, the greater part of evolution, first chemical and then biologic, leading up to the kinds of life that are known today was accomplished in the very distant past. If judgment is based on either fundamental progress in the forms and diversification of living things or on amount of elapsed time since the earliest evidence of organisms, 85 percent probably is a conservative estimate for Precambrian evolution, as compared with 15 percent in more recent time.

The Missing Record

The later Precambrian Era certainly was not so lifeless as the very scanty fossil record indicates. Precambrian carbonaceous strata, including some material that is almost coal, thick and extensive limestone or dolomite formations, and great deposits of sedimentary iron ore almost surely owe their origin to the existence and biochemical activities of organisms that have left no other traces. Why, then, are Precambrian fossils so rare and so indecisive?

Numerous explanations have been offered to account for the abrupt appearance of many kinds of fossils in the rocks of Cambrian age. Most of them can be classified in one or another of three groups: (1) The chemistry of Precambrian seas was unfavorable for life or did not permit the preservation of fossils. (2) Abundant Precambrian life existed, but its record, although still preserved, has not been discovered. (3) Profound changes in life occurred at the beginning of the Cambrian Period, resulting from rapid evolution and perhaps adaptation to new environments.

(1)   Suggestions have been made that Precambrian seas were acid or contained no calcium carbonate that could be built into shells suitable for fossilization. Great masses of limestone or dolomite older than the Cambrian are evidence, however, that this was not so.

(2)   Metamorphosed Precambrian rocks cannot be expected to contain well-preserved fossils. Many late Precambrian formations, however, are not appreciably altered and are very similar physically to younger fossiliferous formations. Careful searching has shown that almost all of them are essentially barren except for structures probably built by algae (Fig. 33). Suggestions have been made that all Precambrian organisms were small and faunas were very local. One version supposes that all observed rocks are nonmarine and therefore unfossiliferous. Another, on the contrary, supposes that life originated in freshwater but that the accessible rocks are all marine. These ideas do not seem to have much merit.

(3)   Several theories relate the appearance of fossils in the Cambrian Period to changes in the kinds and conditions of marine life. There are numerous variations of the idea that all Precambrian organisms were microscopic, planktonic, or without hard parts. Shells did not become useful until bottom-living forms developed which needed protection against early predatory animals. Emphasis commonly has been placed on rapidity of evolution not duplicated in later time. There is, however, no other reason to conclude that rates of evolution were distinctly different before and after the beginning of the Cambrian.

No entirely satisfactory solution to this problem has been reached. A widespread misconception seems to be reflected, however, in most theories concerning the appearance of fossils in the Cambrian. This is that representatives of most of

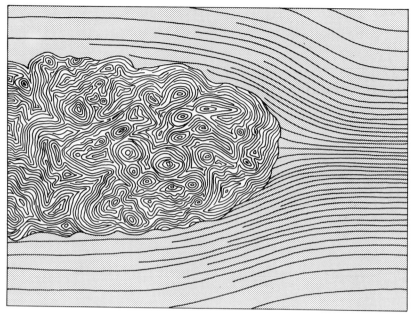

Fig. 33 Vertical section through part of a Precambrian stromatolite mass, presumably formed by algae, 3 ft thick, in the Belt Group of Glacier National Park. Bending of the limestone strata around it shows that this was a resistant mass before the enclosing sediment was compressed and compacted. (*Drawn from photograph; see Fenton and Fenton, 1958, "The Fossil Book," figure on p. 52.*)

the important groups appeared suddenly. When the record is reviewed carefully, it becomes clear that this is not so.

The fossil record shows that marine invertebrates with preservable hard parts became gradually more diverse, from the rare and restricted fossils of the Precambrian to the abundant faunas of the mid-Ordovician, when for the first time such animals occurred in comparable variety to those of modern seas. This interval of time was very long, more than 100 million years, and equal to approximately 20 percent of all time since the beginning of the Cambrian. The biologic changes effected in this interval probably were little if any more remarkable for their suddenness than many later ones.

Each important group of fossils makes its appearance in the record distinct and fully differentiated from the others. There is nothing at all to suggest that any of them at the time of its first appearance was a very recent evolutionary development. Each probably appeared simply because it had acquired the ability to build some kind of hard structure that was preservable. These structures are so different in form, composition, relation to soft parts, and mode of growth that their independent development in the different groups cannot be seriously doubted. Probably all phyla were represented by naked, soft-bodied ancestors long before the beginning of Cambrian time. Then during the course of 100 million years or more, protective or supporting structures became advantageous and the hard parts known as fossils progressively evolved. This development may have been a response to the emergence of relatively large active predatory carnivores, of which the cephalopods probably were the first.

GEOLOGIC TIME

Evolution is a process that has progressed with time. The fact that neither plants nor animals have changed appreciably during the last few thousand years of human observation demonstrates that it has been very slow. Biologic studies provide little evidence as to the speed of evolution; most estimates have been based on the ideas of geologists and paleontologists regarding the duration of geologic time.

Table 34   Estimates of the duration of time, in millions of years, represented by the main divisions of the stratigraphic column

	DURATION	TOTAL TIME
Cenozoic:		
Pleistocene	1	1
Pliocene	9	10
Miocene	15	25
Oligocene	10	35
Eocene	20	55
Paleocene	10	65
Mesozoic:		
Cretaceous	70	135
Jurassic	45	180
Triassic	50	230
Paleozoic:		
Permian	50	280
Pennsylvanian	30	310
Mississippian	35	345
Devonian	60	405
Silurian	20	425
Ordovician	75	500
Cambrian	100	600
Precambrian		

Modified from Kulp (1961), *Science*, vol. 133, p. 1111.

In Europe and some other parts of the world, Mississippian and Pennsylvanian are considered to be subdivisions of the Carboniferous System approximately equivalent to Lower and Upper Carboniferous, respectively. The Tertiary Period includes all the Cenozoic epochs except the Pleistocene.

Estimates of the age of the earth have varied greatly but have tended to increase steadily. In the seventeenth century, Archbishop Ussher of Ireland calculated, on the basis of Biblical studies, that the earth was created in 4004 B.C. In the mid-eighteenth century Buffon in France observed the cooling of differently sized iron balls and by extrapolation estimated the age of an originally molten earth at about 140,000 years. In the mid-nineteenth century Lord Kelvin of England calculated thermal conductivity in the sun and concluded the earth's age to be between 20 and 400 million years but favored a figure of about 100 million years.

Most geologists have been satisfied to consider time in a relative way in terms of a sequence of geologic periods. When estimating time, however, they looked more directly to the earth for evidence. Thus the thickness of sedimentary rocks and the amount of salt in the present sea have been interpreted to indicate ages varying between as little as 3 million to more than a billion years. Early in the present century most geologists accepted an estimate of not more than 100 million years for the time that has elapsed since the beginning of the Cambrian Period.

The discovery of radioactivity provided a new and more accurate method of estimating geologic time. Early measurements of radioactive elements and the products of their decay in minerals furnished several points of reference and demonstrated the unexpectedly great ages of certain igneous rocks. On this basis the beginning of the Cambrian Period was placed at about 500 million years ago. More recent refinements in radiometric methods have made age determinations possible for very minute amounts of radioactive material in sedimentary rocks. Thus a more accurate and somewhat more detailed time scale has been constructed for the geologic periods against which the changing floras and faunas of the past can be calibrated through approximately 600 million years since the beginning of the Cambrian (see Table 34).

The earth, however, is very much older than this last figure, and life certainly began much earlier. Evidence of organisms is believed to have been recognized in rocks whose ages are determined at nearly 3 billion years. Even before that life probably existed. Calculations based on astronomical, physical, and chemical phenomena all suggest that a total time span of about 4.5 billion years may have been available for evolution to progress to its present stage.

BIBLIOGRAPHY

G. W. Beadle (1963): The Place of Genetics in Modern Biology, *Smithsonian Inst. Ann. Rept., 1962,* pp. 399–414.
This popular account briefly traces evolution from its chemical beginnings to the most advanced organisms.

M. Calvin and **G. J. Calvin (1964):** Atom to Adam, *Am. Scientist,* vol. 52, pp. 163–186.
A distinguished chemist outlines the successive steps of chemical evolution.

P. R. Ehrlich and **R. W. Holm (1963):** "The Process of Evolution," McGraw-Hill Book Company, New York.
This is an excellent textbook with principal emphasis on genetics.

C. L. Fenton and **M. A. Fenton (1958):** "The Fossil Book," Doubleday & Company, Inc., Garden City, N.Y.
All types of fossils are illustrated by unusually fine figures and discussed in a popularly written but authentic manner.

P. G. Fothergill (1952): "Historical Aspects of Organic Evolution," Hollis and Carter, London.
The author presents a comprehensive account of the development of evolutionary thought and includes an extensive bibliography.

H. Gaffron (1960): The Origin of Life, in "Evolution after Darwin," S. Tax (ed.), vol. 1, pp. 39–84, The University of Chicago Press, Chicago.
A factual and philosophic consideration of the probable origin of life is presented very clearly.

J. Hadži (1963): "The Evolution of the Metazoa," Macmillan & Co., Ltd., London.
Embryology and larvae are not reliable evidence for evolutionary relations.

J. L. Kulp (1961): Geologic Time Scale, *Science,* vol. 133, pp. 1105–1114.
The time scale is based on recent radiometric determinations.

G. L. La Berge (1967): Microfossils and Precambrian Iron-formations, *Bull. Geol. Soc. Am.,* vol. 78, pp. 331–342.
Reported and described are abundant tiny structures, probably organic, in Precambrian rocks more than 2 billion years old.

C. C. Li (1948): "An Introduction to Population Genetics," National University Press, Peiping; reprinted, **1955,** The University of Chicago Press, Chicago.
This book presents an elementary summary of the principles of population genetics.

G. E. Murray (1964): Indigenous Precambrian Petroleum, *Bull. Am. Assoc. Petrol. Geologists,* vol. 49, pp. 3–21.
See this article for a selected bibliography on Precambrian fossils.

N. D. Newell (1959): Adequacy of the Fossil Record, *J. Paleontol.,* vol. 33, pp. 488–499.
The author optimistically suggests that more than 10 million species of organisms may be preserved as fossils.

_____ **(1959):** The Nature of the Fossil Record, *Proc. Am. Phil. Soc.,* vol. 103, pp. 264–285.
The contributions that paleontology has made to evolutionary theory are critically reviewed.

E. C. Olson (1966): The Role of Paleontology in the Formulation of Evolutionary Thought, *BioSci.,* vol. 16, pp. 37–40.
Paleontology provides no evidence supporting the theory of natural selection but is important to evolutionary theory in other ways.

A. I. Oparin (1961): "Life: Its Nature, Origin and Development," trans. by A. Synge, Academic Press Inc., New York.
This is a revision of Oparin's famous book, "The Origin of Life."

_____ **et al. (1959):** "The Origin of Life on Earth," Pergamon Press, New York.
An international symposium of more than 40 articles is devoted mainly to the chemical and physical origin of organic material and the evolution of metabolism.

P. E. Raymond (1939): "Prehistoric Life," Harvard University Press, Cambridge, Mass.
The author reviews knowledge of Precambrian fossils and theories accounting for their rarity or absence.

A. S. Romer (1949): Time series and Trends in Animal Evolution, in "Genetics, Paleontology, and Evolution," G. L. Jepsen, E. Mayr, and G. G. Simpson (eds.), pp. 103–120, Princeton University Press, Princeton, N.J.
Paleontologic principles such as adaptation, orthogenesis, irreversibility, parallel and convergent evolution, and racial senescence are discussed.

I. S. Shklovskii and **C. Sagan (1966):** "Intelligent Life in the Universe," Holden-Day, San Francisco.
Chaps. 16 and 17, pp. 214–245, are a comprehensive review of the possible physical and chemical conditions and processes that may have resulted in the appearance of life on the earth.

G. G. Simpson (1949): "The Meaning of Evolution," Yale University Press, New Haven, Conn.
A readable semipopular account of evolution is presented by an eminent vertebrate paleontologist.

———— **(1949):** Rates of Evolution in Animals, in "Genetics, Paleontology, and Evolution," G. L. Jepsen, E. Mayr, and G. G. Simpson (eds.), pp. 205–228, Princeton University Press, Princeton, N.J.
Evolutionary rates are discussed in relation to genetics, morphology, and taxonomy.

———— **(1960):** The History of Life, in "Evolution after Darwin," S. Tax (ed.), vol. 1, pp. 117–180, The University of Chicago Press, Chicago.
This article reviews the problem of the scarcity of Precambrian fossils.

R. E. Snodgrass (1963): Some Mysteries of Life and Existence, *Smithsonian Inst. Ann. Rept., 1962,* pp. 517–535.
This popular article briefly reviews ideas concerning the origin of life, movement, instinct, and consciousness.

A. M. Srb and **R. D. Owen (1965):** "General Genetics," 2d ed., W. H. Freeman and Company, San Francisco.
This is a good textbook in its field.

G. F. Stebbins (1950): "Variation and Evolution in Plants," Columbia University Press, New York.
Plants differ considerably from animals in some features of their genetics.

H. H. Swinnerton (1938, 1939): Development and Evolution, *Pan-Am. Geol.,* vol. 70, pp. 161–182; vol. 71, pp. 11–26.
Evidence for and against recapitulation is reviewed, and attention is directed to misunderstandings that have confused conclusions.

P. B. Weisz (1966): "The Science of Zoology," McGraw-Hill Book Company, New York.
This excellent textbook includes sections or chapters relevant to the preceding introduction and also to many subsequent parts of this book.

CHAPTER
TWO

PLANTS

BIOLOGIC GENERALITIES
PLANT DIVISIONS
ORGAN EVOLUTION

One of the main objectives of biologic and paleontologic study is the deciphering of evolutionary relations and reconstruction of the evolutionary history of all organisms.

Evidence of Relationships
The genetic relations of modern organisms are indicated by varying degrees of similarities and differences in: (1) Comparative morphology, including anatomy and histology. (2) Ontogeny, or individual development, including embryology. (3) Physiology, including reproduction and behavior. (4) Biochemistry, including serology. (5) Genetic structure. Evidence of these kinds indicates much about the nearness of relationships that connect the many different organisms inhabiting the world today, but it reveals very little concerning their actual evolutionary history or phylogeny. Evidence that is indispensable in this respect is provided by paleontology in: (6) The temporal relations of groups of organisms which succeeded one another with the passage of geologic time.

Morphology
Comparative morphology of adult organisms provides the principal evidence guiding most phylogenetic reconstructions. It is not adequate, however, to indicate the relations of the main animal phyla, all of which probably had their origins in Precambrian time. It is almost equally inadequate with respect to plants, but this is somewhat less important in paleontology because most plant fossils belong to a

single group roughly comparable to an animal phylum and most, if not all, of the evolution of these plants has taken place since the beginning of Paleozoic time. In the consideration of morphology, however, confusion may result from the misinterpretation of homologous or analogous similarities and the effects of parallel or convergent evolution.

Ontogeny

The comparative morphology and development of the youngest stages of animals, particularly their embryology, have provided much of the evidence upon which theories relating the various phyla have been based. Later ontogenetic development likewise is suggestive of relationships within the phyla. Interpretations of developmental stages and sequences, however, must be made with great caution because not all of them are equally reliable evolutionary guides. Ontogeny is of much less service in reconstructing plant phylogeny.

Physiology

Direct studies of physiology are possible only with living organisms. Very little can be learned about the physiology of fossil animals except by analogy. The higher plants are different, however, because the details of their reproduction are important and the nature of their reproductive processes is indicated by structures that may be found preserved as fossils. Some other structures related to metabolism also are revealed in well-preserved fossils and can be compared with those of modern plants.

Biochemistry

Most features of biochemistry can be investigated only among modern organisms. These studies have been concerned mainly with organic substances that are rarely retained unaltered in the fossil state. The hard parts of animals, however, are of interest because similarities or differences may be indicated by chemical composition, the presence of trace elements, and isotopic ratios. Similar studies find much less application among fossil plants.

Genetics

Genetic studies also are limited to modern organisms. Besides breeding experiments, cytology and the observation of the number and form of chromosomes provide evidence regarding the closeness of relationships. Obviously, all this is impossible with paleontologic material.

Stratigraphic Succession

The time sequence of fossils, as revealed by their presence in vertical successions of layered rocks, must guide all realistic phylogenetic reconstructions. It is clear that no group of organisms could have appeared as a result of evolution before its ancestors were in existence. The fossil record, however, is incomplete, and a group of organisms may have existed for a very considerable length of time before that indicated by its earliest known fossil occurrence. Nevertheless, it is very unlikely that any group which commonly was fossilized and has evolved since the beginning of

Paleozoic time was represented by abundant and well-diversified individuals with preservable hard parts much in advance of the oldest discovered fossils. This, however, is less true of the higher plants than of animals, because the principal centers of terrestrial plant evolution probably were located in regions where sediments did not accumulate and, therefore, plant materials were not preserved as fossils.

No evidence of the foregoing kinds is conclusive with respect to the paths that evolving organisms actually have followed. It must be emphasized that all evolutionary reconstructions are theoretical, although certainly some are founded much more securely than others. Many writers have presented their conclusions in the form of family trees or systems of classification with much greater positiveness than is warranted by the facts. Persons not well acquainted with a group of organisms are likely to accept such conclusions uncritically and without realizing the uncertainties which actually exist in almost every case.

PLANTS AND ANIMALS

The fundamental differences that separate all familiar plants and animals are recognized by everyone. As attention is directed to smaller organisms, however, some of which have been presumed to be relatively very primitive, these differences fade. One group in particular, the single-celled flagellates, includes numerous members which possess both plant and animal qualities. For example, some are equipped with chlorophyll for photosynthesis and produce cellulose as plants do but move about with considerable agility, have eye spots, and feed like animals on other microorganisms. Flagellates are commonly believed to represent the general stock from which all higher plants and all animals evolved.

The existence of animal-like plants or plant-like animals poses a vexing problem in classification. An attempt has been made to resolve it by proposing the recognition of a third kingdom of organisms, the protists, whose members are declared to be neither plants nor animals. This solution has been welcomed by some biologists and paleontologists. It seems, however, to introduce new difficulties, because any division between protists and plants, as restricted on the one hand, and animals, as restricted on the other, is just as arbitrary as that between unrestricted plants and animals. Therefore, an attempt to eliminate one hazy boundary results in the creation of two others that are equally unsatisfactory.

Plants

With the possible exception of viruses, bacteria are the smallest, apparently the simplest, and the most primitive organisms that live today. Although they are classed as plants, most bacteria lack chlorophyll, are motile, and require preformed organic matter for their nourishment. Thus they resemble simple animals. Some groups may have lost plant-like qualities, but others probably never had them. Bacteria undoubtedly are much more complex than the earliest living organisms, but some of them almost certainly are more similar to the earliest forms of life than any other known plants or animals.

Multicellular plants as a whole exhibit much less morphologic and structural diversification than do multicellular animals. Therefore, their main systematic divisions are less sharply differentiated, and more or less transitional forms connect some of them. These higher plants probably evolved from green algae (Fig. 42). They are subdivided into groups mainly on the basis of increasing structural complexity and differences in the details of their reproduction. Botanists classify plants in nearly 20 so-called divisions, or phyla, but there are numerous points of disagreement concerning the phyla that should be recognized. Table 43 shows a conventional arrangement which is somewhat simplified but which, except for this, is widely recognized. Most monerophytes and thallophytes live in water or otherwise damp situations. None produces woody tissues. Among them only the diatoms are found abundantly as fossils. Consequently paleontology provides little information concerning the evolution of these groups.

Animals

At least 10 animal phyla are abundantly represented by fossils (see Table 86). These are distinguished mainly on the basis of more or less fundamental differences in the anatomical and structural plans exhibited by mature individuals. They range from relatively simple to highly complex. The diversity so obvious among them is related to the activities that are characteristic of their ways of life and their adaptations to a great variety of environments that in general are more complicated than those of plants.

Animals are commonly believed to have evolved from flagellates. No sharp division in this group can be made between plants and animals. Some flagellates

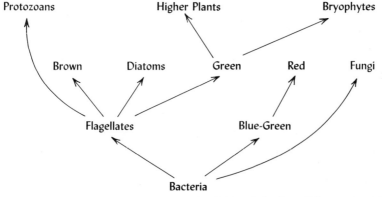

Fig. 42 Diagram showing probable phylogenetic relations of the algae. If the ancestors of these plants were known, they probably would be classified with bacteria.

Table 43 Classification of plants

MAIN DIVISIONS OF PLANT KINGDOM	EXAMPLES
Monerophytes: without organized nucleus	
Schizomycophytes	Bacteria
Cyanophytes	Blue-green algae
Nucleophytes: with nucleus	
Thallophytes: without embryo	
Eumycophytes: without chlorophyll	Fungi
Algae: with chlorophyll	
Rhodophytes	Red algae
Chrysophytes	Diatoms, etc.
Chlorophytes	Green algae
Phaeophytes	Brown algae
Pyrrophytes	Flagellates
Embryophytes: with embryo	
Bryophytes: nonvascular	Mosses, etc.
Tracheophytes: vascular	All higher plants

Most of these groups are variously subdivided or restricted by some botanists. See Table 48 for a classification of the tracheophytes.

TABLE 43

and a few other alga-like organisms are included in both botanical and zoological classifications.

PLANT DIVISIONS

The phylogenetic relationships of the main plant groups are obscure (Figs. 42; 48). As compared with many animals, their geologic record is very incomplete. There are several reasons for this. (1) Few plants produce mineralized hard parts, and their tissues decompose relatively rapidly. (2) The nonvascular plants, that is those without an efficient system for conducting liquids between their different parts, are particularly unsuited to preservation. Most of them are small and many are microscopic. (3) Most of the more highly evolved plants inhabit the land and grow in places where sediments do not accumulate. Therefore, their preservation is unlikely. (4) Most fossils consist of dismembered parts which are uncertainly related to one another. Few fossil plants are known in their entirety. (5) Plant evolution probably has been characterized by parallelism and convergence to a remarkable degree.

Nonvascular Plants

The studies of most nonvascular plants constitute specialties that have interested few paleobotanists. When these plants have been noticed at all as fossils, generally little more than their occurrences have been recorded.

Bacteria

Bacteria must have had a very long geologic history. The Precambrian sedimentary iron ores are presumed to have been produced by their activities. Microscopic objects more or less doubtfully identified as bacteria have been found in Precambrian cherts and other rocks. Bacteria are believed to have been recognized positively, however, in both silicified and coalified plant tissues and in the excrement of some Devonian and younger animals. Very little knowledge of evolutionary value has been gained from these fossil occurrences because only more or less imperfect forms can be observed. This is not sufficient for anything more than the most superficial identification of bacteria.

Algae

The various algal divisions (Fig. 42) are distinguished mainly on the basis of their photosynthetic pigments and the kinds of food substances which they manufacture and store within their cells. The blue-green algae are most primitive and seem to be most closely related to the bacteria because they lack organized nuclei. The red algae probably are related to the blue-greens because they are provided with similar pigments and, unlike the other kinds, both produce spores that do not have whip-like flagella.

Colonial and multicellular algae almost certainly are very ancient plants. They

Fig. 44 Stromatolites as exposed on a glaciated bedding surface, about 4 by 6 ft, in Upper Cambrian limestone near Saratoga, New York. (*Drawn from photograph; see Goldring, 1929, N.Y. State Mus. Handbook 9, fig. 77, p. 290.*)

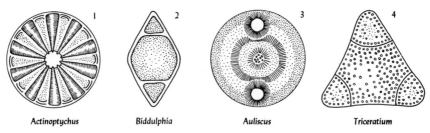

Fig. 45   Fossil diatoms from the Upper Cretaceous of California. (× 250–1,000.) (After Long, Fuge, and Smith, 1946, J. Paleontol., vol. 20, pl. 13, fig. 9; pl. 15, figs. 11, 14; pl. 18, fig. 9.)

are believed to have produced a considerable variety of calcareous structures known as stromatolites (Fig. 44) which include the oldest presumed evidence of life. Various other structures present in limestones and dolomites of many ages have been identified as algal. The cells are rarely preserved, however, or at least they have not commonly been observed. Therefore, many references to the existence of blue-green, red, or green algae are doubtful. Strangely shaped objects or markings termed fucoids are even less certainly identified as the lithified remains of brown algae. Spore-like objects of questionable origin and cellular structures that are almost surely algal have been described from the late Precambrian and occur in strata of younger ages, mainly in black shales and other carbonaceous material. None of these provides much information concerning evolutionary relations.

Diatoms

Diatoms are one important group of algae that are abundantly preserved as fossils (Fig. 45). Their siliceous box-like capsules are so small, however, that they commonly escape notice and are difficult to study. This and their great morphologic diversity have resulted in their restricted investigation by specialists, few of whom have had other paleobotanic interests. The earliest certainly identified diatoms are of Triassic age but there are reports of older ones. They must have had a long antecedent history, because many different kinds appear in the fossil record almost simultaneously. They did not become abundant until Cretaceous time. Practically nothing is known about their evolution.

Charophytes

Charophytes are multicellular plants that commonly have been assigned to the green algae, although the suggestion has been made that they occupy an evolutionary position intermediate between these algae and the bryophytes. Modern species inhabit freshwater, but some of the fossils may have been adapted to brackish environments. Fossils consist mainly of tiny characteristically sculptured, calcified structures in which female spores developed (Fig. 46). They are known possibly from rocks as old as the Devonian and are abundant in some Cretaceous and later strata.

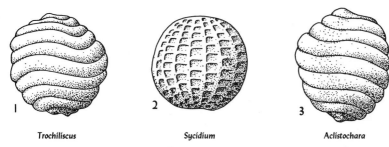

Trochiliscus   Sycidium   Aclistochara

Fig. 46 Fossil charophytes. (Approx. ×50.) 1, Upper Devonian, Ohio; 2, Lower Mississippian, Missouri; 3, Lower Eocene, Utah. (*After Peck, 1934, J. Paleontol., vol. 8, pl. 10, fig. 21; pl. 13, fig. 22; 1948, vol. 22, pl. 21, fig. 25.*)

Nothing is known about the evolution of these plants except that some Carboniferous fossils seems to be more complex than later ones.

Flagellates
Siliceous structures secreted by a variety of single-celled flagellates have been reported from the Paleozoic and are found very rarely in strata as old as the Jurassic (Fig. 46a). Their evolution is unknown.

Coccoliths
Coccoliths are tiny calcareous objects of many types (Fig. 47) borne on the surfaces of certain planktonic organisms that have been classified as either green algae or protozoans. Although they are abundant in some fine-grained calcareous sediments of Jurassic and later ages, they have been very little studied because of their almost submicroscopic sizes.

Fungi
The tubular strands, spore-bearing organs, and spores of fungi have been identified in plant tissues and excrement from strata as old as the Devonian and have been reported from the Precambrian. Tertiary fossils closely resemble modern fungi. They reveal nothing about the evolution of these plants.

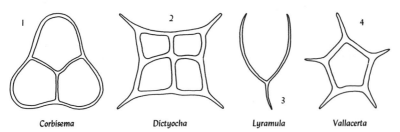

Corbisema   Dictyocha   Lyramula   Vallacerta

Fig. 46a Fossil silicoflagellates from the Cretaceous of California. (Approx. ×300.) (*After Hanna, 1928, J. Paleontol., vol. 1, pl. 41, figs. 1, 3, 4, 7.*)

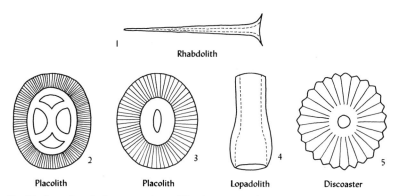

Fig. 47   Examples of a few types of coccoliths, found in the Tertiary strata of California, most of which are widely distributed in other parts of the world. (Approx. × 1,250.) (*After Bramlette and Riedel*, 1954, *J. Paleontol., vol. 28, pl. 38, figs. 1a, 2a, 3, 4, 10.*)

Bryophytes

Bryophytes, the most advanced nonvascular plants, are believed to have evolved from filamentous green algae. Although they are adapted to life on land, they still are dependent on water for their reproduction. The sperms are motile flagellates which make their way to the ova by swimming through films of water. Thus the bryophytes exemplify an adaptive stage among the plants, comparable in a way to that of the amphibians among the vertebrates.

Bryophytes are very rare among the fossils before the Pleistocene. Somewhat questionable specimens have been found at the top of the Devonian, but the earliest seemingly authentic remains of liverworts and mosses are of Upper Carboniferous age. They are so similar to some modern forms that evidently little subsequent evolutionary advancement has been made. Formerly bryophytes were considered to be ancestral to the higher plants. Most botanists are now agreed, however, that they are a side branch not in the direct line of plant evolution. Opinions differ as to whether liverworts or mosses are more primitive.

Vascular Plants

Vascular plants, or tracheophytes, are plants adapted to land life although some have returned to an aqueous existence. They are characterized particularly by the possession of an efficient vascular or circulatory system which conducts fluids between their parts. Most of them have well-differentiated roots, stems, and leaves, all of which are absent or imperfectly developed in other groups. Tracheophytes are essentially plants with woody tissues which resist decay. This accounts for their occurring as fossils more commonly than the nonvascular thallophytes.

The origin of tracheophytes is somewhat doubtful. Most botanists now believe that they evolved from an advanced type of alga similar to some of the brown algae

Table 48   Classification of vascular plants

Tracheophytes: vascular plants	
Pteridophytes: spore-bearing	
Psilophytes	Leafless and rootless
Lycopsids	Scale trees
Sphenopsids	Horsetails
Filicinopsids	Ferns
Spermatophytes: seed-bearing	
Pteridosperms	Seed ferns
Gymnosperms: seeds naked	
Cycadophytes	Cycads
Ginkgoites	Ginkgos
Cordaites	Cordaites
Coniferites	Common cone bearers
Angiosperms: seeds in ovary	
Dicotyledons	Dicots
Monocotyledons	Monocots

except that it was green (Fig. 48). All essential features of the tracheophytes evolved before the end of the Paleozoic Era. The only fundamental new development that has appeared among plants since that time is the angiosperm flower.

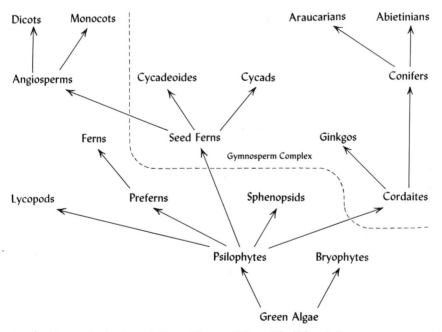

Fig. 48   Diagram showing the probable evolutionary relations of the higher plants.

Psilophytes

A considerable variety of primitive vascular plants has been described from Lower and Middle Devonian strata (Fig. 49). Similar forms have been reported from the late Silurian of Australia, although there seems to be some reason for questioning their greater age. Most of these primitive Devonian vascular plants consist of simply organized, woody stems that forked to produce nearly equal branches and grew only at their tips. These were borne on recumbent stems, or rhizomes, rather than on true roots. Some of them possessed spine-like or scale-like outgrowths or widened shoot tips, but they lacked true leaves. All reproduced by spores. Three species of similar plants still are living.

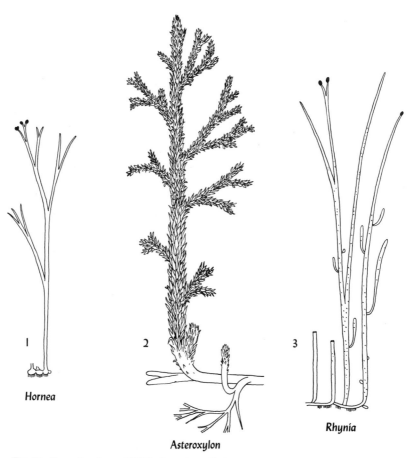

Fig. 49   Reconstructions of Middle Devonian psilophytes from Scotland. (Approx. × ½.) (*After Kidston and Lang, 1921, Trans. Roy. Soc. Edinburgh, vol. 52, pl. 1, fig. 1; pl. 2, figs. 3, 4.*)

Paleobotanists are not agreed regarding the classification of these simplest of all tracheophytes. All or only part of them are referred to the psilophytes. The more primitive species resemble advanced algae in some ways, but their stems contain woody vascular tissues and their surfaces are perforated by respiratory pores, or stomata. These plants probably grew in ponds like reeds, with their stems rising above the water surface. Their structures are such that they might represent the stock which diversified to produce the better-known pteridophytic groups. Other species were more or less intermediate between the simplest psilophytes and the lycopods, sphenopsids, and ferns. Probably none of the known fossils actually was ancestral to the other pteridophytes; they are not of sufficiently great age.

Lycopods

The lycopods were one of the dominant plant groups of the late Paleozoic, when they included trees which grew to heights of more than 100 ft. They have been definitely identified among fossils supposed to be of Silurian age, and some very questionable fragments have been reported from strata as old as the Middle Cambrian. Large tree-like forms did not survive the Permian. Several of the living species are remarkably similar to herbaceous late Paleozoic fossils.

Stems of the large lycopods contained a relatively small vascular cylinder, or

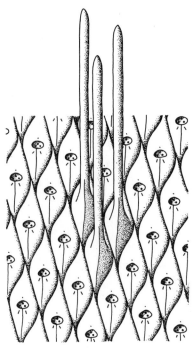

Lepidodendron

Fig. 50   Surface of part of the stem of *Lepidodendron*, an arborescent lycopod, showing leaves and leaf scars. (See Walton, 1953, "An Introduction to the Study of Fossil Plants," fig. 28A, p. 56; Andrews, 1961, fig. 8-11, p. 228.)

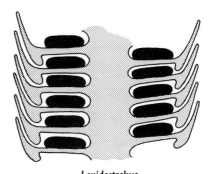

Fig. 51 Section through a cone of *Lepidodendron* (known as *Lepidostrobus*) showing sporangia borne on spirally arranged sporophylls. *(After Felix, 1954, Ann. Missouri Botan. Garden, vol. 41, fig. 1, p. 354.)*

Lepidostrobus

stele, mostly with central pith and wood that was both primary and of later growth. The outer part of the stems consisted of several differentiated layers, some of which were the main strengthening and supporting tissues of these plants. Radial enlargement of the stems seems to have been accomplished at several zones of growth or cambium. The outer layers did not slough away like the bark of modern trees, and therefore scars of the closely set leaves are preserved on the outer surface (Fig. 50). These features set the lycopods apart distinctly from all other plants. The stems show a primitive type of branching which produced nearly equal forks.

The root system of the lycopods also was peculiar. Most of these plants had four large horizontal roots, diverging from the base of the stem, which branched in a very short distance into equal parts two or more times. The roots resemble stems structurally, but some lacked pith and most of the woody tissue was secondary growth. The main roots bore many very regular, spirally arranged rootlets which, unlike those of other plants, were not equipped with root hairs.

The leaves of lycopods are believed to have evolved from surface outgrowths of the stem similar to those possessed by some psilophytes. They were without constricted stalks, or petioles, and generally they were small and narrow, although some are reported to have reached a length of about 3 ft. Most leaves had a single vascular strand, but in some this forked to produce two veins. No leaf gaps occur in the stele (Figs. 68-8; 69), where vascular strands were detached and led outward to the leaves. Lycopods differ in this way from most other plants.

Spore-producing organs, or sporangia, were borne on the surfaces of small specialized leaves, or sporophylls, closely arranged in a spiral order to form loose cones (Fig. 51). These terminated either main or subsidiary branches. The cones of some lycopods produced only a single kind of spore. In others, however, small male spores developed in the upper parts and larger female spores in the lower parts of the same cones. Evolutionary progression is shown by increase in the size of the female megaspores and decrease in the number produced within a sporangium. In several species, only a single megaspore remained. Some have been observed, still enclosed in the sporangium, which had begun to germinate. These are essentially primitive seeds.

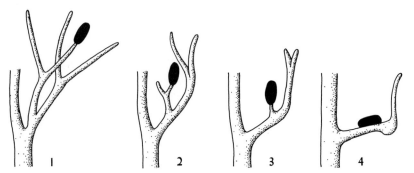

Fig. 52 Sketches showing stages in the supposed evolution of a sporophyll in a lycopod cone. (*After Zimmermann, 1952, fig. 10, p. 468.*)

The lycopods probably evolved from psilophyte-like ancestors. They exhibit a combination of primitive and advanced features different from those seen in any other plants. They definitely do not occupy a position in the main line of plant evolution.

Sphenopsids

The sphenopsids have a history much like that of the lycopods except that such early representatives are not known. Specimens from the Lower and Middle Devonian are somewhat questionable in their relations. Typical sphenopsids appear, however, in the Upper Devonian, and these plants were abundant in the Carboniferous forests, where some grew to heights of 50 ft or more. Such large species did not survive the Triassic. The few later fossils that have been found are small, more or less herbaceous forms, generally similar to some specimens known from the Carboniferous. Evidently little subsequent evolutionary advancement was made by this group of plants. One genus, represented by the modern horsetail rushes, still persists.

Stems of the sphenopsids are jointed, with distinctly marked nodes, where leaves or branches were given off, and internodes (Fig. 53). Those of the larger members of this group were hollow or contained a large central cylinder of pith. The primary wood cells were in small, separate bundles surrounding this central area. Considerable secondary wood was added outward in radially directed sectors. The cortex, or outer stem layers, of these plants was rarely well preserved. It generally was thin, and it was less complex than in the lycopods. Another group of the sphenopsids had a stele whose primary wood formed a solid triangular core surrounded by secondary wood which was denser radially outward from the angles of this structure. The gross branching habit of the sphenopsids is imperfectly known. Some specimens have subsidiary branches that were given off in whorls at the nodes of a main stem.

In some at least of the sphenopsids, the stems arose from horizontal rhizomes, from which grew unjointed roots. The roots commonly contained a solid core of pith surrounded by wood whose structure was generally similar to that present in the stems. The outer cortical tissues of roots were rarely preserved.

Sphenopsid leaves are believed to have evolved from small specialized branches born in whorls at the stem nodes. In one species which may be a primitive representative of this group these structures were stem-like and branched in three dimensions (Fig. 74-2). Typical sphenopsid leaves were flattened and seem to have extended outward approximately in a plane at right angles to the stem. The number of leaves in a whorl varied greatly in different species. In one group the leaves commonly were long and narrow (Fig. 53-1). Each included a single vascular strand, but a corresponding vein is not apparent. Wedge-shaped leaves more or less dissected along their outer margins were characteristic of another group (Fig. 53-2). Each had a single vein that divided into equal branches several times. Reduction in the size of leaves seems to have been an evolutionary tendency. This progressed so far that in modern species the leaves have almost disappeared.

A great variety of sphenopsid cones has been discovered. Some were terminal to main stems, others grew outward at stem nodes, and still others were born on specialized branches. Cones containing either one or two kinds of spores have been described. Although differing in many details, most of the cones had a characteristic

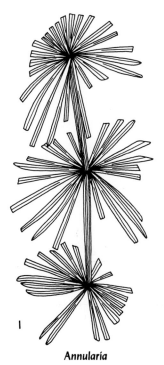

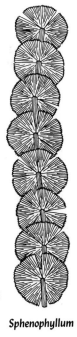

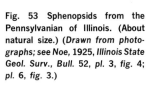

Fig. 53 Sphenopsids from the Pennsylvanian of Illinois. (About natural size.) (*Drawn from photographs; see Noe, 1925, Illinois State Geol. Surv., Bull. 52, pl. 3, fig. 4; pl. 6, fig. 3.*)

1   *Annularia*

2   *Sphenophyllum*

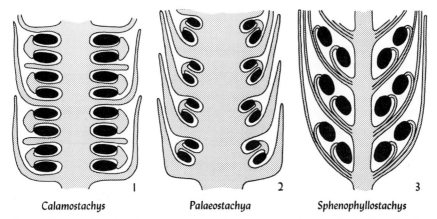

| Calamostachys | Palaeostachya | Sphenophyllostachys |

Fig. 54 Sections through sphenopsid cones showing sporangia enclosed by bracts. (× 3–5.) (See Andrews, 1961, fig. 9-9, p. 270; Baxter, 1955, Am. J. Botany, vol. 42, fig. 2, p. 343; Walton, 1953, "An Introduction to the Study of Fossil Plants," fig. 47, p. 73.)

structure and consisted of alternating whorls of leaf-like bracts and sporangio-phores, or organs composed of several sporangia (Fig. 54). The bracts were commonly united centrally to form a horizontal disk, but their separate scale-like ends turned up abruptly and enclosed the succeeding whorl of sporangiophores. In most cones, the sporangiophores were attached directly to the stem and bore four sporangia, although specimens with a smaller number are also known. A few species had sporangiophores arising from the surfaces of the bracts.

The sphenopsids are almost unique in both their jointed structure and whorled appendages. Their connection with psilophyte ancestors is indicated by some early Devonian plants of intermediate character. Like the lycopods, the sphenopsids are a side branch unrelated to the main line of plant evolution.

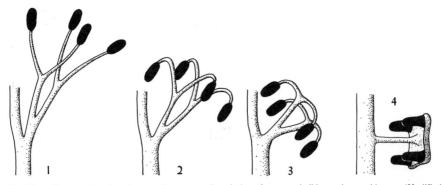

Fig. 54a Sketches showing stages in the supposed evolution of a sporophyll in a sphenopsid cone. (Modified after Zimmermann, 1952, fig. 10, p. 468.)

Ferns and Preferns

From an evolutionary standpoint, the fern-like plants are the most important element of the late Paleozoic floras. As loosely defined, they constitute a complex
group that can be subdivided into the preferns, true ferns, and seed ferns. These
subgroups occur stratigraphically roughly in this order, but their ranges overlap.
Preferns are known from the Devonian to the Permian. True ferns and seed ferns
were both well represented in the Carboniferous. The seed ferns, which are considered in the next section, died out during or shortly after the Jurassic Period.

The stems of ferns show many complications in their structure. Variations in
the fossils are known principally from tree-like plants, some of which attained
heights of more than 50 ft, although many were much smaller. All stems began
with a solid woody cylinder, in the midst of which a pith core appeared as the plant
grew upward. There a stele, consisting of a ring of woody tissue, was separated
into strands by gaps which opened directly above the departure points of bunches
of vascular fibers leading to the leaves. This produced a structure of more or less
regularly anastomosing vascular bundles. The woody tissue in other stems was
primitively lobed. A separation of these lobes probably produced the multiple strands
that did not rejoin each other in some stems. Few if any ferns developed secondary
wood. The outer cortex commonly was surrounded by an armor of broken leaf
bases.

The subterranean roots of fossil ferns are not well known. Some roots probably
grew from horizontal rhizomes rather than directly from upright stems. Several
kinds of tree ferns had abundant adventitious roots which originated in the stem
and passed downward within the cortex. One group possessed trunks composed of
numerous individual stems, all bound together by a complex of small adventitious
roots.

The leaves of fossil ferns were of many types. Most commonly they were
compound structures that branched several times. Some were more than 10 ft long.
They are believed to have evolved from branching shoots that became arranged in
a single plane, thus producing a flattened frond. Consequently they probably had
an origin similar to that of the leaves of sphenopsids but different from those of
lycopods. Primitive stages in the development of such leaves are exhibited by the
preferns. These plants bore minor side branches that divided into more or less
equal parts diverging in three dimensions (Fig. 74-1). Some had widened tips. The
vascular strands in the petioles of many fern leaves and their connections with the
stele had cross sections of several peculiar shapes.

Most fossil ferns seem to have reproduced by means of a single kind of spore.
These were produced in sporangia related to the leaves in several different ways
(Fig. 56). Some leaves bore highly specialized fertile lobes, and some whole appendages were fertile. Most ferns, however, had sporangia situated at the tips of
leaf lobes, along the sides of leaves or on the lower leaf surfaces. Changes in
position in this order seem to have been an evolutionary development.

The ferns rather obviously descended through the preferns from psilophyte

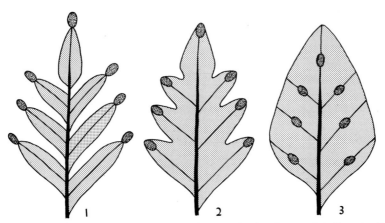

Fig. 56 Sketches showing the theoretical evolutionary migration of sporangia to the underside of the leaves of ferns.

ancestors. The many variations that mark all fern characters suggest that this group of plants may have evolved from several different ancestral lineages. Some modern ferns show little evolutionary advancement over those that existed at the end of the Paleozoic Era.

Seed Ferns

Pteridosperms, or seed ferns, were plants with fern-like foliage which, as their name implies, reproduced by megaspores that developed into seeds. Leaves generally are not sufficient for the recognition of these plants, and most fossil leaves unassociated with reproductive bodies cannot be identified with certainty (Fig. 57). For many years all leaves of fern-like type were believed to have been derived from true ferns.

A considerable variety of both large and small seeds has been discovered in strata of Carboniferous age, but few of them have been observed attached to foliage. The seed ferns certainly were abundant during Carboniferous time. They continued into the Permian, but their numbers seem to have been much reduced during that period. They persisted as a minor element of the floras at least until Jurassic time. Stems have been found in the Upper Devonian which resemble the seed ferns structurally, but no seeds are known to accompany them and their relations are doubtful.

Stems more or less definitely identified with the seed ferns are varied and in general are similar in complexity to those of the true ferns. A feature, however, that serves to distinguish most of them is the presence of well-developed and abundant secondary wood. Some had stems intermediate in structure between those of ferns and cycads. A few specimens have been found which show growth rings that may be annual. Few of the stems attained large size, and the seed ferns probably did not include plants comparable to some of the tall tree ferns.

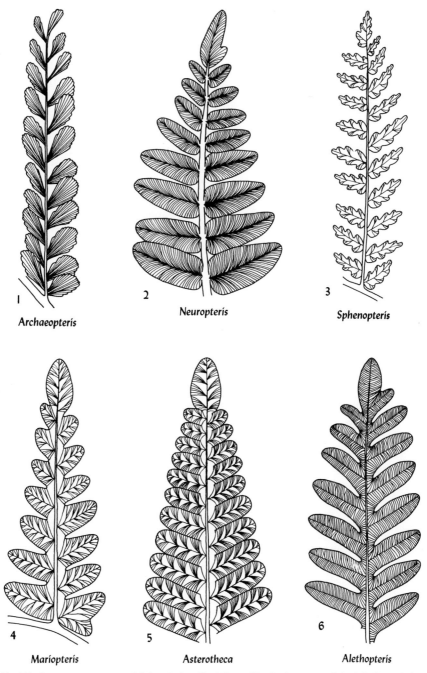

1
Archaeopteris

2
Neuropteris

3
Sphenopteris

4
Mariopteris

5
Asterotheca

6
Alethopteris

Fig. 57 Some common types of Paleozoic fern-like foliage. (Mostly about natural size.) *Archaeopteris*, from the Upper Devonian, bore spores. (*After Andrews, 1961, fig. 3-2c, p. 63.*) The others, from the Pennsylvanian, are all the leaves of seed ferns. *Neuropteris* foliage, however, was also borne by ferns which reproduced by spores. (*After photographs by Noe, 1925, Illinois State Geol. Surv., Bull. 52, pl. 13, fig. 2; pl. 16, fig. 1; pl. 24, fig. 1; pl. 34, fig. 3; pl. 38, fig. 1.*)

The roots which are best known are of the adventitious kind that originated within and grew downward in the stems. Like stems, they include considerable secondary wood.

The leaves were compound fronds of unknown size. Several of them seem to have divided into two equal branches near their emergence from a stem. Some types of foliage are believed to be definitely identified with the seed ferns, but others are so similar to the leaves of true ferns that distinction cannot be made in the absence of reproductive parts. In general, however, certain differences in the epidermal layers of the leaves may be diagnostic.

Some seeds were borne on nonleafy appendages, but mostly they were associated with leaves. The positions of the seeds varied comparably with those of the sporangia of ferns. They were terminal on either specialized or unmodified leaf lobes, marginal at the ends of veins, surficial, or attached to the axis of a frond. A few seeds have been found within loose structures like the husk of a hazel nut. Those of some Mesozoic species were almost enclosed in a fruit-like body and thus nearly resembled the seeds of angiosperms. None of the seed ferns is known to have produced cones. The pollen-bearing organs, or microsporangia, have not been identified certainly, although isolated specimens of such organs have been found in association with the abundant fragments of these plants.

The seed ferns probably evolved from psilophyte-like ancestors but otherwise seem to constitute a group not directly related to the true ferns. This is suggested by the fact that the development of two kinds of spores occurred very rarely if at all among the ancient ferns. This was necessarily antecedent to the seed-bearing condition.

Cycadophytes

Included in the cycadophytes are two groups of plants, the true cycads and the cycadeoids, whose relations to each other are comparable to those of the ferns and seed ferns. The definite record of these plants begins in the Triassic, although some late Paleozoic foliage may be related to them. Most fossils are cycadeoids. This group attained its maximum development in the middle or late Jurassic, continued to be important in early Cretaceous floras, but probably became extinct before the end of that period. The true cycads have persisted to the present time and are represented by nine modern genera which include all known Tertiary species.

Cycadophyte stems superficially resemble those of ferns. They contain a large central area of pith, a discontinuous ring of primary wood in strands which separate at the points where leaves were given off, scanty secondary wood, and a thick cortex commonly enclosed by closely set, spirally arranged leaf bases. Few of these plants were tall, and some were very short, although they seem to have lived for many years. Most of the stems branched irregularly a few times. Roots are very imperfectly known.

Whole leaves are rarely found, but some may have attained a length of 10 ft. Most of them possessed a stout midrib which bore numerous laterally directed

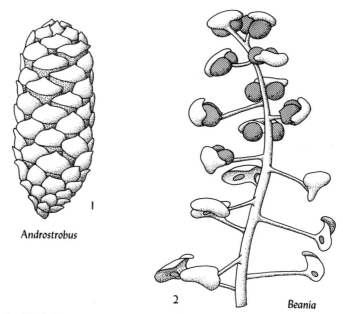

Androstrobus

Beania

Fig. 59   Fruiting structures of mid-Jurassic cycads from England. 1, A male cone with pollen-producing sporangia on the undersides of sporophylls. 2, Seeds in different stages of development. This structure, except for looseness, is similar to the female cones of some modern cycads. Both probably belong to a single species. (*After Harris*, 1941, *Phil. Trans. Roy. Soc. London, ser. B, vol.* 231, *fig.* 3, *p.* 82.)

leaflets with parallel veins, although some were not divided in this way. Leaves of the cycads and cycadeoids differed importantly in details related to their stomata.

From an evolutionary standpoint, the cycadophytes are chiefly interesting because of their reproductive structures. The modern cycads produce male and female cones on different plants (Fig. 59). These consist of spirally arranged sporophylls and arise amidst the cluster of terminal leaves. Several Mesozoic cones of similar structure have been discovered, but the botanical relations of some of them are doubtful. Confusion may be possible between the cones of cycads and fossil araucarian conifers.

The cycadeoids possessed flower-like structures. Some are commonly compared to flowers of the magnolias, which they resemble closely in certain ways. They ordinarily were borne among the leaf bases, and some seem to have been attached in the axils or angles between leaves and stem. Their stalks varied considerably in length. The most perfect flowers were bisexual (Fig. 60). The central seed receptacle consisted of interovular scales whose outer parts were expanded and partly joined together in a cone-like structure. This was surrounded by numerous pollen-producing sporophylls which seem to have developed from small specialized fronds. The

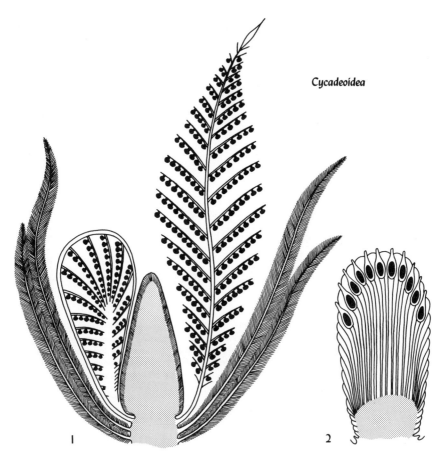

Cycadeoidea

Fig. 60 Reconstructions showing the structure of cycadeoid flowers. 1, A complete flower. (About natural size.) 2, The central seed-bearing, cone-like part of another species. (*Modified after Wieland, 1906, Carnegie Inst. Wash., Publ. 34, vol. 1, figs. 56, 87, pp. 110, 164.*)

sexes were separated in the reproductive structures of other cycadeoids which may or may not have been borne on different plants. Some were cone-like in their form. Most of these structures originated within a bud composed of numerous small leaf-like bracts.

The cycads and cycadeoids were very similar in some respects, but in others they differed greatly. They are believed to represent two groups of plants which evolved more or less concurrently from the same general ancestral stock. This probably was a relatively primitive part of the seed-fern complex.

Angiosperms

The angiosperms are the most advanced group of plants, and they dominate modern floras almost everywhere. They are characterized by a unique type of double fertilization in which two nuclei derived from a pollen grain pass into an ovule to form a seed.

The earliest discovered possible fossil angiosperms are Late Triassic palm-like leaves and other specimens that might be dicot leaves. Angiosperm pollen has been reported from Jurassic coal. The oldest unequivocal fossils occur, however, in the Lower Cretaceous, and some wood of this age has been referred to modern families. The angiosperms expanded greatly in late Cretaceous time. Large diversified floras have been collected in widely separated parts of the world, many of whose species are more or less confidently assigned to modern genera.

The angiosperms are seed-producing plants which bear true flowers with ovules, almost all of which are enclosed completely so that pollen tubes to reach them must penetrate the tissues of a stigma (Fig. 61-2). Their seeds are contained in fruits. All other seed plants have an opening in the integument through which pollen grains must pass to reach an ovule (Fig. 61-1). In addition most angiosperms have wood, including vessels which consist of large elongate cells communicating with one another through perforations in their ends. The presence of vessels, however, is not invariably a distinguishing character because they occur in a few species of lycopods and ferns and some angiosperms lack them. Most angiosperm wood also possesses other less-conspicuous identifying characters, particularly in its radiating structures.

The first angiosperms are believed to have been dicots, i.e., plants whose seeds contain two food-storage bodies which commonly also become the first two leaves. Moreover they probably were tree-like perennials. Such plants have an active cambium, where peripheral growth takes place, and abundant secondary wood as in the older gymnosperms, from some of which they almost certainly evolved. Herbaceous forms were rare before mid-Tertiary time. The monocots, with only one food-storage body in the seed, could have evolved from early dicots by the suppression of cambial activity, the loss of one cotyledon, and the breaking up of vascular

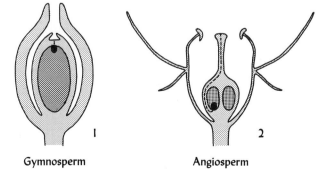

Fig. 61 Generalized drawings showing the ovules and associated structures of gymnosperms and angiosperms. 1, Gymnosperm pollen is received in a pollen chamber, and a pollen tube grows down to the ovum. 2, Angiosperm pollen adheres to the summit of a pistil, which is penetrated by a pollen tube, as indicated by the dotted line.

Gymnosperm   Angiosperm

tissues to produce more or less numerous scattered bundles. Most monocots, except palms, are herbs or herb-like plants. Very few of them produce secondary wood.

Herbs are in general better adapted to withstand severe climates characterized by seasons of cold or drought than are many woody perennials. The theory has been presented that the common development of the herbaceous habit in many different lineages of dicots resulted from deterioration of climatic conditions during Cenozoic time, particularly in the Pleistocene Period. Transformation from trees to vines, shrubs, perennial herbs, biennials, and finally to annuals has been suggested. This climatic explanation is not entirely satisfactory, however, because herbaceous plants are represented in floras much earlier than the Pleistocene. The relative scarcity of dicot herbs and monocots among the older fossils might indicate not that these plants were rare but that they were less likely than the woody ones to be preserved. Relative scarcity seems to be substantiated, however, by pollen studies.

Although the evolution of angiosperms from gymnosperms is generally agreed upon, the opinion is not uniform as to which kind was ancestral to them. One small group of plants commonly assigned to the seed ferns but also possessing similarities to the cycadeoids had flowers remarkably like those of the magnolias and their allies and produced enclosed seeds much like the angiosperms. Because of this the magnolias are considered by many botanists to be the most primitive angiospermous plants. This conclusion is supported by the structure of wood in some plants related to magnolias, which, like gymnosperm wood, lacks continuous vessels and thus differs from the wood of most other angiosperms. In spite of this, however, other botanists have pointed to the oaks, willows, and similar trees, whose simple petal-less wind-pollenated catkins resemble lax cones, as possibly representing the most primitive type of angiosperm, although their wood is not gymnospermous. In any case, the ancestry of the angiosperms is obscure. On the whole, seed ferns seem to be more probable than cycadophytes.

Cordaites

The coniferophytes, including the cordaites and conifers proper, constitute a great group within the plant kingdom that probably evolved entirely separately from the other seed-producing plants. The cordaites were one of the most prominent elements of the late Paleozoic floras, in which they were represented by trees more than 100 ft in height. Plants of this group are reported as ranging from the Upper Devonian to the Triassic, but both the oldest and youngest specimens consist principally of fossil wood. Other parts, with few exceptions, have been identified only from Pennsylvanian and Permian strata.

The stems of some cordaites and related plants grew very large. Fragments of tree trunks 5 ft in diameter have been found. These stems included a moderate to large cylinder of central pith, which commonly separated into imperfect transverse diaphragms. This was surrounded by more or less numerous, small, individual strands of primary wood. The abundant secondary wood was compact, traversed by relatively simple rays, and was without vessels. Thus the wood was essentially similar to that of modern gymnosperms. Possible annual rings occur in specimens from the southern hemisphere, but similar structure is very uncommon in plants which were members of the northern coal-swamp floras. The roots of cordaites are rarely recognized and have been studied by few persons.

Cordaite leaves are abundant in some Pennsylvanian strata. They were mostly relatively long and ribbon-like and superficially resembled large broad blades of grass. Some grew to lengths of 3 ft or more, although commonly they were much shorter. The parallel veins were of equal strength, or strong and weak veins alternated. These leaves were arranged spirally on stems. Their vascular traces in the stems were commonly associated with leaf gaps. Therefore, their origin is believed to have been similar to that of the leaves of ferns rather than the leaves of lycopods.

The male and female reproductive structures of the cordaites were similar loose cones or catkins (Fig. 63). Most of them consisted of sporophylls and intervening sterile bracts spirally arranged on short stems. They were borne in the leaf axils or attached to the stem a short distance above a leaf. Both pollen and seeds commonly were winged. The seeds were much like those of the seed ferns except

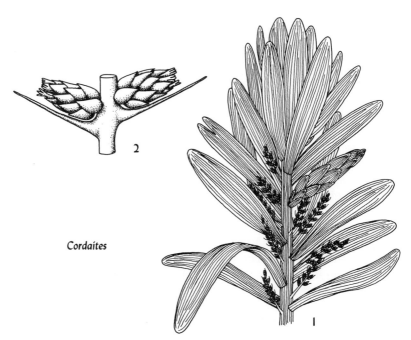

Cordaites

Fig. 63   1, Restoration of a growing tip with spirally arranged leaves, a large vegetative bud, and pollen-producing inflorescences. (*After Andrews*, 1961, *fig.* 11-1, *p.* 318.) 2, Detail of part of a male inflorescence. (× 12½.) (*After Delevoryas*, 1953, *Am. J. Botany*, *vol.* 40, *fig.* 5, *p.* 146.)

that they generally were bilaterally rather than radially symmetrical. No seeds have been discovered associated with the Upper Devonian or Lower Mississippian cordaite-like stems. Some paleobotanists have favored the idea that these possible early representatives of the group were seed-bearing. This is doubtful, and the cordaites surely evolved from ancestors that reproduced by means of spores.

No plants that might have been the immediate forerunners of the cordaites and their allies are known. Perhaps the members of this group are more diverse than present understanding of their structural similarities suggests. Probably, however, all of them were derived from psilophyte-like forms.

Conifers

The coniferites, or conifers proper, are the only ancient seed-bearing plants that persist importantly in modern floras. The earliest known fossils are of late Pennsylvanian age. Several modern families have been identified in the Jurassic. This great group reached its climax in early Cretaceous time, and by the late Cretaceous all modern families were in existence. In the early Tertiary, conifers were widely distributed throughout the world, but subsequent geographic restriction and the extinction of many genera and species probably resulted from competition with the rapidly expanding angiosperms. Numerous genera living at the present time are represented by single species surviving only within very limited areas. Essentially no evolutionary advancement has been made since the Cretaceous, and modern species, although still successful in many regions, are the remnants of a steadily dwindling stock.

The conifers include most modern gymnosperms. This latter large assemblage of plants, however, also embraces the seed ferns, cycadophytes, cordaites, and ginkgos, all of which resemble one another in the nature of their seeds and wood. In spite of these similarities, most botanists believe that evolution occurred independently in at least two great groups of plants and that the seed ferns and cycadophytes were not closely related to the cordaites and conifers.

The stems and roots of conifers contain much secondary wood, which generally is rather homogeneous and lacks the more conspicuous vessels and ray structure of many dicot angiosperms, and possess a thin cortex. Differences in the wood of cordaites and conifers are relatively minor. Some genera of conifers can be recognized but others cannot be distinguished on the basis of their stems alone.

The conifers developed three general types of leaves: (1) narrow, flattened blades generally small and commonly borne alternately on the two sides of a twig, (2) needles arranged spirally on stems or grouped in bundles of 2 to more than 20 arising from short lateral shoots, and (3) small scale-like leaves that more or less overlap each other. More than one kind may be present on a single plant. Reduction in the size of leaves may have been an evolutionary tendency in several lineages of conifers, but some of the oldest species had small leaves not much different from those seen on modern plants (Fig. 65). One group of conifers, the araucarians which are now confined to the southern hemisphere, have larger

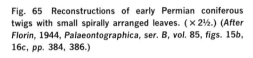

Fig. 65 Reconstructions of early Permian coniferous twigs with small spirally arranged leaves. (× 2½.) (After Florin, 1944, *Palaeontographica, ser. B, vol. 85, figs. 15b, 16c, pp. 384, 386.*)

1 *Lebachia*   2 *Ernestiodendron*

broader leaves very similar in structure to those of the cordaites. The absence of resin-containing canals in the wood of these plants also is suggestive of close relations to the cordaites and is one feature distinguishing them from the abietinians, or the pines and their allies. This has been considered good evidence that the araucarians are more primitive. A different conclusion has been reached by some botanists, however, who recognize supposed close homologies in abietinian and cordaitian cones. They believe that these reproductive structures are more conservative and less subject to environmental adaptation than the leaves and wood.

Many of the conifers differ from the cordaites in the more compact nature of their cones. Pollen and seeds are produced on the same plant in separate cones. These are generally similar in structure, although the seed cones are more complex. Some conifers produce seeds contained in berry-like bodies instead of cones. Opinions differ as to whether these evolved from typical cones or were derived from simpler structures without passing through an ancestral cone stage.

The conifers almost certainly evolved from cordaites.

Ginkgos

The ginkgo tree is one of the most interesting of all modern plants and one of the best examples of a living fossil. It is the last representative of an ancient lineage that can be traced back to the Permian and perhaps to Carboniferous time. Although it has a long geologic history, knowledge of fossil ginkgos is restricted almost entirely to leaves.

Ginkgo wood is typically gymnospermous. Probably it has not been identified in the fossil state because it cannot be distinguished from the wood of conifers. The modern trees grow in a dimorphic way. There are long rapidly extending shoots bearing scattered leaves and also short branches that grow very little and end in clustered leaves.

Leaves of the living ginkgo differ considerably in shape. Probably the leaves of ancient species varied similarly, and consequently they are difficult to classify. Nevertheless, the fossil leaves seem to present a nearly complete evolutionary series (Fig. 66). The earliest ones were narrow blades that divided in a generally evenly forked manner. They were not clearly differentiated from the shoots that bore them.

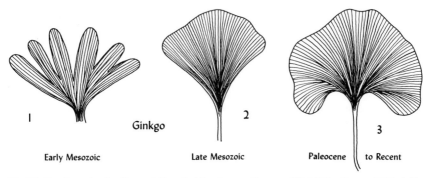

Fig. 66   Drawings showing the evolution of ginkgo leaves. (Approx. × ¾.) (*After Brown, 1943, J. Forestry, vol. 41, fig. 1, p. 863.*)

These were followed by leaves with broader, deeply cut lobes containing parallel veins. The lobes of such leaves seem to have progressively fused laterally until they almost wholly coalesced. At the same time a more and more distinct petiole, or stem, developed. The leaf shape then gradually changed from roughly triangular to semicircular by the lateral broadening of its base.

The modern ginkgo produces pollen-bearing catkins and singly borne seeds within a fruit-like outer coat on different plants. The reproductive structures of very few fossil species have been identified.

Some features of the ginkgo resemble seed ferns, cycads, and cordaites. Although its relations to these other plants are doubtful, a cordaite-like ancestor seems to be most probable.

EVOLUTIONARY TRENDS

The fossil record of ancient plants is so incomplete that few reliable phylogenetic sequences have been established. The great obstacles to an understanding of the origin and diversification of vascular plants are as follows. (1) Evolution seems to have progressed mainly in upland regions where plant remains were not preserved. The fossils, therefore, generally provide a record of accomplished evolution without clear indication of intermediate or ancestral forms that would reveal the actual course of evolution. (2) The fragmentary nature of almost all fossil plants is such that few species can be reconstructed in their entirety. The different parts of plants may be obviously more or less simple or complex, but the relative simplicity or complexity of whole plants is rarely known. Consequently, overall comparisons generally are not possible. In the almost total absence of known sequences of comparable complete fossils, many of the ideas concerning evolutionary relationships are based on the comparative morphology of modern plants. Opinions differ with respect to whether apparent simplicity of structure is primitive or the result of specialization and whether similarities reflect a common ancestry or parallel or convergent evolution. Theoretical phylogenies depend upon how such features are interpreted, and the results, as might be expected, differ greatly.

The mid-Paleozoic psilophytes provide examples of simple and presumably primitive plants representing a developmental stage through which all other vascular plants probably passed in their evolutionary development. In spite of much uncertainty concerning the actual levels at which the major groups diverged from one another, the general relations of most groups seem to have been fairly satisfactorily determined. Although evolution cannot be traced in any detail, separate consideration of the structure of different parts of plants indicates trends in the evolution of these parts from simplicity to more and more structural complexity. Evolution progressed unequally, however, among the associated parts of plants in different lineages, and reversion from relatively complex to more simple structure probably occurred in some of them. Recognition of these trends, therefore, does not necessarily clarify the evolution of particular plant lineages.

Stems

The origin of stems can perhaps be traced from a beginning in the single-celled fllamentous strands of algae or even in the chains produced by some bacteria. Both blue-green and green algae include filamentous forms. Many red algae consist of multicellular branches, but their cells are very little differentiated. Brown algae attain much greater complexity and exhibit the most advanced developments among the thallophytes. In structure, the brown algae range from simple filaments to elaborate growths including root-like holdfasts, branched stem-like stipes, and leaf-like blades. Many have gas-filled bladders which serve as floats. The stipe of such a form consists of (1) a core whose cells do not engage in food production, some of which are sieve-like cells and have vascular functions, surrounded by (2) a cortex capable of growth by the division of cells which contain photosynthetic pigments, enclosed by (3) a nonwaxy epidermis. These algae are buoyed up by the surrounding water and do not require supporting woody tissues. They increase in length partly by intercellular growth and also by the division of a single apical cell, like ferns, or a row of apical cells as in many seed plants.

Structure

The stems of vascular land plants perform three principal functions: they support the chlorophyll-bearing parts, conduct sap both upward and downward, and store a reserve supply of food and water. Although stems vary in structure in many ways, they consist essentially of (1) a central stele, surrounded by (2) a cortex, enclosed within (3) a dermal cover (Fig. 68).

From the center outward, the stele commonly contains (1) a median cylinder of pith-like food-and-water-storage cells, (2) a zone of xylem, or wood consisting of more or less complex conductive and supporting tissue, and (3) an outer member composed of conductive phloem cells that is structurally more homogeneous than

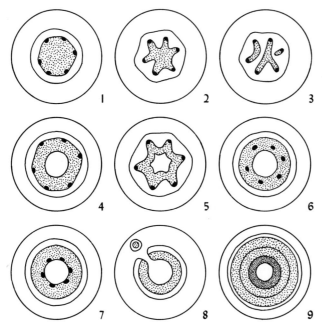

Fig. 68   Diagrams illustrating in transverse section the structures of plant stems showing the centers of woody growth. Xylem (wood) stippled, surrounded successively by phloem and cortex, and enclosing pith. 1 to 5, Xylem matures inward. 6, Xylem matures both inward and outward. 7, Xylem matures only outward. 8, Stem with leaf gap and leaf trace. 9, Stem with two rings of secondary wood surrounding primary xylem.

the xylem. A cambium, or zone of peripheral growth, if one is present, separates the xylem and the phloem. The cortex differs in different plants. In some it consists of food-and-water-storage cells. In others it is transformed into cork and fibers to produce bark. The dermal cover of young stems is a waxy epidermis, one or a few cells in thickness. In most stems which increase progressively in diameter by cambial growth, the epidermis dies and sloughs off, and the dermal cover consists of other dead cells derived from the cortex or phloem.

The simplest and presumably the most primitive stems contain a solid central woody core of xylem (Fig. 68-1 to 3). This is the condition of some fossil psilophytes and also of the ontogenetically earliest formed parts of stems of some plants whose later structure is more complex. Evolutionary advancement is indicated by the presence or appearance of a pith cylinder in the center of such a core (Fig. 68-4, 5). This stage of development was reached in some psilophytes and in almost all higher plants.

All stems grow by the proliferation of cells located at their tips. These cells produce the primary, or first-formed, tissues, which rapidly differentiate. The first xylem cells appear at certain centers and are augmented by the subsequent differentiation of adjacent tissue. Most primitively, the primary wood increases from these points inward toward the center of the stem (Fig. 68-1 to 5). This pattern of development is characteristic of most fossil psilophytes, lycopods, one section of the sphenopsids, and the most primitive ferns both living and fossil. In the next evolutionary stage, primary wood is added on all sides of the original centers of development (Fig. 68-6). This stage is reached in the stems of ferns and seed ferns, and it is characteristic of a few other plants. In other gymnosperms and all angiosperms, differentiation of primary wood progressed only in an outward direction (Fig. 68-7). Secondary wood surrounds the primary in all stems which possess an active zone of growth or cambium (Fig. 68-9).

Other features of the stele have evolutionary implications. A circular cylinder of primary wood (Fig. 68-1) is more primitive than a lobed one (Fig. 68-2) with the first-formed cells located in the protruding parts. Pith may or may not be enclosed in either kind of woody column. Further advancement is shown by the arrangement of primary wood in separate transverse or radiating bands (Fig. 68-3).

Progressive complications of stelar structure also are associated with the relations of wood and phloem, the nature of rays which transect the wood, and the development of leaf gaps. The latter are discontinuities in the primary wood above the points where vascular strands, which continue into leaves, diverge from a stele containing central pith (Figs. 68-8; 69). Leaf gaps occur in almost all vascular plants other than the psilophytes, lycopods, and sphenopsids. They are particularly conspicuous in some ferns whose steles are transformed into a network of anastomosing vascular strands. Furthermore, evolution produced still other variations in stem structure too complex to mention here.

Branching

Patterns of branching exhibit a series of evolutionary advances. The most primitive of all vascular plants probably grew as simple, straight, unbranched shoots. Most of the oldest fossils branched to produce two almost equally strong shoots at each forking (Figs. 49-1; 70-1; 72-1). This stage of development had been reached by

Fig. 69 Generalized sketch of the internal structure of a stem showing xylem (wood) cylinder and pith core, and the relations of a leaf gap to the diverging vascular leaf trace.

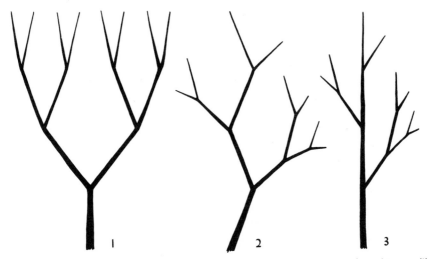

Fig. 70 Diagrams showing how the primitive pattern of equal branching can be transformed to one with a main stem and smaller lateral shoots. This is termed overtopping.

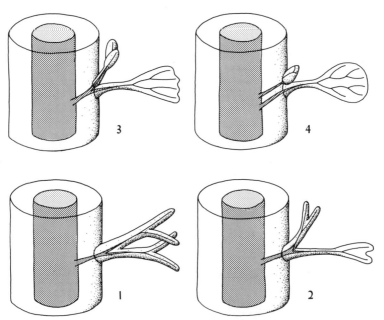

Fig. 70a Sketches illustrating successive stages in the theoretical evolution of the axillary bud. Vascular strands from the stele are shown passing through the cortex of a stem.

at least early Devonian time, and it is particularly characteristic of most fossil lycopods. With only slight modifications it has persisted in some plants to the present time. In the next step, one shoot grew somewhat more rapidly than the other, and the differentiation of a main stem and side branches began (Fig. 70-2). Continued differentiation and the development of a regular system of branching produced stems with more or less equally spaced nodes and internodes (Fig. 70-3). This type of branching occurred in the late Paleozoic and is characteristic of many modern plants. The ultimate stage of differentiation is exemplified by small specialized branches which grow very little in length and bear the reproductive organs or clustered leaves.

The earliest branching resulted from division at the tip of a growing shoot. Branches produced in this way tended to grow upward. They continued to elongate until the shoot tip died or produced a determinate organ such as a leaf or a reproductive structure. The development of, first, small lateral branches and, later, buds in the axils of leaves (Fig. 70a) resulted in a different type of branching. This, of course, was incidental to the transformation of specialized branches into leaves. Shoots arising from axillary buds are younger than the stems which bear them. In many plants they tend to grow laterally rather than upward. Axillary branching is characteristic of modern conifers and angiosperms as well as of some other plants. Its beginning can be traced back to the origin of leaves in mid-Devonian time.

Roots

Algae which live immersed in water require no roots because they are not subject to desiccation and they obtain all their nutritional requirements by absorption from the surrounding medium. Attached forms are held in place most simply by a growth of only slightly specialized cells which adhere to some solid object. Advanced and especially larger algae may develop more elaborate holdfasts, consisting of branched stipe-like structures which penetrate the bottom sediment like roots but which lack other root-like functions.

Land plants must obtain moisture and mineral matter from the soil in which they grow. The most primitive tracheophytes arose as shoots from a recumbent stem, or rhizome (Figs. 49-1, 49-3; 72-1). This was furnished with numerous unicellular or multicellular hair-like appendages or scales which acted as absorbing organs. Plants of this sort could exist only in more or less permanently damp situations.

Roots probably developed from specialized prostrate stems at an early stage in the evolution of vascular plants (Fig. 72-2). The differentiation is evident in all groups except the psilophytes. The first branching of the plant axis which produces the primary root develops during early embryonic growth. In the lycopods and sphenopsids the root is directed laterally, as might be expected of a branch. In ferns, gymnosperms, and angiosperms, however, it is directed downward, in the same axis as the upwardly growing shoot. This evidently is an evolutionary advancement.

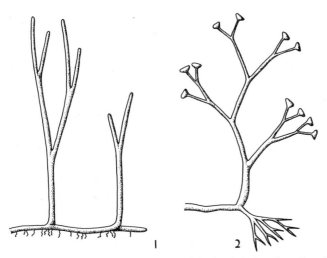

Fig. 72 Diagrams showing the theoretical origin of roots by transformation of a rhizome, or recumbent stem.

The details of growth in roots and shoots are significantly different in several ways. These differences, however, do not produce results that are clearly apparent in the mature areas behind the apical zones of growth. Nevertheless, there are some structural differences, and roots in general are more conservative than their associated stems. Such were the relations even in some psilophytes whose rhizomes possessed a stele that was longitudinally discontinuous. This is a condition more primitive than that known in any stem.

Roots differ most noticeably from stems in the sharper boundary separating stele from cortex. Central pith generally is less developed except in the roots of monocot and herbaceous dicot angiosperms and sphenopsids. Primary wood continues inward from the centers of original differentiation, as it does in primitive stems. Because roots do not bear leaves, there are no nodes and internodes, nor are leaf gaps present.

Leaves

Leaves are specialized organs whose function is the manufacture of food by photosynthesis. This process also is effected by other tissues, e.g., cells underlying the epidermis of stems in plants which do not have a thick outer covering of bark. Most leafless plants have relatively unspecialized cells that perform photosynthesis. These cells generally contain tiny chlorophyll-bearing bodies. A few bacteria and all bluegreen algae possess diffused chlorophyll that is not concentrated in this way.

No alga has leaves, although the blades of brown algae and of some of the larger greens are leaf-like and function much like leaves. These parts, however, do not have the layered structure or vascular tissues of true leaves.

The bryophytes also lack true leaves, although the mosses have leaf-like appendages and the whole upper surface of liverworts has leaf-like functions. Most of the appendages of mosses are only one to a few cells thick, lack the internal structure of leaves, and have no stomata. Liverworts have an upper epidermis pierced by pores opening into air chambers which are lined by cells richly supplied with chlorophyll. Below these is a zone of cells with less chlorophyll, and then a lower epidermis. This superficially resembles the structure in the leaves of flowering plants, but there are no veins.

The most primitive vascular plants were leafless, and photosynthesis was accomplished by cells of the cortex in their stems. Stomatal openings communicating with air chambers where gaseous exchange was accomplished pierced their epidermal layers. Leaves seem to have developed from plants of this kind in two different ways: (1) from local outgrowths of the epidermis (Fig. 49-2), and (2) from the laterally expanded tips of stems (Fig. 75-2).

Some of the psilophytes or their near relatives had small surface outgrowths of several kinds. In one type, strands of vascular cells led from the stele to the bases of spines or scales but did not continue into them (Fig. 73-2). In another, short stems or leaf-like structures were furnished with vascular strands but these did not

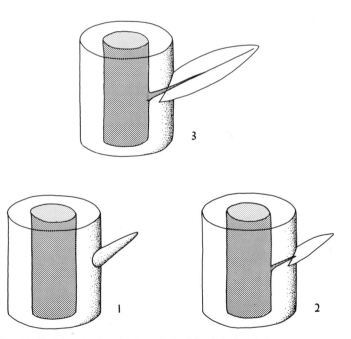

Fig. 73 Sketches showing the theoretical origin of the leaf in lycopods. (*After Smith*, 1955, *"Cryptogamic Botany,"* vol. 2, fig. 86, p. 137.)

connect inward with the stele. The leaves of lycopods probably evolved from structures of the former kind. They were very simple and contained a vascular strand
or vein that generally was single but might fork once. There were no leaf gaps where
these strands left the stele.

One group of psilophyte-like plants was transitional between the typical psilophytes, whose shoot tips were unmodified, and the ferns and seed ferns (Fig. 74-1).
The shoot tips of these preferns exhibit progressive flattening and thus acquired
leaf-like characters (Fig. 75). In the earliest development of this kind, no abrupt
transition separated the stem and flattened shoot, and branching was not restricted
to a single plane.

The derivation of sphenopsid leaves is uncertain. Some botanists believe they
had an origin similar to that of the leaves of lycopods, because they are simply
veined and the vascular strands leading to them are not associated with leaf gaps
in the stem. Others believe they evolved from lateral stems, because seemingly
related fossils have stem-like outgrowths at the nodes which may have been forerunners of the leaves (Fig. 74-2).

In other plants the most primitive leaves derived from stems were narrow and
possessed a single vascular strand continuous with the stele and associated with
a leaf gap. The earliest leaves with multiple parallel veins probably resulted from
the lateral fusion of more simple narrow leaves. Fern-like fronds branched several
times and bore lateral leaflets all lying in a single plane. Leaves of this type grew
at their tips within a tight coil which unrolled as the leaves matured. This is remi-
niscent of the growth of shoot tips in some psilophytes and related plants.

Evolutionary advancement is evident in the increasing complexity of leaf venation. The sequence starts with narrow leaves possessing a single vein and proceeds
to wider leaves with numerous parallel or radiating veins produced by repeated
forking. Primitively such veins end at the leaf margin but in more advanced leaves

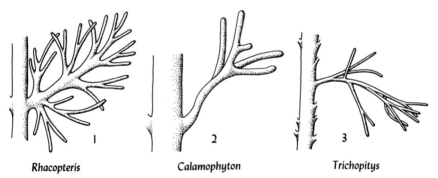

Rhacopteris   *Calamophyton*   *Trichopitys*

Fig. 74 Stem-like appendages considered to be the precursors of leaves. 1, A primitive fern. 2, A
primitive sphenopsid. 3, A primitive ginkgophyte. (*After Leclercq and Andrews, 1960, Ann. Missouri
Botan. Garden, vol. 47, fig. 5, p. 7; Andrews, 1961, figs. 3-130, 11-15, pp. 86, 342.*)

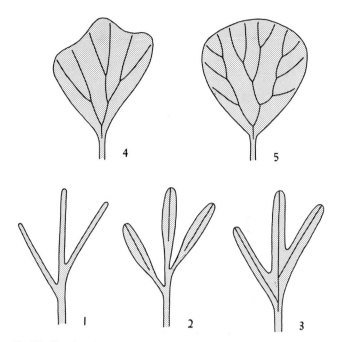

Fig. 75 Sketches showing successive stages in the theoretical evolutionary transformation of shoot tips to expanded leaves.

the veins curve forward and become connected there. The development of a prominent midrib was a structural advancement that strengthened leaves. Veins branch from this rib at increasingly large angles until they extend laterally in some leaves. Anastomosing net-like veins were the ultimate development. Several different patterns were produced. These are particularly characteristic and conspicuous in dicot angiosperms and numerous fern-like leaves, as well as in some other relatively unrelated plants. Obviously these patterns evolved independently several times. Multiple prominent strong veins or ribs occur in many angiosperm leaves, especially those with lobed or coarsely serrated margins.

Not all leaf venation evolved toward more complexity. Most of the conifers have narrow single-veined leaves that suggest a primitive condition. Their small scales and needles, which generally are not shed seasonally, seem to be adapted to reducing the loss of moisture by transpiration.

The constriction of leaf bases and the development of petioles evolved independently in different plant lineages. The leaves of most fossil plants, except the dicot angiosperms and some lycopods and conifers, did not separate cleanly from the stems. Particularly the stems of tree ferns and cycadophytes commonly were encased in a rough layer of broken leaf bases. The shedding of leaves by arborescent angiosperms and some gymnosperms probably was an adaptation to annual climatic changes involving pronounced variations in temperature or moisture, or in both, which do not seem to have characterized the climates of much past geologic time.

Reproduction

Details of the reproductive organs and reproductive processes of plants are extremely varied. Nevertheless, a general sequence can be recognized that represents a series of evolutionary advancements. The simplest and most primitive plants, bacteria and blue-green algae, reproduce by cell division, thus forming new single-celled individuals. Many other plants also reproduce vegetatively by fragmentation or the development of parts that become detached from the parent and grow into new individuals.

Alternation of Generations

The alternation of diploid sporophytic or spore-producing and haploid gametophytic or sexual generations is a characteristic feature in the reproduction of most plants. Bacteria and blue-green algae do not have alternate generations. Their sexual activity seems to be restricted to occasional and perhaps rare conjugation in which two individuals come together and exchange genetic material. Other algae exhibit many variations, not obviously related, in their reproductive processes. Examples may be chosen among these and other plants, however, that seem to mark the main pathway of progressive evolution.

Some filamentous blue-green algae produce large specialized cells that become free and germinate to form new threads. Others have occasional thick-walled cells serving to break the filaments into parts, which then grow separately. Some of them also germinate to produce new filaments. Although cells of these and similar kinds are spore-like in their nature, they are not strictly comparable to the spores of other plants. Their functions are not entirely reproductive, and they play no part in an alternation of generations. Some are important principally because they are adapted to withstanding conditions that would be fatal to ordinary cells.

The most primitive plants possess genetic material, whether concentrated in a nucleus or not, that probably corresponds to the haploid condition in other organisms. Their spores, likewise, are haploid and germinate to produce other haploid plants. No sexual generation occurs. In more advanced plants the sexual process consists of the fusion of two haploid spores to form a fertilized diploid zygospore. Haploid spores of this kind, therefore, are sex cells, or gametes. The zygospores of many green algae almost immediately undergo two successive divisions, as in the process of meiosis, and each produces four flagellate haploid cells that grow to become haploid individuals. This is the first step in the development of alternating generations. The sequence is (1) haploid gametophyte, (2) haploid sex cells, (3) diploid zygospore, (4) haploid nonsexual spores, (5) haploid gametophyte, etc.

In the next step, the second division of meiosis is postponed. The zygospore

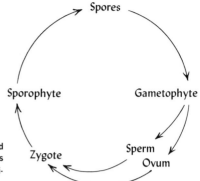

Fig. 77 Diagram illustrating the process of reproduction and alternation of generations in plants. Individual gametophytes may produce both sperms and ova, or they may be unisexual.

grows into a diploid individual or sporophyte. The sequence, therefore, now is (1) haploid gametophyte, (2) haploid sex cells, (3) diploid zygospore, (4) diploid sporophyte, (5) haploid nonsexual spores, (6) haploid gametophyte, etc. Distinct alternation of generations has been achieved (Fig. 77).

Important evolutionary advantages evidently have resulted from sexual reproduction. Something about this process seems to account for the progressive increase in the importance of the diploid sporophyte in the alternation of generations among plants which evolved independently a number of different times. Beginning with the very subordinate position of the zygospore in this sequence as in algae, the diploid generation gradually rose to dominance in the higher plants. Concurrently the haploid generation was progressively reduced until it disappeared (Fig. 77a).

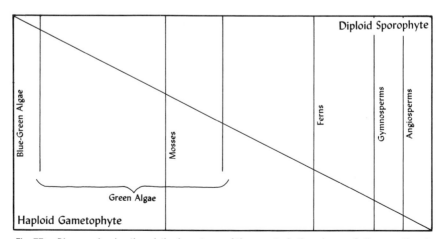

Fig. 77a Diagram showing the relative importance of the gametophytic and sporophytic generations in various groups of plants.

The following six stages may be recognized among modern plants that show increasing evolutionary advancement with respect to reproduction. (1) Zygospores undergo meiosis directly, as in many green algae. (2) Zygospores germinate and grow into small sporophytes, as in some multicellular algae and the bryophytes. The algal sporophytes are relatively small, different in form from the gametophytes, and generally short-lived. Those of the bryophytes are very small and parasitic on the gametophytes. (3) Sporophytes and gametophytes are very similar, if not identical, morphologically and in their way of life, as in some algae of both the single-and multiple-celled types. (4) Gametophytes are much reduced, as in the ferns and some brown algae. The sporophytes are relatively larger, longer-living forms, but the free-living gametophytes are small, inconspicuous, short-lived plants. (5) Gameto-phytes are reduced to only a few cells which are parasitic on the sporophytes, as in the gymnosperms and angiosperms. (6) Gametophytic generation is completely suppressed, as in some brown algae which produce sperms and eggs directly. These plants have reached a reproductive stage exactly comparable to that of most animals.

Spore Differentiation

Most algae produce flagellate spores, known, because of their motility, as zoospores. The most primitive sex cells are smaller but otherwise very similar to them. Male and female gametes, if any differentiation occurs, cannot be distinguished. The sexual process probably began by the meeting and fusion of two undersized spores, neither of which was capable of growing into a new individual by itself. Subsequently such fusion, which results in a zygote or a diploid sexually produced individual, became a common process. Presumably the fusing partners generally were derived from different parents, and thus cross-fertilization was accomplished.

The gametes of other algae are differentiated in such a way that the sexes are distinguishable. In some, the only obvious difference is one of size. As one, the female, becomes relatively larger than the other, its motility generally decreases. The sperms continue to be small flagellate cells, but the ova develop into large nonflagellate ones incapable of any movement. There also is a tendency for the ova which originally had been set free to be retained within receptacles in the body of the parent.

Flagellate sperms are produced by most thallophytes, bryophytes, and the lower spermatophytes. In such land plants as the bryophytes and some pteridophytes, they must swim to the ova through surface films of water. The reproduction of these plants, therefore, can be accomplished only in damp situations or at seasons when moisture is abundant.

Plants living on the land, where they are surrounded by air instead of water, must overcome the problem of desiccation. One phase of their adaptation to ter-restrial life involved the production of gametes within multicellular protective organs. The ova are fertilized within their receptacles, and growth of the sporophyte begins there with the development of a multicellular embryo.

Another very important evolutionary step was the development of pollen and relative reduction of the gametophytic generation. Individual gametophytes may produce gametes of both sexes or of only one. The sexuality of gametophytes is determined by the spores from which they grow. The more primitive plants produce only one kind of spore, and the gametophytes are bisexual, bearing both male and female gametes. Species of this kind also are common in almost every major group of advanced plants, although with some of them this condition may be a reversion rather than a primitive state. The differentiation of sex-determining spores was an evolutionary advancement that seems to have been achieved independently in various plant lineages. If this occurs, the female spores commonly are larger than the male, and distinction is made between megaspores and microspores. Relative size, however, is not entirely conclusive, and the sexuality of many fossil spores cannot be determined (Fig. 79).

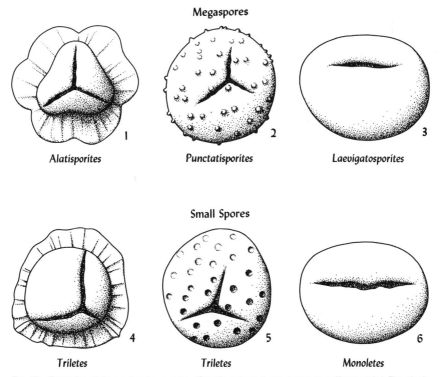

Fig. 79 Examples of Pennsylvanian spores. (Greatly enlarged, not drawn to uniform scale.) The single or three radiating marks are scars where adjacent spores were in contact as they developed in a sporangium. The identification of spores with particular plants generally is very uncertain. (1 to 3 *after Kosanke, 1950, Illinois State Geol. Surv. Bull. 74, pl. 2, fig. 1; pl. 4, fig. 1; pl. 5, fig. 7. 4 to 6 after Schopf, 1938, ibid., Rept. Invest. 50, pl. 1, figs. 8, 9; pl. 6, fig. 5.*)

The spores of most land plants have resistant outer coats as protection against drying out and other hazards. Transformations progress within the cell inside this coat in the higher plants before growth results in bursting of the coat. The nucleus of the spore divides, and several cells are formed. This, of course, is the first stage in the development of a gametophyte. A pollen grain is a microspore in this condition (Fig. 80). Whether fossil microspores actually were pollen grains may be difficult to determine because only the spore coat commonly was preserved. At first un-transformed microspores probably were carried by the wind or other means and deposited adjacent to a megaspore. There they germinated, grew into tiny gameto-phytes, and produced sperms. Spores of this kind might be considered a very primitive type of pollen.

The pollen grains of modern plants, and doubtless of many fossils also, are or were transported in a similar way. After the pollen grain is deposited near an ovum, a tube grows out from the grain and penetrates the tissues leading to and surrounding the ovum (Fig. 61). In most gymnosperms except the conifers motile sperms are freed from the pollen tube. This seems to be an excellent example of recapitulation in which motile sperms have persisted after evolution made their existence unnecessary. Further evolution that was independent in the conifers and angiosperms suppressed the sperm-producing stage; fertilization is accomplished in these plants by male gametes which are little more than nuclei that pass from the pollen tubes directly into the ova.

Seeds

A seed consists of an ovule or an embryo which is a very immature sporophyte in a resting state, a supply of food, and an outer coat. The evolution which led to the development of seeds was one of the most important adaptations of plants to land life. It further freed them from dependence upon water in their reproduction, much as the development of the reptilian egg freed vertebrate animals from a similar

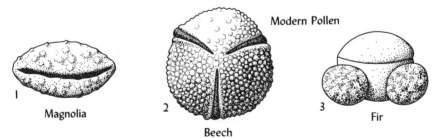

Fig. 80   Examples of modern pollen, greatly enlarged. Many kinds of pollen grains have "furrows" where the coat may expand or contract to compensate for volume changes, depending on the absorption or loss of moisture, and through which the pollen tube emerges. (*After Wodehouse, 1935, "Pollen Grains, Their Structure, Identification and Significance in Science and Medicine," pl. 2, fig. 4; pl. 5, fig. 8; pl. 7, fig. 5; pp. 259, 329, 375.*)

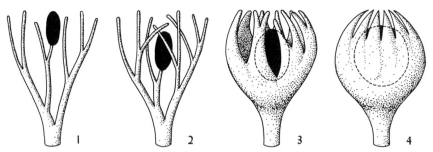

Fig. 81 Generalized sketches showing possible evolutionary stages in the transformation of a mega-spore into an enclosed seed.

dependence. Seed development certainly was accomplished independently in the lycopods, seed ferns, and cordaites. It may also have been independent in some other groups.

The evolution which led to seeds began with the differentiation of megaspores and microspores. It was a gradual process, accompanied by the development of pollen. As finally perfected, the nucleus of a megaspore divides, cells are produced, and a tiny gametophyte is formed. This in turn produces an ovum, which is fertilized by a male gamete of a pollen grain whose tube penetrates what was once the outer coating of the megaspore. The ovum segments into cells and grows, an embryo develops, and a seed has been created.

Cones and Flowers

The spores of most thallophytes are produced within cells that may or may not be specialized and differ from other cells. In all higher plants, spores are produced within multicellular organs, termed sporangia, where the double division of meiosis is accomplished. These organs occur singly and terminate the tips of ordinary shoots in the most primitive pteridophytes. Differentiation, however, seems to have developed very early between fertile and nonfertile shoots. From this beginning, the evolution of reproductive organs in plants largely involved variations in the relations of sporangia to leaves, or structures presumed to be derivatives of leaves, and the development of complex composite structures (Fig. 82).

A rather consistent tendency developed among most plants for the sporangia to become grouped. This seems to have resulted from the stunting of fertile shoots and associated leaves to produce cone-like structures of increasing compactness. Both the sporangia-bearing and seed-bearing parts and the bracts which more or less enclose them probably are leaves much specialized and reduced in size. Although not all stages in the development of compact cones are known in most plant lineages, loose catkin-like structures probably were intermediates. Primitively the cones or their predecessors were borne terminally on main stems, but as the result of unequal branching, they acquired lateral positions and short stems. This or a

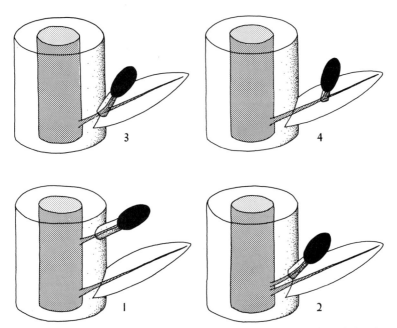

Fig. 82   Generalized sketches showing a possible series of stages in the theoretical evolution of a spore-bearing leaf, or sporophyll.

comparable process did not occur, however, in the ferns or seed ferns, which never produced typical cone-like structures.

Flowers also evolved by the specialization of short fertile leafy shoots. A perfect and complete simple flower consists of a pistil surrounded successively by stamens, petals, and sepals. The sepals and petals are obviously leaf-like. Their primitive spiral arrangement changed to cyclic, and the area of attachment rose from below to above the seed-bearing receptacle at the base of the pistil. Stamens in the most primitive flower-like structures, as in the cycadeoids, were small frond-like foliar organs which bore microsporangia. This resemblance, however, is not retained in the flowers of angiosperms. The pistil contains one or more ovules at its base that have been derived from megasporangia. The whole organ consists of one or more specialized leaves whose edges are fused together.

Almost infinite variation is to be seen in the form, development, and detailed relations of the parts of flowers. Thus some are without petals, some lack either pistils or stamens, and some are compound. Fruits produced by primitive flowers such as the magnolia resemble cones. Others vary greatly in form and structure. Many of these differences are related to the nature of the flowers. The angiosperm flower is the most highly evolved part of any plant.

BIBLIOGRAPHY

H. N. Andrews, Jr. (1960): Evolutionary Trends in Early Vascular Plants, *Cold Spring Harbor Symp. Quant. Biol.,* vol. 24, pp. 217–234.
The earliest plant fossils present a confusing picture which is difficult to interpret in terms of evolutionary relations.
_____ **(1961):** "Studies in Paleobotany," John Wiley & Sons, Inc., New York.
This book is devoted mainly to the descriptive morphology of fossil plants.
C. A. Arnold (1947): "An Introduction to Paleobotany," McGraw-Hill Book Company, New York.
This conventional textbook includes little consideration of evolution.
_____ **(1948):** Classification of Gymnosperms from the Viewpoint of Paleobotany, *Botan. Gaz.,* vol. 110, pp. 2–12.
Two distinct lineages are recognized, pteridosperm-cycadophyte and cordaite-ginkgo-conifer.
_____ **(1948):** Paleozoic Seeds, *Botan. Rev.,* vol. 14, pp. 450–472.
Seed plants were not derived from heterosporous ancestors, and the earliest seeds contain no embryos.
_____ **(1953):** Origin and Relationships of the Cycads, *Phytomorphology,* vol. 3, pp. 51–65.
The relations and ancestry of cycads and cycadeoids are uncertain, but they were not closely related to conifers.
D. I. Axelrod (1959): Evolution of the Psilophyte Paleoflora, *Evolution,* vol. 13, pp. 264–275.
Land plants evolved from algae possibly in the Precambrian, and psilophytes represent an evolutionary stage rather than a group of related plants.
_____ **(1960):** The Evolution of Flowering Plants, in "Evolution after Darwin," S. Tax (ed.), The University of Chicago Press, Chicago, vol. 1, pp. 227–305.
Angiosperms were successful because of their plasticity and ability to adapt to widely different environments.
E. W. Berry (1949): The Beginnings and History of Land Plants, *Johns Hopkins Univ. Stud. Geol.* 14, pp. 9–91.
This article is a simple appraisal of what is known of the main lines of plant evolution and the distribution of fossils.
H. C. Bold (1957): "Morphology of Plants," Harper & Row, Publishers, Incorporated, New York.
This is a comprehensive, systematic, and descriptive text.
D. H. Campbell (1940): "The Evolution of Land Plants," Stanford University Press, Stanford, Calif.
Although mainly morphologic and systematic, this work is much more concerned with evolution than are most botanical texts.
C. J. Chamberlain (1935): "Gymnosperms: Structure and Evolution," University of Chicago Press, Chicago; paperback reprint, Dover Publications, Inc., New York.
See particularly chaps. 20 and 21, pp. 427–445, for a discussion of phylogeny and consideration of progressive stages in the alternation of generations.
W. C. Darrah (1939): "Textbook of Paleobotany," Appleton-Century-Crofts, Inc., New York.
See chap. 23, pp. 405–421, for a brief outline of plant evolution.
T. Delevoryas (1966): "Plant Diversification," Modern Biology Series, Holt, Rinehart and Winston, Inc., New York.
See particularly chap. 5, pp. 96–129, on the evolution and adaptations of angiosperms; this book also discusses other aspects of plant evolution.
A. S. Foster and **E. M. Gifford, Jr. (1959):** "Comparative Morphology of Vascular Plants," W. H. Freeman and Company, San Francisco.
The main morphologic features of modern plants are described in this well-illustrated book.

H. F. Fuller and **O. Tippo (1954):** "College Botany," rev. ed., Holt, Rinehart and Winston, Inc., New York.
Good organization characterizes this comprehensive text.

S. Leclercq (1956): Evidence of Vascular Plants in the Cambrian, *Evolution,* vol. 10, pp. 109–114.
Reported occurrences of pre-Devonian land plants are summarized.

R. C. McLean and **W. R. Ivimey-Cook (1951, 1956):** "Textbook of Theoretical Botany," 2 vols., Longmans, Green & Co., Ltd., London.
This is a recommended textbook with helpful allusions to fossil plants.

R. C. Moore (1954): Kingdom of Organisms Named Protista, *J. Paleontol.,* vol. 28, pp. 588–598.
A taxonomic scheme including a third organic kingdom is outlined.

B. Sahni (1925): Ontogeny of Vascular Plants and the Theory of Recapitulation, *J. Indian Botan. Soc.,* vol. 4, pp. 202–216.
Recapitulation seems to occur in the ontogeny of plant stems and some other structures.

R. F. Scagel et al. (1965): "An Evolutionary Survey of the Plant Kingdom," Wadsworth Publishing Co., Belmont, Calif.
This comprehensive mainly systematic and morphologic botany relates fossils to modern plants excellently and includes sections on the evolution and phylogenetic relations of the main plant groups.

K. R. Sporne (1959): On the Phylogenetic Classification of Plants, *Am. J. Botany,* vol. 46, pp. 385–394.
The present status of plant phylogeny is evaluated, and the problems of phylogenetic reconstruction are considered.

H. H. Thomas (1958): Fossil Plants and Evolution, *J. Linnaean Soc. London, Zool. vol.* 44; *Botan. vol.* 56, pp. 123–135.
Fossil evidence does not reveal evolutionary series, but evolutionary trends are becoming more and more evident.

C. W. Wardlaw (1965): "Organization and Evolution in Plants," Longmans, Green & Co., Ltd., London.
Many aspects of the theoretical and philosophic approach to morphologic and physiologic organization presented in this book are as applicable to animals as they are to plants.

J. M. Weller (1955): Protista: Non-plants, Non-animals? *J. Paleontol.,* vol. 29, pp. 707–710.
The disadvantages of recognizing a protistid kingdom are emphasized.

W. Zimmermann (1952): Main Results of the "Telome" Theory, *Palaeobotanist,* vol. 1, pp. 456–470.
Theoretical modifications in the evolution of plant stems, leaves, and reproductive structures are described.

CHAPTER THREE

NONCOELOMATE ANIMALS

EMBRYOLOGY AND LARVAL DEVELOPMENT
PROTOZOANS—SPONGES—COELENTERATES

The principal animal phyla (see Table 86) are so different in their mature structure that their phylogenetic relations are obscure. They show, however, several grades of increasing structural complexity that seem to indicate corresponding grades of evolutionary advancement. For example, there are (1) the single-celled protozoans, (2) the multicellular sponges, which are essentially without symmetry and have no definite mouth or digestive tract, (3) the radially symmetrical coelenterates, with mouths and stomach-like cavities but without clearly developed internal organs, and (4) all higher animals, most of which exhibit distinct bilateral symmetry and are organized in a variety of more complex ways. (See Table 87.)

All principal animal phyla known as fossils are believed to have differentiated before the beginning of the Paleozoic Era. Consequently the fossil record cannot be expected to reveal much, if anything, about their evolutionary relationships. Practically all ideas regarding the major patterns of animal phylogeny, therefore, are based on the embryonic and larval study of modern organisms. In the absence of other evidence, reliance is placed upon the assumption that similarities in the early ontogenetic stages of different phyla reveal similarities of evolutionary origin (Fig. 86). This accords with the principle of recapitulation which, with some reservations, is widely accepted as valid with respect to general but not necessarily detailed relationships.

Embryology
All sexually derived animals above the protozoan grade of organization, and also those produced parthenogenetically from unfertilized eggs, begin with a single cell.

Table 86   Classification of animals

MAIN DIVISIONS OF ANIMAL KINGDOM	PHYLA
Eozoa: single-celled	1. Protozoans
Metazoa: many-celled	
Parazoa: without enteron	2. Sponges
Enterozoa: with enteron	
Acoelomata: without coelom	3. Coelenterates
Coelomata: with coelom	
Lophophorata: with lophophore	4. Bryozoans
	5. Brachiopods
Schizocoela: coelom opens in mesoderm	6. Molluscs
	7. Annelids
	8. Arthropods
Enterocoela: coelom invaginates from enteron	9. Echinoderms
	10. Chordates

All phyla important in paleontology are listed, but others that are unknown or poorly or doubtfully represented in the fossil record are omitted. Some classifications recognize as many as 40 phyla of modern animals.

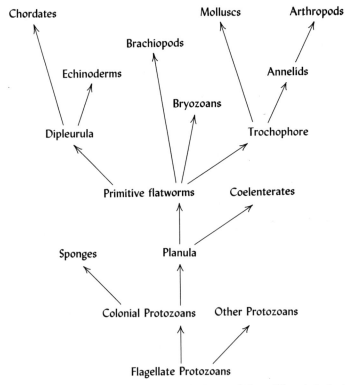

Fig. 86   Diagram showing the probable evolutionary relations of the principal animal phyla known as fossils.

Table 87   The principal animal phyla and their most conspicuous identifying characters

Protozoans: single-celled animals
Poriferans: sponges, the least organized of all multicellular animals, without enteron or stomach cavity
Coelenterates: hydroids, corals, and jellyfishes, radially symmetrical, with mouth and stomach but no anus; many are colonial
Bryozoans: tiny animals composing colonies produced by asexual budding, like corals but more highly organized
Brachiopods: bivalves, with two movable shells differently shaped, but each symmetrical with similar right- and left-hand halves, thus differing from clams
Molluscs: snails, clams, and cephalopods, mostly shelled animals with a highly muscular, so-called foot
Annelids: segmented worms, without legs
Arthropods: segmented animals with a horny external skeleton and jointed legs; includes insects, spiders, crustaceans, and numerous other groups
Echinoderms: starfishes, sea urchins, and crinoids, commonly pentamerously symmetrical with a calcareous skeleton consisting of many plates just beneath or within the skin
Chordates: mostly vertebrates with an internal calcaro-phosphatic skeleton and a few other animals with a different kind of longitudinal stiffening structure or notochord

This cleaves, or segments, and the resulting cells rapidly multiply and differentiate (Fig. 87). Most animals pass through several similar successive developmental stages before they begin to acquire the form or other characters that identify them as members of some particular phylum or more inclusive group.

The Egg

The egg is a cell which contains a nucleus but shows no other very apparent evidence of internal structure. Organization within it does occur, however, because development commonly is definitive and generally reflects the existence of axes or planes of symmetry. The greatest obvious differences between eggs concern the amount of food substance or yolk which they contain. This may affect development importantly, but it does not differentiate major groups of animals. The eggs of some fairly closely related species differ greatly in their yolk content. If yolk is abundant, it ordinarily is concentrated in the center of the egg or in a part adjacent to what is called the vegetative pole. In contrast, the opposite part, where segmentation or

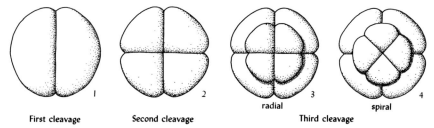

First cleavage   Second cleavage   radial   spiral
                                            Third cleavage

Fig. 87   Generalized diagrams showing the first three cleavages of an egg as seen at the animal pole. Spiral cleavage is characteristic of protostomes, particularly molluscs, annelids, and many modern flatworms. The eggs of most other animals cleave radially.

cellular development begins, is known as the animal pole. Patterns of segmentation differ greatly, but not all of them require detailed consideration here.

The Blastula

Eggs with little yolk commonly pass through a morula stage when the embryo is a solid body composed of only a few cells that extend from the surface to the center. A blastula is produced when the multiplying cells have drawn apart and enclose a central cavity. If the yolk is large and central, it generally is enclosed within the blastula. The comparable stage derived from an egg with much yolk concentrated at the vegetative pole consists of a disk of cells resting upon the yolk.

The Gastrula

In the next, or gastrula, stage, the cells differentiate into external and internal layers, or ectoderm and entoderm. Commonly the surface of the blastula at the vegetative pole turns inward to form a pocket, which becomes a primitive stomach cavity communicating with the exterior through a blastopore, or primitive mouth (Fig. 88). More or less similar double-layered structures, however, also are produced in other ways. For example, cells of the animal region may grow down over and partly enclose the vegetative cells, cells may grow back beneath those of the disk which lies upon a yolk, or new cells may be produced within the hollow blastula. In the last example, there is at first no stomach cavity or mouth but these develop later by rearrangement of the internal cells and the separation of the surface cells at the vegetative pole.

The theory has been widely held that embryonic development, progressing from a single cell to a hollow blastula and an invaginated gastrula, recapitulates the phylogeny of coelenterates and all higher animals. This concept certainly is too simple, but if details and variations are neglected, some such sequence may explain in a very general way the origin of two fundamentally different tissue layers.

Larval Development

The embryos of invertebrate animals become free-living, self-sufficient larvae or young at different stages of development. The presence of abundant yolk permits

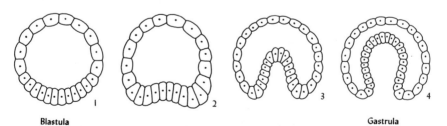

Blastula   Gastrula

Fig. 88   Generalized vertical sections showing how a blastula is transformed into a gastrula by invagination of the cells at the vegetative pole.

embryonic development to proceed further within the egg than would otherwise be possible. All terrestrial invertebrates and most of those inhabiting freshwater produce relatively advanced larvae or almost completely formed young. Many marine invertebrates, however, begin their larval existence as free-swimming flagellated or ciliated organisms that have not progressed much, if any, beyond the gastrula stage.

Blastula Larvae
The simplest larvae of multicellular animals are the ciliated hollow blastulas of some sponges and hydrozoans.

The Planula
A planula is a more or less ellipsoidal or elongated larva that generally swims by means of long whip-like flagella or short hair-like cilia (Fig. 90-1). It consists of an outer layer of cells as in a blastula, but the center is filled with cells that have been produced by or have moved inward from the outer layer (Fig. 89). It lacks a stomach cavity and a mouth. Feeding is accomplished in a protozoan manner. Food particles are captured by the outer, or ectodermal, cells and passed into the interior for digestion, which is intracellular. Most blastulas soon develop into planulas. This kind of larva is typical of many sponges, most coelenterates, and the flatworms. The planula of sponges becomes attached and undergoes rather complex metamorphosis that differs in different groups of these animals. The planula of coelenterates develops into a gastrula by the opening of a central cavity and mouth, either before or after the larva becomes attached.

Gastrula Larvae
Other gastrula larvae are produced by the invagination, or in-pocketing, of a hollow blastula (Fig. 88), as in some sponges and many higher animals, although such a transformation generally is accomplished during embryonic development. The sponge gastrula is peculiar, however, because its invagination takes place at the animal rather than the vegetative pole as in all other animals (Fig. 106).

The gastrulation of a hollow blastula by invagination does not seem to be a

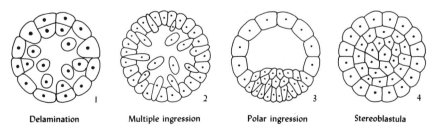

| Delamination | Multiple ingression | Polar ingression | Stereoblastula |

Fig. 89 Diagrams showing, in vertical section, the principal ways in which cells that fill a hollow blastula produce a solid stereoblastula.

primitive process. Instead, it probably is an attainment accomplishing the same result as the gastrulation of a planula but in a more efficient manner. Therefore, invagination is believed to be an example of evolutionary advancement.

Primitive Flatworms

The theory has been widely held that all higher animals evolved from a coelenterate-like ancestor produced by the invagination of a hollow blastula. A serious objection to this theory is now recognized, because the gastrular cavity of almost all coelenterates originates in the other way. The larva of a coelenterate is a planula. The adult condition of these animals results from a rather simple kind of metamorphosis in which the internal cells are reorganized to produce the entoderm and stomach cavity, and a mouth opens through the ectoderm.

Flatworms also develop from planula larvae (Fig. 90). In some respects the nonparasitic species of these animals seem to represent a lower evolutionary grade than do most coelenterates. They have a mouth but no stomach. Food particles are simply absorbed into the internal cells in a protozoan-like manner.

These facts seem to indicate that evolution of the higher animals did not include a coelenterate-like stage. A planula-like ancestor is much more probable. From it evolution proceeded in two directions: (1) Radial symmetry was preserved in one group of animals, the coelenterates, that were adapted to a free-swimming existence and later to an attached one. (2) Bilateral symmetry was acquired in another group, the primitive flatworms, that became adapted to a crawling way of life. If this is so, the stomach cavity evolved independently in these two groups.

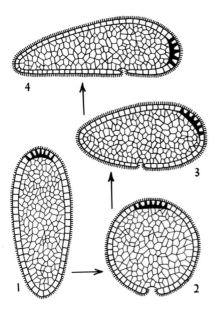

Fig. 90 Generalized diagrams showing in vertical section the theoretical transformation of a ciliated planula (1) into a bilateral organism similar to a primitive flatworm (4). A mouth opens (2), and the nerve center migrates to the anterior end (3). (*Adapted after Hyman, 1951, "The Invertebrates," vol. 2, fig. 2, p. 9.*)

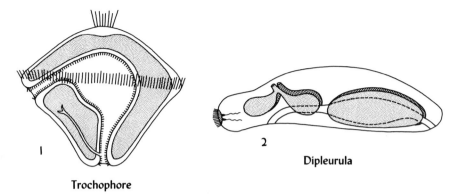

Trochophore   Dipleurula

Fig. 91   Generalized diagrams showing the characteristic features of 1, a trochophore with lateral mouth, and 2, a dipleurula larva, anterior to left. The coelomic cavities of the dipleurula are shaded.

From the theoretical flatworm ancestor, evolution is believed to have produced two great divisions of the animal kingdom, which culminated, respectively, in (1) the molluscs and arthropods, and (2) the echinoderms and vertebrates (Fig. 86). The bryozoans and brachiopods constitute a third group, whose relationships are not so clear.

The Trochophore

A trochophore is a bilaterally symmetrical, oval or pear-shaped larva provided with a stomach which opens through both a mouth and an annus (Fig. 91-1). It bears cilia on its surface in an apical tuft and an equatorial band. It also possesses nephridia, or excretory organs, and both nerve and muscle fibers. The larvae of annelids and molluscs, very similar to each other, are typical trochophores. Some bryozoans and several kinds of worm-like animals develop from larvae of a more or less similar type. This is accepted as evidence that all these animals probably had a common evolutionary origin.

The trochophore generally is believed to have evolved as the larva of a rotifer-like organism whose ancestry probably traces back to a creature resembling a primitive flatworm. A trochophore develops from the gastrula stage of an embryo by an invagination of the ectoderm which reaches and opens into the stomach cavity, thus producing an anus. The phyla presumed to have descended from such an ancestor are known as the protostomes. Most of them share several other note-worthy peculiarities in their embryonic development.

The Dipleurula

The dipleurula is a generalized larval form that follows the gastrula in the ontogeny of echinoderms (Fig. 91-2). It is supposed to represent an ancestor derived from a primitive flatworm-like creature and stands at the base of the other great group of animals, the deuterostomes, consisting mainly of the echinoderms and chordates.

The dipleurula differs importantly from the trochophore, because the blastopore becomes the anus rather than the mouth. The mouth is a new structure developed in the larva by an invagination of the ectoderm. Nephridia are not present in the dipleurula, and the coelom, or intermediate body cavity, has an origin different from that of the protostomes.

Origin of Mesoderm and Coelom

The ectoderm and entoderm are the primary tissue layers that form the outer skin and the lining of the digestive tract of all animals above the grade of sponges and flatworms. A third intermediate layer, the mesoderm, makes its appearance later in ontogenetic development. This is produced by either one, the other, or both of the primary layers in several different ways. Only the hydroids lack a distinct third layer. The other coelenterates are commonly said to be without it, but in them the ectoderm and entoderm are separated by gelatinous material containing several kinds of cells. This material corresponds to an incompletely developed and relatively unorganized middle layer that is produced mainly by the ectoderm. Sponges have a similar type of middle layer.

Most animals above the flatworm grade have well-organized mesoderm that lines the inside of the ectoderm and the outside of the entoderm. It encloses a body cavity, or more than one, known as the coelom. This mesoderm is derived principally from the entoderm. It originates in two main ways (Fig. 93): (1) Cells move into the space between the other layers and there proliferate. The coelom opens in the midst of these cells. Animals which develop in this way are termed schizocoels. They include the annelids, arthropods, molluscs, perhaps the bryozoans, and some brachiopods. (2) A pair of invaginations forms pockets in the entoderm. These invaginations close. The cells thus cut off from the entoderm are the beginning of the mesoderm, and the space which they enclose is the coelom. Animals that develop in this way are known as enterocoels. They include the echinoderms, chordates, and some brachiopods.

All deuterostomes are enterocoels, but only the annelids, arthropods, and molluscs among the protostomes are typical schizocoels. The brachiopods and bryozoans share features with both these groups. They commonly are classed with

Table 92 Classification of animals above the flatworm grade of development according to their embryonic development

---

Protostomes: develop from trochophores, nephridia present. Segmentation spiral. Blastopore becomes mouth. Molluscs, annelids, several other worm-like animals, some bryozoans

Deuterostomes: develop from dipleurula-like larva, nephridia absent. Segmentation radial, Blastopore becomes anus. Echinoderms, lower chordates. All are enterocoels

Schizocoels: coelom opens in midst of early mesodermal cells that move inward from entoderm. Molluscs, annelids, bryozoans, some brachiopods

Enterocoels: mesoderm and coelom originate as invaginations of entoderm. Echinoderms, lower chordates, some brachiopods

---

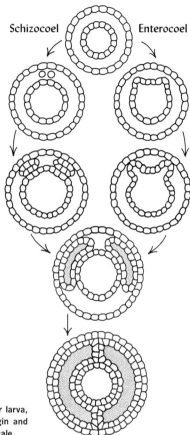

Schizocoel   Enterocoel

Fig. 93 Diagrammatic transverse sections of an embryo or larva, starting with a gastrula, showing the two methods of origin and development of the mesoderm and coelom; not drawn to scale.

the schizocoels among the protostomes because their development from trochophore-like larvae is considered to have greater phylogenetic significance than their possible enterocoelous nature.

The three layers that have been described are the source of all the different parts of the animal body. Nerves are derived from the ectoderm. The function of the entoderm is exclusively nutritional. Muscles, bone, calcareous endoskeleton, other connective tissues, blood vessels, and internal organs all arise from the mesoderm. The only important exception is provided by the coelenterates, some of whose muscles and nerves are of entodermal origin.

Metamorphosis

The larvae of many invertebrate animals bear no close morphologic resemblance to the adult creatures into which they will develop. Transformation is accomplished by a process of metamorphosis when more or less radical reorganization of tissues occurs and a juvenile individual is produced, equipped with structures and organs recognizably similar to those of the ensuing adult. Metamorphosis may so completely interrupt a sequence of gradual and orderly ontogenetic changes that the derivation of adult characters from larval ones is obscured.

Metamorphosis generally takes place during a resting stage of variable duration. At this time the young individual does not feed, growth ceases, and the animal may even decrease in size. Many variations are known, however, ranging all the way from complete reorganization with respect to form and bodily functions accomplished at a single step to more gradual development during which normal growth proceeds.

If eggs are large or contain much yolk, metamorphosis and the development of a juvenile stage may be completed in the embryo before it hatches. Therefore, a larva very different from an adult is not necessarily part of the ontogeny of all animals.

NONCOELOMATE PHYLA

The noncoelomate phyla commonly are regarded as the most primitive of all animals, but this viewpoint certainly can be debated. Some of them, particularly many of the protozoans, obviously are specialized to a degree not exceeded by any of the so-called higher animals. A majority of those that live today, as well as many known as fossils, surely has evolved far beyond the truly primitive creatures that once existed and constituted the evolutionary base and lower levels of the animal kingdom. The gross structural organization of these and other noncoelomate animals, however, is relatively simple. They are not equipped with organs, some lack the quality of individuality that generally is an attribute of animals, asexual reproduction is common, alternation of generations occurs (it is unknown among the higher phyla), and many of these animals are colonial. These are all attributes that variously characterize many plants. They seem to indicate less-fundamental evolutionary divergence from a common ancestral stock than that exhibited by other animals.

PROTOZOANS

The protozoans are an exceedingly varied group made up of species ranging from what seem to be relatively simple single-celled creatures to others that obviously are complex and some that are colonial. Altogether the range of their diversity probably is as great as that of all multicellular animals. The structure of many of them and the ways in which their functions are performed suggest that actually they are as highly organized as some multicelled animals and should, therefore, be considered acellular rather than unicellular.

The opinion has been increasingly expressed that the ''acellular'' plants and

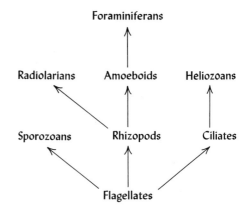

Fig. 95   Diagram showing the possible evolutionary relations of the main groups of protozoans.

animals should be regarded as a third organic kingdom and separated into several independent phyla. These differ among themselves so greatly that this seems to be a logical conclusion. The divisions proposed by different persons, however, reveal numerous disagreements and indicate that present knowledge of these organisms is inadequate as the basis for a phylogenetic classification that is reasonably reliable. Classification, in fact, consists mainly of groupings centered around certain of the better-known species or genera with little or no understanding of actual relationships.

Nothing can be said about the evolutionary origin of protozoans except that most of them emerged from the general flagellate complex that also was the source of the majority of plants. Some theories, however, derive certain groups of protozoans from the fungi or even directly from bacteria. The commonly recognized larger groups may in the main be valid evolutionary branches (Fig. 95), but on the other hand, the diagnostic characters of certain kinds, particularly among the parasites, may reflect similar environmental adaptations rather than common ancestry.

The only protozoans well represented in the fossil record are the foraminiferans and radiolarians. Only these provide an opportunity for actually tracing evolutionary relationships.

Foraminiferans

Foraminiferans are protozoans whose protoplasm can be extended in branching and anastomosing thread-like pseudopods by means of which they capture food and move about (Fig. 96). Most foraminiferans are marine, and also most of them secrete or otherwise form shells of mineral matter that are readily preserved as fossils. The shells commonly range from 0.1 to 1.0 mm in size, although some are smaller and others are much larger. The earliest generally accepted fossil foraminiferans are of Ordovician age, but older ones have been reported, and some of them probably are authentic.

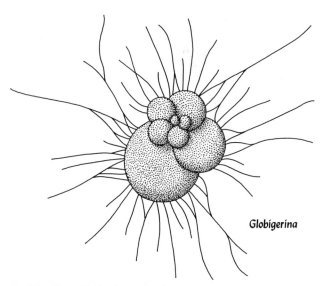

Fig. 96   Living *Globigerina*, a pelagic foraminiferan. (× 60.) Protoplasm streams out through the center of the slender pseudopods and back along their surfaces. The calcareous shell is pierced by many relatively coarse pores.

Shell Composition

Foraminiferan shells are composed of different materials or of the same materials used in different ways. The kinds most commonly recognized are: (1) A thin horny envelope that may or may not be enclosed by gelatinous material. Such shells are very rarely preserved as fossils. (2) Arenaceous or agglutinated. These consist of cemented sedimentary particles of several kinds. Many species are extremely selective in the type and size of the particles utilized. (3) Porcelaneous, consisting of unoriented calcite crystals which scatter light and give the shells an opaque appearance. Pores rarely are evident in either arenaceous or porcelaneous shells, but careful studies have demonstrated that they do occur. (4) Hyaline, consisting of clear glassy calcite in uniformly oriented crystals. Most of these shells are pierced by visible very tiny pores. (5) Granular or fibrous calcite. These shells are mostly Paleozoic. Their present structure has been interpreted as primary, arenaceous, or altered, resulting from recrystallization of calcite that was originally of a different nature. Calcitic shells older than the late Cretaceous generally are more or less recrystallized, so that the different kinds of structure may be difficult to recognize.

Shell Form

Foraminiferan shells present such a wide range of forms that it is difficult to imagine any pattern of growth that does not actually occur (Fig. 97). Some are open tubes (Fig. 100a-1,2), variously sized and shaped, which may or may not branch. Others are single globular chambers (Fig. 97-1). Most, however, consist of a series of chambers added successively as the animals grew. These chambers differ much in size, shape, ornamentation, structure, and their relations to one another. Characteristic patterns of arrangement are (1) coiled in a plane (Fig. 100a-3); (2) coiled in variously blunt or pointed spires (Figs. 97-5,7,8); (3) arranged side by side in two alternating series (Figs. 97-6; 98); (4) linear; and also numerous others more difficult to describe. Many shells change more or less abruptly from one pattern to another, thus recording two or more successive ontogenetic stages (Figs. 98; 101; 102-1). Understanding of how such varied forms can have evolved in response to adaptational advantages is not easy.

Dimorphism

Some species of foraminiferans are dimorphic and are represented by two kinds of shells: (1) microspheric, with a relatively very small initial chamber, and (2) megaspheric, with a relatively much larger initial chamber (Fig. 98). These are presumed to alternate more or less regularly in the reproduction of successive generations. The less common and generally larger microspheric individuals are produced by the sex-like union of two tiny motile spore-like bodies. These individuals, in their turn, reproduce asexually by division to produce megaspheric individuals which generally do not grow so large but are more common. The megaspheric

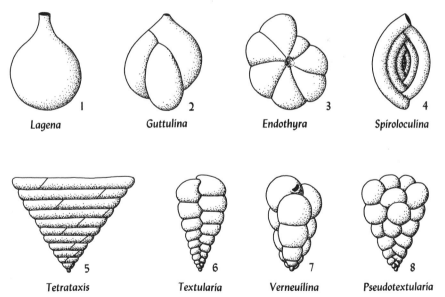

1	2	3	4
Lagena	*Guttulina*	*Endothyra*	*Spiroloculina*
5	6	7	8
Tetrataxis	*Textularia*	*Verneuilina*	*Pseudotextularia*

Fig. 97   Examples of common types of foraminiferan shells, greatly enlarged and not drawn to uniform scale. 1, A single flask-like chamber. 3, 4, 5, and 7, Shells showing various kinds of coiling. 3, The later whorls enclose and hide all earlier chambers. 5, This shell ordinarily has four chambers to a whorl. 6, A biserial shell with regularly alternating chambers. 7, A typical triserial shell with three chambers to a whorl. 8, This shell starts biserially and then changes to an irregular polyserial arrangement.

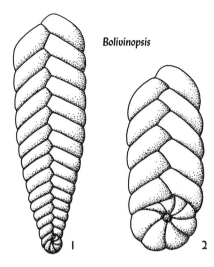

Bolivinopsis

Fig. 98   Example of microspheric and megaspheric shells belonging to the same species. Differences between the generations of most so-called smaller foraminiferans are not nearly so striking. (*See Cushman, 1946, U.S. Geol. Surv., Profess. Papers 206, pl. 44, figs. 11, 13.*)

individuals either produce sexual spores or again reproduce asexually and are followed by other megaspheric individuals which may or may not be slightly different from their single parents. The variations of this reproductive pattern are shown diagrammatically in Fig. 98*a*. This alternation of generations is essentially similar to that which characterizes plants.

The assumption has been common that most if not all foraminiferans repro-

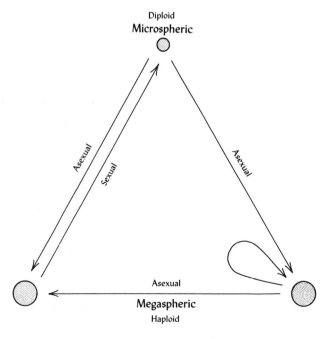

Fig. 98*a*   Diagrammatic representation of variations in the reproductive cycles of foraminiferans.

duce in this way, although both microspheric and megaspheric forms have not been recognized in many groups. This generalization is now known to be unwarranted. The life histories of only a relatively small number of modern species have been observed in detail, and the great range in their reproductive patterns is both remarkable and surprising. The dimorphic species are particularly interesting, however, because the microspheric forms may include early growth stages missing from the megaspheric forms. These suggest that conclusions may be drawn regarding the evolutionary relations of various shell forms, according to the theory of recapitulation, in ways that might not otherwise be considered.

Primitive and Advanced Characters

Foraminiferans have been studied as intensively as any other group of invertebrate fossils, partly because of their great usefulness to the petroleum industry in correlating strata from well to well. This, their great abundance from the late Paleozoic onward, and the presence of well-marked ontogenetic stages in some of them would seem to make the foraminiferans an ideal group of fossils for phylogenetic reconstructions. Ideas regarding which features are primitive and which are evolutionarily advanced, however, and uncertainties in the interpretation of similarities as homologous or analogous have resulted in conclusions that have differed greatly. On the whole no very clear understanding of the larger aspects of foraminiferan evolution has emerged.

Arenaceous foraminiferans have been considered either primitive or advanced types by different persons. Comparable disagreement has marked the identification of either unchambered tubes or single globular chambers as the most primitive form of shell. A linear series of chambers is supposed to have been derived from either a single globular chamber or a coil. Very likely much of this confusion has resulted from attempts to relate all foraminiferans in an orderly way from a single kind of ancestor. Perhaps all these ideas are correct. Probably different lineages evolved from different naked ancestors with subsequent evolution proceeding differently and involving several remarkable examples of convergence.

The arenaceous nature of some shells surely is primitive. Many of the earliest certainly known foraminiferans of Ordovician and Silurian ages are arenaceous, and calcareous forms do not become abundant in the faunas before the Mississippian or more doubtfully the Devonian. It is reasonable to suppose that some foraminiferans began to build shells by cementing sedimentary particles such as fine sand or silt grains derived from the sea bottom where they lived. Gradually the amount of cementing material that they secreted increased, until whole shells were produced in this way and the use of foreign particles became unnecessary. This permitted some foraminiferans to live attached to seaweeds in clear water uncontaminated with suspended sediment. The typical porcelaneous forms, which first appeared sparingly in the Mississippian and did not become common before the Jurassic, seem to have evolved in this way from late Paleozoic ancestors with granular shells. Possibly some lineages reverted to a more arenaceous habit, as modern faunas

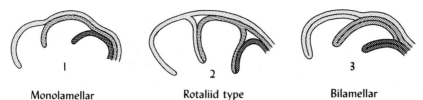

Monolamellar	Rotaliid type	Bilamellar
1	2	3

Fig. 100   Diagrammatic sections showing different types of structure and the relations of successively added shell layers to one another in hyaline foraminiferans; anterior is to left. The evolutionary relations are uncertain, but the monolamellar shell probably is more primitive than the bilamellar.

seem to include a larger proportion of arenaceous species than do many fossil faunas.

The evolution of the hyaline foraminiferans probably did not include an arenaceous stage. They seem to have developed directly from naked ancestors, perhaps in the Permian or somewhat earlier. These animals possibly diverged to produce two or more stocks differentiated by the structure of their shells (Fig. 100). New chambers in most foraminiferans are built relatively rapidly at intervals as the animals grow. A mass of protoplasm assumes the form of a chamber and secretes a thin organic covering, upon which a calcareous layer is then deposited. In one form of shell this layer lies only on the outside of the organic envelope. In another it lines the inside also, producing a bilaminated shell.

One of the more noteworthy results of evolution among the hyaline foraminiferans was the development of planktonic forms. These first appeared in the Jurassic but did not become common until the late Cretaceous.

Evolutionary Trends
Some general evolutionary trends seem to be reasonably well established in one or more groups of foraminiferans. For example, regular coiling, first noted in the Silurian, probably followed irregular wandering of tubes, and the earliest septate

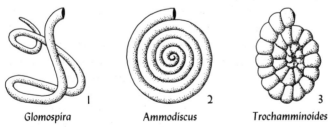

Glomospira	*Ammodiscus*	*Trochamminoides*
1	2	3

Fig. 100a   Evolution in some lineages of foraminiferans is believed to have progressed from 1, an irregularly wandering tube, to 2, a regularly coiled tube, and to 3, a chambered coil. Evolutionary relations of these particular genera are not implied.

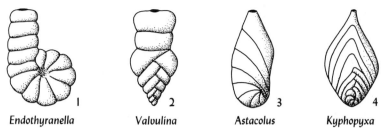

| Endothyranella | Valvulina | Astacolus | Kyphopyxa |

Fig. 101   Examples of different ways in which foraminiferan shells are transformed to a uniserial arrangement of chambers. Compare 1 and 2 with 3 and 6 of Fig. 97.

or chambered shells of the Devonian evolved from nonseptate open tubes (Fig. 100a). The coiling took several forms. After it was attained, uncoiling occurred in a number of different lineages, accomplished in several different ways (Fig. 101). Most commonly shells coiled in a plane simply straightened out. Those coiled in a more or less pointed spire reduced the number of chambers in their whorls, passed from triserial to biserial and finally to a uniserial arrangement. Other trends can be traced with less assurance.

Evolutionary relations are clearer in several groups which exhibit almost complete transitions between shells that differ conspicuously in some ways (Figs. 101a; 102). Particularly if the transitional series progress regularly through successive stratigraphic zones, they can be accepted as demonstrating orderly evolutionary transformation. Two good examples at the generic level are provided by the fusulinids and miliolids (Figs. 101a; 102a). Evolutionary sequences at the species level also have been recognized on the basis of various modifications in shape, size, and ornamentation of shells and form of aperture (Figs. 103; 103a; 103b).

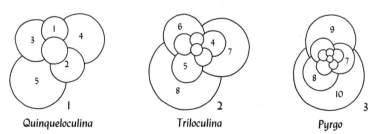

| Quinqueloculina | Triloculina | Pyrgo |

Fig. 101a   Diagrammatic transverse sections showing the relations of chambers in miliolid foraminiferans ranging from the Jurassic to modern time. These shells have two elongated chambers to a whorl (see Fig. 97-4) and the aperture occurs alternately, first at one end and then at the other. The plane of coiling is displaced progressively so that the symmetry is fivefold (1), threefold (2), or twofold (3). The last formerly was known as *Biloculina*. These relations are presumed to characterize evolutionary stages that are distinguished as different genera. The full sequence, which is interpreted as an example of recapitulation, ordinarily is present only in the microspheric generations of triloculine and biloculine species.

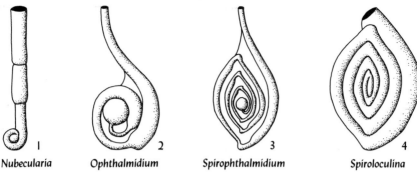

Nubecularia	Ophthalmidium	Spirophthalmidium	Spiroloculina

Fig. 102   In the late nineteenth century, this series of morphologic types of Lower Jurassic foraminif-erans from England suggested the idea that new evolutionary characters first appear early in ontogeny and persist longer and longer in subsequent generations until they become adult characters. According to this idea, an evolutionary trend is illustrated by successive transitions proceeding from left to right. These forms, however, were all contemporaneous and, therefore, cannot be an actual evolutionary sequence. In fact, as far as their evidence is concerned, evolution might just as well have proceeded in the opposite direction. (*After Wood and Barnard, 1947, Quart. J. Geol. Soc. London, vol. 102, fig. 7, p. 103.*)

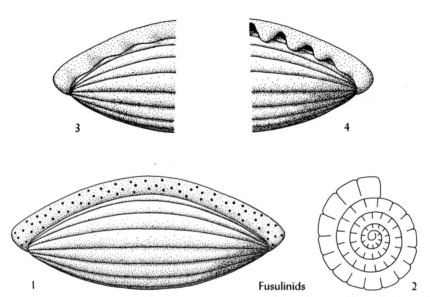

Fusulinids

Fig. 102a   Generalized drawings illustrating some of the evolutionary features of Pennsylvanian and Permian fusulinids. 1, An early typical fusulinid. The axis of coiling is much elongated. The septum closing the last chamber, pierced by numerous small pores, is unfolded. 2, Transverse equatorial section showing the arrangement of internal chambers. These are connected by an equatorial tunnel produced by resorption of parts of the septa. 3, Evolutionary modifications include folding of the septa. This begins at the poles and progresses toward the equator until the septa are deeply folded throughout their length. 4, The folds of adjacent septa are arranged so that the backward folds of one are opposite the forward folds of the preceding septum. Folds increase in depth until they meet at the base. Then the folds of one septum climb up those of the septum behind it. Finally openings are resorbed in the overlapped septal folds a short distance behind the final chamber. Various genera are defined partly on the basis of progressive stages in this evolutionary sequence.

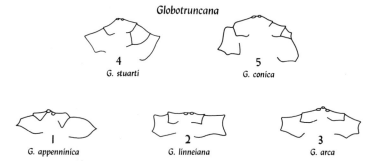

Globotruncana

| 4 | 5 |
| G. stuarti | G. conica |

| 1 | 2 | 3 |
| G. appenninica | G. linneiana | G. arca |

Fig. 103 Vertical sections through coiled specimens characteristic of successive Upper Cretaceous zones which probably represent an evolutionary series. The periphery first becomes vertical; then the spire increases in height and the whole shell thickens. (*After Glaessner, 1945, fig. 33, p. 151.*)

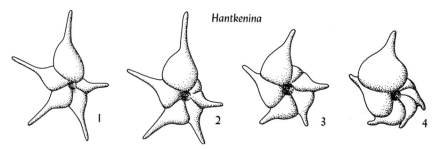

Hantkenina

Fig. 103a Successive morphologic stages in an evolutionary sequence of Eocene foraminiferans. Changes involve shift in position of spines with respect to sutures, shortening of spines, and modification of their inclination. (*Drawings reconstructed and compiled after Thalmann, 1942, A. J. Sci., vol. 240, pl. 1, p. 814); Glaessner, 1945, fig. 32, p. 150; and Bronnimann, 1950, J. Paleontol., vol. 24, fig. 2, p. 406.*)

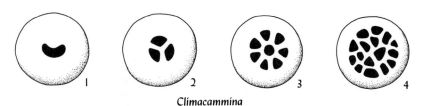

Climacammina

Fig. 103b Ontogenetic series showing increasing complexity of the aperture of a foraminiferan species at successive growth stages. Some other species and genera distinguished by comparable differences in mature specimens probably demonstrate evolutionary advancement. (*After Cushman and Waters, 1928, J. Paleontol., vol. 2, pl. 20, fig. 1.*)

Larger Foraminiferans

Several independently evolved groups of foraminiferans of both granular-porcelaneous and hyaline types acquired shell structures much too complex to be described or illustrated here. Their walls may be connected by pillars or penetrated by delicate canal systems, and their chambers may be subdivided by secondary partitions or connected by passages known as stolons. Shells of these kinds commonly are termed the "larger" foraminiferans, in contrast to the so-called "smaller" ones. Many have attained relatively large size, but it is their complexity rather than their size that is the diagnostic character. These shells require sectioning, and their detailed microscopic study is a specialty. They possess many features that can be compared within the several groups and arranged in what seem to be well-demonstrated evolutionary series.

Radiolarians

Radiolarians are protozoans, mostly much smaller than foraminiferans, that secrete spicules, or delicate lattice-like or coarsely porous skeletons, composed of silica or a horny substance. All are marine, and most of them float or swim in the near-surface zone. Only siliceous skeletons or spicules are known as fossils. These have been reported from rocks as old as the Precambrian, although this age is doubtful. They are certainly known from at least the Devonian onward.

The fossil radiolarians have been inadequately studied, mainly because of their

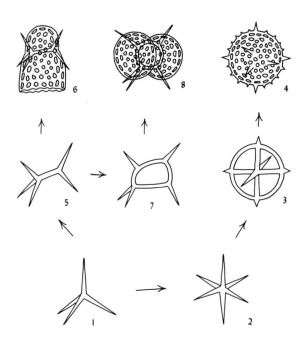

Fig. 104 Diagrammatic representation showing examples of how variously more complex forms of radiolarians may have been derived from one another. (*Mainly after Campbell, 1954, figs. 7, 8, pp. 25, 27.*)

very small size and the great difficulty of extracting them from the hard siliceous rocks in which they commonly occur. Nothing is really known about their evolution. Classifications of radiolarians, which have been presented as close approximations of their presumed phylogenetic relations, are based mainly on series of geometric types that might have been derived from one another (Fig. 104). Some Paleozoic specimens are reported to be larger than more recent ones, but the supposed Precambrian forms are described as being much smaller than modern species.

SPONGES

Sponge spicules are fairly common in many marine sedimentary rocks and have been reported from the Precambrian. They are unsatisfactory fossils, however, because they reveal very little about the animals from which they came. Fossilized whole sponges belong almost exclusively to types in which the spicules are fused to form a rigid framework. Such specimens represent some of the most highly evolved sponges, and paleontology, therefore, provides little evidence concerning the phylogeny of these animals. Almost all ideas regarding sponge evolution are based on the study of modern species. These ideas are vaguely reflected in several variant classifications whose differences indicate much uncertainty.

Sponges almost certainly arose from colonial flagellate protozoans, but evolution among them has progressed very little beyond this grade of development. Sponges have no organs, in the ordinary sense, and their tissues exhibit only a low degree of differentiation. The larval metamorphosis of these animals seems to unite them as a well-marked group and to set them apart as totally unrelated to any other modern animals (Fig. 106). In their transformation, the flagellate outer cells of the larvae, which become the ectoderm in other phyla, move inward, develop collars, and line the flagellate chambers that are one of the most characteristic features of the sponges. These create the water currents that circulate through the sponge body. At the same time, the inner cells of the planulae or the cells at the vegetative end of the gastrulae, which in other animals become entoderm, assume positions on the outer surface. Food particles are captured by the collared cells in a protozoan-like manner from the water currents and are passed inward to amoeboid cells, where digestion is accomplished intracellularly.

Although sponges are classified mainly on the basis of their spicules, evolutionary advancement seems to be indicated most plainly in progressive elaborations of the water circulation system. A definite trend is evident, but similar results probably were attained independently in more than a single lineage.

Three commonly recognized grades of sponge organization are named for well-known modern genera: (1) asconid (Fig. 107-1), (2) syconid (Fig. 107-2), and (3) leuconid (Figs. 107a; 107b). These intergrade morphologically and provide an almost uninterrupted evolutionary sequence. Each flagellate chamber as seen in the simplest sponges, the asconids, may be looked upon as a complete individual exhibiting all the features and performing all the functions necessary for these animals. Most

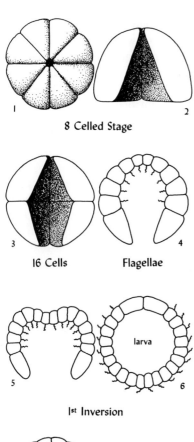

8 Celled Stage

16 Cells **Flagellae**

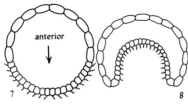

1st Inversion

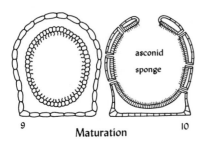

2nd Inversion and Attachment

Maturation

Fig. 106 Stages in the ontogenetic development of one type of calcareous sponge; highly diagrammatic and not drawn to scale. 1, 2, Cleavage of the egg forms an eight-celled plate. 3, 4, This divides longitudinally to produce 16 cells, and the upper ones divide further and develop flagella. 5, 6, The embryo turns inside out; a flagellated blastula forms and becomes a free larva. 7, 8, The larva swims with the flagellated end forward; it invaginates to produce a gastrula, and becomes attached. 9, 10, The flagellated cells develop collars and enclose a central cavity, incurrent pores break through, and an excurrent aperture opens. Other sponges develop in other ways. Many have planula-like larvae which experience complex metamorphosis that is not well understood.

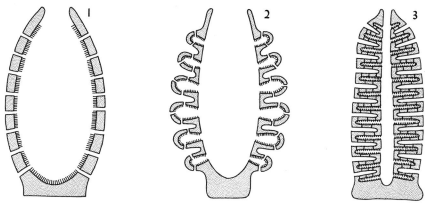

Asconid   Simple Syconid   Advanced Syconid

Fig. 107 Six stages are recognized in the increasingly complex structure of sponges. The first three are known only among the modern calcareous sponges and are named for two characteristic genera. 1, A single chamber lined with flagellated and collared cells. 2, A large chamber without flagellated cells into which open many asconid chambers. 3, The asconid chambers are grown together laterally, leaving tubular incurrent canals between them.

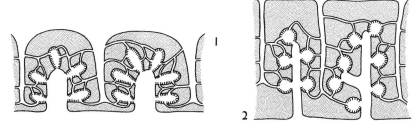

Simple Leuconids

Fig. 107a All siliceous sponges are leuconid, a type of structure named for another well-known modern genus. Three stages advancing in structural complexity are recognized. These simple leuconids have groups of flagellated chambers, like advanced syconids, arranged around excurrent canals into which they open directly. The incurrent canals are more complex than in syconids.

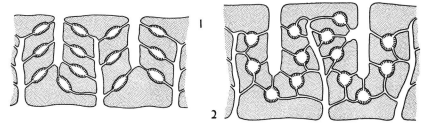

Intermediate Leuconid   Advanced Leuconid

Fig. 107b In intermediate leuconids the flagellated chambers are connected with excurrent canals through narrow tubes. In advanced leuconids the flagellated chambers are connected with incurrent and excurrent canals in more complex ways.

sponges which consist of many chambers may be considered colonial aggregates. The ontogeny of some syconids suggests that their structure was produced by a kind of budding of multiple asconid chambers. The leuconid grade probably was attained by folding of the sponge body, each part of which in its most simple development is similar structurally to a complete syconid. Further evolution obscured this structure, and the sponge became a mass of countless flagellate chambers connected by very complex systems of incurrent and excurrent canals (Fig. 107b).

The sponge skeleton consists of various kinds of spicules composed of hydrous silica or calcite, horny fibers or strands, and, in some, extraneous particles such as sand grains or even the spicules of other sponges. The spicules and their organization have been considered more conservative than the water circulation system and less subject to individual variation than the gross structure of the sponge body. The latter is known to respond to local environmental conditions, particularly water currents, so that a single species may develop different growth forms at different places. Therefore, the spicules are accepted as important distinguishing features of three relatively unrelated groups of sponges: (1) those with calcareous spicules, (2) those with siliceous spicules whose basic hexactinellid form possesses six rays disposed at right angles to one another, and (3) most other kinds, which are known as demosponges (Fig. 108).

Calcareous Sponges

The chemical composition of their spicules differentiates the calcareous sponges distinctly from all others. These sponges also are structurally the simplest. They include all the modern asconid species and all the typical syconids, but some have progressed to the leuconid stage. Advanced forms develop rigid skeletons by the

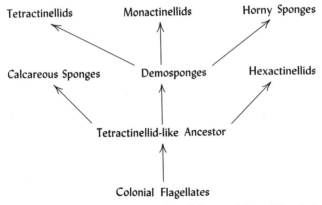

Fig. 108 Diagram showing the possible evolutionary relations of the principal groups of sponges.

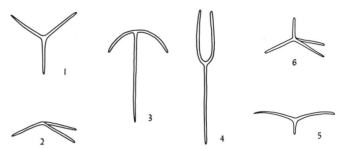

Fig. 109   Spicules of calcareous sponges. These are most commonly three-rayed. 1, 2, Two views of the same symmetrical spicule showing how the rays do not lie in a plane. 3 to 5, Modifications resulting from varying directions and unequal length of rays. 6, An extra ray produces a four-rayed type. Drawn from modern spicules. (Much enlarged.)

fusion of their spicules. These sponges, therefore, seem to present a well-marked phylogenetic sequence. Because this includes grades through which other sponges may have passed in their evolution, the calcareous group might be looked upon as representative of the most primitive or conservative sponges, from which all others were derived.

This conclusion, however, does not seem to be well founded. An evolutionary shift from calcareous to siliceous spicular secretion is unlikely. Moreover, calcareous and siliceous spicules probably are formed in different ways. Calcareous spicules seem to originate as tiny crystal growths within binucleated cells. As the spicules grow these cells divide. One of the daughter cells increases the spicule in length, and the other builds up its thickness. Multirayed spicules (Fig. 109) are produced by the cooperative action of several similar pairs of cells. Siliceous spicules, on the other hand, seem to grow within single cells that may have one nucleus or several. The silica is deposited around an axial organic fiber that is not certainly known to be present in calcareous spicules.

Calcareous and siliceous spicules probably are analogous, rather than homologous. If this is so, the calcareous and siliceous sponges may have arisen from different protozoan ancestors which already had differentiated with respect to the composition of the spicules or other parts which they secreted.

Hexactinellid Sponges

The hexactinellid group seems to stand apart from all other siliceous sponges. Although the spicules are basically six-rayed (Fig. 110), other types result from the distortion or obsolescence of one or more of these rays. The spicules of some are fused into rigid, regularly reticulated supporting structures. The development of more complex spicules from the six-rayed pattern is not considered likely. There is a possibility, however, that the hexactinellid spicule evolved from a tetractinellid form (Fig. 111a).

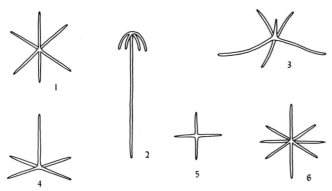

Fig. 110 Siliceous spicules from hexactinellid sponges. 1, A symmetrical spicule. 2, 3, Spicules with distorted unequal rays. 4, 5, Spicules lacking one or two rays, respectively. 6, Spicule with extra rays that may or may not have been derived from a six-rayed form. Drawn from modern and fossil specimens. (Much enlarged.)

The hexactinellids are peculiar in other ways. They are less commonly developed as branching masses than many other sponges. Almost all of them are anchored by a unique kind of root-like tuft or holdfast composed of long, coarse spicules. Most significantly, however, the soft anatomy is different from that of other sponges. The whole body consists of a meshwork of strands whose cellular structure is obscure, enclosing many small irregular open spaces. There is no clear differentiation of surface layers, as in other sponges. The flagellate chambers are arranged in groups resembling those of syconids, but they are not associated with well-developed incurrent and excurrent canals.

Modern hexactinellid sponges are almost all deepwater species; because of the resulting lack of observation very little is known about their ontogeny. Information that might be gained from developmental studies is not available as evidence of possible evolutionary relationships. Hexactinellid and other siliceous sponges may have descended from closely related but different protozoan ancestors or from a single very primitive sponge type. Perhaps the former is more probable.

Demosponges

The remaining modern sponges are a heterogeneous lot of organisms which vary greatly among themselves. Included are forms with skeletons exclusively siliceous, siliceous and horny, or exclusively horny, and some without a skeleton of any substantial kind. The interrelations connecting all or most of them, however, probably are closer than those which exist between this group and either the calcareous or hexactinellid sponges.

All modern demosponges are structurally leuconid. They develop from a juvenile

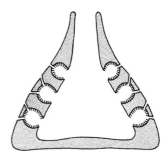

Fig. 111 The rhagonid youthful stage of a leuconid sponge. The flagellated chambers open into the central cavity like a syconid sponge, but mutual incurrent canals are lacking. (*Much generalized after Hyman, 1940, fig. 96A, p. 335.*)

Rhagonid

stage known as a rhagon (Fig. 111), rather than from a typical syconid stage. The rhagon, however, may be a modified syconid. There is some doubt, therefore, that the demosponges evolved from asconid and syconid ancestors after the probable manner of calcareous sponges.

The siliceous demosponges commonly are considered to comprise two groups, (1) the tetractinellids, and (2) the monactinellids, separated mainly for practical reasons. The monactinellids may or may not have horny fibers in addition to their spicules. The nonsiliceous sponges, with or without similar fibers, may have been derived from spiculate forms by loss of spicules, but some perhaps represent lineages in which mineralized spicules never had developed.

(1) As the name suggests, tetractinellid sponges have a basic type of spicule with four rays disposed like the axes of a tetrahedron (Fig. 111a). These generally occur with spicules of several other kinds but without horny fibers. Such an association of spicules has encouraged the supposition that the tetractinellid may be the form from which all other siliceous spicules evolved by either the addition or loss of rays. Some of the tetractinellid sponges have their spicules fused together and overgrown with further siliceous deposits, so that an unusually massive and solid framework is produced.

Fig. 111a Siliceous spicules from tetractinellid sponges. Many modifications develop from the symmetrical four-rayed type. Drawn from fossil specimens. (**Much enlarged.**)

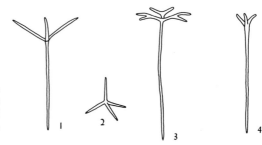

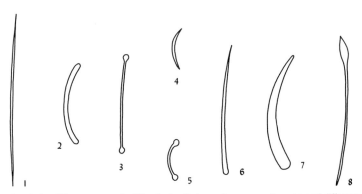

Fig. 112 Siliceous monactinellid spicules that may have come from several different kinds of sponges. These spicules have either similar or different ends, probably indicating that they grew in two directions or in only one. Drawn from fossil specimens. (Much enlarged.)

(2)   The monactinellid sponges possess only single-rayed spicules (Fig. 112), ordinarily bound together by horny fibers. Similar spicules are present in all the other groups of sponges. Such spicules probably have been derived from several other types by the suppression of all but a single ray or two oppositely directed rays. Monaxial spicules of hexactinellid sponges can be identified, if well preserved, by the tiny crossed axial canals of the undeveloped rays which enclosed their organic fibers. Some monactinellid sponges perhaps are more closely related to forms included in other groups than they are to one another.

Heteractinellid Sponges
The so-called heteractinellid sponges are fossils mainly of Paleozoic age. The principal spicules are many-rayed and of either regular or irregular design (Fig. 112a). They are commonly larger and coarser than those of other sponges. Such spicules may have evolved from simpler forms. In the absence of modern species with which these sponges might be compared, and with little or no understanding of their soft anatomy or ontogeny, their relationships cannot be known.

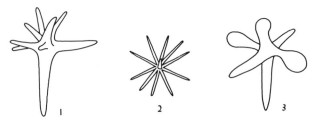

Fig. 112a Heteractinellid siliceous sponge spicules. These and similar kinds may not be related to more symmetrical spicules. They occur mainly in the skeletons of some Paleozoic sponges. Drawn from fossil specimens. (Much enlarged.)

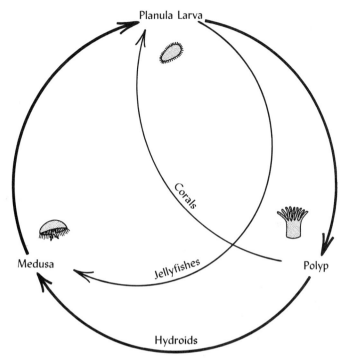

Fig. 113   Diagram showing the relations of polyp and medusa in the reproductive cycle of coelenterates. Most hydrozoans (hydroids) have successive alternating asexually reproducing polyp and sexual medusa generations. Some scyphozoans (jellyfishes) develop through a very brief polyp larval stage. Anthozoans (corals) do not have a medusa stage or generation.

COELENTERATES

The coelenterates include a large number of relatively simply constructed, more or less radially symmetrical animals whose bodies are sack-like and have a stomach cavity and mouth but lack an anus. They consist of ectoderm and entoderm, and most of them have an intermediate layer which is largely noncellular gelatinous material. Tentacles surround the mouth, and these and other parts of the body are provided with stinging cells, which are both protective and an aid in the capturing of prey. The nervous system is diffuse, but simple muscles are fairly well developed. Organs generally are lacking, but primitive sense organs occur in some of the more motile forms. Digestive enzymes are secreted by specialized stomach cells; some solid particles also are absorbed into the entodermal cells, where they are digested in a protozoan-like manner. Reproduction is accomplished both sexually and asexually. Many coelenterates produce more or less extensive colonies by asexual budding.

Three main divisions of the coelenterates are recognized (Figs. 113; 114; Table 114): (1) hydrozoans, or hydroids; (2) scyphozoans, or jellyfishes; and (3) anthozoans,

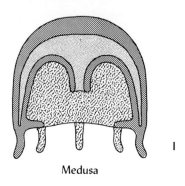

Medusa

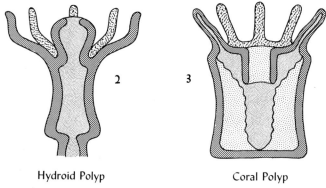

Hydroid Polyp   Coral Polyp

Fig. 114 Generalized vertical sections showing the body forms of polyps and medusa. The mouth region of the coral polyp is turned inward, and the stomach cavity is compartmented by fleshy partitions, or mesenteries. The medusa is essentially a free upside-down polyp of umbrella shape.

or corals and sea anemones. These groups differ anatomically in ways that do not need to be detailed here. They are believed to have evolved their distinguishing characters before the beginning of the Paleozoic Era.

The conspicuous dimorphism and alternation of polyps and medusae in the reproduction of many hydrozoans is quite different from the alternation of haploid

Table 114 Most conspicuous distinguishing characters of the three main groups of coelenterates

---

Hydrozoans: hydroids. Mostly regularly alternating generations of sessile colonial polyps which bud from each other asexually, and free-swimming medusae that reproduce sexually. Hydrocoralines build supporting calcareous skeletons

Scyphozoans: jellyfishes. Medusae which reproduce sexually and do not alternate with a generation of colonial polyps. Exclusively soft-bodied, without skeletons

Anthozoans: corals with supporting skeletons and sea anemones without supporting skeletons. Exclusively sessile polyps reproducing both sexually and asexually to produce colonies. No alternate generation of medusae

---

and diploid generations which is characteristic of most plants and many protozoans. Both generations of hydrozoans normally are diploid. The asexual reproduction of hydrozoan polyps is of the vegetative type, in which new individuals bud off from old ones. It is duplicated in plants and other groups of animals, but with these dimorphism is not involved. This type of dimorphism, therefore, is peculiar to the coelenterates and it presents a unique problem with respect to the evolution of this group of animals.

Opinions have differed sharply concerning which form, polyp or medusa, is more primitive and, therefore, which points to the kind of ancestor required for all coelenterates. One view that has become popular regards the medusa as the an-cestral form and the polyp as a derived larva. If this were so, the evolutionary sequence would seem to be from scyphozoan first to hydrozoan and then to antho-zoan, with the corals and anemones providing an example of development arrested at a larval stage. Despite its uncertainty, this theory of arrested development has been applied to an explanation of evolution in other groups of animals, particularly the chordates and even, in a minor way, among the primates to man himself.

The polyp is the exclusive form in the anthozoans. It is at least equal in impor-tance to the medusa in the hydrozoans, and many modern scyphozoans develop from an abbreviated polyp-like stage in which asexual reproduction may occur. These facts strongly suggest that the polyp is a form, inherited from the earliest and most primitive coelenterates, which subsequently has been totally eliminated from the development of only a few of them. This conclusion does not necessarily imply that the earliest mature coelenterates were polyps, although this possibility cannot be dismissed with confidence. Perhaps the larval planula became attached and in its metamorphosis produced a polyp as a preliminary stage in the develop-ment of a free medusa. If so, subsequent evolution resulted in (1) specialization of the scyphozoan medusa for a free-living existence; (2) postponement of the final metamorphosis in hydrozoans, with lengthening of an intermediate polyp stage and specialization of asexually growing colonies without loss of primitive anatomical characters; and (3) complete suppression of the medusa stage in anthozoans and specialization of the polyps. In such an evolutionary history (Fig. 115) the polyp would

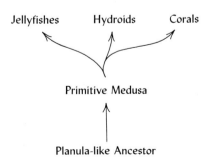

Fig. 115   **Diagram showing the possible evolutionary relations of the three coelenterate groups.**

not be more primitive than the medusa except that it represents an earlier ontogenetic stage.

Many coelenterates build calcareous or horny external skeletons which serve mainly supporting functions, and a few of them secrete calcareous spicules, some of which are much like those of sponges. Of these, only the calcareous skeletons are likely to be preserved. The fossil record of the coelenterates is extensive, but none other than the corals is represented by specimens that provide much evolutionary information.

Hydroids

Many hydrozoans secrete a thin outer horny cover and a few, known as the hydrocorallines, build a basal calcareous skeleton much like that of some corals (Fig. 116-1,2). Indistinct carbonaceous films from rocks as old as the Cambrian have been described as compressed hydrozoans of the former type. Calcareous structures recording the existence of hydrocorallines have been found sparingly in strata of Mesozoic and Cenozoic Ages.

In addition to the dimorphic polyp and medusa generations of the hydrozoans, most asexual colonies of these animals exhibit another kind of polymorphism, with the development of three kinds of individuals specialized for different functions: (1) ordinary polyps or feeding individuals; (2) individuals without mouths or stomachs but liberally supplied with stinging cells which aid in the capture of food and ward off enemies; and (3) individuals that bud off small medusae, which in turn produce

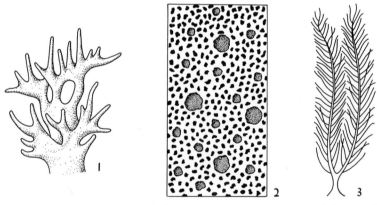

Hydrocoralline Hydrozoan   Gorgonid Octocoral

Fig. 116 Coralline hydrozoans (1, 2) build external calcareous skeletons with surface pits of two sizes. The larger ones contain feeding polyps; the smaller are occupied by protective individuals. (*After Boschima, 1956, "Treatise on Invertebrate Paleontology," Pt. F, figs. 76A, 77-1b, pp. 92, 93.*) A sea pen (3) builds a calcified horny supporting skeleton. Impressions of structures of this general type have been found in presumed late Precambrian rocks of Australia.

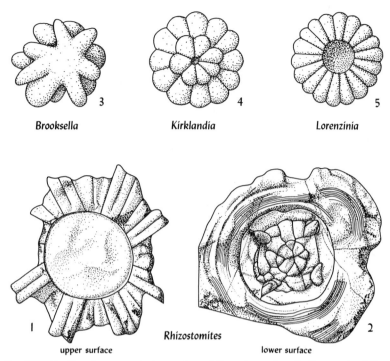

Fig. 117   Fossil medusae may be impressions of the outer surface (1, 2) or casts of the stomach cavity (3 to 5). Most of them show radial lobed structures, and those that are most surely identified exhibit fourfold symmetry. The assignment of fossil medusae to the commonly recognized groups of coelenterates ordinarily is doubtful. (*After Harrington and Moore,* 1956, *"Treatise on Invertebrate Paleontology," Pt. F, figs.* 11-2d, 32-2b, 37, 38-2, 54-1b, *pp.* 22, 42, 48, 49, 71.)

the eggs and sperms of the sexual generation. The hydrocoralline skeleton has two sizes of pits in which the individual animals live (Fig. 116-2). The larger pits house the feeding polyps, and the smaller ones contain the protective individuals. Medusae are produced by highly specialized individuals which develop within the communal tissue on the outer surface of the skeleton that connects the other kinds of individuals. This polymorphism is a feature requiring careful consideration when comparisons are made between the hydrozoans and some other colonial animals, particularly the fossil graptolites.

Jellyfishes

Of all multicellular animals, the soft-bodied scyphozoans, or jellyfishes, seem least well suited for fossil preservation. Nevertheless, a considerable variety of markings or objects found in fine-grained sediments as old as the late Precambrian has been supposed to represent medusae (Fig. 117). These specimens have been considered to be either external impressions or casts of the stomach cavity, which became filled with mud or sand, of individuals stranded probably in the intertidal zone. Most of them show a kind of lobed radial symmetry suggesting stomach pouches separated by fleshy partitions, or mesenteries. Some of these specimens probably are correctly interpreted, but others may be no more than concretions or markings of inorganic origin.

Corals and Anemones

The anthozoans are polyps considerably more complex in their anatomy than the hydrozoans (Fig. 114-3). For example, they have an inturned gullet, and the stomach cavity has its surface much increased by the presence of numerous thin inwardly directed fleshy folds, or mesenteries. Many of these animals reproduce asexually by budding to form colonies (Fig. 121-2,3). Eggs and sperms are produced just beneath the cells that line the stomach. The gullet generally is an elongated oval or a slit with a ciliated groove at one or both ends that drives a water current into the stomach. This has imposed a secondary bilateral symmetry upon the fundamental radial symmetry of the body that is more or less reflected in the structure of a calcareous exoskeleton if one is formed.

The anthozoans are conveniently divisible into two practical groups: (1) Anemones, which lack hard external skeletons. These are abundant in modern seas, but they are ill suited for preservation and the only reported fossils are carbonaceous films in Middle Cambrian shale. (2) Corals, with preservable skeletons that are commonly calcareous although some consist of more or less calcified horny material. Corals surely evolved more than once from anemone-like polyps by the acquisition of a skeleton.

Differences in the structural patterns of the skeleton permit the recognition of four general groups of fossil and modern corals: (1) tabulates, (2) tetracorals, (3) hexacorals, and (4) octocorals (see Table 118, Fig. 119).

Tabulate Corals

The first known tabulates, with one possible Cambrian exception, appear in Middle Ordovician rocks slightly older than those which contain the earliest tetracorals. Tabulates were important elements in Ordovician to Devonian faunas and are especially prominent in the great Silurian reefs that were built in several parts of

Table 118   Most conspicuous distinguishing characters of the four main groups of corals

Tabulate corals: with well-developed horizontal skeletal partitions, or tabulae, and poorly developed or few vertical partitions, or septa. Exclusively colonial; extinct

Tetracorals: with many vertical septa, arranged in four quadrants producing bilateral symmetry. Solitary or colonial; extinct

Hexacorals: septa exhibit sixfold symmetry, although polyps are bilateral. Colonial or solitary

Octocorals: with eight fleshy mesenteries but without skeletal septa. Skeleton horny or spicular. Polyps dimorphic. Exclusively colonial. Rare as fossils

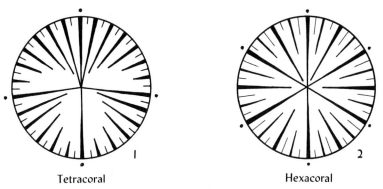

Tetracoral   Hexacoral

Fig. 119   Diagrams showing the symmetry and order in which septa or vertical calcareous partitions appear in the growth of tetracorals and hexacorals. The four or six first-formed, or primary, septa are shown meeting at the center. The order in which other septa appear is indicated by progressive decrease in length. Septa of a first cycle are shown by thick lines, those of a second cycle by thinner lines. More than two cycles are uncommon in tetracorals. More than three cycles may be present in hexacorals. The lengths of lines in these diagrams indicate the order of insertion, not the relative lengths of septa in actual coral specimens, although those of successive cycles generally are progressively shorter. Growth of septa in actual corals ordinarily is somewhat irregular in number, order, and length.

the world. As reef builders, these corals have been considered good indicators of paleoecology and climate. Parallels drawn between their distribution and that of modern reef corals, however, are not so secure as has commonly been assumed, because they are not known to have been adapted to the same conditions. The tabulate corals began to decline during the Devonian Period, and only one group may have survived the Paleozoic and lived on to Eocene time.

The tabulates include about half a dozen groups of fossil corals, which are distinctly different from the contemporary tetracorals but whose interrelationships are not clear (Figs. 120; 120a). Each possibly evolved from a different lineage of skeletonless anthozoans. Most ideas regarding evolution within the tabulates as a whole seem to have been influenced by observation of tetracoral development, which may not have much application here.

Tetracorals

The earliest tetracorals are present in strata of Middle Ordovician age; by Middle Silurian time individuals had become locally abundant. Their decline began in the Mississippian Period. This group is not recognized to have survived the Paleozoic Era, although the hexacorals, first recorded in the Mesozoic, may have evolved from some of them.

The soft anatomy of the tetracorals is not known with any certainty. Their skeletons exhibit a wide range in the variation of almost all their structures and the combinations in which these variants occur, suggesting that parallel and convergent evolutionary developments probably were common. Much uncertainty re-

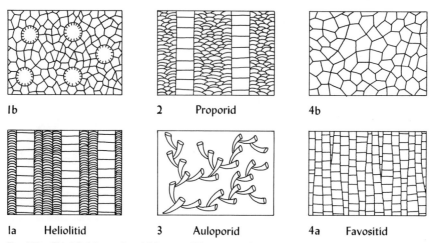

1b   2   Proporid   4b

1a   Heliolitid   3   Auloporid   4a   Favositid

Fig. 120   The tabulate corals exhibit many differences in form and structure. Tabulae, or transverse calcareous diaphragms, ordinarily are well developed. Septa, if present, mostly are represented by inwardly directed spines. The polyp-bearing tubes of some are separated by skeleton laid down by communal tissue (1, 2). The tubes of others are closely crowded and not separated by structures of that kind (4). The most simple colonies consist of individuals directly budded from one another (3).

garding what features are most reliable in identifying relationships and tracing evolutionary lineages is evidenced by the various classifications that have been proposed. Ontogenetic studies and stratigraphic distribution, however, reveal a number of evolutionary trends that seem to have been more or less duplicated in what probably are different lineages. These include: (1) The development of colonial aggregates by the budding of solitary corals to form, first, loose associations of branching tubes (Fig. 121) and then solid masses composed of closely appressed polygonal corallites (Fig. 121a). (2) The appearance and accentuation of an upright

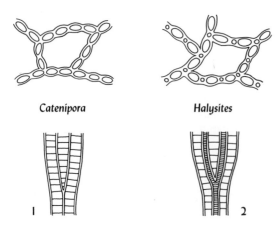

Catenipora   *Halysites*

1   2

Fig. 120a   Tubes of the earlier chain corals are directly in contact laterally (1). Later ones with smaller intermediate tubes (2) probably evolved from them, although transitional types have not been discovered. The small tubes have been interpreted as either the loci of diminutive modified polyps or structures built by communal tissue connecting the polyps of the colony.

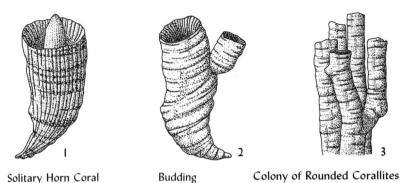

Solitary Horn Coral   Budding   Colony of Rounded Corallites

Fig. 121   The most primitive form of a tetracoral skeleton probably was a solitary horn-like structure (1). All stages in colony formation are known, from a simple bud arising laterally on a horn coral (2) to more and more compact and crowded tubes that budded from one another (3, see also Fig. 121a). The depression, or cup, at the end of the skeleton (2) was occupied by the polyp, which gradually built up the skeleton beneath itself. A central projection, or columella, in the center of the cup (1) evolved independently in several different lineages. Columellae did not become common until Mississippian time.

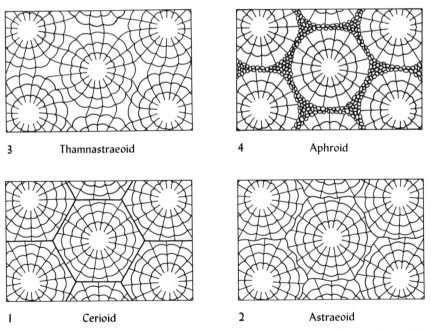

3   Thamnastraeoid   4   Aphroid

1   Cerioid   2   Astraeoid

Fig. 121a   Crowding of coral tubes in colonies produced solid structures. The individual tubes may be polygonal and well differentiated (1), but the tubes lost their distinctness (2, 3) and became separated by thin zones of communal tissue (4), or the polyps merged with each other laterally. (*Modified after Easton, 1944, Illinois State Geol. Surv., Rept. Invest. 97, pl. 2, figs. 21 to 24.*)

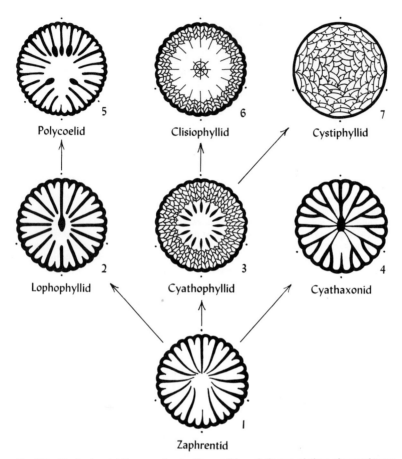

Fig. 122   Idealized septal diagrams showing the possible evolutionary relations of several types of tetracorals from a simple and generalized zaphrentid. The arrows indicate a morphologic series but are not intended to suggest that these genera are related in such a detailed way. (*Modified after Easton, 1960, fig. 5.16, p. 187.*)

axial column in the center of the coral cup (Fig. 121-1) as a solid shaft, by the junction of enrolled or thickened ends of the vertical septs, by uparching of the horizontal tabulae, by the piling up of small vesicular chambers, or perhaps in other ways. (3) Decrease in the distinctness of the septal quadrants, resulting from the more uniform development of septa, increase in their number, or various complications in their arrangement (Fig. 122). (4) The thickening of septa until they make lateral contact with one another and produce rings of solid skeletal material. (5) The shortening of septa so that they do not reach to near the center of the cup (Fig. 123). (6) The replacement of regular and complete tabulae by less-continuous somewhat inclined and overlapping partitions or by small curved calcareous plates

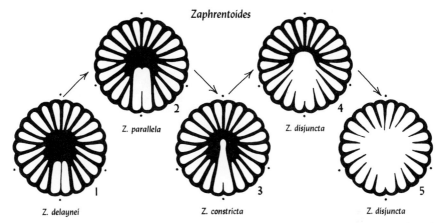

Fig. 123   Generalized diagrams of the cup seen from above in an evolutionary sequence of small Carboniferous corals from Scotland. (× 3½.) Specimens collected through rocks 4,000 ft thick show gradual changes from a type with septa connected in the center to another with shortened unconnected septa. (*Adapted after Carruthers, 1910, Quart. J. Geol. Soc. London, vol. 66, pl. 37.*)

that enclose numerous irregular chambers between the septa (Fig. 123*a*). The latter developed particularly in an outer zone of increasing width where they might disrupt the septa. There is some evidence that all these trends may have been reversed in certain instances.

A search for more conservative features, less subject to variation than those

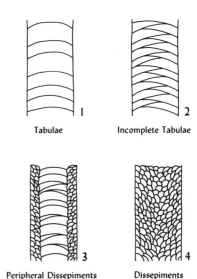

Fig. 123*a*   The simpler and presumably more primitive tetracorals raised themselves periodically in their tubes and laid down complete tabulae beneath themselves (1). A tendency developed for the production of incomplete or overlapping tabulae (2). The formation of a peripheral zone filled with small arched plates followed (3), and finally regular tabulae were eliminated entirely (4).

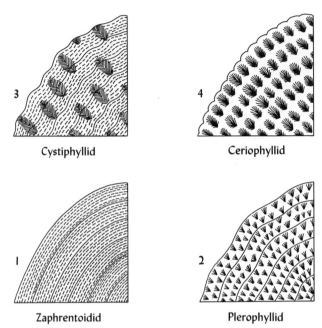

Cystiphyllid   Ceriophyllid

Zaphrentoidid   Plerophyllid

Fig. 124   Generalized diagrams showing the minute structure of tetracoral septa as seen in median longitudinal sections. (Greatly enlarged.)
Evolution seems to have progressed from lamellar structure (1) in two
directions (2, 3) to produce a structure which, with numerous modifications, is characteristic of many corals (4). (*Modified after Wang,* 1950,
fig. 76, *p.* 182.)

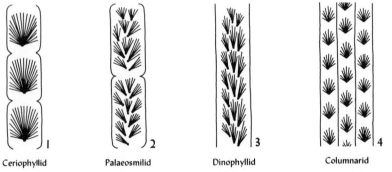

Ceriophyllid   Palaeosmilid   Dinophyllid   Columnarid

Fig. 124a   Generalized diagrams showing the minute structure of tetracoral septa as seen in
transverse sections. (Greatly enlarged.) Differences of these kinds are believed to indicate genetic
relations of corals more accurately than grosser features of the skeleton. (*Modified after Wang,*
1950, *fig.* 77, *p.* 183.)

just mentioned, has directed attention to minute structures of the septa (Figs. 124; 124a). These partitions, which arise from the floor and project inward from the sides of the coral cup, were secreted by restricted parts of the ectoderm where the base of the polyp was folded into narrow troughs alternating with the fleshy mesenteries. The most primitive septa are lamellar and consist of minute crystal flakes arranged parallel to the growing surface. Tiny crystal fibers directed approximately perpendicular to the growing surface appeared here and there within the lamellar septa. These increased in abundance and either completely replaced the flaky structure of the lamellae or became grouped in short columns surrounded by the still flaky lamellae. Septa of both types developed into long columns consisting of bundled crystal fibers radiating outward from closely spaced points of origin that completely obliterated the lamellar structure. The bundles of fibers are arranged in several characteristic ways in different groups of corals.

Hexacorals

The relatively abrupt and complete replacement of tetracorals by hexacorals after the ending of the Paleozoic Era rather naturally suggests that the younger group may be the direct evolutionary descendant of the older one. This conclusion has been encouraged by the many obvious similarities of the two kinds of coral skeletons, and some paleontologists have believed that hexacoral ancestors can be recognized in certain late Paleozoic tetracorals. Nevertheless, there are serious objections to this view. The principal ones are (1) that it is not clear how the change in symmetry and the even more important change in the order of septal insertion (Fig. 119) could have been accomplished, and (2) that sudden transition in the crystal nature of the skeleton, which is calcite in tetracorals and aragonite in hexacorals, does not seem likely. Therefore, most paleontologists who have studied corals are convinced that relations are not close and that hexacorals evolved from skeletonless ancestors similar in a general way to modern anemones. Furthermore, the hexacorals, like the tetracorals, probably acquired their skeletons as new structures in more than a single lineage.

The tetracoral and hexacoral skeletons are so similar in most ways, however, even in some of their minutest details, that confident inferences perhaps can be made concerning the soft anatomy of the older group from observations of living members of the younger one. If so, this would add greatly to an understanding of the tetracorals.

Relations among the hexacorals are believed to be revealed most accurately by the patterns of crystal fibers in the septa. Grosser similarities mostly reflect parallel and convergent evolutionary trends that were repeated in different lineages. Several of the trends listed in connection with the tetracorals also are recognized among the hexacorals. Some of these, e.g., colonial growth form and relations of individual polyps, seem to demonstrate evolutionary sequences in groups of corals whose other features show them to be related closely (Fig. 126). At least two trends continued further among the hexacorals: (1) the tendency for septa to be perforated

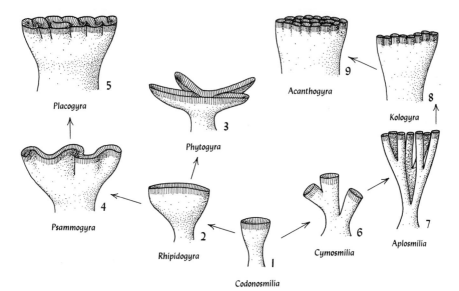

Fig. 126 Sketches showing presumed evolutionary progression in the gross skeletal structure of a group of closely related hexacorals. One sequence starts with a solitary coral (1). Budding of new polyps inside the ring of tentacles produced first an elongated narrow trough (2). More complex forms resulted from folding of the trough (4, 5) or branching (3). Another sequence starts with the same solitary coral (1) and proceeds by lateral budding (6) to branching colonies of increasing compactness (7, 8) and finally to a massive colony (9). (*Modified after Wells, 1956, "Treatise on Invertebrate Paleontology," Pt. F, fig. 260, p. 364.*)

by small openings through which the stomach compartments communicated, and (2) the intimate lateral association of adjacent polyps. In forms like the brain coral (Fig. 126a-1), budding occurs within the ring of tentacles, and individuals develop as a series of mouths bordered on two sides by rows of tentacles.

Fig. 126a The opening of new mouths within the ring of tentacles of some hexacorals produced compound polyps that occupied curved and branching grooves in the surface of the skeleton (1). This is an evolutionary development more advanced than any seen in the massive colonies of tetracorals. In other corals (2) new polyps budded from older ones and grew up between them. (*After Vaughan and Wells, 1943, pl. 36, fig. 2; pl. 38, fig. 8.*)

Octocorals

Octocorals are abundant in parts of the modern seas (Fig. 116-3), but they are relatively poorly known as fossils. Their record seems to begin in the early Mesozoic, although a few Silurian specimens have been assigned doubtfully to this group and impressions found in presumed late Precambrian strata of Australia resemble some modern octocoral colonies.

All known octocorals are colonial. The skeletons consist of variously calcified horny material and are mostly spicular, although more solid structures are built by some of them. The stomach cavity of the polyps is divided by eight mesenteries, but intervening septa are not developed in the skeleton. Polyps are connected by communal tissue, and the stomach cavities communicate through open tubules. Dimorphism of the polyps is the rule. Feeding individuals possess eight feathery tentacles, but the other type has only one and seems to function mainly by producing water currents that circulate through the tubules.

Octocorals are very different from the contemporary hexacorals and must have had a different evolutionary origin. The suggestion has been made that they evolved from tabulates, but this does not seem likely. Their ancestors probably were skeletonless colonial polyps unlike any that now are living. The dimorphism of their polyps suggests that they may be more closely related to the hydrozoans than are any other corals. Relationships among the octocorals are inferred mainly on the basis of the forms and arrangement of their spicules.

BIBLIOGRAPHY

INVERTEBRATES IN GENERAL

J. R. Beerbower (1968): "Search for the Past," 2d ed., Prentice-Hall, Inc., Englewood Cliffs, N.J.
This is a more generalized but broader consideration of paleontology than is presented in most other books.

R. B. Clark (1964): "Dynamics in Metazoan Evolution: Origin of the Coelom and Segments," Oxford University Press, Fair Lawn, N.J.
See final chapter, pp. 205–261, for phylogenetic conclusions based on the interpretation of the mechanics of locomotion.

W. H. Easton (1960): "Invertebrate Paleontology," Harper & Row, Publishers, Incorporated, New York.
This is a good intermediate textbook.

L. H. Hyman (1940–1959): "The Invertebrates," 5 vols., McGraw-Hill Book Company, New York.
These books are an excellent source for invertebrate zoology; molluscs and arthropods are not yet covered.

E. Marcus (1958): On the Evolution of the Animal Phyla, *Quart. Rev. Biol.*, vol. 33, pp. 24–58.
Relationships are indicated by similar details of embryology and early ontogeny; understanding of fossils is imperfect.

R. C. Moore (ed.) **(1953–):** "Treatise on Invertebrate Paleontology," 16 pts. in 20 vols. published, 8 pts. in preparation, Geological Society of America, Boulder, Colo., and University of Kansas Press, Lawrence, Kans.
This profusely illustrated work, to which more than 100 specialists are contributing, is devoted mainly to classification.

————, **C. G. Lalicker,** and **A. G. Fischer (1952):** "Invertebrate Fossils," McGraw-Hill Book Company, New York.
This excellent morphologic and systematic text includes brief considerations of the evolution of some groups of fossils.

T. J. Parker and **W. A. Haswell (1940):** "A Text-book of Zoology," 6th ed., vol. 1, revised by O. Lowenstein, Macmillan & Co., Ltd., London.
This is a well-known and widely consulted work; vol. 2 is devoted to vertebrates.

R. R. Shrock and **W. H. Twenhofel (1953):** "Principles of Invertebrate Paleontology," 2d ed., McGraw-Hill Book Company, New York.
This is a comprehensive textbook devoted mainly to morphology and classification.

H. H. Swinnerton (1947): "Outlines of Paleontology," 3d ed., Edward Arnold (Publishers) Ltd., London.
This book, covering both invertebrates and vertebrates, is less concerned with classification and presents more data on evolution than American texts.

P. B. Weisz (1966): "The Science of Zoology," McGraw-Hill Book Company, New York.
This book contains much information on the morphology and ontogeny of modern animals and some conclusions regarding their phylogeny.

NONCOELOMATE ANIMALS

F. M. Bayer et al. (1956): Coelenterates, in "Treatise on Invertebrate Paleontology," R. C. Moore (ed.), pt. F, Geological Society of America, Boulder, Colo., and University of Kansas Press, Lawrence, Kans.

See pp. 5–8, 181, 256, and 362–368 for brief consideration of the evolutionary relations and trends of various groups of coelenterates, chiefly corals.

A. S. Campbell (1954): Radiolaria, in "Treatise on Invertebrate Paleontology," R. C. Moore (ed.), pt. D, Protista 3, pp. 11–163, Geological Society of America, Boulder, Colo., and University of Kansas Press, Lawrence, Kans.
Attempts to reconstruct phylogeny are based entirely on increasing complexity of the siliceous skeleton.

J. A. Cushman (1948): "Foraminifera, Their Classification and Economic Use," 4th ed., Harvard University Press, Cambridge, Mass.
Arenaceous tubes are considered to be the most primitive type of foraminiferans.

A. Dendy (1911): Sponges, in "Encyclopaedia Britannica," Encyclopaedia Britannica, Inc., Chicago, 11th ed., vol. 25, pp. 715–732.
A brief discussion of sponge phylogeny is presented on p. 730.

R. M. Finks (1960): Late Paleozoic Sponge Faunas of the Texas Region, *Bull. Am. Museum Nat. Hist.*, vol. 120, pp. 3–160.
Some conclusions regarding evolutionary relations, particularly of Paleozoic sponges, are stated; see pp. 10–13.

J. J. Galloway (1933): "A Manual of Foraminifera," Principia Press, Inc., Bloomington, Ind.
Evolution is presumed to have progressed from a primitive spherical form with a gelatinous envelope; see pp. 18–20.

M. F. Glaessner (1945): "Principles of Micropalaeontology," Australia University Press, Melbourne; U.S. ed. **(1947),** John Wiley & Sons, Inc., New York.
The text includes brief general discussions of the phylogenetic relations within superfamilies.

H. J. Harrington and **R. C. Moore (1956):** Protomedusae, in "Treatise on Invertebrate Paleontology," R. C. Moore (ed.), pt. F, pp. 21–23, Geological Society of America, Boulder, Colo., and University of Kansas Press, Lawrence, Kans.
These peculiar fossils are presumed to be casts of the stomach cavities of medusae.

D. Hill, J. W. Wells, et al. (1956): Hydrozoa, in "Treatise on Invertebrate Paleontology," R. C. Moore (ed.), pt. F, pp. 67–106, Geological Society of America, Boulder, Colo., and University of Kansas Press, Lawrence, Kans.
Most types of hydroids are poorly represented among fossils.

L. H. Hyman (1940): "The Invertebrates," vol. 1, McGraw-Hill Book Company, New York.
This thorough descriptive work includes some consideration of evolutionary relations among the protozoans.

M. W. de Laubenfels (1955): Porifera, in "Treatise on Invertebrate Paleontology," R. C. Moore (ed.), pt. E, pp. 21–112, Geological Society of America, Boulder, Colo., and University of Kansas Press, Lawrence, Kans.
This taxonomic work includes nothing on evolution.

A. R. Loeblich, Jr., and **H. Tappan (1964):** Protista 2 (foraminiferans, etc.), in "Treatise on Invertebrate Paleontology," R. C. Moore (ed.), pt. C, 2 vols., Geological Society of America, Boulder, Colo., and University of Kansas Press, Lawrence, Kans.
This work includes good sections on the morphology and biology of foraminiferans but almost nothing on evolution.

R. C. Moore and **H. J. Harrington (1956):** Scyphomedusae, in "Treatise on Invertebrate Paleontology," R. C. Moore (ed.), pt. F, pp. 38–53, Geological Society of America, Boulder, Colo., and University of Kansas Press, Lawrence, Kans.
The impressions of jellyfishes are unusual fossils.

V. Pokorný (1963): "Principles of Zoological Micropaleontology," vol. 1, trans. by J. W. Neale, The Macmillan Company, New York.
Taxonomic problems of foraminiferans are considered in relation to their ontogeny and presumed evolutionary development.

R. E. H. Reid (1962): Hexactinellida or Hyalosponges? *J. Paleontol.*, vol. 37, pp. 232–243.
The author's objections to de Laubenfels's classification have evolutionary implications.

T. W. Vaughan and **J. W. Wells (1943):** Revision of the Suborders, Families, and Genera of the Scleractinia, *Geol. Soc. Am. Spec. Paper* 44.
See pages 90–99 for conclusions concerning hexacoral evolution.

I. M. van der Vlerk (ed.) **(1963):** "Evolutionary Trends in Foraminifera," Elsevier Publishing Company, Amsterdam.
This symposium consists of 17 papers devoted to various groups of foraminiferans.

H. C. Wang (1950): A Revision of the Zoantharia Rugosa in the Light of Their Minute Skeletal Structures, *Phil. Trans. Roy. Soc. London, Ser. B,* vol. 234, pp. 175–246.
This important article describes tetracoral structure and traces the history of these corals from the Ordovician to the Permian.

J. W. Wells, D. Hill, et al (1956): Anthozoa, in "Treatise on Invertebrate Paleontology," R. C. Moore (ed.), pt. F, pp. 161–477, Geological Society of America, Boulder, Colo., and University of Kansas Press, Lawrence, Kans.
Brief considerations of coral evolution are given on pp. 162–164, 181, 256, and 362–368.

FOUR

LOPHOPHORATE ANIMALS

BRYOZOANS—BRACHIOPODS

A lophophore is a structure surrounding the mouth and bearing ciliated tentacles which differ in their origin and organization from those of the coelenterates (Fig. 132). The possession of a lophophore characterizes three phyla: (1) bryozoans, (2) brachiopods, and (3) phoronids. These animals share other characters in various degrees, suggesting that they probably had a common evolutionary origin. They have intermediate body cavities, or coeloms, within the mesoderm and thus are more advanced in their organization than are the coelenterates. None possesses any structure that might be considered as a head; in this way they rank below most of the other common coelomate phyla. Otherwise the lophophorates differ greatly among themselves.

The phylogenetic relations of the lophophorates are puzzling; in some respects their embryology seems to ally them with the protostomes and schizocoels, and in others with the deuterostomes and enterocoels. For example: (1) Cleavage of the eggs generally is radial—deuterostomous. (2) The blastopore becomes the mouth—protostomous. (3) The coelom of phoronids, bryozoans, and one group of brachiopods opens in the mesoderm—schizocoelous—but in the other brachiopods it invaginates from the stomach cavity—enterocoelous. (4) The larvae of phoronids and some bryozoans resemble trochophores—protostomous—but those of other bryozoans and brachiopods do not. Perhaps the lophophorates evolved from the same group which produced the typical schizocoels and enterocoels before these became fully differentiated (Fig. 132a). The lophophorates commonly are classed with the schizocoels, but they may constitute a third and separate branch of the animal kingdom.

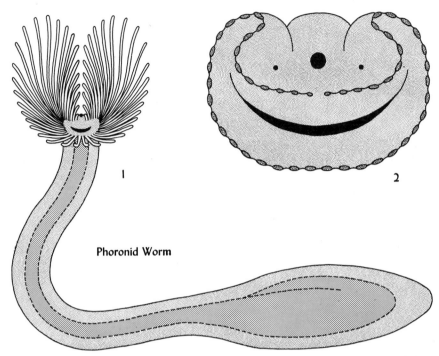

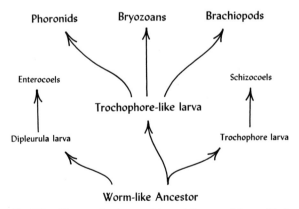

Phoronid Worm

1   2

Fig. 132   1, Complete animal showing long looped gut in worm-like body and tentacle-bearing lophophore. 2, Generalized form of anterior end showing mouth surrounded by tentacle bases, overlying
anus, and paired nephridial openings. The lophophore grows by the addition of pairs of new tentacles
inserted at the midline above the mouth. Modern phoronids range from 6 to 200 mm in length. Many
individuals commonly live in dense association. (*Adapted after Hyman, 1959, figs. 82C, 83A, pp. 231, 234.*)

Phoronids   Bryozoans   Brachiopods

Enterocoels   Schizocoels

Trochophore-like larva

Dipleurula larva   Trochophore larva

Worm-like Ancestor

Fig. 132a   Diagram showing possible evolutionary relations of lophophorate phyla. Compare with Fig. 86.

Table 133   **Principal distinguishing characters of the lophorate phyla**

Bryozoans: tiny colonial animals, budding asexually, some polymorphic, commonly build calcareous skeletons. Alimentary system complete. Nephridia and blood circulatory system lacking. Ordovician to present
Phoronids: worm-like animals without skeletons. Mostly inhabiting tubes in bottom sediments. Some reproduce asexually by transverse fission. Alimentary and circulatory systems complete. Have nephridia. Blood red. Unknown as fossils
Brachiopods: mostly sessile animals with bivalved shells. Alimentary system with or without an anus. Circulatory system imperfect. Have nephridia. Blood colorless. Cambrian to present

Ranking of the lophophorates is uncertain, but the bryozoans seem to be the most primitive and the phoronids the most advanced. Bryozoans are colonial animals with a complete alimentary system leading to an anus but no blood circulatory system or nephridial excretory tubes. Perhaps lack of these features is related to the animals' minute size rather than being an indication of a primitive condition. Brachiopods have nephridia and an imperfect circulatory system with a colorless type of blood. One group has a blind gut that does not open through an anus. Phoronids possess nephridia and complete digestive and circulatory systems, the latter with red blood.

Formerly the bryozoans and brachiopods were classed together in a single phylum and were termed molluscoids. Most zoologists, however, now believe that bryozoans and phoronids are more closely related to each other than either is to the brachiopods.

Bryozoans and brachiopods occur abundantly as fossils. Phoronids have left no ancient record unless some vertical worm-like borings in sediment were made by animals of this kind.

BRYOZOANS

A bryozoan colony originates from a sexually produced individual which attaches itself to some suitable object or surface and buds off other individuals that retain close connection with their parents. Each tiny individual secretes an outer horny covering that commonly is strengthened by an inner wall of calcite. The skeleton so built consists of small chambers in the form of tubes or cells occupied by the living animals. Some skeletons closely resemble those of certain corals. Most of the bryozoan individuals are constituted much like the phoronids, with a crown of tentacles surrounding the mouth and a strongly looped gut opening through an anus outside the lophophore. When a bryozoan animal is active, this crown is protruded from the skeletal chamber, but disturbance and adverse conditions cause it to be withdrawn so that the animal is completely enclosed by its outer skeleton (Figs. 134; 147). A hinged operculum, or cover, may close the opening in the skeleton when the animal is withdrawn.

Some bryozoan colonies consist only of ordinary feeding individuals, but others are polymorphic. Specialized individuals are of several types: (1) Those that produce

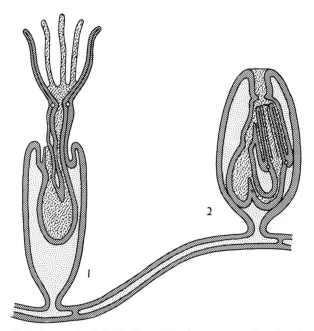

Fig. 134   Simplified sketch of uncalcified bryozoans, greatly enlarged, showing two individuals of the ctenostome type arising from a stolon, one with lophophore extruded and the other with lophophore withdrawn. Contraction of muscles in the body wall reduces the internal volume and forces the lophophore outward. Contraction of a long muscle extending from the base of the tentacles to the posterior body wall withdraws the lophophore, and the wall bulges outward.

slender stolons, from some of which feeding individuals are budded off (Fig. 138-2). (2) Those in which the operculum is transformed into a movable structure of more or less extreme form. There are two kinds; (*a*) avicularia, whose opercula may resemble the beak and lower jaw of a tiny bird's head (Fig. 145), and (*b*) vibracula, in which the operculum is a long slender flexible bristle. The functions of these individuals are uncertain. They may protect the feeding individuals, prevent the settling of sediment or larvae of other animals on the colony, or aid in respiration. (3) Those that are variously transformed as breeding individuals or become chambers in which the larvae are brooded (Fig. 135). (4) Finally, dwarf or aborted individuals which lack some or all of the structures of normal ones. Among these, (*a*) some occupy mesopores, or slender tubes between the larger chambers (Fig. 135*a*), and (*b*) others presumably were associated with acanthopores, known only in fossils, which are narrow tubes in the walls of normal chambers identified by their cone-in-cone structure and projecting as short spines (Fig. 135*a*). The functions of these individuals are not known. Mesopores may simply fill in space between

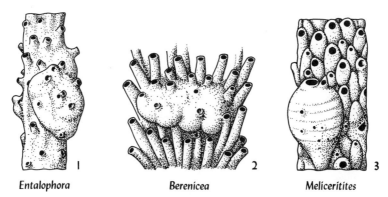

Entalophora   *Berenicea*   *Meliceritites*

Fig. 135   Brood chambers of cyclostomes. (Approx. × 15.) (*After Gregory, 1899, "Catalogue of Cretaceous Bryozoa in the British Museum," vol. 1, pl. 5, fig. 7a; pl. 10, fig. 12; pl. 14, fig. 10.*)

feeding individuals. Acanthopores may mark the sites of individuals comparable to avicularia or vibracula. Almost all gradations between the ordinary feeding individuals and these specialized forms are known. Many of the avicularia and vibracula arise from very small chambers located on the frontal surface of a colony and bear almost no resemblance to ordinary individuals (Fig. 145-4 to 7).

The individual bryozoan animals periodically degenerate and then are reorganized to produce new individuals. In this process the animal withdraws into its chamber, its tissues disintegrate, much of the resulting material is consumed by scavenging cells, and a new individual grows from a bud that appears within the old chamber. The physiology of this process is not understood. Bryozoans lack excretory organs, and degeneration may be related to the accumulation of nitro-

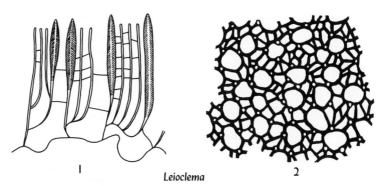

Leioclema

Fig. 135a   Trepostomes, vertical and horizontal sections. (Approx. × 28.) Mesopores and acanthopores surround ordinary tubes. The immature region of this bryozoan is very short. (*After Ulrich, 1890, fig. 1, p. 301. See also Fig. 143a.*)

genous waste or some other condition associated with aging. There is also the possibility that degeneration follows the sexual activity of individuals. Several successively regenerated animals may in turn occupy the same chamber, or a new individual may move forward, build an extension to a tube, and lay down a tabula behind itself (Figs. 135a-1; 143-5; 143a-2).

The skeletons of bryozoan colonies take many forms. For example, there are (1) chains of successive chambers or tubes, (2) thin encrusting expanses, (3) thick compact structures composed of closely appressed tubes, (4) erect leaf-like fronds, (5) erect more or less slender branching forms, and (6) very delicate lace-like growths. Colonies of more than one type occur within a single group, and those of similar types occur in different major groups of bryozoans. Form is influenced to some extent by the environment within which a colony grows.

Evolutionary Trends

Ideas concerning major subdivisions of the bryozoans have remained almost unchanged for nearly three-quarters of a century. This seems to be good evidence that evolutionary concepts have played a negligible part in the classification of these animals. The recognized groups are (1) phylactolaems, (2) cyclostomes, (3) trepostomes, (4) ctenostomes, (5) cryptostomes, and (6) cheilostomes (see Table 136). These groups are not so clearly distinguishable as commonly is assumed; forms

Table 136   Characters of major bryozoan groups

Phylactolaems: U-shaped lophophore, without calcareous skeleton, restricted to freshwater and brackish water environments; almost no fossil record

Cyclostomes: circular lophophore, skeleton delicate to massive, generally composed of cylindrical tubes with *minutely porous walls*, rounded apertures, and *rare tabulae*, may be separated by mesopores or vesicular tissue, *brood chambers present*; marine, late Cambrian to Recent

Trepostomes: *skeleton massive* to delicate, generally composed of long polygonal to circular tubes, with contiguous but *separate walls*, divisible into *immature and mature regions*, mostly with *abundant* complete or incomplete *tabulae* or curved partitions, separated by mesopores and with *acanthopores* in their walls, surface with *monticules*, or elevated areas, and *maculae*, or depressed areas, *without brood chambers*; marine, Ordovician to Permian (Triassic ?)

Ctenostomes: circular lophophore, individuals generally bud from a *stolon*, skeleton only rarely or partly calcified; marine, Ordovician to Recent, scant fossil record

Cryptostomes: colonies *delicate*, many lace-like forms, skeleton consists of *short tubes or cells* mostly *without tabulae* except *one or two that are incomplete* which are believed to mark *sunken aperture* probably closed by an operculum, *without brood chambers*; marine, Ordovician to Permian

Cheilostomes: circular lophophore, skeleton delicate, variably calcified, generally consisting of closely appressed more or less regularly arranged cells with restricted anterior apertures closed by *opercula, avicularia common, brood chambers present*; marine, Jurassic (?) to Recent

These groups as generally recognized cannot be diagnosed in any way that is consistent and mutually exclusive. Characters considered especially important are italicized, but mostly they are neither invariably possessed by nor restricted to the bryozoans assigned to any single group. Combinations of characters, rather than the presence or absence of single characters, are regarded as significant. Some bryozoans might be classified in either of two groups depending on how the relative importance of their morphologic features is assessed.

are known that in one way or another are morphologically intermediate between most of them, and the groups probably do not accurately represent evolutionary divisions of the bryozoans.

Evolutionary speculation has been concerned mainly with the relations of these groups to one another, and very little consideration has been given to evolution within any of them except the cheilostomes. On the whole, morphologic evidence is scanty and stratigraphic succession has not been carefully assessed.

All but the phylactolaems and cheilostomes are represented by fossils in Ordovician faunas, and a few more or less doubtful Cambrian occurrences have been reported. The almost simultaneous appearance of a considerable variety of bryozoans probably indicates that the early fossils evolved from nearly a dozen different ancestral stocks which had not previously constructed preservable calcareous skeletons. Perhaps much of the major differentiation among them dates back to Precambrian time.

Several general trends that seem to be evolutionarily progressive can be recognized with respect to a number of different features of the bryozoans. Among the most obvious are (1) calcification of the skeleton, (2) colony formation, (3) form of chambers, (4) development of polymorphism, and (5) variation in the mechanism of lophophore extrusion and retraction. There is no reason to believe, however, that any of these trends was constantly progressive. Numerous reversions almost certainly occurred. Also parallel and convergent types of evolution probably were common.

Calcification of Skeleton

All bryozoans surely evolved from uncalcified ancestors, and all groups probably have included uncalcified representatives in Ordovician and later time. With very few exceptions, bryozoans of this type have left no fossil record. Uncalcified forms such as the phylactolaems probably belong to lineages that never acquired hard structures, although some other bryozoans may have reverted to this seemingly primitive condition.

Only the cheilostomes among modern bryozoans show much variation in the relative amounts of horny and calcareous material in their skeletons. Therefore, only in this group can observations be made concerning progressive calcification. Perhaps conclusions drawn from them are not applicable to other groups, because of some important differences in their structures.

Modern cheilostomes exhibit all gradations from wholly uncalcified skeletons to those that are heavily mineralized. Certain fossils, like some living species, clearly were incompletely calcified (Fig. 138-1). Different conditions of the skeleton show that secretion of calcite began in the side walls of the cells or perhaps on the back surface. Subsequently calcified front walls were formed in several different ways. Lastly, front walls were thickened, depressions between adjacent chambers were filled in, and collars were erected around some apertures.

Fossils demonstrate that the ability to secrete more or less tubular walls evolved

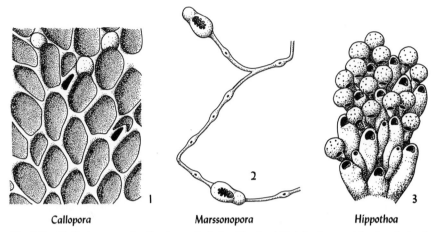

Callopora   Marssonopora   Hippothoa

Fig. 138   Cheilostomes, contrasting types. 1, Cells without calcified fronts except for brood chambers. 2, Breeding individuals connected by reduced stolon-like individuals. 3, Colony consisting of small male individuals, large sterile individuals, and female breeding individuals with hooded brood chambers. (Approx. ×20.) (*After Canu and Bassler, 1928, Proc. U.S. Natl. Museum, vol. 72, pl. 3, fig. 2; pl. 26, fig. 2; and Hyman, 1959, fig. 120A, p. 324.*)

independently in more than one type of early bryozoans. Tabulae and other kinds of transverse partitions probably were later developments in some lineages; they also suggest parallel evolution in several different stocks.

Colony Formation

All bryozoans are colonial. They probably evolved, however, from animals that lived as separate individuals. Colony formation presumably began with chains of individuals produced by asexual budding, and multiple budding from an individual resulted in branching growth. Two forms probably developed (Fig. 139): (1) erect colonies of more or less bushy type, and (2) recumbent colonies that spread over the sea bottom or adhered to other surfaces.

The chains of individuals in early colonies probably continued to grow independently of one another. More compact organization perhaps was accomplished in two ways: (1) When different chains by chance made contact, they fused together locally in an anastomosing manner (Fig. 139-3). (2) Branching chains adhered to each other from the start and continued to grow closely side by side (Fig. 139-4). Both developments would have strengthened erect colonies, and the latter would have conserved the space available to recumbent colonies. More frequent budding in recumbent forms produced compact colonies that grew peripherally and advanced in all directions (Fig. 139a-2).

Tendencies seem to have developed in some recumbent colonies for upward as well as outward growth. This probably was initially accomplished by the lengthening of upwardly directed tubes (Fig. 139a-3). Promiscuous budding of these tubes produced hemispherical or biscuit-shaped massive colonies, some of which attained

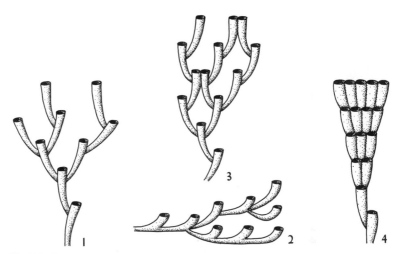

Fig. 139   Diagrammatic representations of primitive types of bryozoan colonies. 1, 2, Upright and recumbent simple branching colonies. 3, Anastomosing branches. 4, Laterally adherent branches.

large size. The concentration of budding in certain areas, however, resulted in the beginning of localized erect growth.

Budding in all but the simplest erect colonies generally either occurred in a plane or it was more or less restricted to a central axis. The first type is characteristic

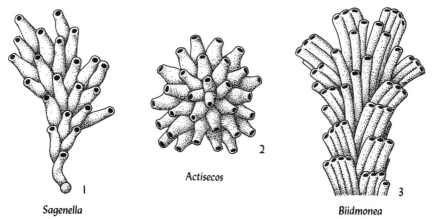

Fig. 139a   Colonies representing presumed successive evolutionary stages. (Approx. × 10.) 1, Unidirectional encrusting growth. 2, Peripheral encrusting growth. 3, Upright growth of elongated tubes. (*After Bassler, 1953, figs. 14-3a, 15-7, 174-8, pp. 44, 45, 232.*)

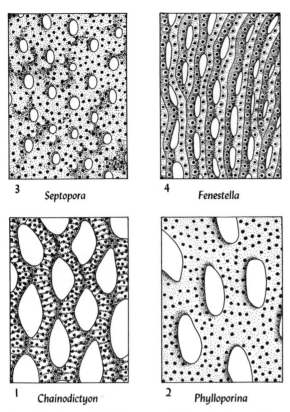

3 *Septopora*   4 *Fenestella*

1 *Chainodictyon*   2 *Phylloporina*

Fig. 140   Fenestellid-like bryozoan colonies, miscellaneous types.
(× 10.) 1, This structure may have developed by the rather regular
periodic fusion of curved branches with variably arranged pores. 2,
Areas without apertures at the ends of perforations suggest the
beginning of polyporoid or septoporoid kinds of structure. Compare
Figs. 141-5 and 142-5. 3, This type may have originated by the
opening of perforations in a continuous foliaceous expanse in adaptation to the force of water currents. 4, This specimen may show
transition from anastomosing branches (upper left) to straight
branches connected by cross bars. (*After Ulrich,* 1890, *pl.* 44, *figs.*
5a, 6b; *pl.* 51, *fig.* 4; *pl.* 62, *fig.* 3a.)

of foliaceous, or laminar, growth with the animals facing all in one direction, or in
both directions if budding was bilateral from a median plane. Budding from a
common axis produced more or less cylindrical rising stalks.

The anastomosing type of growth already mentioned possibly became more
regular, and a variety of delicate lace-like patterns might have developed from it
(Figs. 140 to 142). These patterns, however, may have originated in at least three
other ways: (1) by the appearance of perforations in a continuous lamellar frond,

(2) by the fusion of converging side branches, and (3) by the development of crossbars which connect adjacent branches.

Other features also must be considered in assessing the mutual relations of bryozoans. These include minute structural details that can be observed only by the microscopic study of thin sections, and the gross form of colonies. For example, among the lace-like colonies and their allies, the earliest forms were funnel-shaped growths. Fan-like colonies evolved later. They appeared in the Silurian and became common in the Devonian faunas. Still later in the Mississippian Period, several kinds of specialized supporting structures developed. Although morphologic series can be

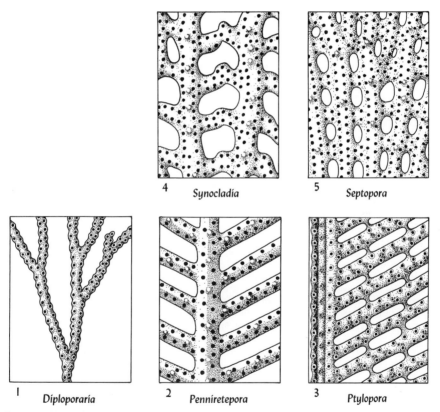

Fig. 141   Fenestellid-like bryozoan colonies showing possible successive evolutionary stages. (× 10.) 1, A simple branching ribbon. 2, Perhaps derived from the former type by the regular arrangement of side branches. 3, Possibly derived from the last type by the development of crossbars. 4, Perhaps derived from 2 by the fusion of side branches converging from adjacent main branches. 5, Possibly derived from 4, although origin from a colony similar to that of Fig. 140-2 but with more regularly arranged apertures is equally likely. This sequence is not intended to suggest direct evolutionary relationships of these genera. (*After Ulrich, 1890, pl. 62, fig. 12a; pl. 64, figs. 4, 1; pl. 66, figs. 6, 2.*)

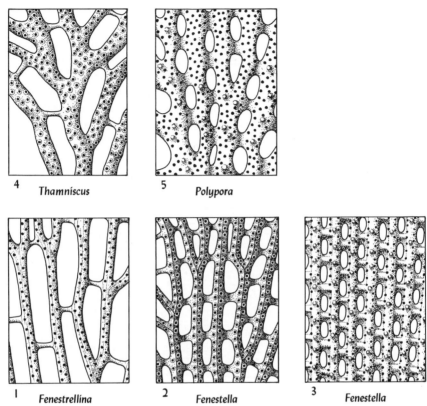

Fig. 142 Fenestellid-like bryozoan colonies showing possible stages of evolutionary advancement. (× 10.) Fenestellids with more than two rows of apertures are rare before the Devonian and did not become common before the Mississippian Period. They seem in general to be more advanced than the two-rowed kind. 1, This type might have evolved from one like Fig. 141-1 by the acquisition of crossbars. 2, 3, These show increasing perfection in the regularity of fenestellid structure. 4, This type may have evolved from a form like 1 by increase in the number of rows of apertures or from one like that in Fig. 141-1 but with three rather than two rows of apertures. 5, More regular structure may have evolved by modification of colonies like either 2 or 4. All such developments would have been advantageous in strengthening colonies and permitting them to grow in disturbed water or strong currents. (*After Ulrich,* 1890, *pl.* 44, *fig.* 1; *pl.* 49, *fig.* 20; *pl.* 51, *fig.* 5b; *pl.* 60, *fig.* 8; *pl.* 62, *fig.* 4a.)

recognized, the evolutionary relations of these bryozoans are confused, and more than one original stock and several evolutionary lineages seem to be represented among them. This group almost surely provides an example of convergent evolution.

Form of Chambers

All bryozoans might have had their origin in colonies consisting of moderately

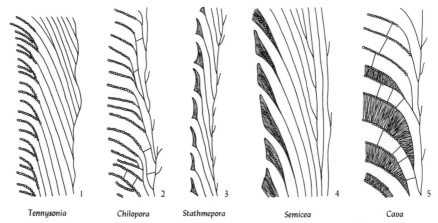

Fig. 143   Cyclostomes; longitudinal sections showing long tubes, apertures to left, separated by mesopores or laminated calcareous deposits. Tabulae are few or absent, and distinction between immature and mature regions is not so conspicuous as in trepostomes (compare Fig. 143a). (Approx. × 15.) (*After Canu and Bassler, 1922, Proc. U.S. Natl. Museum, vol. 61, art. 22, figs. 6D, 8C, 9C, 24C, 29F, pp. 36, 40, 51, 96, 106.*)

elongated, slightly expanding tubes. These may have been erect, with simple terminal apertures, or recumbent, with upturned apertures or openings located anteriorly in their sides. From a colony of either kind, evolution may have progressed in two directions: (1) lengthening of tubes, which were extended by a succession of regenerated and advancing individuals (Figs. 143; 143a); and (2) shortening of tubes, finally producing more or less oval or polygonal cells that were occupied by a

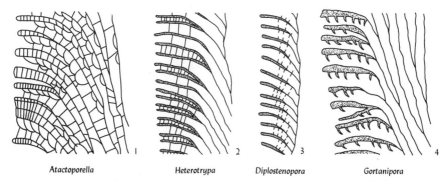

Fig. 143a   Trepostomes; longitudinal sections showing long tubes, apertures to left, with complete or incomplete tabulae and curved cystiphragms, separated by mesopores and acanthopores. Distinction between immature and mature regions is more conspicuous than in cyclostomes (compare Fig. 143). (Approx. × 15.) (*After Bassler, 1953, figs. 55-2c, 56-2b, 59-1c, 66-3b, pp. 92, 93, 96, 103.*) See also Fig. 135a-1.

succession of regenerated individuals without further skeletal growth (Figs. 138; 144-6; 145).

Short cell-like chambers almost certainly evolved in several different stocks. Some cells seem to have been a secondary development that followed a preliminary lengthening of tubes, and many may have had this evolutionary history. In these the cells correspond to the immature portions of longer tubes, and their extensions in some forms probably were not occupied by advancing individuals. Similar longer chambers also may have developed from short cells by the upward growth of skeleton around their apertures (Fig. 144-4).

Polymorphism

Most modern cheilostomes are polymorphic, i.e., the colonies consist of individuals of more than a single kind. Some of the individuals are so extremely altered that their nature would not be known if almost complete morphologic transitions could not be traced connecting them to normal individuals. This is best exemplified by the avicularia, which are highly specialized for some unknown function (Fig. 145). A similar clear transition to vibracula is not so evident.

Other individuals degenerated and built small tubes. Some of these are slender recumbent stolons that connected or budded off normal individuals and thus accelerated the spreading of a colony (Fig. 138-2). Some served as anchoring holdfasts. Still others produced various types of strengthening or supporting structures.

Many fossil bryozoan colonies contain small tubes located either between the larger ones or on the backs of expanded colonies, all of whose ordinary animals faced in one direction. The assumption is general that they were occupied by specialized individuals of several kinds. Tubes termed mesopores are common in the mature regions, particularly of trepostomes (Figs. 135a; 143a-1,2). Perhaps they served only to separate the normal feeding individuals. There is some indication that the individuals which built these tubes degenerated completely in certain stocks and formed a kind of communal tissue that ceased to construct tubes but laid down curved partitions beneath itself (Fig. 144-5).

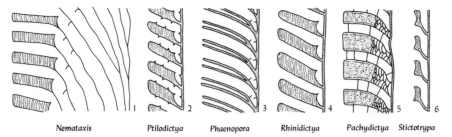

Nemataxis   Ptilodictya   Phaenopora   Rhinidictya   Pachydictya   Stictotrypa

Fig. 144   Cryptostomes; longitudinal sections showing tubes or chambers, opening to left, mostly with short immature regions, sunken apertures, and incomplete tabulae (hemisepta). (Approx. × 15.) (After Bassler, 1953, figs. 93-3b, 100-1a, 102-3, pp. 133, 140, 142; and Ulrich, 1890, figs. 11a, 12b, 13b, pp. 391, 392, 394.)

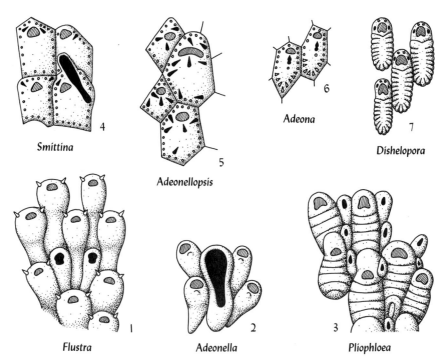

Fig. 145   Cheilostomes with pores indicating the locations of avicularia, or specialized opercula, variously enlarged. Evolution seems to have progressed through several stages, from pores on chambers little different from ordinary ones (1) or rarely of larger size (2), to smaller ones associated with interstitial cells (3), those borne in the surfaces of ordinary cells, first of variable size (4), then smaller ones randomly arranged and oriented (5), finally to those that are definitely oriented and located generally adjacent to the apertures of ordinary cells (6, 7). (*After Hyman, 1959, figs. 121B, D, 122F, 124C, 125H, pp. 326, 329, 344, 346, and Lang, 1921, pl. 6, fig. 1; pl. 7, fig. 7.*)

Some modern bryozoans release their eggs into the surrounding water, but many more brood them and shelter the developing larvae. Brood chambers of several different kinds evolved among both fossil and recent cyclostomes and cheilostomes. Some are structures built over or around the apertures of normal individuals, either a single aperture, as in many cheilostomes (Fig. 138-3), or several apertures, as in some cyclostomes (Fig. 135). Some are sunken between the normal chambers and either may enclose specialized breeding individuals that degenerate after the eggs are produced or are outgrowths from ordinary chambers. Some are large and do not differ much from their neighbors except in size (Fig. 145-5). Others seem to change in form as brooding progresses. The larger tubes that occupy some elevated areas, or monticules (Fig. 146-1), or surround some depressed areas, or maculae (Fig. 146-2), in the fossil trepostomes may have housed breeding

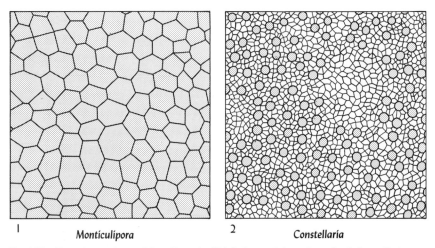

Monticulipora	*Constellaria*

Fig. 146   Trepostomes, tangential sections. (× 18.) 1, Larger tubes of an elevated monticule near center. 2, Tubes surrounded by mesopores and a depressed macula without tubes to upper right. (*After Ulrich, 1895, Geol. of Minn., vol. 3, pt. 1, pl. 15, fig. 3; pl. 21, fig. 11.*)

individuals, but there is no indication that these bryozoans brooded eggs or larvae.

Lophophore Extrusion and Retraction

When a broyozoan animal is disturbed, muscles draw the tentacled lophophore back completely within its chamber. Even though the stomach and gut collapse, the volume of material in the chamber is increased and compensation is required. In uncalcified bryozoans the horny chamber walls bulge outward (Fig. 134). Many calcified cheilostomes have a horny frontal wall which can accommodate volume changes in a similar way (Fig. 147-2,3). This is believed to identify a primitive condition.

Other bryozoans with completely calcified skeletons are faced with a different problem. This is met simply by the cyclostomes. Fluid filling the posterior part of the coelom moves forward into a space in the body just within the aperture. There inward swelling completely closes the passage through which the lophophore was retracted (Fig. 147-1).

A somewhat similar mechanism probably operated in the trepostomes and perhaps in many cryptostomes. The cells of some of the latter bryozoans are so restricted, however, that complete withdrawal of the lophophore may have been impossible. If this is so, an explanation may be provided for the evolution of the trellis-like superstructures that rise above and extend across the surfaces of some fossil colonies in a way that seems to have been protective (Fig. 147a).

Three different kinds of calcified frontal walls evolved in the cheilostomes. One

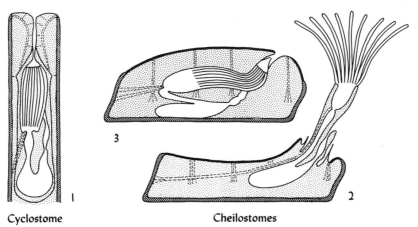

Cyclostome   Cheilostomes

Fig. 147   1, Cyclostome; when retracted, coelomic fluid flows forward and produces swelling that closes aperture. When muscles contract in region just inside the aperture, the fluid flows backward into the posterior part of the coelomic cavity and the lophophore is extruded. 2, 3, Cheilostome, with extended and retracted lophophore, showing flexible frontal membrane and muscles that pull it downward. (*After Borg, 1923, Arkiv. Zool., vol. 15, art. 11, fig. 1, p. 3; and Bassler, 1922, Smithsonian Inst. Ann. Rept., 1920, figs. 17G, H, p. 74.*)

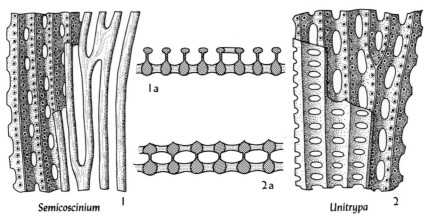

Semicoscinium   1   *Unitrypa*   2

Fig. 147a   Fenestellid-like colonies with superstructures shown in front view and transverse cross sections to illustrate relations to underlying parts occupied by the bryozoan animals. (×7.) Bryozoans of this type present a seemingly nearly complete evolutionary series of stages. 1, Ribs with smoothly convex surfaces. (2) With regularly spaced, small, acanthopore-like spines between the rows of apertures. (3) With larger acute spines. (4) With narrow median longitudinal ridges (5) Ridges surmounted with lateral expansions (1). 6, Adjacent expanded ridges connected by crossbars (2). Perforations in superstructures of the last kind may correspond to those in the underlying colonial skeleton; they may be twice as numerous in the longitudinal series (2), or two longitudinal series may occur between adjacent supporting ridges. (*After Bassler, 1953, figs. 85-2b, 4a, p. 124.*)

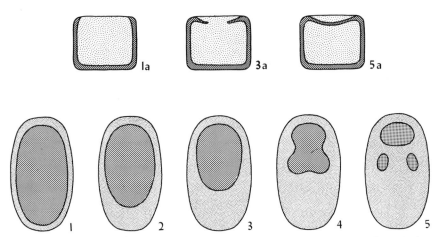

Fig. 148   Cheilostomes. Diagrammatic representation of evolutionary stages in the development of a calcified frontal wall beneath a flexible membrane. The wall began to form posteriorly and progressed forward and inward. Paired openings pierce the complete wall, as shown in 5, for passage of muscles attached to the membrane. Transverse sections show relations of wall to membrane.

grew inward from the side walls below the horny frontal membrane (Fig. 148), and another developed by the fusion of spines above it (Fig. 149). These provide protection for the animal within its cell, but the membrane remains free and flexible and accommodates volume change by moving in or out.

The third kind of frontal wall is attached directly to the membrane, which thereby loses its accommodating function. Compensation was provided by the evolution of a sac within the coelomic cavity opening to the exterior either in the aperture of the cell or through a pore in the calcified frontal wall (Fig. 149a). Seawater passes in and out of this sac, changing its volume as this is required by the position of the animal.

The withdrawal mechanism typical of the cyclostomes probably evolved far back before the Ordovician Period and before calcareous skeletons began to be secreted. The mechanism of the typical cryptostomes is not known, but if it differed, it also must have had a similar very early origin. Cheilostomes with a compensation sac probably evolved from ancestors with a flexible frontal membrane. Their mechanism is the most highly evolved of any bryozoans.

Evolutionary Relations

Evolutionary relations of the main bryozoan groups are not clear. Part of the difficulty seems to be related to the probability that fossils are classified in a way that does not accurately reflect their genetic relationships. Ontogeny and adult structure of modern bryozoans suggest the existence of two branches of this phylum. Similar comparisons, however, cannot be made for the extinct forms, and

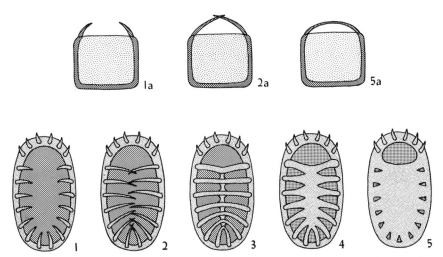

Fig. 149   Cheilostomes. Diagrammatic representation of evolutionary stages in the development of a calcified frontal wall above a flexible membrane. Overarching spines lengthened, fused, and widened. The complete wall (5) is pierced by pores that permit the passage of water into and out of the chamber above the membrane. Transverse sections show relations of wall to membrane.

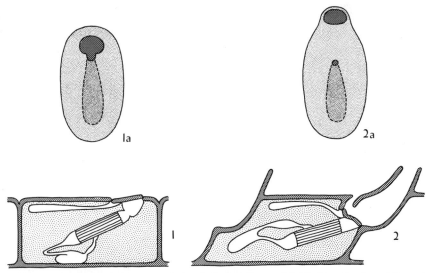

Fig. 149a   Cheilostomes. Diagrammatic longitudinal sections and front views showing relations of compensation sac to aperture or pore in frontal wall. (*After Bassler, 1922, Smithsonian Inst. Ann. Rept.,* 1920, *figs.* 11, 13, *pp.* 370, 371.)

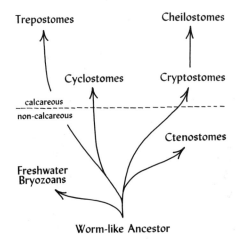

Fig. 150   Diagram showing possible evolutionary relations of the main bryozoan groups.

judgments regarding them are based solely on the nature of their calcified skeletons. Reasonable conclusions, nevertheless, can be reached if consideration is restricted to the more typical representatives of the groups (Fig. 150).

Phylactolaems

Zoologists believe that the phylactolaems, which are freshwater bryozoans, are the most primitive existing forms. Features considered to be significant are (1) elongated worm-like body, (2) muscular body wall, (3) lack of complex mechanism for lophophore extrusion and retraction and closure of the orifice, (4) horseshoe-shaped lophophore, and (5) absence of polymorphism. These bryozoans almost certainly evolved from marine ancestors. Peculiarities of their embryology are explained if this was so.

Cyclostomes

The simpler living cyclostomes resemble phylactolaems, and these groups are believed to be related. Similarities include (1) body form, (2) lack of conspicuous polymorphism, and (3) embryonic development, although the larvae are quite unlike. On the other hand cyclostomes differ importantly in possessing (1) a coelomic sac that aids in lophophore extrusion, and (2) brood chambers. Divergence of these groups must have occurred far back in time, probably in the Precambrian.

Trepostomes

Trepostomes probably evolved from the same ancient stock as the cyclostomes. The tubular skeletons of these two groups are generally similar (Figs. 143; 143a) and both lack opercula. Mesopores are present in both groups, and both have variously differentiated monticules and maculae (Fig. 146). These last features, however, are especially characteristic of the trepostomes, which also are more complex in other ways. They are provided, for example, with acanthopores (Fig. 135a) and have much more abundant tabulae as well as other kinds of partitions in their tubes.

Ctenostomes

The ctenostomes differ importantly in several ways from the two modern groups already mentioned, and relations with them seem to be remote. The embryonic development of ctenostomes is more primitive, and the larvae, which resemble trochophores with bivalved horny shells (Fig. 151), probably are phylogenetically significant. These bryozoans also seem to exhibit an early stage in the evolution of opercula.

Cheilostomes

Cheilostomes of Paleozoic age have not been surely recognized. Therefore, it is possible that they evolved directly from one of the older bryozoan groups. Some specialists in the study of these animals have believed that cheilostomes differentiated from the cyclostomes, but this derivation is no longer considered probable. Several similarities suggest that the cheilostomes are most closely allied with the ctenostomes, among modern forms. For example: (1) Embryology is similar. (2) Larvae are almost identical in structure. (3) The muscular system of the less-calcified and presumably more primitive cheilostomes closely resembles that of the ctenostomes. (4) Cheilostome opercula are foreshadowed in the ctenostomes.

Cryptostomes

The cryptostomes bear some resemblance to trepostomes in their incomplete tabulae and structures that may be acanthopores. It is generally agreed, however, that they are most closely related to the cheilostomes. This is indicated by (1) similar budding patterns in the beginning stages of colony growth, (2) similarity of cell-like chambers in many forms, and (3) the probability that cryptostomes were provided with opercula. A few late Paleozoic cryptostomes look much like cheilostomes and might easily be mistaken for them. The possibility is reasonable that cheilostomes may have evolved from an uncalcified cryptostome stock that left no fossil record.

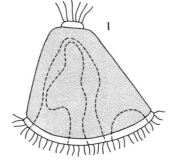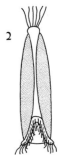

Fig. 151   Cyphonaut larva, greatly enlarged, of ctenostome and cheilostome bryozoans which do not brood their eggs. Generalized side and end views showing horny bivalved shell, apical sensory tuft, looped gut (mouth is to left), posterior adhesive organ by which larva attaches itself when it settles to bottom, and ciliated band with which the tiny creature swims. This larva is similar in a general way to a trochophore with its lower part missing and without nephridial tubes. (*After Hyman, 1959, figs. 133C, B, p. 353.*)

Entoprocts

In the past a small group of animals known as entoprocts generally has been regarded as a separate branch of the bryozoans. These creatures have left no fossil record. Their name refers to the way in which they differ most conspicuously from typical bryozoans, i.e., their anus is located within a circle of tentacles, which are arranged very similarly to those in a lophophore. Although they develop from a trochophore type of larva, careful studies have shown that their embryology and metamorphosis are different from those of typical bryozoans and that the adults are without a coelom. This last fact places them lower than the bryozoans in the evolutionary scale. The entoprocts probably are an offshoot of a primitive stock from which the annelid worms evolved.

BRACHIOPODS

Brachiopods are marine lophophorate animals enclosed in a bivalved shell (Fig. 152). Most of them lead a completely sedentary existence, attached posteriorly by a muscular stalk, or pedicle. The viscera are located in the rear of the shell, and the forward part, lined by the fleshy mantle, encloses the lophophore.

Phylogenetic placement of the brachiopods poses several unsolved problems. Possession of a lophophore and overall anatomical construction seem to ally them definitely with the bryozoans and phoronids. Their circulatory system resembles that of annelids, and the blastopore becomes the mouth. These features may indicate relation to the protostomes. On the other hand, brachiopods do not develop from trochophore larvae, and the coelom of one of the great divisions, the articulates, originates in modern forms by invagination of the entoderm. This suggests affinity with the deuterostomes.

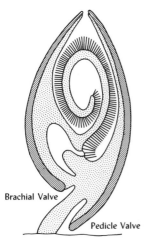

Brachial Valve

Pedicle Valve

Fig. 152   Generalized longitudinal section through a brachiopod. Most shells are hinged by paired teeth and sockets (not shown) at the sides of the posterior fleshy pedicle by which the animal is attached. Valves open anteriorly (above). The larger part of the shell, on right, is the pedicle valve (also termed ventral), and the other is the brachial (dorsal) valve.

The brachiopods constitute one of the most important groups of Paleozoic fossils. They make their first appearance at the very base of the Cambrian System and achieved their maximum differentiation during Devonian time. Thereafter decline set in. Some resurgence occurred in the middle Mesozoic, when modern kinds became temporarily abundant. Brachiopods now, however, are a relatively insignificant element in most marine faunas, persisting particularly in parts of the Pacific Ocean.

Fossil brachiopods have been studied by many paleontologists because of their abundance, diversity, and usefulness in stratigraphic correlation. In general, however, these animals have been neglected by zoologists. Much of the knowledge of modern species is based on studies reported in publications many years ago.

Classification

Two great divisions of the brachiopods are almost universally recognized: (1) the inarticulates, whose more or less phosphatic shells do not have hinges provided with interlocking structures; and (2) the articulates, with calcareous shells whose hinges are furnished with paired teeth and sockets. The lack of articulation and the phosphato-organic nature of their shells generally are considered to mark the inarticulates as the more primitive division, although many of these animals are highly specialized in other ways.

Modern brachiopods of these types seem to differ importantly. For example, their embryology and larval development are not the same, particularly with respect to the origin of the coelom and the pedicle (Figs. 154; 155). Furthermore, inarticulates have a digestive system ending with an anus; the articulates do not. Whether ancient brachiopods were similarly different cannot of course be known.

Ideas regarding the evolutionary relations of brachiopods have been expressed mainly in systems of classification. The one most widely accepted in the past, consisting of four main groups, to which a fifth was later added, is outlined in Table 154. It was based primarily on the nature of the pedicle opening and its modifications. The atremates were considered to represent the most primitive stock, from which both neotremates and telotremates evolved. The protremates were considered by different persons or at different times to have evolved either from the neotremates or directly from the atremates. These ideas implied that the division between inarticulates and articulates was unrealistic.

The later-distinguished palaeotremates are believed to be very primitive. They have been considered to represent the stock from which all other brachiopods evolved, or they were inserted in the evolutionary sequence between atremates on the one hand and protremates and telotremates on the other. The palaeotremates have been variously classed with either the inarticulates or the articulates.

This classification is now recognized to be unsatisfactory from an evolutionary point of view, but the several mentioned groups still provide a convenient basis for description, comparison, and general discussion.

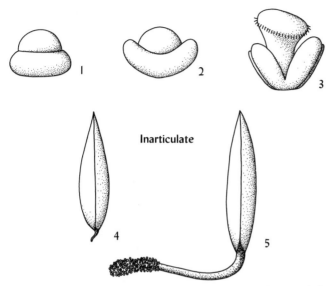

Inarticulate

Fig. 154   Diagrammatic representation of the ontogeny of an inarticulate brachiopod like modern *Lingula*. 1, The embryo segments into anterior and posterior parts. 2, The posterior part grows into the mantle lobes, which begin to enclose the anterior part. 3, The shell begins to form on the outside of the mantle lobes, and the lophophore begins to develop on the anterior part. 4, The pedicle appears in the young shelled animal as an outgrowth of the brachial mantle lobe. 5, The mature pedicle is a long muscular organ which becomes anchored in the bottom by the cementation of sedimentary grains.

Table 154   Main groups of brachiopods as formerly recognized

Atremates: hinge without teeth and sockets, shell generally horny and phosphatic, pedicle emerging through a little-modified opening shared by both valves, lophophore unsupported

Neotremates: hinge without teeth and sockets, shell generally horny and phosphatic, pedicle emerging through an opening in only one valve that may be more or less closed, lophophore unsupported

Protremates: hinge with teeth and sockets, shell calcareous, pedicle emerging through opening in one valve only that may be more or less closed by a single transverse plate, calcified lophophore supports absent or little developed

Telotremates: hinge with teeth and sockets, shell calcareous, pedicle opening indenting both valves in early youth, later confined to one valve and more or less closed by paired laterally advancing plates, lophophore supported by calcareous rods, spires, or loops

Palaeotremates: articulation lacking or rudimentary, shell horny and phosphatic or calcareous, pedicle emerging through an opening shared by both valves, lophophore unsupported. This group, consisting of only a few species ranging from the Lower Cambrian to the Middle Ordovician, was distinguished much later than the other groups. It is considered to be very primitive.

More complete knowledge of the brachiopods, particularly of the nature of their pedicle openings, seems to have demonstrated that the protremates and telotremates are not so clearly differentiated as was formerly believed.

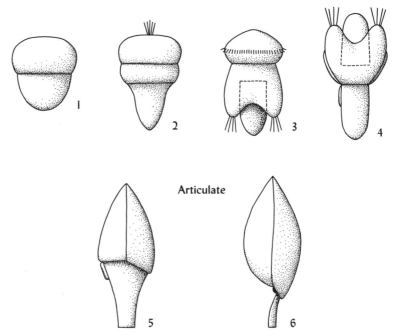

Articulate

Fig. 155   Diagrammatic representation of the ontogeny of a modern articulate brachiopod. 1, The embryo segments into anterior and posterior parts. 2, A third middle segment appears. 3, The middle segment which will later become the mantle grows back around the posterior segment or beginning of the pedicle. 4, The mantle segment turns forward and partly encloses the anterior or body segment. The shell which had begun to form in the preceding stage now lies on the outer sides of the mantle lobes. 5, 6, The animal becomes attached by its pedicle and grows to its mature form. Stages 4 and 5 as illustrated here show a third shelly element that was supposed to develop on the surface of the pedicle. This is now believed to be a misinterpretation as there probably is no such element.

Evolutionary Trends

Any consideration of the comparative morphology of brachiopods becomes complex because these animals and their fossils at the same time resemble and differ from each other in so many ways that no clear overall pattern of relationships is evident. Nevertheless, if different features are considered separately, various probable or possible evolutionary developments can be traced.

Shell Composition

Brachiopods evolved from unknown ancestors whose probably thin unmineralized shells were not preserved. The horny and phosphatic shell of the inarticulates is believed to represent a relatively primitive condition. The first-appearing shells in the larvae of modern species are horny, but as growth proceeds mineral matter is added to this substance. The original composition of many early fossil brachiopods is uncertain, but there seems to have been a tendency for shells of the inarticulates to become more calcareous as they increased in size and thickness. Some specialized representatives of both the atremates and neotremates had shells almost wholly calcareous.

Shells of the protremates and telotremates consist of calcite. These brachiopods may also have evolved from ancestors with unmineralized horny shells. There is no clear evidence that they passed through a phosphatic stage.

Shell Structure

The shells of most articulate brachiopods are distinctly layered, but this structure is likely to be destroyed in fossils that are recrystallized. Two layers generally are recognizable (Fig. 156) although the outer one may be very thin or even nonpersistent in some species. In addition, a third layer of somewhat coarser texture may line part of the interior of a shell, particularly in or near areas of muscle attachment Some early brachiopods, however, seem to lack this structure entirely and have shells that are wholly granular or amorphous. A few specialized forms, particularly some that are cemented to other objects rather than attached by a pedicle, also do not exhibit this structure and differentiation of layers. Possibly, therefore, evolution of articulates progressed from granular to fibrous layered shells and then reverted in some instances to what seems to be a primitive condition.

The shells of those inarticulates whose structure has been studied are differently constructed and are not composed of corresponding layers or mineral fibers comparable to those of articulates. Instead, they consist of successive very thin laminae, either of alternating horny and phosphatic material or of these substances

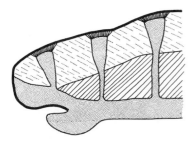

Fig. 156   Cross section through the edge of one type of articulate brachiopod shell and its adherent mantle. (Greatly enlarged.) A very thin surficial horny layer, not preserved in fossils, is formed by cells within the groove near the mantle edge to left. The outer calcareous layer is laminated and consists of very finely crystalline or fibrous calcite. This layer grows only at the mantle edge. The inner calcareous layer consists of inclined crystalline calcite fibers, each secreted by a single mantle cell. This layer continues to grow and thicken during the life of the animal and is the main part of most brachiopod shells. Tiny canals or punctations of the shell occupied by slender extensions of the mantle are characteristic of several groups of brachiopods. (*After Williams, 1956, fig. 2, p. 247.*)

intimately combined. There is no clear evidence that one of these shell types was evolutionarily antecedent to the other. Finally, it seems unlikely that the calcareous shells of articulates evolved from shells of typical phosphato-organic composition.

Punctation

Many brachiopods have been described as being punctate, i.e., as having shells perforated by tiny pores. Present knowledge indicates that three types of "punctate" structure are distinguishable: (1) endopunctate, (2) exopunctate, and (3) pseudopunctate.

Endopunctae are tiny open tubes extending perpendicularly through the shell from the interior but ending with somewhat enlarged diameter just beneath the surface of the outer calcareous layer (Fig. 156). They communicate with the exterior through a bundle of very minute tubules. These punctae ordinarily are not visible on the exterior of a shell unless a thin surface lamina is removed. The study of modern brachiopods shows that the tubes are occupied by slender strands of living tissue continuous with the underlying mantle. These are formed by the budding off of a few cells at the outer edge of the mantle. They maintain their connection with the mantle as the shell thickens and grows peripherally.

The so-called exopunctae are tiny pits in the outer surface of a shell which are not continued by tubes extending to the interior like endopunctae. Their origin is not understood. Perhaps they were produced by groups of cells budded off at the outer mantle edge which did not maintain connection with the mantle as the shell grew and thickened. The pits may be mistaken for endopunctae, but the shells that possess them are not truely punctate.

Pseudopunctate shells are traversed by slender rods of calcite that have been misidentified as tube fillings. They end internally as low elevations which may have been attachment areas for ligaments of the mantle. These structures do not seem to be related in any way to endopunctae. They occur in only one type of brachiopods, the strophomenoids and their near relatives. Pseudopunctate shells almost certainly evolved from nonpunctate ones.

The minute structure of many fossil brachiopods has not been investigated adequately. Typical endopunctae, however, that seem to be morphologically identical characterize several different groups of articulate brachiopods. The similarity of the punctae suggests that all endopunctate forms may be related in their evolutionary origin and that perhaps an endopunctate condition was inherited from a single primitive ancestral stock. Such a possibility seems to be decisively denied, however, because the different groups of endopunctate brachiopods appeared, without exception, later in geologic time than nonpunctuate ones that are practically identical in all other essential features. This is compelling evidence that endopunctae evolved independently more than once in different impunctate brachiopod lineages.

The relations, if any, of exopunctae to endopunctae are very doubtful. Either condition might have evolved from the other. It is even possible that some exopunctate forms were ancestral to endopunctate ones whereas others are reversions to a seemingly more primitive state.

The shells of some inarticulates are pierced by tiny canals which also contained extensions of the mantle. These seem to be different, however, from the endopunctae of articulates. Some of them branch outward into progressively more slender tubes, but others branch inward without reduction in diameter. Their evolutionary significance is unknown.

Shape

The forerunner of the brachiopod shell appears as a pair of delicate horny plates on the mantle lobes of the larvae. These tiny shells ordinarily are more or less semicircular with straight posterior edges. Growth is accomplished by the addition of shelly material around their margins and, at the same time, the shell is thickened by the deposition of material on its inner surface. The shape attained by any shell depends on the rate at which it was built outward at various places about its margin. Progressive changes in shape from a youthful to a mature condition commonly are shown by concentric growth lines, each of which records a temporary pause in the building of a shell.

Each valve of most brachiopods has the shape of a distorted cone, the apex of which is the starting point of growth. Distortion results from differential growth. This is much more rapid anteriorly and laterally than posteriorly, except in some of the neotremes and a few highly specialized articulates. The straight posterior edge of the larval shell becomes the hinge line. Posterior growth along it generally is relatively very slow and produces a narrow hinge area.

One reason for considering the palaeotremates very primitive is that they deviate little from the semicircular shape of the larval shell. Other brachiopods assume a variety of shapes. In a general way two groups of articulates can be distinguished, although all intermediate grades occur (Fig. 158). These are: (1) Long-hinged forms with the hinge equaling or nearly equaling the greatest width.

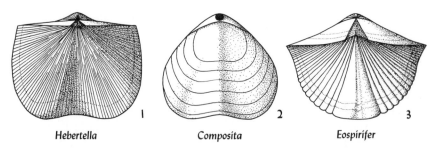

Hebertella   *Composita*   *Eospirifer*

Fig. 158   Three types of articulate brachiopods, showing variation in length of hinge line and surface features. (About natural size.) 1, With narrow ribs. 2, With growth lines. 3, With plications. Brachiopods are customarily illustrated with the posterior end up. (*After Hall and Clarke, 1892, Paleo. N.Y., vol. 8, pl. 5A, fig. 8; pl. 22, fig. 1; pl. 84, fig. 31.*)

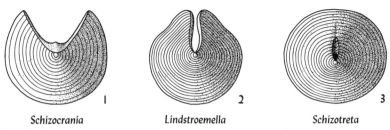

Schizocrania   *Lindstroemella*   *Schizotreta*

Fig. 159   Neotremate brachiopods, showing progressive reduction of the pedicle opening. (*After Hall and Clarke, 1892, Paleo, N.Y., vol. 8, pl. 4E, figs. 27, 10; pl. 4G, fig. 27.*)

In the more extreme examples the posterior part of the shell adjacent to the hinge line is extended to form more or less acute points. (2) Short-hinged forms with the hinge much less than the maximum width of the shell. Some hinges are so short that there seems to be no real hinge line. A brachiopod of this kind commonly has valves one or both of which has a more or less sharply pointed apex that may be strongly curved over the hinge area. The observation of modern brachiopods shows that most long-hinged forms have short pedicles which hold them close to the surface of attachment and that most short-hinged forms have longer ones that permit some movement on their stalks.

Evolution with respect to shape has progressed in two main directions, viz., increase or reduction in the length of hinge line. Although reversals in these trends occurred, few very short-hinged brachiopods seem to have evolved from lineages that had previously acquired increasingly long hinges. Also long hinge lines probably never have evolved from ancestors with very short hinges.

Many neotremates evolved in an entirely different way. In them growth posterior to the apex steadily increased until enlargement was almost equal in all directions. This produced low conical valves of nearly circular outline (Fig. 159). These have neither an apparent hinge line nor a hinge area.

Contour

The shells of most brachiopods are biconvex; a lenticular form with both valves moderately and almost equally arched in profile undoubtedly represents a primitive condition. Some brachiopods have deviated little from this form, but many groups show evolutionary trends of several kinds that are most obvious with respect to their anterior-posterior curvature. These trends involve: (1) Increase in the curvature of a valve or of both valves. Curvature generally is greater in the pedicle than in the brachial valve. It also ordinarily is greatest posteriorly and gradually becomes less as a shell grows. (2) Decrease in curvature of one valve or of both. A trend toward decreasing curvature in one valve may be accompanied by increase in the other. (3) Reversal of curvature in one of the valves. Reverse curvature can only have followed a trend that first reduced the ordinary curvature of a valve.

Similar evolutionary trends of these kinds are evident in different groups of brachiopods, although some are more characteristic of certain groups than others. Some trends have been carried to extremes, and some have been reversed. There is little or no evidence, however, that noteworthy reversal has followed any extreme development.

Plications

Plications are folds, commonly beginning near the apex of a shell and radiating to its anterior and lateral margins (Fig. 158-3). They reproduce the form of the mantle edge where shell growth occurs. Upfolds in one valve generally correspond to downfolds in the other which interlock along the margin when a shell is closed. This strengthens the shell and seems to be functionally advantageous.

Plicated shells evolved from smooth ones. Plications probably originated in two different ways: (1) from flat bands separated by shallow grooves which deepened and widened until a folded surface was developed, and (2) from a few broad gentle undulations of the shell margin that first appeared at a late ontogenetic stage. In both cases the plications were accentuated in derived forms and appeared at younger and younger growth stages until their beginnings are seen very near the apex of a shell.

The earliest plications develop along the midline of a shell between smooth lateral areas. Subsequently in both ontogenetic development and evolutionary sequences of species, successive new plications appear in more and more lateral positions until the entire shell is folded. The plications of some brachiopods are constant in number, except for lateral additions, and they increase in size proportional to growth. In others, which seem to have descended from ancestors of this kind, the plications do not exceed a certain width and the number increases as the shell enlarges. The new plications arise by bifurcation of the upfolds or downfolds in different groups of brachiopods. No sharp distinction can be made between fine plications and certain kinds of radiating ribs. The latter corrugate shells around their edges, but the depressions between them on the inside are obliterated near the margins by the deposition of shell material. Coarser plications may be similarly obliterated internally in specimens that have attained old age.

Fold and Sulcus

Particularly prominent undulations along the midline of many brachiopods produce a raised fold in one valve (Fig. 158-3), most commonly the brachial, and a corresponding depressed fold or sulcus in the opposite valve. Such folds may develop without any others and become evident only as the shell approaches mature size, or they may be flanked by plications, be no more than central relatively large members of this series, and originate very near the apex. These features evolved in two ways: (1) As parts of the original folding that resulted in plication. Such medial folds may become secondarily plicated in derived forms. (2) As secondary folds that followed the development of plications. The final results of these two processes are identical. Secondary folding has rarely affected the lateral plicated parts of a shell.

The development of fold and sulcus seems to be related functionally to water

Fig. 161   Anterior view of brachiopod with prominent fold and sulcus, showing lateral and central passages adapted to the inward and outward movement of water currents. (*After Hall and Clarke, 1892, Paleo. N.Y., vol. 8, pl. 61, fig. 31.*)

Eatonia

currents created by the lophophore. Prominent folds of this kind, when the shell is open, provide two lateral passages for inward flow and a more or less separated central passage for the departing current (Fig. 161).

Spines

The shells of most brachiopods grew continuously except for brief interruptions caused mostly by external environmental influences. These produced irregularly spaced concentric growth lines which commonly are shallow grooves where one shell lamina ends and another emerges from beneath it. From brachiopods with shells of this type, others evolved whose mantle edge withdrew slightly in a rather regular periodic way. Each withdrawal left a delicate shelly frill extending outward, and the repetition of this process resulted in a succession of emergent frills arranged much like the shingles on a roof.

In modifications of this trend, the periodically advancing mantle edge which grew out after each withdrawal was divided into many narrow, closely spaced extensions. Each extension secreted shelly matter, and thus a row of delicate spines was produced, instead of a continuous frill. The mantle extensions of some brachiopods secreted shell only on their outer surfaces and formed flattened scale-like spines. Others surrounded themselves with shell and produced hollow tubular spines. Spines of both kinds commonly are broken away in fossils, but their remaining bases are clearly evident and account for the characteristic surface patterns of some shells.

Other hollow spines, either regularly or sporadically arranged, grew out approximately at right angles to the surface of a shell. They formed around extensions of the mantle that first appeared at the growing edge and seem to have lengthened rapidly. The living tissue within many of these spines soon degenerated, and the shelly tubes were sealed off by deposits laid down on the inner surface of the shell. Some spines of this kind served as anchors, holding their possessors in loose or soft sediment. Others grasped foreign objects and helped to secure some brachiopods which lacked a functional pedicle.

The evolution of brachiopod spines has not been studied carefully. When it is, related trends in the development of these structures probably will be recognized.

Pedicle Opening

Much emphasis was placed in the past on the nature of the pedicle opening and its confinement as a guide to the classification and probable phylogenetic relations of various brachiopods (Fig. 162). Some of the conclusions reached are now known to be unreliable, if not erroneous, and agreement has not been reached regarding the significance of all pedicle structures. Continuing interest in these features, however, and their possible evolutionary relations make them worthy of attention, particularly among the articulates.

The pedicle of primitive brachiopods, and in the early ontogenetic stages of advanced ones, simply emerges from between the two valves, whose posterior edges are accommodated to its shape. As the articulate shell begins to grow, paired teeth and sockets appear upon the hinge line, closely flanking the pedicle on both sides, with teeth on the pedicle valve and sockets in the brachial. Subsequently these structures gradually diverge laterally in each valve proportionally to growth. As the hinge area becomes higher, their traces bound triangular pedicle areas generally much better developed in the pedicle than in the brachial valve. These areas may be unmodified openings, they may be variously enlarged by resorption of the shell adjacent to the pedicle, or they may be more or less closed by various kinds of shelly growth (Fig. 163).

Distinction among articulate brachiopods formerly was made between those whose pedicle areas are partly or wholly closed by (1) a single plate that was believed to originate on the pedicle (Fig. 155-5), and (2) paired plates that grew inward from

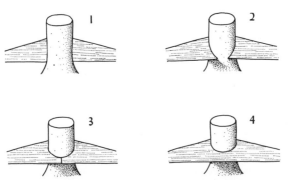

Fig. 162   Diagrams showing relations of pedicle of articulate brachiopods to that portion of the mantle lobe which secretes hinge area. 1, Pedicle area entirely open. 2, Pedicle area partly closed from sides. 3, Area completely closed below pedicle but mantle extensions not fused. 4, Mantle extensions fused. Shell corresponding to condition 3 has pedicle area partly occupied by a pair of lateral plates. That corresponding to condition 4 has area partly occupied by a single plate; see Fig. 163. These stages may occur in ontogenetic sequence, or each may characterize the mature development of a brachiopod.

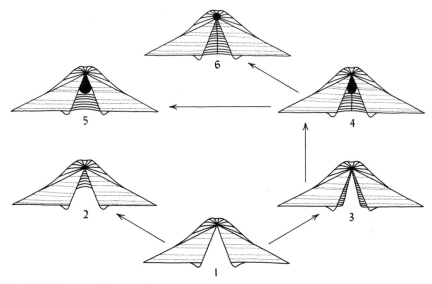

Fig. 163   Diagrammatic representation of the hinge areas in pedicle valves of articulate brachiopod shells showing projecting hinge teeth and central pedicle area. The probable evolutionary relations of different structures are indicated. 1, Open pedicle area. 2, Area partly closed posteriorly, either by thickening of the shell in the apex region or by the forward growth of a plate. 3, Partial lateral closure by two plates. 4, Complete closure in front of pedicle opening by two plates. 5, Complete closure by a single plate. 6, Complete closure of pedicle area and migration of pedicle opening to the apex by resorption of shell.

the sides of the pedicle opening. These brachiopods were designated protremates and telotremates, respectively. More modern studies seem to have demonstrated that the single plate was misinterpreted, that it did not originate on the pedicle but is homologous with the paired plates of other brachiopods.

The assumption has been general that some progressive evolutionary trends among the articulates are indicated by (1) increasingly more complete closure of the pedicle area, and (2) migration of the pedicle opening out of the pedicle area by resorption of the shell. There seem to be some reasons, however, for concluding that a completely unobstructed pedicle area does not necessarily indicate a primitive condition.

Muscles and Their Attachment

The musculature of most brachiopods is relatively simple. Articulates have three kinds of muscles: (1) those that open the valves, (2) those that close them, and (3) those that activate the pedicle (Fig. 164). Inarticulates have a fourth kind that serves to move the valves laterally with respect to each other or to rotate them. The areas of muscle attachment commonly are evident as so-called scars on the inner surface of the shell. Several kinds of internal shelly structures are more or less closely related to muscles or their areas of attachment.

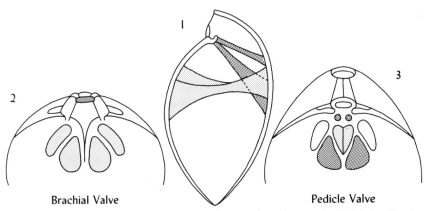

Fig. 164   Generalized diagrams showing the musculature and muscle scars of articulate brachiopods. The opening muscles and their scars are indicated by heavier shading and the closing muscles and their scars by lighter shading. Pedicle muscles not shown; Their areas of attachment are unshaded.

The teeth and sockets of the articulate hinge are strengthened in many shells by buttresses which extend from the inner surface of the shell to the underside of the hinge area and divide the interior apex region into three compartments (Fig. 165-2). Some of these buttresses continue anteriorly for variable distances, rising from the inner surface of the shell as thin vertical septa. There may be a pair of laterally diverging septa extending from beneath the teeth of a pedicle valve and a similar diverging pair related to the sockets in the brachial valve. In addition, median septa may be developed, more commonly in the brachial than in the pedicle valve. Some of these septa only separate areas of muscle attachment, but the muscles of many brachiopods were attached partly or almost wholly to their sides.

The teeth and sockets of the most primitive brachiopods are not supported by buttresses. These strengthening structures probably evolved in several different lineages. They were subsequently lost in more than one group of advanced brachiopods, which thus reverted to a seemingly more primitive state.

Muscles pass through the body of a brachiopod and reduce the space available for reception of the viscera. Large and powerful muscles evidently resulted in a seriously crowded condition. This difficulty was overcome by the evolution of various modifications of the buttresses and their continuations as septa (Fig. 165-4 to 6). Raised platforms developed in the pedicle valves of several groups of brachiopods that, by advancing stages, became spoon-shaped areas of muscle attachment beneath which space was provided for the viscera. Somewhat similar structures also evolved in brachial valves.

Lophophore

The brachiopod lophophore is a tentacled organ whose cilia create water currents

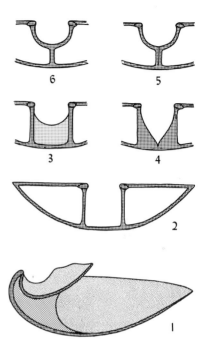

Fig. 165   Diagrammatic representations of structures showing evolution of a spondylium in the pedicle valves of some articulate brachiopods. 1, Longitudinal section showing spondylium supported by a median septum. 2, Transverse section showing hinge teeth supported by vertical buttresses that compartment the space beneath the hinge area. 3, Area between buttresses partly filled by thickening of shell. 4, Buttresses thickened inward until they meet. 5, Spondylium supported by buttresses which have come together medially. 6, Spondylium supported by a median septum.

which carry food particles to the mouth. Reports describing the origin and early growth of the lophophore in embryo and larva are somewhat confused and contradictory. This organ starts as one or a pair of tentacles to which other pairs are added laterally until a circle is produced which surrounds the mouth (Fig. 166-1, 2). Further growth and the addition of more tentacles in the restricted space available result in the development of lobes and coils. These are the so-called arms from which brachiopods derive their name.

Modern brachiopods exhibit three final lophophore patterns (Fig. 166-5,7,9). The ontogenetic stages leading up to them almost certainly recapitulate their evolutionary history. Mature brachiopods in all stages of lophophore development, from the circular form onward, are known among modern species and inferred among the fossils. Some Paleozoic brachiopods undoubtedly were relatively primitive as far as the lophophore was concerned. Others may have had lophophores different from those of any living species, and perhaps more complicated (Figs. 168; 168a). Some recent brachiopods with simple lophophores probably are examples of arrested development which have retained the lophophore in an immature rather than a primitive condition.

Lophophore Supports

Primitively the brachiopod lophophore was unsupported by any shelly structure. No support ever was developed by the inarticulates, and the evolution of such a struc-

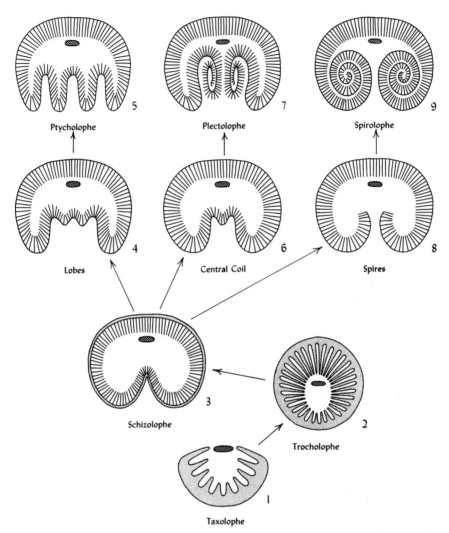

Fig. 166   Diagrammatic representation of the ontogenetic development and probable evolutionary relations of the principal types of lophophores known in modern articulate brachiopods. The lophophore grows by the introduction of new tentacles in the middle, and gradually encircles the mouth (1, 2). Thereafter it is thrown into folds (3 to 5) and may develop a central coil (7) or lateral spires (9).

ture did not proceed far in several groups of the articulates. The absence of supports, however, is no indication of a primitive lophophore. Observation of modern brachiopods shows that some elaborate lophophores are wholly unsupported. Likewise these animals demonstrate that essentially identical lophophores may be borne on calcareous supports of different kinds (Fig. 167). Therefore the form of lopho-

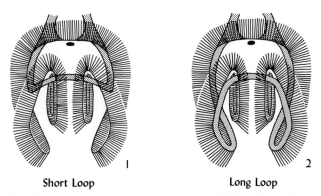

Short Loop   Long Loop

Fig. 167   Diagrams showing the relations of practically identical lophophores to the short and long calcareous loops of modern terebratuloid brachiopods. (*After Stehli*, 1956, *fig.* 3, *p.* 189.)

phore is not a reliable guide to the nature of a possible supporting structure or vice versa.

On the other hand, a shelly support must conform to the requirements of the type of lophophore with which it is associated. Thus some limiting indications of the kind of lophophore are provided by the supports possessed by many fossil brachiopods (Fig. 167a). The possibility must not be overlooked, however, that some

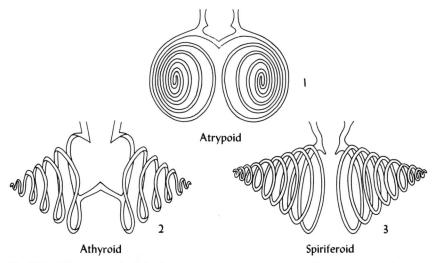

Atrypoid

Athyroid   Spiriferoid

Fig. 167a   Three types of spiral calcareous lophophore supports in fossil brachiopods. 1, The spiral ribbons extend outward and more or less parallel the edges of the shell. 2, The ribbons turn down abruptly into the curvature of the shell and are connected by a crossbar. 3, The ribbons do not depart so abruptly from the rods connecting them to the shell and are not joined by a crossbar.

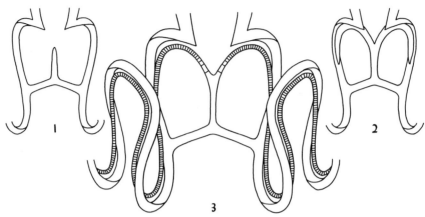

Fig. 168 Athyroid spires may be complicated in various ways by growths continuing the crossbar. 1, The first stage develops a posteriorly directed upgrowth from the bar. 2, This divides, and two calcareous ribbons curve outward. 3, These may continue and produce complete secondary spires inside the primary ones. (*Compare Williams, 1956, fig. 5-9, p. 262.*)

fossils had lophophores different from any known in modern species (Fig. 168). Ordinarily if a fossil brachiopod lacks more than rudimentary supports, nothing can be determined about its lophophore. A few exceptions are provided by species whose lophophores left impressions on the inner surface of the brachial valve (Fig. 168a). These mostly are shells of flattened form with little space between the valves. The lophophore in brachiopods of this kind probably was attached to the mantle surface rather than extending freely within the mantle cavity.

Evolution seems to have progressed independently in the lophophores of brachiopods and their calcareous supports, although obviously some restrictions were imposed on the form that either might assume because of their intimate relations to each other. Evolution of the supports probably can be traced morphologically in

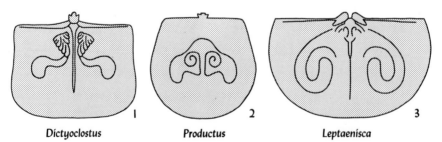

| *Dictyoclostus* | *Productus* | *Leptaenisca* |

Fig. 168a Somewhat generalized sketches showing internal markings of brachial valves, probably indicating the forms of lophophores attached to the dorsal mantle and not provided with calcareous supports. (*After Hall and Clarke, 1894, pl. 22, fig. 5; Williams, 1956, fig. 5-6, p. 262; Muir-Wood and Cooper, 1960, pl. 133, fig. 9.*)

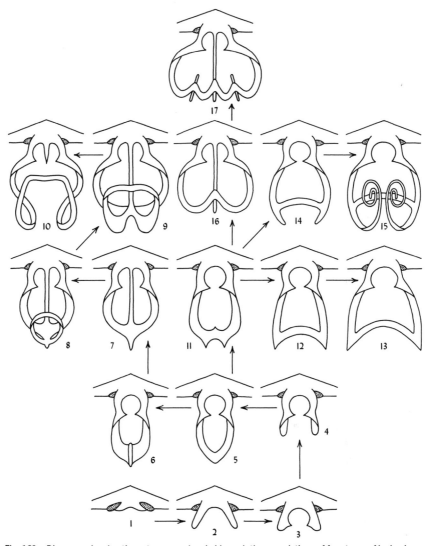

Fig. 169 Diagrams showing the ontogeny and probable evolutionary relations of four types of lophophore supports (10, 13, 15, 17). 1, Central part of the articulate hinge line with no structures except sockets for reception of the hinge teeth. 2 to 4, Stages in the development of crura, or lophophore bases. 5, A simple loop is formed bearing a trocholophe (see 2 of Fig. 166). 6 to 10, Ontogeny of the long tere-bratuloid loop, which is related to a median septum. There are many variants of the completed loop. 11 to 13, Ontogeny of the short terebratuloid loop. Like the last, it bears a plectolophe (see Fig. 167). 11, 14, 15, Ontogeny of lateral spires bearing a spirolophe. 11, 16, 17, Ontogeny leading to ptycholophe-bearing structure.

a general way on the basis of the ontogeny observed in both modern and fossil species and the successive time relations of the fossils (Fig. 169). A reconstruction of this kind, however, obscures parallel trends in evolution that almost certainly occurred, and it may be otherwise misleading if there has been convergence.

Hinge Process

The hinge process is the area or projecting structure near the apex of a brachial valve to which the opening muscles are attached. The nature of this process has been noticed in connection with the classification of some brachiopods, but little attention has been devoted to its evolution. Its features, however, certainly are related to the characters of the hinge and pedicle area.

The hinge process serves as one limb of a lever acting on a fulcrum provided

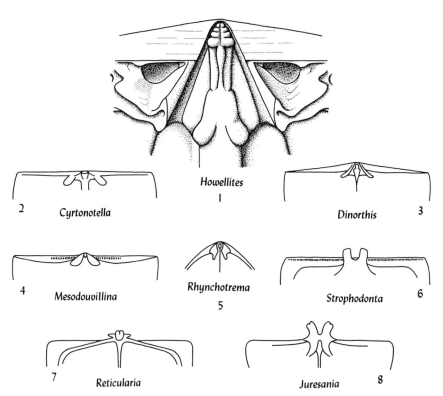

Fig. 170   Sketches showing relations of the hinge process in brachial valves to other articulating features of the hinge line. 1 to 5, Hinge process short, not extending beyond edge of shell. Sockets for reception of hinge teeth present on either side of diverging structures which, in other brachiopods, become the bases of the calcified lophophore supports. 6 to 8, More prominent hinge processes project beyond shell margin, providing more efficient levers for the opening muscles. Many shells of this type lack sockets, which indicates that hinge teeth have been lost by the other valve. 4 and 6, Tiny tooth-like crenulations evolved and spread laterally along the hinge line in one group of brachiopods. (*After Williams, 1963, J. Paleontol., vol. 37, fig. 1a, p. 4; Cooper, 1956, Smithsonian Inst. Misc. Collections, vol. 27, pl. 41, fig. 25; pl. 127, fig. 26; pl. 130, fig. 8; Williams, 1953, pl. 12, fig. 1; pl. 7, fig. 12; Muir-Wood and Cooper, 1960, pl. 79, fig. 15; pl. 105, fig. 8.*)

by the teeth and sockets. The efficiency of this lever depends, of course, on the distance between the area of muscle attachment and the hinge. The process rises above the hinge line and is contained within the space beneath the pedicle area of the pedicle valve. Its length is limited by the interior space available to it. This is much restricted in shells with low hinge areas and in those whose apex regions are occupied by large functional pedicles.

Primitively the process is located very close above the hinge line, and the lever is relatively inefficient (Fig. 170-1 to 5). Lengthening of the process surely was advantageous. Longer processes evolved in several different forms (Fig. 170-6 to 8). They achieved their maximum development in brachiopods with greatly reduced or nonfunctional pedicles. Perhaps reduction of the pedicle was related to crowding by an enlarged process, or conversely, reduction of the pedicle permitted the process to grow longer. Perhaps, also, migration of the pedicle opening out of the triangular pedicle area by resorption of the shell (Fig. 163-6) resulted from crowding by an enlarged process.

Mantle Sinuses

Extensions of the brachiopod coelom penetrate the mantle lining the interior of the shell and branch repeatedly. They provide passages for circulation of the coelomic fluid. These sinuses may leave impressions on the inner surface of the shell which are much more likely to be evident in some kinds of brachiopods than in others. The various patterns of branching are commonly believed to be characteristic of various groups of brachiopods, although the significance of pattern variations is not clearly understood.

If minor differences in sinus patterns are neglected, a series of general evolutionary modifications can be traced (Fig. 172). The principal advance features the progressive appearance of new sinuses which radiate outward from the areas occupied by the sex glands.

Inarticulates

Approximately half of the main recognized groups of inarticulates are represented in the Lower Cambrian. It is evident, therefore, that much of the evolutionary diversification of these brachiopods probably was completed before the beginning of the Paleozoic Era. In the absence of much difference in the time of their first appearance, ideas regarding the relationships of these early fossils are based on the ways in which they had diverged from the theoretical form of the most primitive brachiopod shell. This prototype is believed to be indicated by the earliest ontogenetic stage seen in most later species. From such a beginning, evolution seems to have progressed mainly in one or the other of two ways in the modification of a semicircular shell whose transverse posterior edge embraced the pedicle and had not yet become a hinge line.

The more conventional evolutionary trend produced shells that grew mainly in

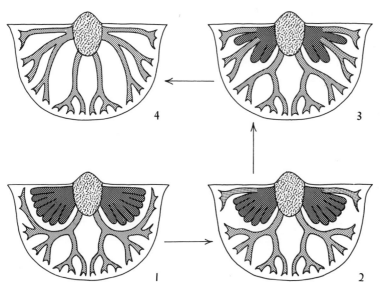

Fig. 172 Diagrammatic representation of evolutionary progression in the mantle sinuses of pedicle valves, as shown by internal shell markings. 1, Two main trunks send branches along the whole shell margin and surround areas occupied by the sex glands. 2, 3, New trunks extend outward from these glands, replacing branches of the main trunks. 4, Pockets occupied by the sex glands have been eliminated, and these glands probably extended into the mantle sinuses. Patterns differing only in some details occur in brachial valves. (*Adapted and generalized from Williams, 1956, fig. 6, p. 274.*)

both anterior and lateral directions, with anterior growth gradually accelerating (Fig. 173). Posterior growth largely was restricted to thickening of the shell in the apical region, which became more or less pointed. Shells of this type were transformed gradually from transverse ovals to triangular shape and finally to more and more elongated form. The pedicle continued to emerge from between the two valves. These are the atremates. Many of them seem to have had strong pedicles, much like those of modern *Lingula,* adapted to burrowing in muddy or silty sediments.

This trend differentiated into a number of evolutionary offshoots whose advanced characters are seen in (1) internal features, such as increasingly thick or raised platforms to which the muscles were attached, (2) surface features, such as delicate projecting frills or spines, and (3) change in shell substance from phospho-organic to calcareous.

The other evolutionary trend involved posterior as well as anterior and lateral growth in only the pedicle or in both valves. This produced conical shells of more and more circular form which varied considerably in their height. Posterior growth served to enclose the pedicle first in a notch and later in an opening restricted to the pedicle valve (Fig. 159). These are the neotremates. Most of them probably had

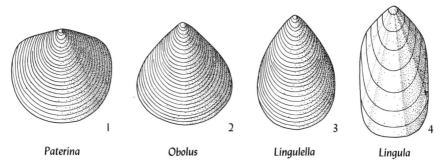

Paterina	Obolus	Lingulella	Lingula
1	2	3	4

Fig. 173   Atremate brachiopods, showing shape transitions from a Cambrian form, considered to be primitive because it resembles a larval shell, to modern *Lingula*. (*After Walcott, 1912, U.S. Geol. Surv., Mon. 51, pl. 2, fig. 2; pl. 8, fig. 1; pl. 20, fig. 6; Hall and Clarke, 1892, Paleo. N.Y., vol. 8, pl. 4K, fig. 5.*)

short pedicles by which the animals were attached to seaweeds or other objects on or above the bottom sediment.

Brachiopods of this kind probably evolved from more than a single atremate type. They differentiated further into a variety of evolutionary offshoots in which several different tendencies became increasingly conspicuous. In some the pedicle maintained a position posterior to the apex, although the opening may have been partly closed by shelly growth (Fig. 159). In others the pedicle migrated forward by resorption of the shell so that the external opening is situated in front of the apex (Fig. 173a). As in the atremates, evolutionary trends also are exhibited by development of (1) internal features, (2) surface modifications, and (3) the acquisition of calcareous shells. The pedicle atrophied in brachiopods of the last type, which preserve no indication of a pedicle area or opening. These animals lived much like barnacles, cemented to convenient surfaces by the pedicle valve.

The inarticulates attained their maximum diversification during the Ordovician Period. Thereafter they declined rapidly, probably as a result of competition from the rising articulates. Only a few persistent types have survived and are represented in modern faunas.

Fig. 173a   Neotremate brachiopods showing pedicle opening which has migrated anteriorly by shell resorption. (*After Walcott, 1912, U.S. Geol. Surv., Mon. 51, pl. 83, fig. 1; pl. 84, fig. 4.*)

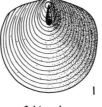
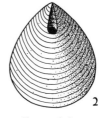

Schizambon	*Trematobolus*
1	2

Articulates

The fossil record of the articulates begins in the Lower Cambrian. Consequently evolutionary diversification that produced the two great brachiopod divisions must have occurred well back in Prepaleozoic time. The conclusion that close relationships connect the articulates and inarticulates, as these are known from their fossil representatives, may not be justified.

The articulate brachiopods are divisible into about fifteen morphologic groups, each group consisting of forms resembling one another in such ways that close relationships within each group seem to be reasonably assured. These groups in turn represent six or seven types of brachiopods that differ from one another in what seem to be fundamental ways (see Table 174). The phylogenetic interrelationships of some of the groups are far from clear. The chronologic order of their appearance, however, limits to some extent the evolutionary probabilities. Only one group is present in the Lower Cambrian, and conceivably all others, which appear successively in younger strata, evolved from members of that kind. On the other hand, the possibility cannot be excluded that some of the later-appearing articulates evolved from different unknown Precambrian and early Paleozoic ancestors. Figure 175 shows in a general way the possible evolutionary relations of the principal and better-known articulate groups.

Parallel and convergent types of evolution involving most features of the artic-

---

Table 174   Characters of the main groups of articulate brachiopods

---

I. *Orthoids:* generally long-hinged, radially ribbed, lophophore supports lacking or rudimentary; L. Camb.-U. Perm.

    Dalmanelloids: like orthoids but endopunctate; M. Ord.-U. Perm.

II. *Syntrophioids:* generally smooth with fold and sulcus, spondylium in pedicle valve, no similar structure in brachial valve; M. Camb.-L. Dev.

    Pentameroids: large, like syntrophioids but with spondylium-like structure in brachial valve; M. Camb.-U. Dev.

    Triplesioids: short-hinged, forked hinge process; M. Ord.-M. Sil.

III. *Strophomenoids:* long-hinged, radially ribbed, pseudopunctate, lophophore supported only basally by calcareous rods; L. Ord.-Rec.

    Productoids: like strophomenoids with spines; U. Ord.-U. Perm.

IV. *Atrypoids:* generally short-hinged, lophophore supported by calcareous spires whose ribbons parallel shell edges; M. Ord.-M. Miss.

    Rostrospiroids: short-hinged, with calcareous spires whose ribbons turn down abruptly into curvature of shell, connected by crossbar; M. Sil.-Jura.

    Spiriferoids: generally long-hinged with fold and sulcus, with calcareous spires whose ribbons turn down less abruptly, not connected by crossbar; M. Sil.-Jura.

    Punctospiroids: endopunctate, lophophore supported by calcareous spires; U. Sil.-Jura.

V. *Rhynchonelloids:* mostly short-hinged, plicate with fold and sulcus, spiral lophophore supported only basally by calcareous rods; M. Ord.-Rec.

    Rhynchoporoids: like rhynchonelloids but endopunctate; Miss.-Perm.

VI. *Terebratuloids:* mostly short-hinged, endopunctate, lophophore supported by calcareous loop; U. Sil.-Rec.

Subdivisions accounting for about double this number of main groups are recognized in some classifications.

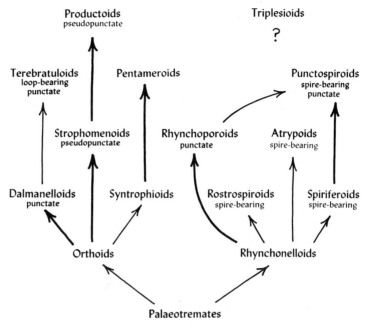

Fig. 175   Diagram showing the possible evolutionary relations of the main groups of articulate brachiopods. Heavier arrows indicate the more reliably established evolutionary sequences. Vertical arrangement corresponds approximately to the successive stratigraphic appearance of these groups from the Cambrian to the Mississippian. The relations of triplesioids are uncertain.

ulate brachiopods almost surely have occurred. Therefore, the significance of various degrees of similarity or difference is doubtful, and certain examples of close similarity may be misleading as indications of close relationships. Nevertheless some conclusions regarding evolutionary derivations seem to be fairly satisfactorily established. For instance, the direct descent of strophomenoids and dalmanelloids from orthoids is reasonably certain. The similar derivation of other groups, however, is probable or only possible in various degrees.

Few attempts have been made to trace evolution in any detail among the brachiopods. One example is furnished, however, by some of the strophomenoids whose hinge teeth and sockets degenerated, disappeared, and were replaced by an articulation consisting of small denticles which became more and more numerous and spread laterally along the hinge line (Fig. 170-4,6). Such developments were accompanied by other progressive evolutionary modifications that seem to distinguish several different lineages, all believed to have evolved from a common ancestor. These particular fossils, like most strophomenoids, have a convex pedicle valve and a flat or concave brachial valve. They are accompanied by other shells similar in almost all respects except that the curvature of the valves is reversed.

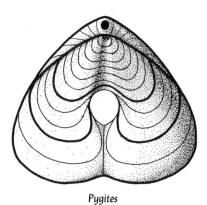

Pygites

Fig. 176   The "keyhole" brachiopod, whose growth stages rather clearly recapitulate its evolution. This began in the middle Mesozoic and progressed with adult shells whose forms correspond to the growth lines as these are accentuated in the drawing.

Three explanations for these associations are possible. (1) The pedicle of these brachiopods atrophied at a youthful stage, and the animals lived lying on the sea bottom in an unattached condition. When they became free, the curvature of further growth may have been determined by which valve happened to lie uppermost. (2) Two main lineages may have evolved, characterized by opposite curvature that developed in closely similar ways. (3) Reverse curvature may have evolved independently at different times in different lineages and, in this respect, these fossils are examples of convergent evolution. When all features of the shells and their stratigraphic succession are considered, the last possibility seems to be most probable.

Another example of an evolutionary lineage seems to be provided by a succession of brachiopods that developed a reentrant in the forward margin. The two lateral parts grew outward and then converged and joined, leaving behind them a peculiar perforation through the shell (Fig. 176).

BIBLIOGRAPHY

R. S. Bassler (1953): Bryozoa, in "Treatise on Invertebrate Paleontology," R. C. Moore (ed.), pt. G, Geological Society of America, Boulder, Colo., and University of Kansas Press, Lawrence, Kans.
This well-illustrated systematic work includes no consideration of evolution.
C. E. Beecher (1891, 1892): Development of the Brachiopoda: pt. 1, Introduction; pt. 2, Classification of the Stages of Growth and Decline, *Am. J. Sci.*, ser. 4, vol. 41, pp. 343–357; vol. 44, pp. 133–155.
These articles shaped ideas concerning brachiopod classification and evolution for many years.
F. Borg (1926): Studies on Recent Cyclostomatous Bryozoa, *Zool. Bidrag., Uppsala,* vol. 10, pp. 181–504.
See pp. 479–490 for discussion of bryozoan relationships.
P. E. Cloud, Jr. (1942): Terebratuloid Brachiopoda of the Silurian and Devonian, *Geol. Soc. Am. Spec. Paper* 38.
The origin and evolution of early loop-bearing brachiopods are discussed; see pp. 32–36.
G. F. Elliot (1948): The Evolutionary Significance of Brachial Development in Terebratelloid Brachiopods, *Ann. Mag. Nat. Hist.,* ser. 12, vol. 1, pp. 297–317.
The author relates evolution of the lophophore and its calcareous supports to intraspecific competition and to competition with pelecypods.
_____ **(1953):** Brachial Development and Evolution in Terebratelloid Brachiopods, *Biol. Rev.,* vol. 28, pp. 261–279.
This article traces the ontogenetic development of brachial loops in modern brachiopods.
C. L. Fenton (1931): "Studies of Evolution in the Genus *Spirifer,*" Wagner Free Institute of Science, vol. 2, Philadelphia.
Supposedly evolutionary trends are traced in great detail, but the main conclusions are questionable.
T. N. George (1933): Principles in the Classification of the Spiriferidae, *Ann. Mag. Nat. Hist.,* ser. 10, vol. 11, pp. 423–456.
The author points out the difficulty of recognizing genetic lineages because many morphologic characters evolved similarly in different groups.
J. Hall and **J. M. Clarke (1894):** "An Introduction to the Study of Brachiopods Intended as a Handbook for the Use of Students," pt. 2, Report of New York State Geologist, 1893.
This old but important work includes a chapter devoted to general evolutionary considerations; see pp. 907–930 (pp. 159–182 in separate edition).
L. H. Hyman (1959): "The Invertebrates," vol. 5, McGraw-Hill Book Company, New York.
See chaps. 20 and 21, pp. 275–609, for accounts of the bryozoans and brachiopods.
W. D. Lang (1919): The Pelmatoporinae, an Essay on the Evolution of a Group of Cretaceous Polyzoa, *Phil. Trans. Roy. Soc. London, ser. B,* vol. 209, pp. 191–228.
Increased structural complexity is correlated with the supposed necessity for the secretion of more and more calcium carbonate.
_____ **(1921):** "Catalogue of the Fossil Bryozoa (Polyzoa) in the Department of Geology, British Museum (Natural History); The Cretaceous Bryozoa (Polyzoa)," vol. 3, British Museum, London.
The introduction, particularly pp. xxxiii–li, presents ideas regarding cheilostome evolution based on principles now largely outmoded.
H. Muir-Wood and **G. A. Cooper (1960):** "Morphology, Classification and Life Habits of the Productoidea (Brachiopoda)," *Geol. Soc. Am. Mem.* 81.
This book contains a brief account of the probable evolution of the productoids; see pp. 47–50.

J. P. Ross (1964): Morphology and Phylogeny of Early Ectoprocta (Bryozoa), *Bull. Geol. Soc. Am.,* vol. 75, pp. 927–948.
The fossil record begins essentially with 11 lineages of unknown relationships.
C. Schuchert and **G. A. Cooper (1932):** "Brachiopod Genera of the Suborders Orthoidea and Pentameroidea," Memoirs of the Peabody Museum of Natural History, vol. 4, pt. 1, Yale University Press, New Haven, Conn.
The introductory part of this work includes considerations of brachiopod evolution.
F. Stehli (1956): Evolution of the Loop and Lophophore in Terebratuloid Brachiopods, *Evolution,* vol. 10, pp. 187–200.
This article compares the ontogeny of modern species and the geologic record of ancient species.
J. A. Thomson (1927): "Brachiopod Morphology and Genera (Recent and Tertiary)," New Zealand Board of Science and Art, Manual 7, Dominion Museum, Wellington, New Zealand.
This book contains much information pertinent to the consideration of brachiopod evolution.
E. O. Ulrich (1890): "Palaeozoic Bryozoa," Geological Survey of Illinois, vol. 8, pp. 283–688.
Chap. 2, pp. 323–366, devoted to classification, contains observations on probable evolutionary trends and relations.
A. Williams (1953): "North American and European Stropheodontids: Their Morphology and Systematics," *Geol. Soc. Am. Mem.* 56.
The author reconstructs the evolution of a group of brachiopods ranging from the Upper Ordovician to the Upper Devonian.
———— **(1956):** The Calcareous Shell of the Brachiopoda and Its Importance to Their Classification, *Biol. Rev.,* vol. 31, pp. 243–287.
This important article discusses and interprets morphology as it probably reflects evolution.
———— **et al. (1965):** Brachiopoda, in "Treatise on Invertebrate Paleontology," R. C. Moore (ed.), pt. H, 2 vols., Geological Society of America, Boulder, Colo., and University of Kansas Press, Lawrence, Kans.
See pp. 164–199 for a consideration of brachiopod evolution and phylogeny.

CHAPTER FIVE

SCHIZOCOELATE ANIMALS 1

MOLLUSCS

Schizocoelates

The schizocoelate phyla include several of the most abundant and intensively studied groups of fossils. Most of the members of these phyla exhibit a much higher degree of cephalization than any of the groups considered in preceding chapters. For this reason, among others, they seem to stand definitely higher in the evolutionary scale.

The coelomic cavities of schizocoelate animals develop in the larvae or embryos as openings which appear within the mesoderm after this begins to form. Included here are (1) molluscs, (2) annelids, and (3) arthropods. Although bryozoans and some brachiopods also are schizocoelate, these animals are excluded. They constitute the lophophorate phyla, which were discussed in the foregoing chapter.

Many marine molluscs and annelids develop from typical trochophore larvae (Fig. 91-1). Modern marine arthropods produce eggs with a relatively large amount of yolk. Most of these hatch in a more advanced state of development as tiny creatures, known as nauplii (Fig. 262), which already exhibit the typical characters of arthropods. In spite of their very different appearance when mature, the molluscs and annelids almost certainly sprang from a common stock, and convincing evidence indicates that arthropods evolved from annelid-like ancestors.

MOLLUSCS

The molluscs are, by any comparative standards, one of the greatest of all animal phyla. Their geologic record is unsurpassed, and they are represented by a profusion of species in modern marine, freshwater, and terrestrial faunas. Although they exhibit a high degree of superficial morphologic diversity (Fig. 180) and vary vastly in their habits, the essential structural similarity of the main molluscan groups has been recognized for more than a century and a half. These animals are unsegmented. Parts of the body generally are well differentiated into a head with nerve ganglia grouped around the esophagus, a so-called foot specialized in various ways

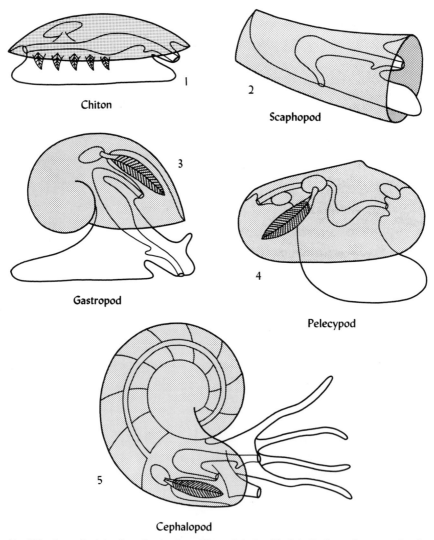

Fig. 180   Generalized drawings showing the relations of shell and body in the five main groups of molluscs.

Table 181   Principal distinguishing characters of the main groups of molluscs

Placophorans: relatively primitive marine molluscs with multiple paired gills. They crawl on a so-called foot in much the same way as a gastropod. Ordovician to Recent

  Monoplacophorans: with cap-shaped dorsal shell of one piece

  Polyplacophorans: chitons, with dorsal shell composed of eight overlapping pieces

  Aplacophorans: without shells

Gastropods: snails, marine, freshwater, and terrestrial molluscs, commonly provided with a spirally coiled shell. The body has been twisted so that generally the main nerve trunks are thrown into a figure 8 and one member of some of the originally paired organs has been eliminated. These animals crawl on a flattened so-called foot in a worm-like manner. Cambrian to Recent

Pelecypods: clams, marine and freshwater molluscs enclosed in a two-pieced shell whose lateral halves are mirror images of each other. They are without heads. Many move about on or in soft sediment by the action of an anterior so-called foot that is thrust forward and then pulls the animal after it. Ordovician to Recent

Cephalopods: marine molluscs equipped with prehensile tentacles, commonly provided with sucker disks, and a funnel through which water from the mantle cavity can be forced in a jet that propels the animal backward. Fossils consist of chambered shells, straight, curved, or coiled, ordinarily in a plane. Cambrian to Recent

Scaphopods: marine molluscs with slightly curved, gradually expanding tubular shells open at both ends. They are without gills, and the so-called foot is similar to that of pelecypods. Silurian to Recent

These groups of molluscs were well differentiated at the time of their first-known fossil occurrence. All probably had evolved from a common ancestor in Precambrian time before shells had been acquired.

for locomotion, and a mantle which commonly secretes a calcareous aragonitic shell. Most molluscs possess a peculiar tongue-like rasping organ, termed a radula.

The phylogenetic relationships of the main groups of molluscs are obscure. Differentiation almost certainly was completed before the opening of the Paleozoic Era. All that can be said is that these groups probably evolved from a common definitely molluscan ancestor. This creature in its turn probably derived from the same stock that produced the annelids. This is indicated by similarity of the trochophore larvae of both molluscs and annelids, the occurrence of generally similar nervous systems with paired ventral nerve trunks, and other evidence. The chief difference seems to be that evolution in the annelids involved the development of increasingly perfect segmentation, whereas in the molluscs an initial tendency to segment mostly declined and disappeared.

Placophorans

The placophorans are interesting mainly because some of them come nearer than any other animals to exemplifying the primitive mollusc. A single-shelled group of monoplacophorans is known as fossils (Fig. 182) extending from the Cambrian to the Devonian, and several modern species have been discovered recently living on the deep sea bottom in the western Pacific Ocean. The cap-shaped shells of these animals resemble those of some limpet-like gastropods (Fig. 187) but are marked internally by a series of paired muscle scars rather than by a single strongly crescentic scar (Fig. 182-8*a*). The modern representatives probably are similar in a

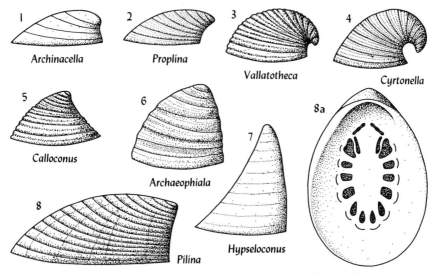

Fig. 182   Representative monoplacophorans, showing their cap-shaped shells and paired muscle scars. Notice that the shell apex generally points forward; compare with Fig. 187. (Not drawn to scale; mostly about natural size.) (*After Knight and Yochelson, 1960, figs. 46–50, pp. 78–82.*)

general way to the ancestors of gastropods, before torsion deformed the bodies in this latter group, if not to the ancestors of all molluscs. There are some reasons to suspect, however, that the ancient placophorans had only one or two pairs of gills and that the greater number in modern species resulted from subsequent evolution.

The chitons, or polyplacophorans, with shells consisting of eight low, posteriorly overlapping plates (Fig. 183), are a conservative and compact group, dating back to the late Cambrian in the fossil record. Fossil specimens are rare except in a few faunas that seem to represent very near-shore environments, and they are relatively unimportant paleontologically. The only general evolutionary trend that has been noted is the progressive development of improved articulation, which increased the effectiveness of defensive enrollment when these animals were disturbed. Chitons are rather obviously related closely to the single-shelled placophorans. Other some-what similar creatures without shells, the aplacophorans, known only in modern faunas, probably evolved from chitons.

Gastropods

Gastropods are abundant in many fossil faunas; their shells exhibit a great variety of modifications in form and ornamentation. They are the most numerous and widely adapted molluscs and the only ones that have made the transition to terrestrial life. Their record begins in the Lower Cambrian, and their relative impor-

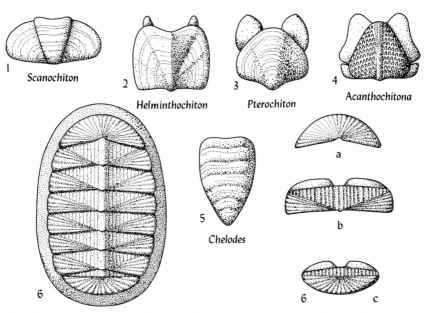

Fig. 183   Representative fossil chiton plates. Evolutionary advancement is shown mainly by the form and structure of the forward extensions that are overlapped by the next anterior plates. (Mostly about natural size; not drawn to uniform scale.) (*After Smith, 1960, figs. 34, 36, 43, pp.* 51, 53, 68.) 6, Modern specimen showing arrangement of the eight plates. 6a-c, Anterior, middle, and posterior plates of a similar specimen. (About natural size.)

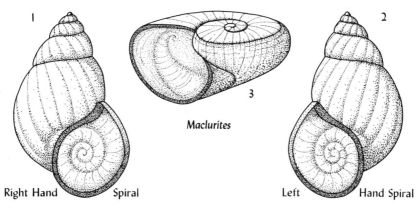

Fig. 183a   Most gastropod shells are right-handed, or clockwise, spirals (1), but some are left-handed (2). Many gastropods are equipped with an operculum, which closes the aperture when the animal withdraws into its shell, and this may also show spiral structure. If so, its spiral is opposite to that of the shell. Gastropods ordinarily are illustrated with the apex of the shell (posterior) pointing upward. When the Ordovician *Maclurites* (3) is oriented in what seems to be the conventional way, it appears to be left-handed. The counterclockwise structure of its operculum demonstrates, however, that this shell actually is a right-handed coil with the true apex hidden by the enlarging whorls, and the specimen is illustrated wrong side up.

tance has increased steadily, especially in post-Paleozoic time, when they seem to have effected adaptations to wider environmental opportunities.

The most remarkable peculiarity of gastropods is the torsion accomplished in larvae or embryos, by which the mantle and visceral parts of the body rotate to the right through an angle of about 180° with respect to the head and foot (Fig. 184). This results in an unsymmetrically organized body with the paired nerve trunks thrown into a figure 8, a looped gut with a forwardly directed anus, and generally the reduction in size or loss of one member of several originally paired organs (Fig. 190). This transformation may have been initiated by a single genetic mutation which affected the symmetry of the larval shell muscles. If so, it was immediately advantageous, because it shifted the mantle cavity from a posterior position above the main part of the foot to an anterior one above the head. In this position, the head could be withdrawn easily into the shell where it would be protected by the subsequently retracted foot.

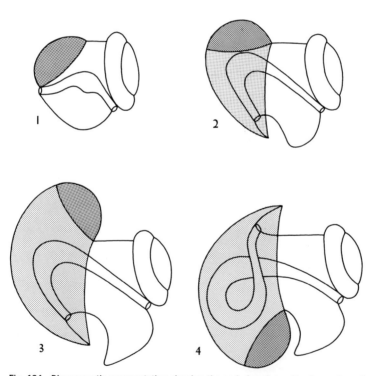

Fig. 184 Diagrammatic representation showing the early larval growth of a gastropod and torsion of the visceral mass and shell with respect to the head and foot. This carries the mantle cavity containing the gills and anus to an anterior position above the head.

Table 185   Some noteworthy distinguishing characters of the main groups of gastropods, recognized by zoologists

---

Archaeogastropods: snails with bifoliate gills, primitive forms with a pair, advanced forms with only one gill. Aperture not notched for the reception of a tubular inhalant siphon. Cambrian to Recent

Caenogastropods: includes both meso- and neogastropods

Mesogastropods: snails with a single monofoliate gill; many have an apertural notch for the reception of an inhalant siphon. Pennsylvanian(?) to Recent

Neogastropods: like mesogastropods but commonly with the shell aperture drawn out into a canal for the reception of an inhalant siphon. Cretaceous to Recent

Opisthobranchs: nerve trunks shortened, not twisted as in the foregoing groups. Single gill, if present, in a lateral or posterior rather than an anterior position. Shell may be reduced, internal, or absent. Pennsylvanian(?) to Recent

Pulmonates: nerves not twisted, without gills, mantle cavity transformed into an air-breathing lung. Pennsylvanian(?) to Recent

---

Archaeo-, meso-, and neogastropods are successive evolutionary stages derived from one another that probably progressed concurrently in several different lineages. Opisthobranchs and pulmonates are highly specialized and of doubtful origin. All these groups, set up for modern gastropods, are based mainly on features such as nerve patterns, nature and location of gills, type of radula, and details of genital organs that cannot be determined certainly for fossils. Consequently, the assignment of many fossils to these groups is difficult if not impossible.

Most theories regarding phylogenetic relations among the gastropods have been formulated by zoologists, whose ideas have been based mainly on the comparative anatomy of modern species. The features believed by them to be important are provided by the soft parts and radula, which are not preserved in fossils. Paleontologists on the whole have contributed very little to the solution of evolutionary problems. They have attempted mainly to fit their fossil species into the system of classification set up by zoologists. Unfortunately the fossil shells tell disappointingly little about the animals that once inhabited them. Therefore many doubts remain concerning the relations of fossil to living gastropods and of the fossils among themselves. The most reliable approach at present involves three steps: (1) to infer the course of water currents which circulated through the mantle cavity, as indicated by the form of the shell aperture, (2) to infer the positions and relations of the gills and anus that this suggests, and (3) on this basis to relate fossils to modern gastropods.

Some minor evolutionary sequences have been recognized among fossil gastropods on the basis of progressive changes in shape and ornamentation of closely similar forms. A few more ambitious attempts have been made to trace gastropod lineages on the assumption that conclusions regarding recapitulation and accelerated development can be relied upon. In its larger aspects, however, the reconstruction of gastropod phylogeny is most highly theoretical. It is based on ideas concerning (1) the probable consequences of torsion of the molluscan body and coiling of the shell, and (2) steps that seem necessary thereafter to account for the diversity observed in modern gastropods.

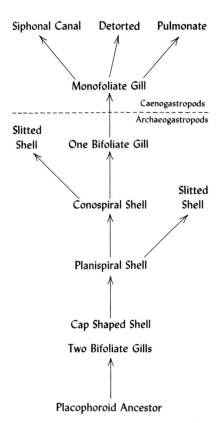

Fig. 186 Diagram showing probable stages in the evolution of gastropods. The caenogastropods include both meso- and neogastropods.

Primitive Archaeogastropods

Among other features that are equally unrecorded in fossil shells, the modern archaeogastropods are identified by twisted visceral nerves, paired bifoliate gills, and a heart with two auricles (Fig. 190-1). The twisted nerves are the result of torsion; the other features presumably are part of the inheritance from a pregastropod ancestor.

Gastropods are believed to have evolved from a bilaterally symmetrical mollusc, similar in some ways to the monoplacophorans, with a posterior anus and mantle cavity, probably enclosing two gills, and a flattened foot adapted to crawling. Torsion brought the mantle cavity forward. Water to bathe the gills entered it at the sides and left flowing forward in the middle, so that fecal matter was discharged near the gills and in the vicinity of the head. Opportunities for contamination were reduced first by the development of a median reentrant, or sinus, in the aperture of the shell, and later by a deepening slit that permitted the anus to be directed upward (Fig. 189-4,5).

The earliest shells probably were low, relatively flat, and cap-like (Fig. 187), but higher ones, in proportion to their size, which provided more complete protection for the body rapidly evolved. Most molluscs seem to have had a common tendency to coil, and gastropod shells are essentially variously coiled, more or less regularly enlarging tubes. Shells coiled symmetrically in a plane soon developed (Fig. 188). Fossils illustrating different stages in this presumed succession occur in the Cambrian, and some planispiral forms with a slit band continue into the Triassic.

The planispiral shell of a crawling snail must be held erect, and a narrow one would tend to topple over sideways. Better stability was achieved by the development of a shell drawn out excentrically in a more or less conical spire which extended first upward, if low, or sideways if slender and then backward where it lay upon the posterior part of the foot when the gastropod was active. Shells of this type appear in the Upper Cambrian. They probably progressed in the development of sinus to slit concurrently with the planispiral shells. Gastropods with slit-bearing shells of this kind are still living in the sea, and their anatomy is known. By analogy,

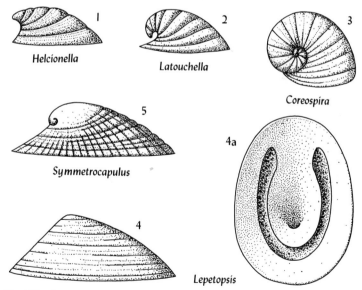

Fig. 187   Representative cap-like gastropods. 1 to 3, Primitive Cambrian archaeogastropods. 4, Late Paleozoic, probably primitive form; note horseshoe-shaped scar. 5, Jurassic form that may be a reversion to a seemingly primitive condition. (*After Knight and Yochelson, 1960, figs. 89, 143, 144, pp. 172, 232.*) (All enlarged; not at uniform scale.) Compare with Fig. 182; note that apex points to rear.

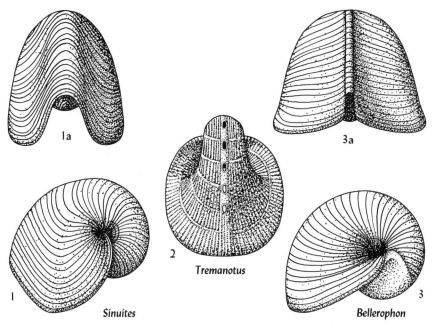

Fig. 188 Representative planispiral gastropods. 1, An early Paleozoic shell with a sinus in its aperture. 2, A middle Paleozoic shell with a perforated band. 3, A late Paleozoic shell with a well-developed slit and band. (× ½–2.) (*After Knight and Yochelson, 1960, figs. 93-6a, b, 97-1b, 102-4b, c, pp. 176, 179, 185.*)

fossils with a reentrant near the middle of the aperture's outer lip are believed to have had two gills (Fig. 190-1) and to have been similarly organized in other ways.

Advanced Archaeogastropods

A recumbent conispiral shell tended to drag behind a crawling gastropod. Deflection of the shell axis in this way caused some distortion of the mantle cavity. The anus was displaced somewhat to the right of the midline of head and foot, and the right gill was crowded into a smaller sector than the left. This gill consequently was reduced in size and eventually disappeared. With only the left gill remaining, water currents entered from the left and flowed outward from the right side of the mantle cavity. Gastropods of this type did not require a marginal slit. Fossils that are interpreted as being advanced archaeogastropods organized in this way appear in the Lower Ordovician, and many species are present in modern faunas.

So far as known, none of the archaeogastropods had copulatory organs, and their eggs and sperm were discharged into the sea, where fertilization was accomplished. All of them presumably have been marine.

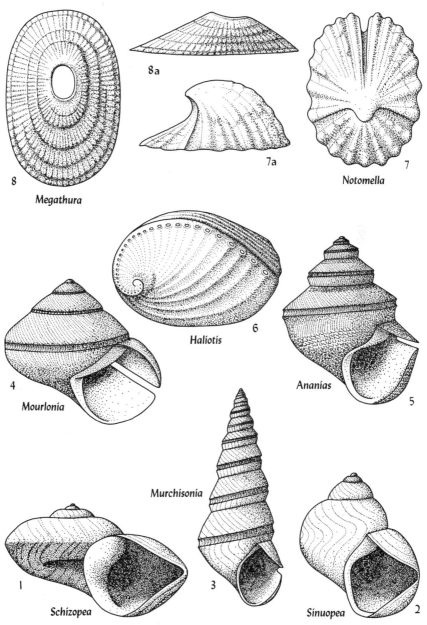

Fig. 189   Representative primitive archaeogastropods. 1, 2, Early Paleozoic shells with sinus in outer lip of aperture. 3 to 5, Late Paleozoic shells with well-developed slit and band. 6 to 8, Modern shells with perforations, slit, and summit opening. (×½–3.) (*After Knight and Yochelson, 1960, figs. 199, 200, 204, 205, 224, 228, 229, 292, pp. 112, 113, 117, 118, 137, 141, 142, 190.*)

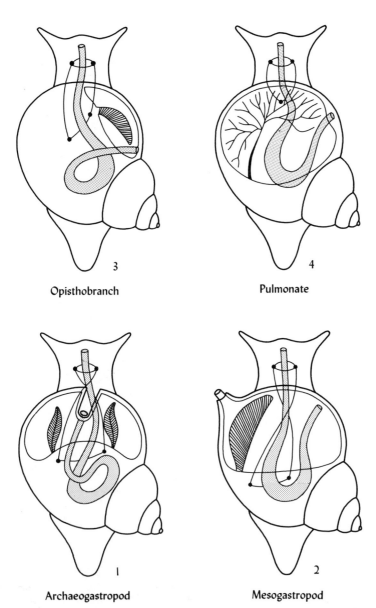

Opisthobranch

Pulmonate

Archaeogastropod

Mesogastropod

Fig. 190 Highly diagrammatic representations showing relations of mantle cavity, gills, and anus in different types of gastropods. 1, Primitive form with two gills; exhalant current discharges centrally through a slit. 2, Left gill only; inhalant current enters through a siphon, and exhalant current leaves at right. 3, Gill and anus shifted to right and backward; nerves detorted. 4, Mantle cavity supplied with blood capillaries becomes a lung. (*Adapted in part from Cox, 1960, "Treatise on Invertebrate Paleontology," pt. I, figs. 87, 88, pp. 154, 155.*)

Mesogastropods

The mesogastropods surely evolved from advanced archaeogastropods, but the transition probably occurred in more than a single lineage. These groups both have twisted nerves, no right gill, and a heart with a single auricle (Fig. 190-2). Modern mesogastropods are recognized by their monofoliate gill, better-developed nerve ganglia, and characteristic type of radula. These gastropods acquired an inhalant fold or tube in the mantle, termed a siphon, whose presence generally is indicated first by a notch at the anterior end of the aperture and later by a moderately projecting enrollment of the shell in line with its axis. Some of them also have a small notch, adjacent to the suture where the last whorl overlaps the one behind it, that marks a place where the water current left the mantle cavity.

Some mesogastropods copulate and produce young in an advanced stage of development. This has permitted them to invade freshwater environments. A few have adapted to land life. These lost their gill, and the mantle cavity was converted to an air-breathing lung.

The mesogastropods appear in Triassic faunas, although some Paleozoic fossils are uncertainly identified with this group. Fossil shells which present no evidence for the existence of a siphon cannot be distinguished surely from archaeogastropods. Those that are tentatively assigned to the mesogastropods exhibit other shell characteristics which are only suggestive of this relationship.

Neogastropods

No very sharp line of demarcation separates the neogastropods from the mesogastropods. The newer group shows evolutionary advancement in the further development of nerve ganglia and the presence of a more pronounced siphonal canal (Fig. 192-3). It also is distinguished by a special type of radula. Neogastropods are less numerous in species and less varied in their adaptations than the mesogastropods. They appeared in the early Tertiary, and almost all of them have been marine.

The evolution of an inhalant siphon probably was related to increased gastropod activity. When well developed it conducts a water current to a smelling organ within the mantle cavity as well as to the gill. The siphon can be moved in various directions; it tests the surrounding environment and directs the animal to a source of food. Many siphonate gastropods are carnivorous.

Opisthobranchs

Some of the characteristic effects of gastropod torsion have been lost in the opisthobranchs. The anus and gill have migrated backward to the right, the auricle lies behind the ventricle of the heart, and the visceral nerves are no longer twisted (Fig. 190-3). The results of torsion still persist, however, because these animals possess only a single gill, as in their probable mesogastropod ancestors. Otherwise almost complete practical bilateral symmetry in the soft anatomy may be regained. There is a tendency for the nerve ganglia to become concentrated around the esophagus.

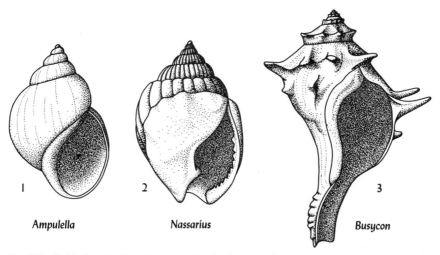

I	2	3
Ampulella	*Nassarius*	*Busycon*

Fig. 192   Illustrations to show the progressive development of a siphonal canal in gastropod shells. 1, The aperture is rounded and gives no indication of a siphon. 2, An anterior notch provides an opening for the egress of a siphon. 3, A conspicuous projecting shell fold forms a protective cover for a siphon. If soft parts were not available for observation, shells of these types might be classed as archaeogastropod, mesogastropod, and neogastropod, respectively. (*After Cox, 1960, "Treatise on Invertebrate Paleontology," pt. I, fig. 72, p. 117.*)

Modern opisthobranchs are all hermaphroditic, a condition that was first attained by some of the mesogastropods. They are all marine and show a wide range of environmental adaptations. A tendency for the shell to be reduced in size, become internal, or disappear entirely is evident. Many of the modern shell-less forms have lost their mantle cavity and original gill and have developed secondary respiratory organs. Although supposed opisthobranchs have been reported from the Devonian and younger Paleozoic strata, their reasonably certain range does not antedate the Mesozoic.

Pulmonates

The pulmonates also are believed to have evolved from the mesogastropods. They are mainly land snails and have differentiated into a multitude of species. The original gill was lost, and the mantle cavity, which does not contain the anal opening, became a lung (Fig. 190-4). Some, however, have reverted to an aqueous existence and have developed secondary gills different from the old ones. Pulmonates have not undergone detorsion like the opisthobranchs, and the single auricle retains its position anterior to the ventricle of the heart. The nerves are not twisted, however, because the ganglia are all grouped closely around the esophagus. Like the opisthobranchs, modern pulmonates are all hermaphroditic.

Many of the modern pulmonates are without shells, and of course snails of this kind have left no fossil record. Some Carboniferous nonmarine species have been classed with the pulmonates, but this assignment is doubtful. Most of the questionable forms probably were adapted to freshwater rather than terrestrial habitats. Most pulmonate fossils are of Cenozoic age.

Convergence

The shells of gastropods have been so plastic, especially with regard to ornamentation and spiral form, and examples of convergence are so abundant that such features cannot be relied upon in many instances to indicate relationships at any taxonomic level. Furthermore there seem to be good reasons to conclude that the groups discussed briefly in the preceding sections represent little more than successive grades of evolutionary advancement. The mesogastropods and neogastropods, especially, probably consist of numerous unrelated lineages, each of which was derived from a different stock of much more ancient origin. If this is true, these lineages owe their likenesses to evolution progressing along parallel pathways that produced closely similar results.

Convergent evolution is demonstrated by some known modern species with soft parts very differently organized but with shells which are such exact duplicates of each other that they cannot be distinguished. Among the very similar-appearing limpet-like gastropods, various species must be classed with both primitive and advanced archaeogastropods and mesogastropods. There can be little doubt that likenesses among these unrelated forms have resulted from convergence and adaptation to a similar way of life. Except for this example, however, many of the more obvious features of gastropod shells do not seem to be related closely to environmental adaptations. On the one hand, very differently appearing modern species inhabit almost identical environments. On the other, very similar-appearing species inhabit environments of the most diverse kinds. All this poses a most difficult problem in the paleontologic study on an evolutionary basis of this whole group of molluscs.

Ontogeny

Well-preserved spiral gastropod shells present an easily observed and almost perfect ontogenetic record, and in this respect they are superior to many other kinds of fossils. Details of ornamentation and type of coiling may change abruptly in the early whorls (Fig. 194), and the distinguishing characters of adults are likely to appear gradually in the course of growth. Distinction cannot be made among the young stages of some easily differentiated genera and species. Features of these kinds have been used to a limited extent in the recognition of close relationships. They have not, however, been much employed in attempts to reconstruct evolutionary lineages on the basis of possible recapitulation.

Many gastropods that hatch from large yolky eggs lack the larval trochophore stage and emerge as juveniles already provided with an embryonic shell. This shell does not exhibit growth lines, and it may be remarkably different from the main part of the shell that is subsequently added (Fig. 194-1,2,4). There seems to be

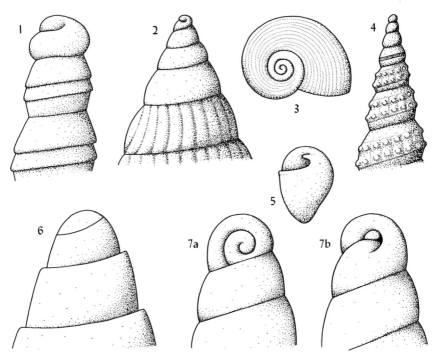

Fig. 194   Early growth stages of several Tertiary gastropods, drawn from specimens. 1 and 6, Abrupt changes in the shape of whorls. 2 to 4, Gradual change in ornamentation. 7, Abrupt shifting in axis of coiling and relation of whorls.

no reason to presume that this tiny beginning of a shell evolved in adaptation to its embryonic environment, because a shell would have no useful function while still within the egg. If this is so, the supposition might be entertained that the embryonic shell recapitulates ancestral characters to some degree. Such a possibility, however, seems to be decisively denied, at least in those instances where extremely sharp morphologic change, which may include displacement of the coiling axis (Fig. 194-7a), marks the beginning of postembryonic growth. A break in continuity of this type certainly does not harmonize with the concept of gradual evolutionary transformation. Although this feature of gastropod ontogeny remains unexplained, later modifications, particularly of ornamentation, probably have some evolutionary significance.

Pelecypods

Pelecypods vary greatly in the shape and ornamentation of their shells. Their general anatomy, however, is remarkably uniform, suggesting that little fundamental evolutionary change has occurred since their differentiation from other molluscs in late Precambrian or earliest Paleozoic time. Their conservatism probably is related to the fact that most pelecypods lead comparatively sedentary lives and that they early adapted to the three general types of habitat which they still occupy: (1) on the sea bottom, where they move about slowly, (2) in bottom sediments, where they live quietly in permanent burrows, and (3) attached to objects on or above the sea bottom, either by an adhesive byssus or by cementation of their shells. All pelecypods inhabit aqueous environments; most of them are marine, but some have adapted to living in freshwater.

The earliest certainly identified pelecypods have been found in Lower Ordovician strata. Some Cambrian bivalved fossils may be either crustaceans or molluscs, more probably the former. After the Cambrian, however, the history of the pelecypods parallels that of the gastropods rather closely. They gradually increased in abundance and variety during Paleozoic time but thereafter expanded greatly in relative importance in many marine and some freshwater faunas. This expansion may have resulted from the perfection of siphons that permitted pelecypods to prosper particularly as burrowers.

The pelecypods are believed to have evolved from placophoroid ancestors. The early ontogeny of modern species shows that the beginning of a shell is secreted

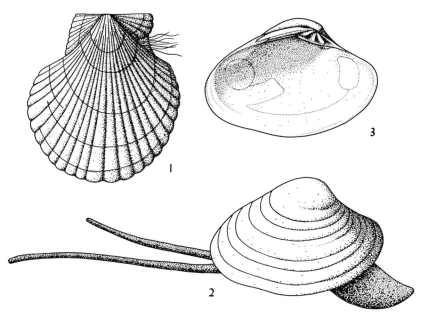

Fig. 195   1, Pectenoid pelecypod, showing where the byssus, an adhesive outgrowth from the foot, emerges through a notch between the valves. 2, Burrowing pelecypod with extended foot and siphons. 3, Inside of left valve of same, showing hinge structure, muscle scars, and sinus in pallial line, indicating the possession of retractable siphons.

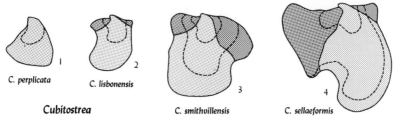

C. perplicata   C. lisbonensis

Cubitostrea   C. smithvillensis   C. sellaeformis

Fig. 196   Four species of Tertiary Gulf Coast oysters from successive stratigraphic zones have been interpreted as stages in a single evolving lineage. Changes in these left valves include more or less regular increase in size, appearance and enlargement of ears, and twisting of shell. Strength of ribbing decreases on right valve. (After Stenzel, 1949, fig. 8, p. 44.)

by a single dorsal gland. At first the tiny shell consists only of horny material. Later calcium carbonate is added on the two sides, but a central zone remains uncalcified. This is the beginning of the hinge and ligament. The sides bend downward, enclosing the body, and two pairs of adductor, or closing, muscles develop from parts of the pallial muscle that parallels the edges of the mantle. At this stage the typical form of a pelecypod has been achieved. This ontogenetic sequence probably indicates very closely how the transformation of a monoplacophoroid to a pelecypod shell was accomplished.

Clues to patterns of evolution among the pelecypods have been sought in (1) structure of the paired gills, (2) nature of the hinge dentition, (3) form, structure, and disposition of the elastic ligament, (4) relative position and size of the closing muscles and their scars, (5) presence or absence and type of siphons, (6) shape and ornamentation of shells, (7) crystalline shell structure, (8) occurrence of a byssus, (9) nature of the stomach, (10) general mode of life, as well as some other features. Undoubtedly all are significant, but there has been much disagreement as to relative significance, which probably has varied greatly among various pelecypod stocks. Short evolutionary lineages seem to have been recognized fairly satisfactorily in a few restricted groups (Figs. 196; 209). No consensus has yet been reached, however, concerning overall pelecypod evolution, probably because parallelism and convergence have been common.

Differences in each type of feature mentioned can be separately considered

Table 196   Pelecypod gill types

Protobranch: primitive gills consisting of paired leaf-like structures similar to those of chitons and gastropods
Filibranch: gills consisting of two parallel rows of filaments hanging downward on each side of the body. In most forms the ends of the filaments are recurved upward, producing W-shaped structures.
Eulamellibranch: ends of the recurved filaments attached on the outside to the mantle and on the inside to the body, thus closing off channels through which posteriorly flowing exhalant water currents pass.
Septibranch: gills of very different type, transverse muscular partitions separating upper and lower parts of the mantle cavity

The first three types are successive evolutionary stages, probably derived from one another in several different lineages. The relations of the last are problematic. Gill types, considered important in the classification of modern pelecypods, cannot be known for fossils.

theoretically on an evolutionary basis with some degree of confidence. Unfortunately little systematic attempt has been made to correlate more than a very few of these evolutionary sequences with one another. Until recently many zoologists have emphasized gill structure and dismissed most features of the shell as having slight importance. On the other hand, paleontologists necessarily have concentrated attention on the shells, and almost unanimously they have accepted hinge structure as being most revealing. It is not surprising, therefore, that presumed evolutionary relations, generally expressed in systems of classification, have differed greatly.

Gills

The most primitive type of pelecypod gill is somewhat similar to that possessed by gastropods, and it probably reflects the structure of this organ in the placophoroid ancestor (see Table 196). It consists of two series of parallel, transversely leaf-like, ciliated filaments hanging downward in the mantle cavity (Fig. 197-1). Originally

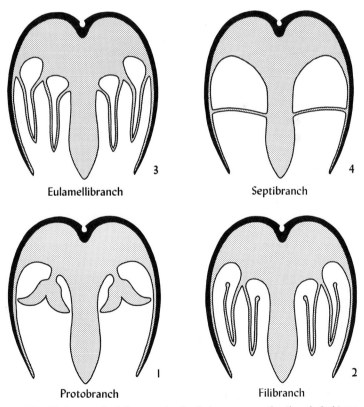

Eulamellibranch   Septibranch

Protobranch   Filibranch

Fig. 197   Much-generalized diagrams showing in transverse section the principal types of pelecypod gills. There are several variations of these types.

the two gills, one on either side of the body, probably were located posteriorly and their function was exclusively respiratory. Their cilia created a gentle water current that entered the mantle cavity at the front and flowed out behind. At an early stage the pelecypod ancestor probably had a fairly well-developed head and radula and crept about and fed in much the same way as a gastropod.

No modern pelecypod has a well-developed head or radula. The degeneration of these parts probably was related to the evolution of the foot, which became a powerful digging organ. It developed in such a way that it extended forward and interfered with the functioning of the mouth. This was raised and no longer could come in contact with the source of food, the head was permanently withdrawn inside the shell, and ciliary feeding replaced direct ingestion. Such a change created a typical pelecypod.

Modern *Nucula* and its allies are primitively organized. Their gills are of the type described, produce a backwardly directed current, and serve for respiration only. Feeding is accomplished by the action of ciliated lip-like flaps and tentacles thrust outward from the shell. These animals are particularly interesting because fossils, very similar to their shells, occur among the most ancient pelecypods. They also exhibit other features which almost certainly are primitive. For example, their hinge dentition is of a distinctive simple type, and the foot has a flattened sole, although reports that these animals can creep about like gastropods probably are mistaken. In spite of their seemingly very primitive characters, these pelecypods continue to live successfully and in considerable variety in modern marine faunas.

Most other types of gills are, in theory, readily derivable from the simple type possessed by *Nucula*. In the first stage of evolutionary transformation,, the leafy filaments seem to have been reduced to hanging threads. Freshwater clams with gills of this kind are living now, and some other pelecypods pass through a comparable ontogenetic stage in their early development. Next the filaments lengthened, became abruptly recurved upward (Fig. 197-2), and were held together by entangled cilia. Thus each gill consisted of two cohesive V-shaped, sieve-like lamellae hanging from a common longitudinal axis. Further evolutionary advancement involved first the development of interfilamentary and interlamellar tissue connections and then the extension of blood vessels through these junctions. Finally the recurved lamellae became joined on the outside to the mantle and on the inside to the animal's body (Fig. 197-3). Posteriorly the inner margins of the two gills became adherent.

These progressive gill developments were correlated more or less directly with changes in respiration and feeding methods and perhaps with living habits. Burrowing must have made respiration more difficult because the surrounding sediment interfered with the flow of water directly through the mantle cavities and past the gills. Compensating improvements included transformation of the gills into more efficient filtering organs, consisting of lamellae that separated increasingly perfect incurrent passages below from excurrent passages above. The direction of incurrent flow was reversed so that water entered from the rear. Fusion of the posterior edges of the mantle produced first an excurrent opening, tube, or siphon, through which deoxygenated water and feces were expelled, and then a corresponding incurrent siphon (Fig. 199). The siphons lengthened and permitted burrowers to maintain direct contact with water above the sediment surface. Food particles borne by the incoming current were filtered out in the gills and conducted forward toward the mouth by ciliary action. The lip-like flaps and tentacles no longer were required to search for food particles in the external sediment but obtained them from the gills. Further fusion of the mantle edges ventrally enclosed the body, leaving only an anterior opening through which the foot could be protruded.

The most specialized type of gill-like structure is a fleshy and muscular unciliated partition extending horizontally from the body to the mantle (Fig. 197-4). Water circulation is maintained by a kind of pumping action, forcing a current upward through numerous perforations. How this may have evolved from another kind of gill is doubtful. Opinions vary, but certain differences in stomach structure and function, which need not be detailed here, suggest that pelecypods with these organs probably did not evolve from ancestors whose gills were transformed in the way that has been described.

Gill structures rather obviously exhibit several grades of progressive evolutionary advancement. Perhaps each improvement originated in a single lineage and then radiated through many subsequently differentiating forms. More probably, however, parallel evolution characterized the development of multiple lineages. Whatever its evolutionary status, gill structure can be determined only for modern pelecypods. No feature of a shell or fossil indicates the kind of gills that an animal possessed. The siphons, however, commonly left traces. They are muscular and can be re-

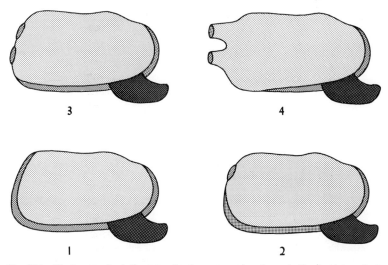

Fig. 199   Much-generalized diagrams showing progressive stages in the development of siphons in pelecypods by fusion of the posterior mantle edges.

tracted to some extent. Most modern pelecypods with short siphons can withdraw them completely within their shells, but others with long siphons cannot do so. The existence of siphons generally is indicated by a posterior reentrant, or sinus, in the pallial line impressed on the inside of a shell by the muscle which encircles the mantle where it becomes free (Fig. 195-3). Some modern siphonate pelecypods do not show this feature, but if it occurs in fossils the existence of siphons is assured.

Dentition

The most primitive pelecypods presumably possessed thin valves whose dorsal edges, joined by a flexible membrane or ligament, were not in actual close contact. A hinge of this type must have been very weak, and obvious advantage was to be gained by the development of some kind of closely interlocking structure that would, at the same time, prevent most slippage or extreme rotation of the valves but not interfere with their opening and closing.

Many fossil pelecypods of the early Paleozoic seem to have had relatively primitive and simple hinge dentition; complexity increased in later forms. By Cretaceous time most of the hinge types recognized among modern pelecypods had developed. This progress, however, has not been traced in detail, so that the actual evolution of these structures is more or less uncertain. Hinges are difficult to study in many fossils because specimens commonly adhere to the solid matrix of enclosing rock by their concave sides. Freeing a hinge from the matrix is not easy, and little effort has been made to uncover this part of the shell of many fossil species.

Table 200   Principal named types of pelecypod hinge dentition

Actinodont: hinge teeth of taxodont type diverging downward
Anodont: without hinge teeth
Anomalodont: teeth very small or absent
Asthenodont: teeth obsolete in burrowing pelecypods
Ctenodont: teeth of taxodont type converging downward; a very primitive type
Cyclodont: strong cardinals curving outward from below beaks. Without a hinge plate
Desmodont: teeth weak or nearly absent, with a large internal ligament. Probably derived from heterodont
Diagenodont: three or fewer cardinals and one or two laterals on a hinge plate
Dysodont: teeth very weak or absent below a narrow external ligament
Heterodont: two or three well-differentiated cardinals with or without laterals before and behind them. Ligament external. Probably derived from diagenodont or pantodont. A very common type
Isodont: two nearly equally sized cardinals below beaks of both valves. Ligament internal, central
Pachyodont: massive, thick teeth, as in rudistids
Pantodont: like diagenodont but with more than a single pair of laterals
Schizodont: two strong cardinals in right valve and a Y-shaped tooth in left valve that fits between them
Taxodont: numerous small hinge teeth in a row. See actinodont and ctenodont
Teleodont: heterodont without laterals

Pelecypod hinges vary in so many ways that classification is difficult and evolutionary relations are uncertain. Several of the types of dentition that have been named are not well characterized or differentiated. Many hinges have two kinds of teeth, strong short cardinals that are vertical or steeply inclined below the beaks and long laterals extending nearly parallel to the hinge line before and behind the cardinals.

It is conceivable that hinge structures developed in at least two different ways: (1) by the appearance of longitudinal ridges along the dorsal margin of one valve which fit into corresponding grooves along the other; and (2) by the appearance of more or less numerous transverse ridges along the dorsal margins of the valves, which fit together in an alternating fashion. Both these types of hinges are known in modern as well as fossil pelecypods, and gradations connect them with more elaborate kinds. Many modern pelecypods characterized by the first type of hinge are thin-shelled burrowers whose mode of life does not require a strong hingement. The hinges of some may actually be primitive, but others probably are the result of regressive evolution. There are also pelecypods whose valves have no interlocking elements but only slightly thickened edges that rotate against each other. Determination of the evolutionary status of fossils of these kinds is practically impossible. Poorly developed hinges of this type are known as dysodont.

The other type of primitive hinge can be interpreted with more assurance. The short transverse teeth (Fig. 202-3,4) probably originated as extensions of surface ribs or thickenings where these reached the valve margin. Many pelecypods have ventral margins with interlocking ribs, and some show similar structure along part of the dorsal margin. Possibly dorsal interlocking of this kind persisted and was transformed to alternating teeth and sockets as a hinge area developed by slow growth of the valve below its pointed beak. Taxodont dentition of this kind seems to have evolved twice. It is believed to be primitive in *Nucula* and its near relatives, and this type of hingement has persisted to modern times with only minor modifications.

The shells of many pelecypods in their earliest ontogenetic stages are nearly circular with very short hinge lines. As they grow, greater and greater longitudinal asymmetry commonly develops. If the dorsal margin of the shell continues to be curved, however, efficiency of the hinge cannot be increased by elongation. Improvement is possible either by enlargement of the teeth and deepening of the sockets immediately below the beak (Fig. 202-5,6) or by the enlargement of more lateral teeth so that articulation can be maintained even though these teeth are partly disengaged from their sockets. In the first case the more lateral teeth are likely to degenerate and disappear. In the second they may decrease in number as they rotate toward a longitudinal orientation and lengthen.

An efficient hinge also may be produced by elongation. If a long straight hinge develops extending both ways from the beak, large teeth and sockets are unnecessary because the outer ones are not disengaged when the shell opens. Such a hinge has developed in *Arca* and its allies (Fig. 202-4). The teeth in modern species are taxodont, much like those of *Nucula,* but they had a different evolutionary origin. As the rather obvious ancestors of *Arca* are traced backward, it can be seen that the teeth become fewer and that the outer ones, particularly, are longer and rotate toward parallelism with the hinge line. In this lineage a seemingly primitive type of hinge dentition is believed to have resulted from convergent evolution.

Most commonly, elongation of the pelecypod shell is greater in one direction,

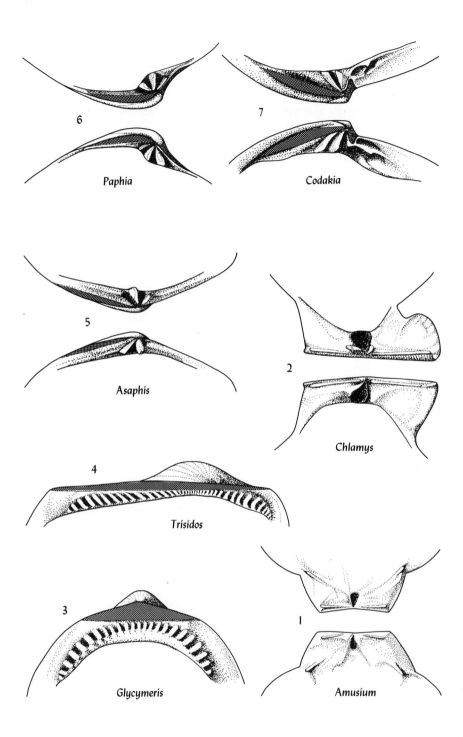

6 *Paphia*

7 *Codakia*

5 *Asaphis*

2 *Chlamys*

4 *Trisidos*

3 *Glycymeris*

1 *Amusium*

Fig. 202   (Opposite page.) Hinge structures of pelecypods. 1, 2, Two examples of pectenoid pelecypods with very weak dysodont hinges. There are no interlocking teeth, an external ligament is poorly developed and does little more than hold the valves together, but an internal resilium is functional. These pelecypods are highly evolved in some respects, and their hinges probably are examples of reversion to a seemingly primitive condition. 3, 4, Two examples of the arcoid type of taxodont dentition which probably resembles the nuculoid type as a result of convergent evolution. Some of these pelecypods have long, straight hinges and external ligaments extending both ways below the beaks. The ligament areas are stippled. 5 to 7, Three examples of heterodont dentition have two or three strong cardinal teeth. The ligament area extends only posteriorly from the beaks.

ordinarily posteriorly. Hinges in this kind of shell generally are specialized, and longitudinal teeth and sockets are likely to be prominent. The anterior dentition may degenerate or consist of a few large elements that maintain their articulation when partly disengaged. The most advanced, or heterodont, type of hinge consists of various combinations of strong teeth of two different kinds. Two or three short more or less transverse cardinals commonly are present below the beaks. These maintain proper positioning of the valves when the shell is open. They are flanked on one side or both sides by long diverging laterals which guide the valves together when they close. Both kinds of teeth may have evolved from taxodont dentition. Some modern pelecypods seem to show this transition in their ontogeny. Other heterodont hinges, however, probably had a different origin. For example, cardinal teeth may be structures independently developed in association with the receptacle of a short internal ligament, and laterals may have evolved from a weak kind of dysodont dentition.

Most paleontologists believe that pelecypod dentition has been relatively stable. Some different patterns of hinge structure probably can be relied upon to distinguish different evolutionary stocks. The converse conclusion, however, is far less certain because closely similar structures almost surely have resulted from parallel or convergent evolution in different lineages.

Ligament

The ligament holds the two valves of a pelecypod shell together dorsally and, with the adductor muscles, governs their opening and closing. The most primitive ligament probably was no more than an uncalcified band containing the axis of rotation about which movement of the valves occurred (Fig. 204-1). Such a ligament was not strained when the shell was open by a natural amount. When closed, however, the upper, or outer, part was stretched and the lower, or inner, part compressed. With the evolution of articulated hinges, the two valves were brought close together and the axis of rotation lay along or near their line of contact. Such a relationship served to divide the functional ligament into more or less separate outer and inner parts (Fig. 204-2) subject to opposite elastic strains; this was one of the influences that conditioned alteration and further evolution of the ligament. A second and equally important influence was the shape of the dorsal margin and hinge, which

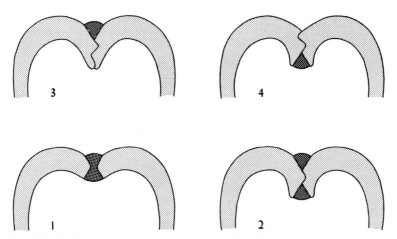

Fig. 204   Highly diagrammatic transverse sections illustrating various types of pelecypod ligaments. 1, The presumably very primitive condition, with ligament both external and internal. 2, The development of a denticulated hinge has divided the ligament into external and internal parts. 3, Only an external ligament remains. This is the most common condition in modern pelecypods. 4, An external ligament has essentially disappeared and is little more than a connection between the horny outer coverings of the valves. An internal ligament, or resilium, serves to open the valves. It ordinarily is confined to a small spoon-like depression in the shell below the beaks.

is arcuate in most pelecypods. This has limited the possibilities of ligamental modification and development.

Although the axis of valve movement may wander somewhat, it must be a straight line, and a ligament cannot function successfully far on either side of this line because of excessive strains. Consequently if a straight hinge is not produced, the main ligament, as well as the dentition, tends to be best developed in, or entirely restricted to, an area immediately beneath the beaks or in one direction along the curved shell margin as this centers below the beaks (Fig. 202-5 to 7).

Evolution, proceeding presumably from a single primitive type of ligament, has produced a variety of different structures (Fig. 205). Some parts of the primitive ligament degenerated or disappeared; others were enlarged and assumed most, if not all, of the elastic functions of this organ. Commonly the external part, or ligament proper, is more prominently developed than the inner part, termed the resilium, which ordinarily is short.

Pelecypods with valves nearly symmetrical in an anterior-posterior direction may have a short external ligament between the beaks (Fig. 206-1). More commonly, however, and especially in the pectenoids, they have only a centrally located resilium (Fig. 202-1,2). Pelecypods with a long straight hinge, like that of *Arca* (Fig. 202-4), have what are fundamentally long external ligaments, but these may be separated into a series of segments (Fig. 206-4), especially if growth of the valves below the beaks produced wide areas of ligament attachment. Most commonly of all, in pelecypods that are asymmetrical because of greater growth posteriorly, a long ligament occupies the hinge line lengthened in this direction (Fig. 202-5 to 7). Generally it is external in shells with heterodont dentition, but this may be accompanied by a small central or anterior resilium. Dysodont pelecypods, however, may have a long ligament located partly or wholly below the weakly articulated hinge.

Altogether, ligament structures are so diverse that not all varieties can be mentioned here. All of them can be derived, however, by functional modifications of the presumed primitive ligament in accordance with the type of articulation and shape of the dorsal margin of the shells. Unfortunately the areas of ligament attachment cannot be identified certainly in all fossils. Finally, because of parallel

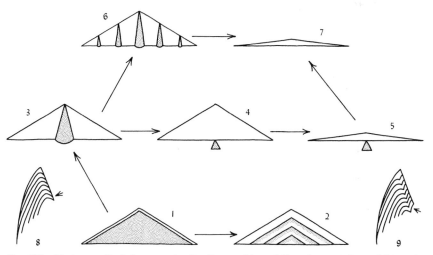

Fig. 205   Much-generalized diagrams showing the possible evolution of some pelecypod ligaments. They consist of two kinds of material: (a) fibrous and partly calcified, which can be compressed but has little tensile strength, shown shaded, and (b) lamellar, which can be stretched considerably, occupying all or parts of the unshaded areas. 1, One type of ligament that seems to be relatively primitive consists mainly of fibrous material with an outer covering of lamellar material. 2, Derived from this is a type consisting of alternating chevrons of fibrous and lamellar material. 3, Probably also derived from 1 is a type in which the fibrous material occupies a much narrower area. 4, Reduction in the amount of fibrous material and its confinement to an area below the hinge axis produced a characteristic resilium. 5, Reduction of the lamellar material left only the resilium as the functional ligament. 6, Possibly derived from 3 is a type provided with several areas of fibrous material which were added successively as the shell grew. 7, Perhaps a ligament consisting only of lamellar material evolved from 5. 8, As a hinge area grew below the beaks, the ligamental material, particularly the fibrous kind, attached to its upper part degenerated, and only the part near the hinge axis remained functional. Thus the functional ligament migrated down the hinge area, as indicated by the arrow. 9, An internal resilium, however, migrated in advance of the growing hinge area. Although these diagrams are symmetrical, most pelecypod hinges are not, and the ligaments of many are limited to an area on only one side of the beaks. (*Adapted in part from Newell, 1938, fig. 13, p. 33.*)

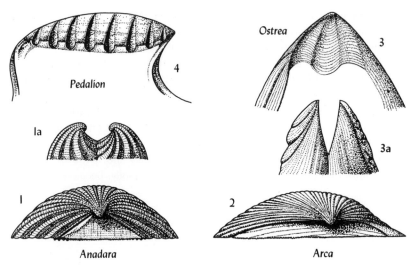

Fig. 206 Ligamental areas of pelecypods. 1, The ligament of this shell is mainly of the compressional kind, with a narrow layer of lamellar ligament extending down both sides of the triangular area. 2, This shell has a ligament of chevron-like bands of alternating compressional and lamellar material. 3, The oysters have a narrow central band of compressional material flanked on both sides by lamellar material. 4, This shell has segments of compressional material, separated by lamellar material, that were added successively as it grew. Compare with Fig. 205.

and convergent evolution, similar ligamental structure probably is no better guide to close relationship than is similarity of dentition.

Muscle Scars

Most pelecypods are equipped with two transverse muscles, of about equal strength, that counteract the elastic ligament and by contraction close the shell. An anterior muscle is located above the mouth, and the other is present below the posterior part of the intestine. Their areas of attachment to the shell commonly are shown by scars on the upper parts of the internal valve surfaces (Figs. 195-3; 207-1,2). Muscles of this kind and scars in these positions undoubtedly indicate a relatively primitive or conservative condition. Evolution, related primarily to modification of the foot and to the inhalant and exhalant structures, has resulted in profound changes which rather obviously reflect different modes of life.

The posterior muscle and its scar in many pelecypods tends to be larger than the anterior, and in some it has moved downward to a position where muscular action is more efficient (Fig. 207-3). Enlargement of this scar commonly has accompanied posterior lengthening of the shell, which generally is related to greater development of siphonal structures. Large retractable siphons, however, inhibit much downward migration, because space is required for reception of the siphons when retracted, as indicated by the pallial sinus below the scar. Relation of the anterior muscle to the mouth and underlying foot likewise prevented much downward shifting of this scar.

Relatively large posterior scars commonly are associated with correspondingly reduced anterior ones; the ultimate is reached by some pelecypods in which the anterior scars are not apparent. Modern species are known in whose ontogeny the anterior muscle progressively degenerates and finally disappears. Oysters and pectenoids are not elongated, but they lack anterior scars, and the very large posterior ones have moved downward conspicuously (Fig. 207-4). This migration has resulted in the rotation of their bodies in the anterior-posterior plane through about 90°, so that the originally dorsal hinge line shifts to an anterior position. Such pelecypods do not have siphons, and some of them lack a functional foot. Other pelecypods, e.g., some of the mytilids, have the foot reduced to little more than

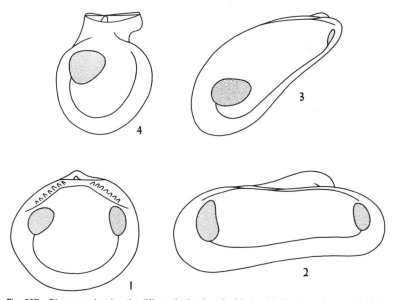

Fig. 207   Diagrams showing the differently developed adductor (closing) muscle scars. 1, The scars are almost equally strong. This is a relatively primitive condition. 2, As the shell lengthens, the posterior scar tends to become the larger one. Here the difference in a burrowing pelecypod is not great. 3, The anterior scar is very small. This is a pelecypod that lived permanently attached by its byssus. 4, Only the posterior scar remains. This pelecypod lived on the sea bottom, unattached most of the time, and probably was able to swim by the opening and rapid closing of its valves.

a byssus-producing organ but retain a fair-sized anterior muscle in spite of considerable reduction in height of the forward part of the shell.

Shell Shape

The shape of a pelecypod shell reflects the form of the enclosed body, and this in turn is determined largely by the animal's way of life. In general the anterior part mainly houses the foot, and the posterior part accommodates most of the circulatory apparatus, consisting of the gills and siphons, if the latter are developed. The basic shape from which all others have evolved probably was moderately convex and symmetrical transversely, slightly oval, and with approximately the same height in front and in back, but with the posterior somewhat longer than the anterior. An unspecialized pelecypod of this form seems relatively well adapted to active movement at or shallowly below the sediment surface.

The foot of such a supposed prototype probably was well developed. Deeper burrowing, therefore, involved mainly the problem of respiration, and ordinarily the shells of animals living in this way lengthened posteriorly so that more room was provided for gills and siphons. Pelecypods with greatly elongated shells probably lived in a nearly vertical position. Some shells gap behind because of siphons too large to be retracted. Evolution generally produced relatively thin smooth shells in active burrowers of this kind, whose hingement may be weak. Surface ribbing commonly indicates more sedentary habits, and thicker shells were required for adequate protection in very shallow water habitats.

The foot atrophied in some pelecypods. This, of course, generally accompanied restriction in burrowing activity. Evolutionary trends of this kind proceeded mainly in two directions and produced forms that lived on or above the sea bottom. In one group, the mytilids, the shell decreased in height anteriorly (Fig. 207-3) and became diagonally more or less triangular, with the beak very near the pointed anterior extremity. In the other group, which includes the pectenoids, the shell shortened and approached a compressed form, nearly symmetrical longitudinally and with central beaks (Fig. 207-4). An adhesive byssus commonly provided attachment for both groups. This was secreted by a gland on the reduced foot and passed out between the valves through an anterior notch close below the hinge line (Fig. 195-1). Some of these pelecypods, particularly those of the first type, lived permanently attached. Others, especially those of the second kind, could free themselves, and some were able to swim agilely by clapping their valves together. When at rest, these lay with one side upon the bottom, and many of them became laterally unsymmetrical, with one of the valves much flattened. The lower valve of these pelecypods commonly is more conservative than the other in ornamentation or shell structure, or in both. The lower valves of some different species did not evolve sufficiently to be distinguishable from one another.

Some restricted evolutionary lineages among pelecypods can be traced in gradually changing morphologic series (Fig. 209). Similarity of form, however, is not

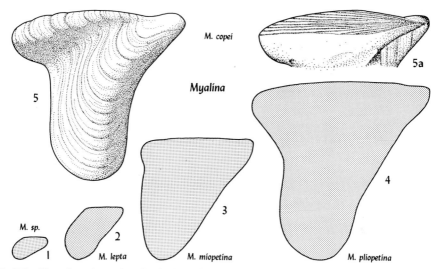

Fig. 209   These five pelecypod species, beginning in the Mississippian and extending to the Lower Permian, have been interpreted as a direct evolutionary series. Rather regular increase in size is conspicuous, but the shells also show successive change in obliquity and the growth of a posterior ear. (*After Newell, 1942, Kansas Geol. Surv., vol. 10, pt. 2, fig. 12, p. 47, pl. 5, fig. 3b.*)

reliable evidence of close relationship, because adaptations to similar ways of life in different lineages produced numerous examples of convergence in the shapes of shells.

Shell Structure

The shells of many modern pelecypods consist of aragonite in two layers, the outer one formed of tiny prismatic crystals directed at right angles to the surface, and an inner one of thin laminations parallel with the surface, which commonly shows the nacreous irridescence of mother-of-pearl. Many fossil shells have been so altered that their original composition and structure are not preserved. Nevertheless most of them, and almost all those older than the Mesozoic, were composed of aragonite, which is more soluble than calcite, the other common form of calcium carbonate. The pelecypod shells in many fossil faunas have been dissolved, leaving only molds or other imperfect specimens, although associated calcitic fossils such as corals, brachiopods, and crinoids remain intact. This is true also for other molluscs.

The composition of well-preserved pelecypods of different ages shows that most of the earlier fossils were predominantly nacreous and that the prismatic structure, which characterizes others, evolved independently in a convergent way in several different lineages. Furthermore, the substitution of calcite for aragonite, as in the oysters, was a later process that evolved more than once. Calcitic pelecypods are not known to have appeared in any lineage in pre-Mesozoic time.

Cephalopods

The cephalopods include the largest of all invertebrates and also probably those with the most advanced nervous and muscular organization. All are marine. They are represented in modern faunas by about 300 species, but almost all of these are quite different from the cephalopods commonly known as fossils. Only the modern *Nautilus* is similar to the great majority of fossil species, which were shell-bearing animals. The squids, octopuses, and most of their allies without shells have in general left a very meager record.

Zoologists divide modern cephalopods into two very unequal groups, characterized by the possession of four gills and two gills, respectively. The assumption commonly has been made that because *Nautilus* has four gills, all the similar shelled fossil species were similarly provided. This is far from certain, however, because the fossils furnish no evidence of the number of their gills. *Nautilus* rather obviously is primitive in some respects, and probably all the two-gilled cephalopods evolved from creatures of this general type. The stage at which evolution resulted in reduction of the gills, if indeed four gills were primitive, or in their increase, if two was the original number, cannot be known. Therefore the number of gills is not a suitable character for paleontologic differentiation. A more satisfactory system recognizes three groups: (1) nautiloids, with a very long fossil record; (2) ammonoids, which became extinct at the end of the Cretaceous Period; and (3) coleoids, which include the modern two-gilled forms, their immediate ancestors, and the fossil belemnoids (see Table 210).

More attention has been devoted by paleontologists to evolution among some of the cephalopods than to that of any other invertebrates. The ammonites have been favored because they are well preserved in great abundance and variety in the Mesozoic rocks of western Europe, where they were early recognized for their unrivaled usefulness in stratigraphic zonation and correlation. Late-nineteenth-

Table 210   Conspicuous distinguishing characters of some of the main groups of cephalopods

Nautiloids: cephalopods with external shells. The sutures where internal partitions meet the outer shell generally are unfolded.
　Orthocones: shells straight. Cambrian to Triassic
　Cyrtocones: shells more or less strongly curved. Ordovician to Devonian
　Nautilicones: shells coiled in a plane. Ordovician to Recent
Ammonoids: cephalopods with external shells. The sutures are folded into backwardly directed lobes and forwardly directed saddles.
　Goniatites: sutures in simple folds. Devonian to Cretaceous
　Ceratites: lobes secondarily folded. Mississippian to Cretaceous
　Ammonites: both lobes and saddles folded. Permian to Cretaceous
Coleoids: cephalopods without external shells
　Belemnoids: shells internal, greatly lengthened, and thickened posteriorly. Mississippian to Eocene
　Sepioids: shells internal, reduced to little more than a spongy dorsal remnant. Jurassic to Recent
　Octopoids: without internal shells. Jurassic to Recent

The three subgroups in each main group are, in general, successive stages in evolutionary sequence. Some reversals, however, are believed to have occurred.

century studies of the patterns of their sutures, which mark the junction of internal partitions with the external shell, account for much of the emphasis and reliance on the recapitulation theory and its popularity particularly among paleontologists.

The shelled cephalopods, unlike gastropods and pelecypods, are extinct except for a single genus. Therefore very little knowledge of modern representatives is available as a possible aid in interpreting the fossil record. This ignorance is advantageous from a practical point of view, because conclusions concerning evolution based on shell structure alone are not disturbed by the possibly conflicting testimony of soft anatomy. The uncertainties that are so evident in the consideration of gastropod and pelecypod evolution, however, suggest that similar complications might be introduced if living cephalopods of ancient type were known in comparable detail. As it is, no general accordance of opinions has been reached regarding certain important aspects of presumed evolutionary relations among cephalopods. Parallelism and convergence may obscure these relations, and some interpretations of the fossil record have varied greatly.

Agreement is general that both ammonoids and coleoids evolved from nautiloids. The kinds of ancestors from which these groups arose, however, have not been identified with any certainty. Divergent views that are still current seem to be strongly influenced by ideas concerning (1) whether straight shells were ancestral to or derived from those that are curved or coiled, and (2) whether or not ontogeny is a reliable guide to evolutionary progress, because it is conceivable that new characters may have appeared at early rather than at late growth stages.

Nautiloids

Cephalopods probably evolved from a generalized placophoroid ancestor by the lengthening of a conical cap-like shell, downward looping of the gut into a U-shaped course, and development of a hydrostatic system that contributed to greater facility of movement. The hydrostatic system is evidenced by differentiation of the shell interior into a tubular siphuncle partly or wholly surrounded by transverse partitions, or septa, defining a series of chambers that presumably were occupied mainly by gas. Only the anterior part of the shell beyond the last partition was the living chamber that contained the major portion of the animal's body.

A few early Cambrian fossils have been doubtfully referred to the cephalopods, but the most ancient certainly identified specimens are of late Cambrian age. These and the earliest Ordovician species are small nautiloids with simple uncrenulated sutures and marginal siphuncles that are believed to be ventral in position. Most of the shells are nearly straight, and the septa are closely spaced. Therefore the gas chambers are short and probably only compensated for density of shell material without providing excess buoyancy.

The fossil record of the nautiloids expands greatly in the Lower Ordovician. Evolutionary transitions from the older and presumably more primitive cephalopods, and relations among the considerable variety of newer forms, however, are not completely understood. Either the primitive cephalopod stock evolved rapidly in

Fig. 212 Probable evolutionary relations of various gross forms of nautiloid cephalopods. Starting with a presumed primitive cap-like shell (1) the principal trend passed through endogastric curvature (3), straight (4), exogastric curvature (5), open coil (6), and closed coil (7), with more and more completely embracing whorls (9). Several other trends branched off from different stages in this sequence. (*Adapted in part from Flower, 1955, Evolution, vol. 9, fig. 1, p. 248.*)

several divergent ways, or a number of different stocks almost simultaneously acquired the habit of secreting shells. Perhaps such parallel evolution is more likely.

The tracing of actual evolutionary lineages among later nautiloids also is no simple matter. Evidence of close relationship is variously seen in (1) the size, position, and detailed structure of the siphuncle (Fig. 216), (2) the presence and nature of deposits within the surrounding chambers (Fig. 220), (3) the degree of shell curvature or coiling (Fig. 212), (4) the cross-sectional shape of the expanding shell, (5) external ornamentation, etc. Apparent trends of these different features, however, do not all point in the same direction. Consequently the conclusions reached in any instance depend on the weighting given to various kinds of evidence according to their supposed relative importance. Such conclusions are all fairly speculative and are almost sure to vary.

Evolutionary changes in different groups of nautiloids produced some conspicuously peculiar forms. For example, among them are shells characterized by (1) very rapid expansion in diameter (Fig. 213-1), (2) greatly restricted apertures (Fig.

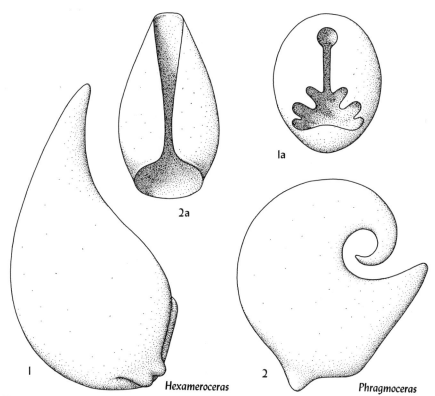

Hexameroceras   *Phragmoceras*

Fig. 213   These two Silurian nautiloids have peculiarly restricted apertures, and probably only the eye-bearing portion of the head and the tentacles could be extruded from the living chamber. Their general form suggests that these animals were rather sluggish bottom-living creatures. The orientation of these shells is somewhat doubtful. If the nature and location of the gas chambers are considered, buoyancy relations seem to demand that dorsal and ventral directions are the reverse of those commonly assumed. (*After Shimer and Shrock, 1944, "Index Fossils of North America," pl. 230, figs. 12, 13; Eastman-Zittel, 1913, "Text-book of Paleontology," 2d ed., vol. 1, fig. 1136, p. 613.*)

213-1a,2a), (3) the shedding of immature parts so that adult specimens consist of only the living chamber with perhaps a few restricted overlying gas chambers (Fig. 214), and (4) the development of ammonoid-like sutures (Fig. 215).

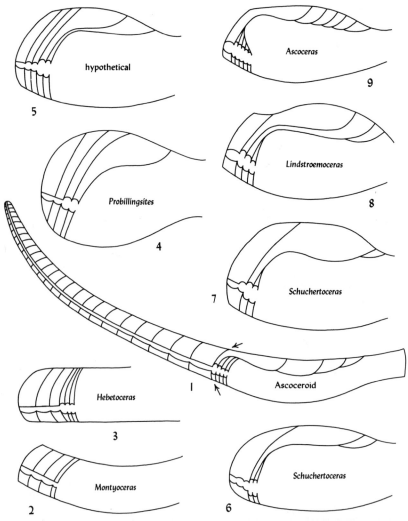

Fig. 214 Some nautiloids solved the problem of posterior buoyancy by shedding part or nearly all of the unoccupied portion of their shells. 1, At maturity the ascoceroids developed a peculiar living chamber overlain by gas chambers, and shed the septate portions of their shells (arrows). 2 to 9, Longitudinal vertical sections showing relations of septa and siphuncle to living chambers in a series of genera that may be a direct evolutionary lineage. Notice transitions in the shape of the siphuncle and the abrupt shortening of the gas chambers which increasingly overlie the living chamber. (*Adapted from Flower, 1941, J. Paleontol., vol. 15, figs. 1–20, pp. 526–530.*)

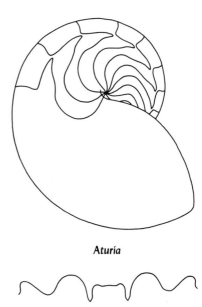

Aturia

Fig. 215   This early Tertiary nautiloid has a sutural pattern that might cause it to be mistaken for an ammonoid, but the siphuncle is dorsal and has backwardly directed septal necks. The septa between chambers are internal, and the sutures do not show on the outside of a shell. These patterns are evident only on exfoliated specimens or the internal molds of shells.

Siphuncular Structure

In recent years the siphuncle has been singled out by most paleontologists as that part of a nautiloid shell of greatest importance in indicating taxonomic and evolutionary relations. Many of the earliest and presumably most primitive nautiloids have relatively very wide siphuncles, some equaling half the diameter of the external shell (Fig. 220-2). They are so large that they are supposed to have been occupied by a considerable part of the animal's viscera, including perhaps the liver and sex glands. All other cephalopods are believed to have evolved from nautiloids with these large ventrally located siphuncles. Few if any fossils of this type have been found in rocks younger than the Ordovician.

The early nautiloids show a persistent evolutionary tendency for reduction of space within the siphuncle, probably indicating the development of a more compact cephalopod body. This was accomplished in four quite different ways: (1) The wide siphuncle was closed off behind, except for a small central tube, by a series of nested calcareous cones (Fig. 220-2). These show that parts of the siphuncle were periodically abandoned as the animal grew and moved forward in its expanding shell. (2) The siphuncle was more or less filled by the growth of longitudinal, radially converging calcareous plates. (3) Massive calcareous deposits were laid down inward from the ends of the septa, nearly filling the siphuncle except for a narrow central passage (Fig. 220-5). (4) The siphuncle decreased in width to become a relatively slender tube.

Walls of the siphuncle are formed partly by backwardly directed necks of shelly matter continuous with the septa, and partly by rings, probably originally of horny membrane, containing abundant calcareous spicules, which connect the necks. Septal necks are not developed in a few of the early nautiloids (Fig. 216-1), but later ones show complete gradations between those that are very short (Fig. 216-2) and others long enough to reach to or beyond the next posterior septum (Fig. 216-3). Gradations also connect very short necks with those that are strongly recurved and appear hook-like in cross section (Fig. 216-4). These series undoubtedly are evolutionary, but parallelism seems to have been common, and the series do not distinguish well-marked lineages. Reversion may not have been exceptional.

The connecting rings are preserved in many specimens. They range from thin to thick, are variously shaped, and exhibit some differences in structure that cannot be detailed here. Evolutionary trends among them probably exist but have not been clearly recognized.

Curvature and Coiling

During an earlier period of paleontologic study, nautiloid shells were classified primarily on the basis of their shapes. Thus they were separated into groups that were straight, or orthoconic; curved, or cyrtoconic; loosely coiled, or gyroconic; and closely coiled, or nautiliconic. The belief was natural that this was an evolutionary sequence. These relations, however, are not so simple as they seemed. Internal structures demonstrate that similar shape modifications evolved in numerous different lineages.

The shapes of most cephalopod shells seem to be directly related to stability requirements. The gas chambers provided buoyancy, and depending on other factors, changes in shell shape commonly were required to permit the animal inhabiting a shell to maintain itself in a position suitable to its way of life.

The earliest fossil cephalopods include shells that are somewhat curved, with the concavity facing downward, a condition termed endogastric (Fig. 212-1 to 3). These are the most primitive of all nautiloids. Their form probably reflects the same influences that resulted in the downward U-shaped bending of the cephalopod gut,

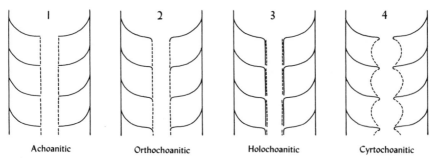

| Achoanitic | Orthochoanitic | Holochoanitic | Cyrtochoanitic |

Fig. 216   1, Most primitively, the septa of nautiloids do not turn backward to produce necks that form part of the tubular siphuncle. 2, 3, In more advanced shells, necks commonly develop and lengthen. 4, Another type of siphuncle has necks that hook out backward, partly enclosing an annulated siphuncle.

but the shell lagged behind the body in its responsiveness. The siphuncles of these shells are large, and the surrounding chambers, which may not have contained gas at this early evolutionary stage, are small and short. At any rate, buoyancy of the posterior part of these shells does not seem to have been particularly effective; the animals probably were bottom-dwelling creatures.

The ability of cephalopods to dart backward rapidly by forcefully expelling water from the mantle cavity surely contributed to the successful existence of these animals. The development of this kind of jet propulsion may have been an important influence in the evolution of straight orthoconic shells (Fig. 212-4,10) from the primitive ones with endogastric curvature. A straight shell directed horizontally would be more efficiently activated in this way than a curved one with the aperture facing somewhat downward. These first orthoconic shells were similar to the curved ones in having large siphuncles and small chambers that probably did not buoy up the hind end.

The detailed workings of the hydrostatic system possessed by fossil cephalopods and *Nautilus* are not well understood. Probably varying the pressure in its gas chambers enabled an animal to modify its density so that it was able to rise toward the surface or sink to the sea bottom as it desired. Wider spacing of the septa and enlarging of the gas chambers perhaps was necessary for such a function. If so, this probably had its evolutionary origin in some of the early orthoconic nautiloids whose internal structures were different from those of the older type.

Buoyant gas chambers posteriorly located in a straight shell, if not compensated in some way, would cause this end to rise. Reorientation of a cephalopod to an inclined or vertical position seems generally to have been disadvantageous, and evolutionary developments counteracted such a tendency in three ways: (1) The shell began to curve exogastrically, or with its concavity facing upward (Fig. 212-5). In such a shell the buoyant posterior was raised but the living chamber and the animal within it remained approximately horizontal. (2) The posterior part of the shell was weighted, or ballasted, with internal calcareous deposits, so that an animal with a straight shell was able to maintain its horizontal position (Fig. 220). (3) The posterior part of the shell, consisting mainly of gas chambers, broke away and was discarded, so that it no longer exerted an uplifting influence (Fig. 214).

The foregoing explanation of evolution progressing from endogastrically curved shells to exogastric cyrtoconic ones and its continuation to coiled forms seems reasonable. No lineage connecting these morphologic types, however, has been surely recognized. Furthermore some uncertainty in such a sequence is provided by shells that start with a tightly coiled spiral and then suddenly become straight (Fig. 218). These have been interpreted as examples of reversion to a much earlier morphologic state, with the initial coil recapitulating a more recent ancestral form. If this is so, not all orthoconic shells represent a single evolutionary grade. A contrary view holds that initially coiled shells connect orthoconic forms and nautiliconic coils in a direct evolutionary sequence. If this idea is correct, cyrtoconic and gyroconoic forms are not necessary intermediates. The explanation offered is that new characters, in this case tight coiling, first appeared at an early ontogenetic stage and persisted longer and longer in descendant populations until they became characters of adults.

Perhaps both these processes actually operated in different lineages of nautiloids. It is obvious, of course, that phylogenetic reconstructions depend on which view is accepted with reference to a particular group of fossils. If the relations are misinterpreted, the supposed phylogeny is sure to be exactly the reverse of the true evolutionary sequence. Several examples of progressive uncoiling are believed to have been established with considerable assurance. These begin with shells ranging all the way from cyrtoconic forms to tightly embracing nautiliconic coils. The reverse process does not seem to have been established with comparable certainty.

Ballasting
Advantages gained by cephalopods as the result of evolution of their hydrostatic system might have been nullified if the horizontal position believed to have been

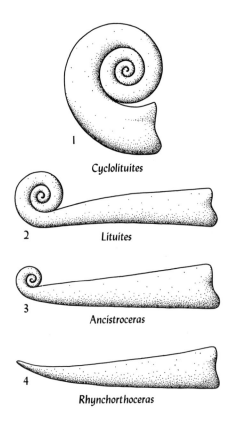

1   *Cyclolituites*

2   *Lituites*

3

Ancistroceras

4

Rhynchorthoceras

Fig. 218   These Middle Ordovician nautiloids have been presented as successive stages in an evolutionary sequence. They supposedly show that a new character, in this case coiling, first appeared at an early ontogenetic stage and then persisted longer and longer in later stages until it became an adult character (4 to 1). Unless these forms are closely related and certainly come from successively younger zones, this interpretation seems questionable because the evolutionary sequence might be exactly opposite to the postulated one and show uncoiling beginning at earlier and earlier stages (1 to 4). (*After Schindewolf, 1950, "Grundfragen der Paläontologie," figs. 164–167, p. 148.*)

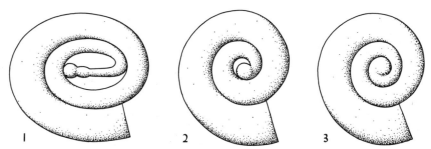

Fig. 219 Very young stages in the ontogeny of three nautiloid cephalopods. (Much enlarged.) 1, The straight initial part has been considered an example of recapitulation indicating that coiled forms evolved from straight ones. 2, The initial part of this coil is not quite closed. An umbilical perforation of this kind is typical of most coiled nautiloids. 3, This coil is without a central perforation and shows the condition typical of most ammonoids. These three specimens may illustrate a general evolutionary trend. (*After Barrande, 1865, "Système silurien du Bohème," etc., vol. 2, pl. 11, figs. 2–4.*)

natural for most of these animals was disturbed too greatly. The problem of excess posterior buoyancy and the consequent tendency for the rear end of the shells, especially straight ones, to rise was solved in several different ways. One of the most successful solutions involved calcareous ballasting within the shell. This remedy evolved along two different but related paths: (1) Calcareous deposits were laid down within the siphuncle (Fig. 220-1,2). Such deposits are of several types, including the nested cones (Fig. 220-2) and radially converging plates already mentioned. Another conspicuous kind consisted of rings growing inward from the septal necks until they nearly filled the siphuncle (Fig. 220-4,5). (2) Similar deposits laid down within the chambers that surround the siphuncle (Fig. 220-3). These demonstrate that chambers, although abandoned successively by the animal's body as it grew and moved forward in its shell, were not entirely empty. They must have retained a lining of tissue capable of secreting calcium carbonate.

Deposits of these two types may occur separately or together in the same shells (Fig. 220-5). Ordinarily they are best developed in straight or only gently curved cyrtoconic shells. Their ballasting functions are rather obvious, because they are most prominent posteriorly and decrease forward in thickness and extent (Fig. 220-4,5). In some shells they are developed much more fully ventrally than dorsally.

Ammonoids

Ammonoids as a group are much less diversified in their detailed structures than are the nautiloids. They are distinguished by a slender siphuncle, at first central but soon becoming ventral and located adjacent to the external part of the outer shell (Fig. 221), and septa whose edges are thrown into wave-like folds (Fig. 226). The sutures consist of undulating lines with so-called saddles, which are convex toward the aperture, and backwardly extending concave lobes. The earliest known ammonoids occur in the Lower Devonian strata of Europe. They evolved slowly at first and then more rapidly, but during the remainder of Paleozoic time they generally were subordinate to nautiloids in the cephalopod faunas.

Ammonoid history is marked by three great crises. At the end of the Paleozoic Era and again at the end of the Triassic Period these animals suddenly approached extinction. Both times a comparatively few types survived, but these few subsequently multiplied and evolved rapidly, so that the previous abundance of these molluscs was soon restored. In Triassic time the ammonoids overcame the former dominance of the nautiloids. The third crisis, at the end of the Cretaceous Period, resulted in the complete extinction of these animals.

Although ammonoids surely evolved from nautiloids, no fossils have been found that demonstrate this transition clearly. One group of late Silurian and early

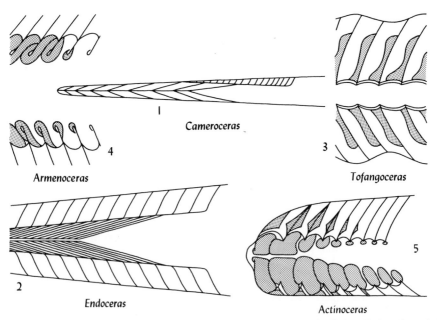

Fig. 220   Examples of ballasting in nautiloid cephalopods shown in longitudinal vertical sections. 1, A large siphuncle with nested cones. This siphuncle is ventral, so that this section does not cut the gas chambers except dorsally. Gas chambers were not formed until the shell diameter became greater than that of the siphuncle. 2, Nested cones in a large central siphuncle. 3, One type of calcareous deposits laid down within the gas chambers. 4, Calcareous deposits enclose the recurved septal necks. 5, Here deposits are present within both siphuncle and chambers. They are best developed posteriorly and ventrally. (*After Ruedemann, 1905, Rept. N.Y. State Geol., 1903, fig. 18, p. 316; Eastman-Zittel, 1913, "Text-book of Paleontology," vol. 1, 2d ed., fig. 1105, p. 595; Kobayashi, 1934, J. Fac. Sci. Univ. Tokyo, sec. 2, vol. 3, pt. 8, pl. 27, fig. 14; pl. 28, fig. 4; 1937, Japan. J. Geol. Geography Trans., vol. 14, fig. 1, p. 6.)*

Fig. 221   Section through a coiled ammonoid. Septa in the first two and one-half whorls have necks turning backward into the siphuncle, as they do in nautiloids. The structure then changes to forwardly directed septal collars, which are typical of ammonoids. This is an example of recapitulation. (*After Miller et al.,* 1957, *"Treatise on Invertebrate Paleontology," pt. L, fig.* 4, *p.* 17.)

Devonian straight cephalopods, the bactritids, is uniquely characterized by slender ventral siphuncles. Some paleontologists suppose that ammonoids evolved from ancestors of this kind by the rapid attainment of a coiled form. On the other hand, some coiled nautiloids of similar age resemble the early ammonoids very closely in details of shape and ornamentation. It is equally likely, if not more probable, that nautiloids of this latter kind produced the first ammonoids by shifting of the siphuncle from a central to a ventral position.

General trends in ammonoid evolution are traced through the progressive developments of their sutural patterns (Fig. 229). These trends, of course, are checked and supplemented by other evidence, such as ontogenetic development, similarity in form of shells and ornamentation, stratigraphic sequence, etc. In some instances reversion from complex to more simple sutures seems to have been established. These instances commonly are accepted as examples of precocious development in which sexual maturity was attained at progressively earlier onto-genetic stages and morphologic development to more advanced stages thereby was cut short.

Detailed studies of abundant specimens have shown that more variability may appear in the sutural patterns of a species or in their development in an individual than is commonly supposed. Within the limitations so imposed, however, practically all paleontologists accept ontogenetic changes in the sutures (Fig. 225) as evidence of ancestry according to the theory of recapitulation. Similar significance cannot be attached in all instances to other features of the shell. For example, some sequences have been interpreted as demonstration that new developments in shape and ornamentation may appear first in early ontogenetic stages and then persist

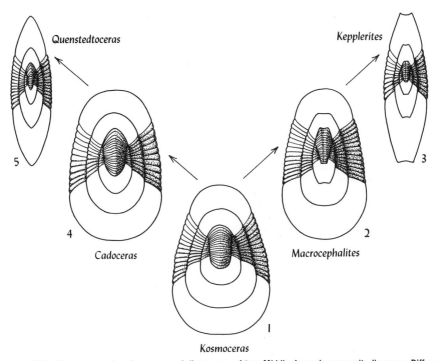

Fig. 222   Diagram showing the presumed divergence of two Middle Jurassic ammonite lineages. Differentiation in whorl cross section first appears in young stages and then continues until it becomes typical of mature shells. (*After Schindewolf, 1950, "Grundfragen der Paläontologie," figs. 216–221, pp. 257, 258.*)

longer and longer in later populations until they become adult characters (Figs. 222; 223).

The last great evolutionary radiation of the ammonoids that began in early Jurassic time produced a considerable variety of bizarre shells, particularly during the Cretaceous Period. These include (1) variously uncoiled forms, (2) straight or coiled shells that abruptly changed growth directions, (3) spired to irregularly wandering coils, and (4) shells with apertures peculiarly restricted or provided with lateral and ventral extensions (Fig. 224). Such fossils have been considered to indicate declining racial vigor, prophetic of extinction soon to come. This, however, cannot be a valid view. The numbers of specimens and their distribution show that some of these species must have been relatively successful for considerable intervals of time. All these developments probably were responses to some kind of environmental stimulus, and they are evidence of rapid and effective evolutionary adaptation. The resulting specializations, however, seem to have unfitted these forms to continued existence when conditions changed.

One small group of late Devonian ammonoids, the clymenids, evolved in a peculiar way that sets them apart from all the others. At an early ontogenetic stage their siphuncles are ventral, as in other ammonoids. The siphuncles then moved to a completely dorsal position. Rather obviously more normal early ammonoids were the ancestors of these fossils. Their sutures did not develop great complexity.

Sutures

The sutural patterns of ammonoids have been studied intensively by many paleontologists for more than a century. It has been known for many years that sutural complexity increases in individual shells from early ontogenetic stages to later ones

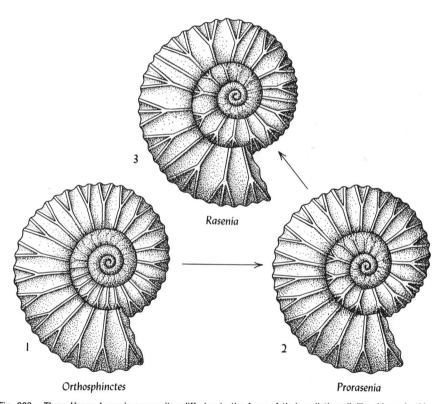

Fig. 223   Three Upper Jurassic ammonites differing in the form of their radiating rib-like ridges. In this supposed evolutionary sequence the change from bifurcating to trifurcating ridges first appears in the early whorls and then persists and becomes a mature character. (*After Schindewolf, 1950, "Grundfragen der Paläontologie," figs. 222–224, p. 259.*)

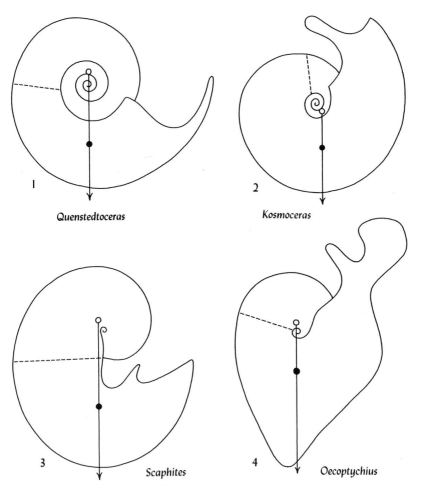

1 *Quenstedtoceras*

2 *Kosmoceras*

3 *Scaphites*

4 *Oecoptychius*

Fig. 224 These ammonite shells with peculiarly formed apertures probably belonged to planktonic animals. The dashed lines show the extent of the living chambers, and the solid and open dots locate the approximate centers of gravity and buoyancy, respectively. The farther these are separated, the more difficult it would have been for the animal to alter its orientation. The stable position of these shells shows that the animals probably faced upward. (*Compare Trueman, 1941, figs. 14, 15, pp. 372, 373; Arkell, 1957, fig. 157, p. 121.*)

(Fig. 225). Also sutural patterns in general became more complicated with the passage of time (Figs. 226; 229). These facts certainly are significant from an evolutionary point of view. The observation that immature sutures in some shells can be matched with the mature sutures of others coming from older stratigraphic zones was largely responsible for acceptance of the recapitulation theory as a valid

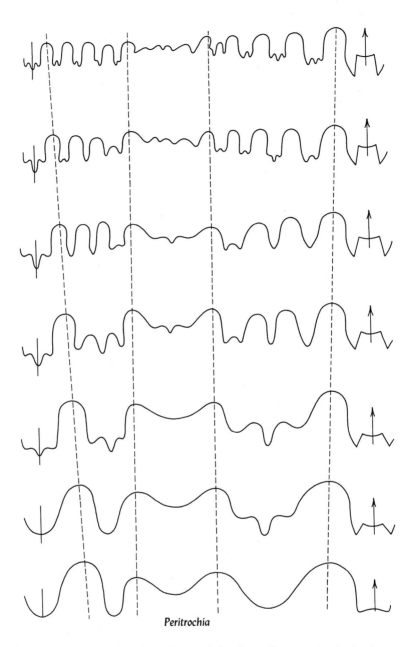

Peritrochia

Fig. 225   An example of the increasing complexity of sutural patterns that develop in the growth of an individual ammonoid of the late Paleozoic. Notice that new saddles develop within the lobes. The illustrated sutures are not successive ones. (*Modified after Schindewolf, 1954, fig. 20., p. 233.*)

evolutionary principle by most paleontologists. Many rather elaborate presumed lineages have been reconstructed on this basis, although in recent years, skepticism regarding some interpretations of this kind has been increasing, especially among students of Mesozoic ammonites.

Three general types of sutures are recognized: (1) The simplest, or goniatitic,

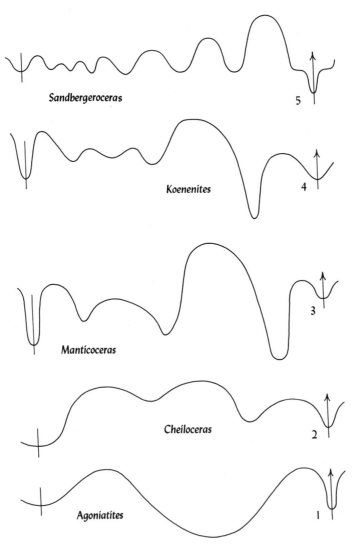

Sandbergeroceras   5

Koenenites   4

Manticoceras   3

Cheiloceras   2

Agoniatites   1

Fig. 226 Goniatitic suture patterns. 1, This suture of a primitive Middle Devonian goniatite has a single lateral lobe. Its pattern is almost the simplest one known. 2 to 5, Sutures of Upper Devonian goniatites showing increase in the number of lateral lobes. *(After Miller et al., 1957, "Treatise on Invertebrate Paleontology," pt. L, figs. 13c, 25c, 27a, 29c, 32f, pp. 29, 32, 33, 34, 35.)*

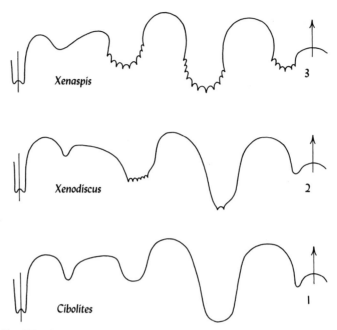

Fig. 227   Suture patterns of three closely related Upper Permian ammonoids. These illustrate transition from a goniatitic to a ceratitic pattern by the development of crenulations in the lobes. (*After Arkell et al.*, 1957, figs. 162-2c, 163c, d, pp. 131, 132.)

type consists of evenly rounded saddles and simple rounded-to-pointed lobes (Fig. 227-1). (2) Next comes the ceratitic type, distinguished by the appearance of minor irregularities or serrations within the lobes (Figs. 227-3; 228-1). (3) This is succeeded by the ammonitic suture, whose saddles as well as lobes are intricately subdivided (Fig. 228-4 to 6). Such a progression may be considered the result of successive cycles of folding of the septal edges. Goniatitic sutures are representative of the first cycle of simple folding. The beginning of the second cycle appears in ceratitic sutures, in which the goniatitic folds are themselves folded. This cycle is completed in the simplest ammonites. The folding process does not cease here, however, because as many as five cycles can be recognized in the more intricate sutures of some Mesozoic ammonites (Fig. 229-3).

A morphologic sequence of more and more complex sutures is certainly evolutionary, but a classification made only on this basis may cut across phyletic lines, because similar results were produced by parallel evolution in different lineages. Reversion also is believed to have occurred in some groups of ammonoids in which species with creatitic or even goniatitic sutures seem to follow typical ammonites. Stratigraphically, goniatitic sutures are particularly characteristic of the Paleozoic,

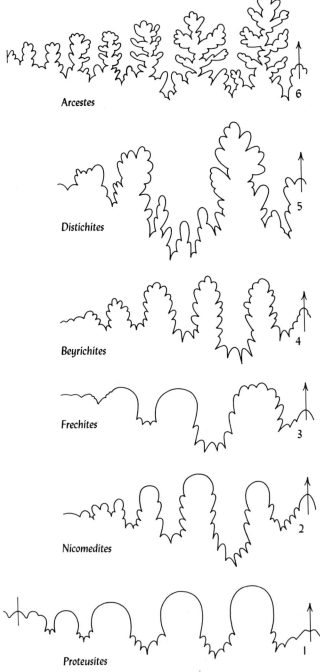

Arcestes 6

Distichites 5

Beyrichites 4

Frechites 3

Nicomedites 2

Proteusites 1

Fig. 228   Suture patterns of six Triassic ammonoids. The fossils belong to several different lineages, but the increasing complexity of patterns illustrates transition from ceratitic to ammonitic sutures. (*After Arkell et al., 1957, figs. 182-3c, 4c, 185-2c, 189-5c, 199-4c, 208-1e, pp. 150, 153, 157.*)

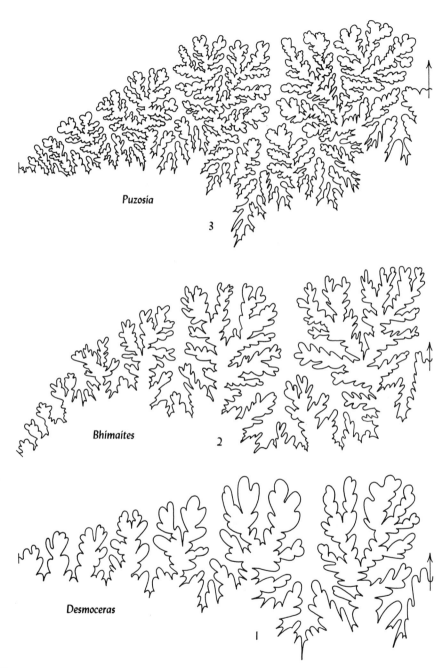

Puzosia

3

Bhimaites

2

Desmoceras

1

Fig. 229 Suture patterns of three related Cretaceous ammonites. These clearly show progressive increase in the complexity of closely similar patterns. (*After Arkell et al.,* 1957, figs. 476-5c, 477 1c, 482-5c, pp. 364, 365, 369.)

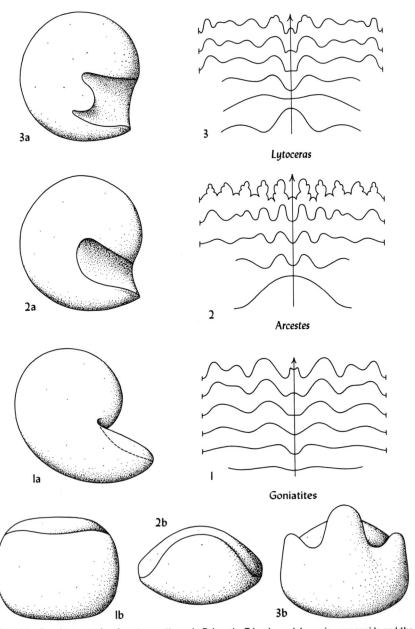

3a

3 *Lytoceras*

2a

2 *Arcestes*

1a

1 Goniatites

2b

1b

3b

Fig. 230 Development of early suture patterns in Paleozoic, Triassic, and Jurassic ammonoids and the initial shell chambers of comparable forms showing the surface and suture of the first septum. 1, Paleozoic, first, second, eighth, and sixteenth sutures and those occurring at shell diameters of 4 and 9 mm. 2, Triassic, first and fifth sutures, two present in the second whorl and one occurring at a shell diameter of 10 mm. 3, Jurassic, first, second, fourth, and tenth sutures and those occurring at shell diameters of 2 and 4 mm. (*After Branco, 1879, 1880, Palaeontographica, vol. 26, pl. 5, fig. 2; pl. 6, figs. 2a, 5a; pl. 8, fig. 4a; pl. 10, figs. 4g–o; vol. 27, pl. 6, figs. 1c, g–n, 3b.*)

although they persisted well into the Triassic and perhaps beyond. The first ceratitic sutures are early Mississippian, but they are especially abundant in the Triassic. Well-developed ammonites make their appearance in the Middle Permian and dominate the Jurassic and Cretaceous cephalopod faunas.

The first suture of an ammonoid shell differs from all others (Fig. 230) and does not require detailed consideration here. The ontogenetic sequence of gradual development begins with the second suture. Most Lower Devonian goniatites possess four lobes—dorsal, ventral, and a pair of laterals—and some show no development of a more complicated pattern (Fig. 226-1). These sutures do not differ fundamentally from the gently undulating sutures of some nautiloids, and distinction is based on the location of the siphuncle. The septa of these early ammonoids are less conspicuously folded than those of some Tertiary nautiloids (Fig. 215). Most other Paleozoic ammonoids in their very early stages have sutures with another pair of lobes in addition to those already mentioned (Fig. 226-2). Located between the dorsal and lateral lobes, these are known as umbilical lobes. Further development of the suture in more advanced ammonoids resulted in the proliferation of lobes in lateral positions and then in the introduction and compounding of additional complications.

Ontogenetic studies of Paleozoic goniatitic sutures indicate that increase in the number of lateral lobes was accomplished in two ways, whose end results are so similar that they cannot be distinguished with certainty (Fig. 232): (1) Commonly the saddles adjacent to the ventral lobe were subdivided by the appearance of new lobes. Ammonoids of this type evolved to the ammonitic stage, but all became extinct before the end of the Paleozoic Era. (2) In a smaller group that seems to have begun its development in late Devonian time, the saddles adjacent to the dorsal lobe were subdivided in a comparable way. Only a few ammonites of this latter type survived into Triassic time; they provided the base for the evolutionary radiation from which all Mesozoic ammonoids were derived.

The second suture of Triassic ceratites has a third pair of laterally located sutural lobes (Fig. 230-2). This condition is interpreted as an example of accelerated evolution in which an early developmental stage has been eliminated from the ontogenetic sequence. Jurassic and Cretaceous ammonites begin with four pairs of lobes of this kind and seem to record a continuation of this process (Fig. 230-3). All these Middle and Upper Mesozoic cephalopods are believed to have evolved from a single Triassic group whose sutural pattern was relatively simple and conservative.

Coleoids

The coleoids are cephalopods without external shells. The earliest authentic specimens have been collected from Mississippian strata, although a possible Devonian example has been reported. These and later fossils are evidence of a persistent evolutionary trend that began by the expansion of the mantle which enclosed an orthoconic shell. Perhaps enlargement of the mantle which had this eventual result originated in a group of especially active nautiloids. If so, they achieved advantage

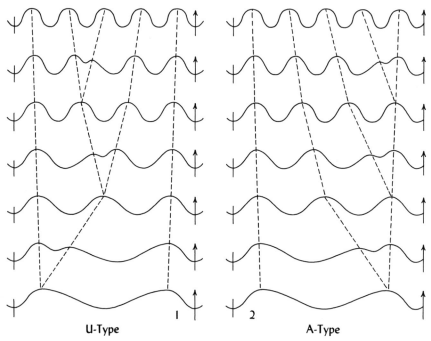

U-Type   A-Type

Fig. 232   Diagrams showing two ways in which lateral lobes of ammonoids increase by subdivision of the saddles to produce identical sutural patterns. These ways differ depending on whether the new lobes originate by division of an internal saddle (in which case they are termed umbilical lobes) or of an external saddle (in this case they are termed adventitious lobes). The umbilical lobes appear alternately first on one side of their area and then on the other. The adventitious lobes appear regularly on the ventral side. The U-type pattern did not survive the close of the Paleozoic Era. (*Data from Schindewolf, 1954, pp. 223–227.*)

in the development of lateral fins, or planes, that grew backward along the body and served to guide their darting movements through the water. After the shell ceased to be an outer protective covering, it was modified to produce a supporting and stabilizing structure, which subsequently degenerated (Fig. 233) until it disappeared entirely, as in the modern octopus.

It cannot be known when envelopment of an external shell began. The existence of an outer tissue covering is demonstrated, however, by the presence of calcareous deposits on the outside of the shell. In the belemnoids, which are the earliest coleoids, such a deposit was built backward in a solid structure termed the guard. The outer wall of the living chamber no longer served a protective function, and it degenerated except for the middorsal part, which projected forward and is known as the proostracum. The siphuncle of belemnoids, located ventrally, suggests that the ancestors of these creatures had one similarly positioned. The only fossils which meet this requirement are the straight bactritid nautiloids.

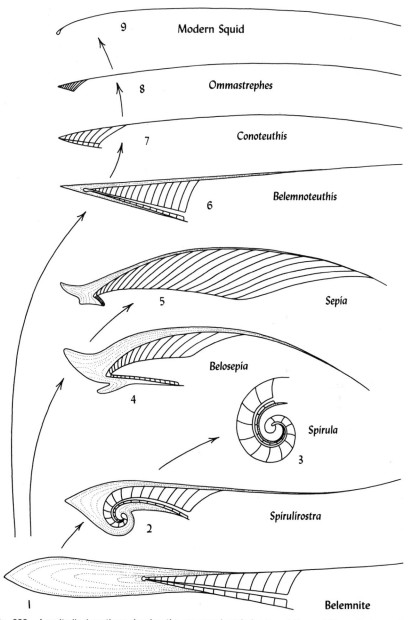

Fig. 233   Longitudinal sections showing the presumed evolutionary relations of the main types of coleoid cephalopods that have retained some vestige of a shell. (*After Bather, 1911, "Guide to Fossil Invertebrates in British Museum," fig. 85, p. 155.*)

Possibly the shells of some bactritids were more or less enclosed within an expanded mantle, but, lacking outer calcareous deposits, they cannot be recognized as being different from other nautiloids in this respect. The possibility also deserves consideration that part or all of those characters which differentiate modern two-gilled cephalopods from living *Nautilus* had already evolved in a branch of the nautiloids before the appearance of the first coleoids. These characters include (1) change in the number of gills from four to two or from two to four, depending on which was the more primitive condition; (2) perfection of the arms whose suckers may be homologous with the many tentacles of *Nautilus;* (3) development of an ink sac; and (4) appearance of lensed eyes. Most of these features of modern coleoids seem to be definite evolutionary advancements as compared with *Nautilus.*

The belemnoids are the only coleoids that are well known in the fossil state. Their cigar-shaped guards (Fig. 233-1) are abundant, especially in some Jurassic and Cretaceous sediments. Their last representatives are of Eocene age. Much earlier, however, perhaps before the beginning of the Mesozoic Era, divergent evolution led to the forerunners of the sepias, squids, octopuses, and a few other of their less-familiar allies. Some fossils of these kinds have been found in Jurassic and younger strata, but the record is too scanty to tell much about their history. In each lineage, however, it is evident that the septate shell and guard degenerated. The sepias have the remnant of a guard that is much reduced, and only the dorsal part of the shell remains as a porous calcareous structure. The squids retain only what corresponds to the proostracum. Octopuses have lost all traces of both shell and guard.

Scaphopods

The shells of scaphopods are gradually expanding, slightly curved tubes (Fig. 234) found in strata as old as the Ordovician. Fossils are nowhere abundant, and

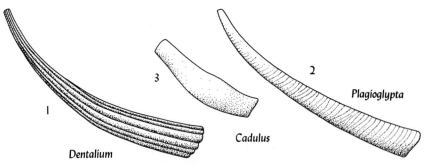

Fig. 234   Scaphopod shells. Fossils present little in the way of diagnostic features other than surface ornamentation and shape of opening at smaller end. (*After Ludbrook, 1960, fig. 30-4.*)

pre-Cretaceous ones are relatively rare. These shells tell so little about the animals which made them that they probably would not be recognized as molluscs if representatives did not occur in modern faunas. Nothing is known about their evolution, but the similarity of specimens of all ages suggests that they have been a remarkably stable and conservative group since at least mid-Paleozoic time.

BIBLIOGRAPHY

W. J. Arkell et al. (1957): Mollusca 4 (Ammonites), in "Treatise on Invertebrate Paleontology," R. C. Moore (ed.), pt. L, Geological Society of America, Boulder, Colo., and University of Kansas Press, Lawrence, Kans.
See pp. 20–22 and 102–118 for general observations concerning ontogeny and evolution.

L. R. Cox (1960): Thoughts on the Classification of the Bivalvia, *Proc. Malacol. Soc. London,* vol. 34, pp. 60–88.
This article attempts to trace evolution among pelecypods and shows how greatly conclusions have varied among previous taxonomists.

W. H. Dall (1895): Tertiary Mollusks of Florida, pt. 3: A new Classification of the Pelecypoda, etc., *Trans. Wagner Free Inst. Sci.,* vol. 3, pp. 485–560.
See pp. 485–504 for an account of the origin, function, and development of pelecypod hinges and subsidiary structures.

W. K. Emerson (1962): A Classification of the Scaphopod Molluscs, *J. Paleontol.,* vol. 36, pp. 461–482.
Important features for classification and presumably for phylogenetic reconstruction are shape and ornamentation of shells and form of posterior apertures.

R. H. Flower (1946): Ordovician Cephalopods of the Cincinnati Region, pt. 1, *Bull. Am. Paleontol.,* vol. 29, pp. 83–738.
The importance of siphuncular structure and its bearing on evolutionary relations are discussed, see pp. 68–90.

R. T. Jackson (1890): Phylogeny of the Pelecypods: The Aviculidae and Their Allies, *Mem. Boston Soc. Nat. Hist.,* vol. 4, pp. 277–400.
This work is out of date, but see pp. 375–400 for an attempt to trace evolution by morphology as related to growth and function.

J. Johansson (1952): On the Phylogeny of the Molluscs, *Zool. Bidrag Från Uppsala,* vol. 29, pp. 277–292.
The author is concerned with the functional relations of a segmented coelom and methods of locomotion, and concludes that molluscs evolved from segmented ancestors adapted to life on a hard bottom.

J. B. Knight (1952): Primitive Gastropods and Their Bearing on Gastropod Classification, *Smithsonian Inst. Misc. Collections,* vol. 117, no. 13.
This paper traced the theoretical origin and early development of gastropods from monoplacophorans before modern examples had been discovered.

———— **et al. (1960):** Mollusca 1 (Paleozoic Gastropods, etc.), in "Treatise on Invertebrate Paleontology," R. C. Moore (ed.), pt. I, Geological Society of America, Boulder, Colo., and University of Kansas Press, Lawrence, Kans.
This volume contains good introductory accounts of molluscs in general and gastropods in particular; see pp. 141–146 for a consideration of gastropod evolution.

———— and **E. L. Yochelson (1960):** Monoplacophora, in "Treatise on Invertebrate Paleontology," R. C. Moore (ed.), pt. I, Mollusca 1, pp. 77–84, Geological Society of America, Boulder, Colo., and University of Kansas Press, Lawrence, Kans.
These animals are believed to exemplify the earliest and most primitive group of molluscs.

N. H. Ludbrook (1960): Scaphopoda, in "Treatise on Invertebrate Paleontology," R. C. Moore (ed.), pt. I, Mollusca 1, pp. 37–40, Geological Society of America, Boulder, Colo., and University of Kansas Press, Lawrence, Kans.
Fossil scaphopods possess few characters indicating evolutionary relations.

A. K. Miller (1949): The Last Surge of the Nautiloid Cephalopods, *Evolution,* vol. 3, pp. 231–238.
Nautiloids evolved rapidly after extinction of the ammonoids at the end of the Cretaceous but declined to only two genera after the Eocene.

J. E. Morton (1958): "Molluscs," Hutchinson & Co. (Publishers), Ltd., London; paperback reprint **(1960),** Harper Torchbooks, no. 529, Harper & Row, Publishers, Incorporated, New York.
This is an excellent semipopular book, mostly zoologic but with an evolutionary accent.
_____ **(1958):** Torsion and the Adult Snail: A Re-evaluation, *Proc. Malacol. Soc. London,* vol. 33, pp. 2–10.
Form is related to function in several evolutionary grades of gastropods.
N. D. Newell (1938): "Late Paleozoic Pelecypods: Pectinacea," University of Kansas Publication, Geological Survey, vol. 10, pt. 1.
See pp. 26–35 for consideration of pelecypod ligaments and their possible evolution.
W. G. Ridewood (1903): On the Structure of the Gills of the Lamellibranchia, *Phil. Trans. Roy. Soc. London,* vol. 195, pp. 147–284.
Evolution progressed from leaf-like filaments to lamellae joined by cilia and lastly to the development of organic cellular connections.
O. H. Schindewolf (1954): On Development, Evolution, and Terminology of the Ammonoid Suture Line, *Bull. Museum Comp. Zool.,* vol. 112, pp. 217–237.
The evolution of sutures is traced, and similar patterns are shown to have developed in different ways in different lineages.
A. G. Smith (1960): Amphineura, in "Treatise on Invertebrate Paleontology," R. C. Moore (ed.), pt. I, Molluscs 1, pp. 41–76, Geological Society of America, Boulder, Colo., and University of Kansas Press, Lawrence, Kans.
Fossil chitons provide very little evidence concerning their evolution.
B. Smith (1945–1946): Observations on Gastropod Protoconchs, *Palaeonto. Am.,* vol. 3, pp. 221–268, 285–302.
Opinions are reviewed regarding the significance of protoconchs in determining the relations of gastropods; see pp. 229–242.
L. F. Spath (1936): Phylogeny of the Cephalopoda, *Palaeontol. Zeit.,* vol. 18, pp. 156–181.
The author contends that adult similarities are more convincing evidence of close relationship than early ontogenetic similarities.
H. B. Stenzel (1949): Successional Speciation in Paleontology: The Case of the Oysters of the *Sellaeformis* Stock, *Evolution,* vol. 3, pp. 34–50.
Four species of Eocene pelecypods are related as stages in the evolution of a single lineage.
C. Teichert (1967): Major Features of Cephalopod Evolution, pp. 162–210 in "R. C. Moore Commemorative Volume," C. Teichert and E. L. Yochelson (eds.), University of Kansas Press, Lawrence, Kans.
Presumed evolutionary relations are traced among families and genera without much morphologic explanation.
_____ **et al. (1964):** Molluscs 3 (Nautiloid Cephalopods), in "Treatise on Invertebrate Paleontology," R. C. Moore (ed.), pt. K, Geologic Society of America, Boulder, Colo., and University of Kansas Press, Lawrence, Kans.
See particularly pp. 106–113 for interpretations of evolutionary relations.
J. Vagvolgyi (1967): On the Origin of Molluscs, the Coelom and Coelomic Segmentation, *Syst. Zool.,* vol. 16, pp. 153–168.
Molluscs and annelids evolved from a common flatworm-like ancestor.
G. E. G. Westerman (1958): The Significance of Septa and Sutures in Jurassaic Ammonite Systematics, *Geol. Mag.,* vol. 95, pp. 441–455.
Analogous sutures may be produced by different types of septal folding.
C. M. Yonge (1939): The Protobranchiate Molluscs: A Functional Interpretation of Their Structure and Evolution, *Phil. Trans. Roy. Soc. London, Ser. B,* vol. 230, pp. 79–147.
See pp. 132–139 for consideration of the evolution and phylogenetic relations of the most primitive pelecypods.
_____ **(1947):** The Pallial Organs in the Aspidobranch Gastropoda and Their Evolution throughout the Mollusca, *Phil. Trans. Roy. Soc. London, Ser. B,* vol. 232, pp. 443–518.

See pp. 480–507 for discussion of the evolution of asymmetry in gastropods and of the gills in other molluscs.

_____ **(1953):** The Monomyarian Condition in the Lamellibranchia, *Trans. Roy. Soc. Edinburgh*, vol. 62, pp. 443–478.

This article traces the evolutionary reduction of the anterior adductor muscle in pelecypods and its relation to life habits and byssal functions.

CHAPTER SIX

SCHIZOCOELATE ANIMALS 2

ANNELIDS
ARTHROPODS

The annelids and arthropods are distinctly segmented and constitute two great phyla which, more than any others, exhibit such obvious anatomical similarities, in this and other ways, that they are believed to be very closely related to each other. The most conspicuous difference is that the arthropods are equipped with jointed appendages activated by systems of special muscles in place of the much more simple structures possessed by many annelids. In addition, the arthropods have a stiffened horny external skeleton, lack cilia, have muscles which differ in their detailed structure, and their main body cavities are parts of the blood circulatory system rather than coelomic. Surely the arthropods are evolutionarily more advanced.

ANNELIDS
Annelid worms are an important element of many modern marine faunas (Fig. 240). Locally they live in such profusion that individuals outnumber all other nonmicroscopic creatures, and probably they were equally abundant in many faunas of the past. Unfortunately these animals generally did not produce hard parts suitable for fossilization. Although the occurrence of worm-like creatures is indicated in strata as old as the late Precambrian, the fossil record furnishes no information concerning their evolution.

Except for rare impressions, positive evidence for the existence of annelids is provided by jaw parts, known as scolecodonts, and calcareous tubes. The tiny jaws

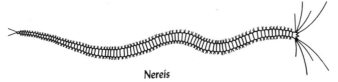

Nereís

Fig. 240   A modern marine annelid worm. Creatures of this kind are soft-bodied, and the chances that they may be preserved as fossils are exceedingly slight. (*After Parker and Haswell, 1940, "A Text-book of Zoology," 6th ed., vol. 1, fig. 277A, p. 310.*)

consist of horny material, but only those which are partly composed of silica or calcium phosphate are likely to be preserved. Fossils of this kind have been discovered most commonly in strata ranging from the Ordovician to the Devonian. Calcareous tubes of variable abundance and different forms are known from the Ordovician onward. Many of them grew attached to seaweeds and to shells and other hard objects. They are relatively numerous in some late Paleozoic and younger limestones. Closely coiled worm tubes are easily mistaken for tiny snails.

Worm burrows and trails are present in rocks of many kinds, but they are commonly overlooked. Most of the burrows are either straight and vertical or U-shaped with two openings at the surface. The nature of the animals which inhabited them cannot be known; some of them probably were made by worm-like animals other than annelids. Identification of the kinds of animals which made the trails (Fig. 614) is even more uncertain.

The larvae of most marine annelids are typical trochophores. The main part of each of these tiny creatures is destined to become the head of a developing worm, and the posterior part elongates to become its body. As the posterior part lengthens, openings appear in the mesoderm and exterior constrictions begin to form (Fig.

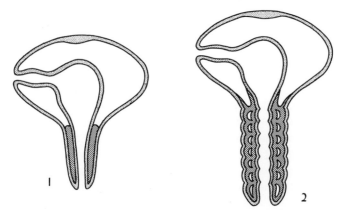

Fig. 240a   Generalized longitudinal sections through annelid trochophore larvae showing beginning of mesoderm formation (shaded), appearance of coelomic cavities, and origin of metameric segmentation. (*Modified after Parker and Haswell, 1940, "A Text-book of Zoology," 6th ed., vol. 1., figs. 335B, 336B, p. 373.*)

240a-2). The parts distinguished in this way become the segments of the worm. New segments originate successively at the posterior end, and growth continues until the full number of segments has been attained. This method of development resembles in a general way the budding of coelenterates and bryozoans, but the new parts, instead of becoming complete individuals, remain in an immature condition and in intimate association. A somewhat similar but more primitive kind of segmentation is characteristic of several types of flatworms, and a few of these actually multiply by budding. The annelids probably evolved from creatures of this general kind. They differ from all the flatworms, however, by possessing coelomic cavities and a more distinct head where the mouth is located, and many of them have bristles or primitive leg-like structures borne by each of the segments behind the head.

ARTHROPODS

The arthropods have left a long, complex geologic record. The base of the Cambrian System commonly is identified by the first appearance of trilobites; fossil arthropods believed by some paleontologists to have been terrestrial are reported to occur in strata of Silurian age; and arthropods are by far the most abundant of all modern multicellular animals. Most arthropods are active creatures, and they have diversified in a multitude of ways, adapting to a great variety of environments and modes of life. Nevertheless, they appear to be a well-unified group, identified by their segmented structure, horny exoskeleton, jointed legs, and several features of their internal organization.

Table 241   Important distinguishing characters of the main groups of arthropods

Trilobitoids: early marine arthropods that cannot be classified with any of the modern or better-known fossil groups. Some share characters of trilobites, chelicerates, and crustaceans. Cambrian

Trilobites: marine arthropods with a head shield probably consisting of six fused and specialized segments, a thorax of two to many articulated segments, and a tail shield whose fused segments vary greatly in number. Antennae, one pair; other appendages are legs with two branches generally similar on all following segments. Cambrian to Permian

Chelicerates: aquatic and terrestrial arthropods of many forms, with a cephalothoracic shield probably consisting of eight fused segments. More posterior segments may be articulated or fused. Without antennae. First pair of appendages equipped with pinching claws, next pair differentiated from the four following pairs of legs, all of which may or may not be clawed. None of the appendages has a second branch. Cambrian to Recent

Crustaceans: mostly aquatic arthropods of extremely varied form with two pairs of antennae. Carapace of head and thorax may be fused to produce a cephalothoracic shield. Appendages behind antennae fundamentally two-branched but functionally modified in many ways. Cambrian to Recent

Myriapods: centipedes, terrestrial arthropods with a head, probably consisting of six fused segments, bearing one pair of antennae and mouthparts. Remainder of body composed of many similar articulated segments, each bearing a pair of jointed legs. Devonian to Recent

Insects: terrestrial arthropods, mostly with two pairs of wings on thorax. Head probably formed of six fused segments with one pair of antennae and mouthparts. Thorax consists of three segments, each with a pair of legs. Abdomen without appendages. Devonian to Recent

These groups seem to belong to two divisions, the myriapods and insects probably not being closely related to the others.

Basic similarities of annelids and arthropods have been recognized for more than 150 years. These include (1) metameric segmentation with serially repeated appendages, (2) paired ventral nerve trunks, and (3) a dorsal heart-like blood vessel in which the flow is forward. Many of the differences separating the arthropods from annelids seem to be almost wholly specializations that contributed particularly to the efficiency of muscular action. These have conferred enormous advantages upon the arthropods, and more than any other phylum these animals have exploited almost every situation available to living creatures.

In spite of essentially universal agreement that arthropods probably evolved from polychaete-like annelids, attempts at reconstruction of phylogeny within the phylum have varied greatly. The onychophores have played an important part in much of this speculation. These caterpillar-like animals (Fig. 271-1) resemble the annelids in certain ways but also exhibit more features allying them with the arthropods. They have variously been placed at the base of all arthropod evolution or viewed as the prototype from which all arthropods that breathe by means of delicate internal tubes, or tracheae, have been derived. When detailed comparisons are made, however, they seem to find their most natural position as the representatives of a group from which the myriapods and insects evolved.

One of the greatest obstacles to the assessment of evolutionary relations among the arthropods has been the reluctance of some biologists to accept the possibility that parallelism or convergence may have been important. The features relied upon most commonly to demonstrate close relationships are (1) the compound arthropod eye, and (2) the tube-like tracheae, which are possessed by such divergent groups as the insects and spiders. All schemes based on such premises, however, are subject to objections which have increased in seriousness as the anatomy and ontogeny of the several living arthropod groups have become better known. A reappraisal of this problem suggests that almost certainly both eyes and tracheae evolved independently at least twice, and that two great groups, each exhibiting reasonable evolutionary continuity, are indicated (Fig. 243).

Division of the arthropods in this way carries with it the implication that the two groups may have evolved from different annelid stocks. The suggestion even has been made that the ancestors of one or both may not have been annelids, although this does not seem likely. In any case the acquisition of a stiffened exoskeleton would have imposed similar restrictions on subsequent evolution. However these groups originated, the development of movable segments and jointed legs was a necessity for active creatures, as also were better cephalization and more efficient vision. Finally, a respiratory system consisting of tracheal tubes seems to be the simplest and most effective adaptation of small, originally aquatic animals with impervious integuments to a terrestrial air-breathing existence.

In spite of their long geologic history, great diversity, and abundance in modern faunas, most of the main groups of arthropods are relatively poorly represented among the fossils. Only the trilobites and the ostracod crustaceans are abundant at many places. Most of the other groups consist of creatures too small or too fragile

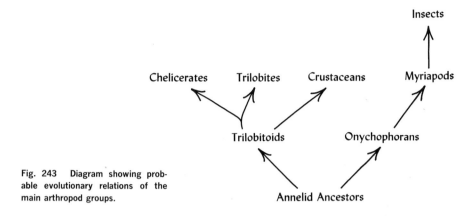

Fig. 243   Diagram showing probable evolutionary relations of the
main arthropod groups.

to be preserved commonly, or they inhabited environments where preservation was unlikely. Trilobite evolution has attracted the attention of paleontologists, but, with a few exceptions, problems of evolution between and within the other groups have been considered almost exclusively by zoologists.

Trilobitoids

The abundant occurrence of a varied lot of trilobites beginning in the Lower Cambrian has led some evolutionists to postulate that all arthropods descended from creatures of this kind, but such a theory no longer has much following. Some light seems to be thrown upon the problem of arthropod evolution by a number of peculiar and almost unique fossils collected from the Middle Cambrian of British Columbia (Fig. 244). They include species that resemble in several different ways, and variously combine, certain characters of trilobites, chelicerates, and crustaceans. The relations of these fossils among themselves and to the better-known groups just mentioned are uncertain. They seem to furnish evidence, however, of considerable evolutionary radiation from a single group ancestral to them all. For want of better understanding, these fossils and other comparable creatures are known collectively as trilobitoids. The discovered fossils could not, of course, have been the actual ancestors of the more familiar groups of aquatic arthropods, although these probably evolved from animals similar to some of them.

The trilobitoids all have several anterior segments solidly fused together to produce a cephalon, or head, or a cephalothoracic shield. The number of these segments is uncertain, but several quite different species seem to have had four pairs of legs originating in this region, as in the trilobites. Most of them appear to have had antennae, some one pair as in the trilobites and others two as in the crustaceans. As far as is known, all the appendages except antennae consisted of two branches, one of which was leg-like. They resembled each other and were similar to those of trilobites. Some species had legs on all the articulated segments, but

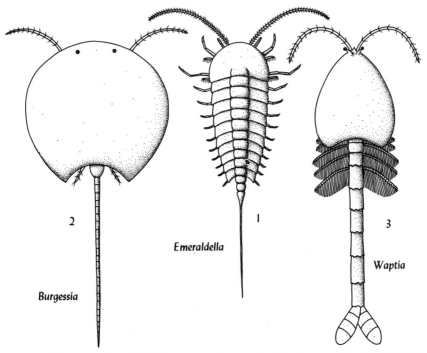

2   Burgessia

1   Emeraldella

3   Waptia

Fig. 244   Three examples of Middle Cambrian arthropods from the Burgess Shale. These peculiar trilobitoids cannot be assigned to any well-known group of arthropods. They are evidence of evolutionary radiation that produced stocks intermediate between trilobites, crustaceans, and chelicerates. 1, This creature resembles a trilobite in its longitudinal trilobation, a crustacean in its two pairs of antennae, and a chelicerate in its long posterior spine. 2, 3, These fossils resemble branchiopod crustaceans in their general appearance, but their legs seem to be of the trilobite type. (*After Størmer, 1959, "Treatise on Invertebrate Paleontology," pt. O, figs. 17, 21-1, 6, pp. 29, 33.*)

in others these structures seem to have been missing from the posterior part of the body.

Trilobites

Trilobites are intriguing fossils because they so obviously resemble, and at the same time so conspicuously differ from, modern arthropods. They have been studied intensively by paleontologists because (1) well-preserved specimens are abundant, (2) many features of their relatively complex skeletons are easily compared, (3) a great variety of forms occurs, and (4) many of them are excellent index fossils, useful for stratigraphic correlation. Although trilobites range through the entire Paleozoic section, their diversity was greatest during the Cambrian and Ordovician periods, after which their importance in the faunas steadily declined.

In spite of great variability in all their details, the trilobites were a conservative group as far as their general structure is concerned. The variable feature that has attracted most attention is the facial suture where the cephalon, or head shield, split when these creatures molted (Fig. 245). Each half of this bilateral suture generally begins somewhere along the anterior margin of the cephalon and extends backward around the eye. In the commonest type, which is termed opisthoparian, it continues to the posterior margin inside the corner of the cephalon that ordinarily is produced as a backwardly directed genal spine. In another type, known as proparian, the suture extends outward from behind the eye and meets the lateral margin in front of the corner of the cephalon. A third type, which is theoretical because no suture opened on the upper surface at the time of molting, is suggested by surface markings. A groove possibly indicating a vestigial suture both begins and ends at the posterior margin of the cephalon; it is referred to as metaparian. Finally, some trilobites have nothing suggestive of a facial suture. The name applied to them is hypoparian, and most of them were blind (see Table 247).

Paleontologists have not agreed regarding the evolutionary significance of the trilobite facial suture, and certainly other features need to be considered in assessing trilobite relations. Parallel and convergent evolution surely have occurred (Fig. 246), and deciding what features are most important is not easy in many cases. Most trilobites can be arranged in family or superfamily groups consisting of rather obviously closely related forms, but the interrelations of such groups generally are questionable. A particularly puzzling problem is presented by numerous Ordovician forms which seem to have no possible ancestors in any of the known Cambrian faunas.

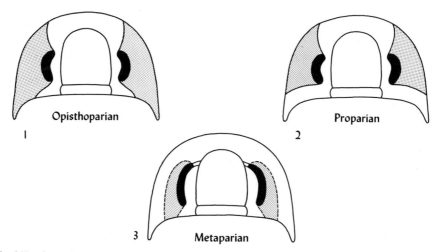

Fig. 245   Generalized diagrams showing nature of the free cheeks (shaded) in three groups of trilobites characterized by differences in the course of their facial sutures.

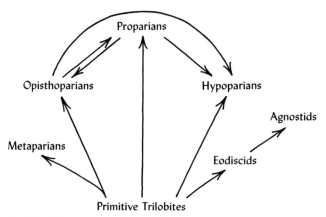

Fig. 246   Diagram showing the possible evolutionary relations of the major groups of trilobites as defined mainly by their facial sutures. Several of these groups seem to include representatives whose similarities are the result of convergent evolution.

Evolutionary Trends

Although the tracing of evolutionary lineages among the trilobites in any detail is very difficult and uncertain, some general evolutionary trends are evident which do not seem to have been adaptively related to particular environments. If trilobites evolved from annelids, the most primitive forms presumably should (1) be most worm-like and consist of numerous, similar, and clearly marked-off segments, (2) have small tail shields, or pygidia, and (3) bear strong furrows on the head shields, revealing their segmentation. Progressive changes, not confined to single lineages, can be followed with respect to all these features, from older to younger strata.

(1)   As far as segmentation is concerned, the most primitive-appearing trilobites are metaparians from the Lower Cambrian (Fig. 247-1). Some have more than 40 segments behind the head, and the last are very worm-like. The number of articulated thoracic segments between the head and tail shields in some species seems to have been variable, but in other trilobites the number of these freely articulating segments is constant in most species. Several general lineages show progressive reduction in the number of free thoracic segments.

In their most primitive condition, the segments of trilobites end laterally in spines, but in more advanced forms the ends generally are rounded. Spines, however, developed secondarily on certain segments in several not closely related lineages.

(2)   The metaparian trilobites also are most primitive with respect to the development of pygidia. In some there seems to have been no fusion of posterior segments to produce even the most rudimentary compound pygidium, and others have only one joined to the final segment. Trilobites in other groups have pygidia of variable

Table 247 Distinguishing characters of the principal morphological groups of trilobites, based mainly on structures of the head

Eodiscids: tiny trilobites of conventional form with two or three articulated thoracic segments, with or without eyes. Cambrian

Agnostids: tiny trilobites much modified from conventional type, without eyes or dorsal facial sutures, with only two articulated thoracic segments. Cambrian to Ordovician

Metaparians: trilobites with long crescentic eyes, without dorsal facial sutures, having many articulated thoracic segments and a few fused to produce a tail shield. Cambrian

Opisthoparians: trilobites with dorsal facial sutures which extend backward from eyes to the posterior border of the head shield inside of posterolateral corners. Cambrian to Permian

Proparians: trilobites with dorsal facial sutures which extend outward from eyes to the lateral borders of the head shield in front of the posterolateral corners. Ordovician to Devonian

Hypoparians: blind trilobites without dorsal facial sutures, having five or more articulated thoracic segments. Ordovician to Devonian

These groups seem to be related to one another evolutionarily in complex ways. Several of them surely consist of two or more unrelated lineages.

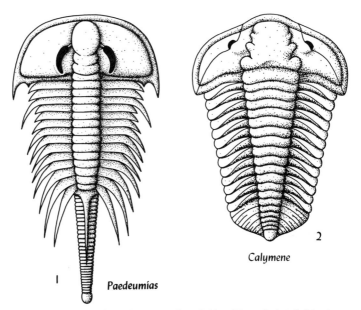

1   *Paedeumias*

2   *Calymene*

Fig. 247   1, A Lower Cambrian metaparian trilobite without distinct facial sutures and with a body ending in a series of worm-like segments. 2, A Silurian trilobite whose facial sutures cut the genal angles. It cannot be classed strictly as either an opisthoparian or a proparian. (*See* Burling, 1916, *Ottawa Natur., vol.* 30, *pl.* 1; Harrington et al., 1959, *fig.* 353-1a, *p.* 451.)

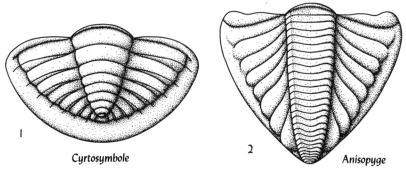

1   *Cyrtosymbole*   2   *Anisopyge*

Fig. 248   Two trilobite pygidia, contrasting the primitive nature of a Devonian specimen (1) whose axial and lateral segments correspond, and a highly advanced Permian form (2) with many more segments in its axis. (*After R. & E. Richter, 1912, Preuss. Geol. Landes., N.F., vol. 99, pl. 2, fig. 18a; Harrington et al., 1959, fig. 307-6b, p. 403.*)

complexity and size (Fig. 248). As more segments became fused, the pygidia grew relatively larger until many of them equaled the cephalon in size and shape. This almost certainly was an advantageous adaptation related to enrollment, which seems to have been the trilobites' principal means of protection from their enemies. Species with large pygidia could bring these parts into apposition with the under-surface of their heads and thus completely enclose their vulnerable ventral struc-tures (Fig. 248a).

In the evolution of pygidia consisting of more and more fused segments, there was a tendency for the segmentation of the longitudinal central axis and the lateral parts to become increasingly unequal. Most late Paleozoic pygidia have appreciably more segments axially than are indicated laterally. The most extreme observed discrepancy occurs in some Middle Permian trilobites with segments nearly four times as numerous in their axes (Fig. 248-2).

Primitive pygidial segments terminate in spines exactly like those of the thorax.

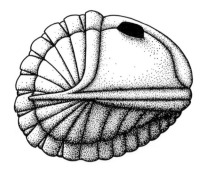

Fig. 248a   This trilobite is enrolled in such a way that its vulnerable ventral parts are completely enclosed and protected. Possession of a pygidium approximately equal to the head in size and shape seems to have been distinctly advantageous.

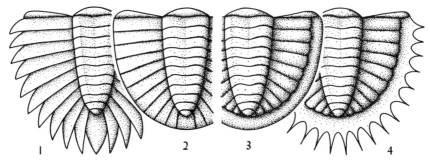

Fig. 249   Much-generalized drawings showing a series of progressive evolutionary modifications of the borders of trilobite pygidia.

The division between thorax and pygidium may be difficult to identify in trilobites of this kind. These spines are lost in advanced forms, whose pygidia attain a smooth and even outline. Further advancement is indicated by the development of a flattened rim which is not encroached upon by the furrows defining pygidial segments. Spines reappear secondarily, however, on the margins of some evolutionarily advanced pygidia (Fig. 249).

(3)   The cephalon of a trilobite consists of about five fused segments with an additional anterior lobe. In primitive forms these segments are plainly indicated by bounding transverse furrows which cross the raised axial portion, or glabella, of the head (Fig. 249a-1). These furrows commonly steadily degenerated. At first the lateral parts ceased to meet across the center of the glabella, then they became progressively less well impressed, until finally they disappeared. These changes first affected the most anterior furrows and gradually spread backward. The last furrow that marks off the final segment or so-called neck ring is by far the most persistent one

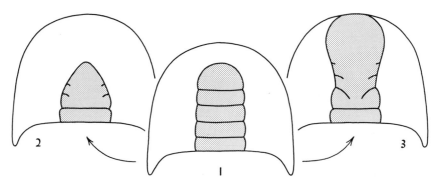

Fig. 249a   Much-generalized drawings showing: 1, The most primitive type of trilobite glabella (shaded), parallel-sided and with continuous transverse furrows. 2, 3, Advanced types that evolved in almost opposite directions.

and may remain as a continuous well-impressed groove after all the others have faded out completely. In only a few highly specialized trilobites has the neck furrow disappeared.

The primitive glabella probably was parallel-sided and did not reach to the anterior margin of the cephalon. From this condition evolution seems to have progressed in two almost opposite directions (Fig. 249a-2,3). In one, the forward part of the glabella widened, became swollen and longer, until it reached or even overhung the anterior margin. In the other, it shrank, i.e., shortened and narrowed between converging sides. The trilobite's stomach underlay the glabella, and these changes probably were related to different feeding habits.

Several other evolutionary trends produced similar effects in a number of trilobite groups not closely related to one another. One involved the weakening of all segmentation in both cephalon and pygidium and also, to a less extent, of the longitudinal furrows which separate the median axis and lateral parts. This resulted in the development of nearly smooth and very untypical-appearing trilobites. In another, the free cheeks, which ordinarily do not meet anteriorly, approached each other on the underside of the margin and finally fused to produce a single piece that separated at the time of molting (Fig. 250).

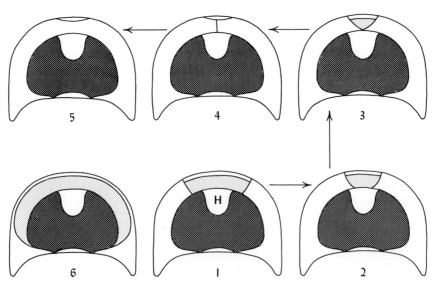

Fig. 250   Much-generalized sketches of the undersides of trilobite head shields, showing relations of the hypostomal plate *H* and epistomal plate (lighter shading) to the free cheeks. 1, Primitively a large epistomal plate occurs in front of the hypostome and separates the anterior parts of the free cheeks. 2, 3, In advanced forms the epistome is reduced in size. 4, It is eliminated, but the free cheeks are separated by a suture. 5, The free cheeks are joined anteriorly. 6, The relations of structure in metaparian trilobites to this evolutionary sequence are uncertain.

Opisthoparians and Proparians

Most trilobites have either opisthoparian or proparian facial sutures. Theoretical considerations and ontogenetic developments reported in a few Cambrian species have led to the conclusion that the proparian suture is primitive and that opisthoparians may have evolved independently several times from proparian ancestors. That there is some connection between these groups certainly is suggested by a few trilobites whose sutures extend exactly to the genal angles so that they cannot be classified definitely in either group (Fig. 247-2).

Paleontologists have become more and more inclined to doubt the accuracy of older ontogenetic observations and to reject the proparian-to-opisthoparian theory. The facts that opisthoparian trilobites appeared earlier and are much more numerous are not conclusive but seem to favor the opposite opinion. Opisthoparians also have a much longer stratigraphic range than any other group.

Metaparians

The metaparians, as already mentioned, are believed to be very primitive trilobites insofar as their postcephalic segmentation is concerned (Fig. 247-1). Likewise their eyes seem to exhibit a primitive condition, which is considered in the next section. Their chief peculiarity, however, is the nature of their facial sutures (Fig. 245-3), if indeed certain surface markings are correctly interpreted as indicating the course of nonfunctional sutures of this kind.

Some paleontologists believe that all trilobite facial sutures, regardless of their positions, are homologous and mark the division between two of the segments which are fused to form the cephalon. The validity of this belief has not been established, but if it is correct, a theoretical explanation of the differences in the course of these sutures can be devised.

The posterolateral genal angle of the cephalon, which commonly is produced into a spine (Fig. 252), serves as a fixed point in describing the general course of a facial suture. Thus the posterior part of the suture reaches the cephalic margin in advance of the genal angle only in the proparians, and the possible anterior part leads to the margin behind this angle only in the metaparians. Obviously if these sutures are homologous, the genal angle or its spine cannot also be homologous and, therefore, it is not a true fixed point.

Some Cambrian trilobites, particularly in their early ontogenetic stages, possessed a cephalon bearing three pairs of marginal spines. If these were to migrate posteriorly and carry the facial sutures with them, the proparian, opisthoparian, and metaparian conditions would follow in this order (Fig. 252).

This theory is ingenious, but the postulated sequence of facial sutures seems to be contradicted by stratigraphic evidence. Proparian trilobites are rare in the Cambrian. They become relatively abundant in the Ordovician and Silurian but exhibit other features that certainly are not primitive. The metaparians are restricted exclusively to the Lower Cambrian.

The relations of facial sutures and genal spines may be viewed, however, in a

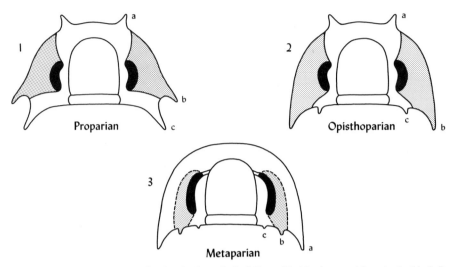

Fig. 252   Generalized diagrams showing the theoretical relations of facial sutures and free cheeks (shaded) to spines that are possessed by some Cambrian trilobites and have been supposed to migrate. Whether or not the genal spines of most trilobites are homologous is an unsettled question.

different way. Starting with an opisthoparian disposition, proparians might have been derived by the forward migration of the posterior spines. The metaparians, nevertheless, remain as aberrant and presumably the most advanced forms if they are to be explained by spine migration. To the primitive nature of their segmentation and eyes, already mentioned, however, may be added the composition of their exo-skeletons. Unlike the skeletons of most trilobites, these were very little calcified. Such a condition almost certainly was primitive. Probably the supposedly nonfunctional facial suture of the metaparians is a misinterpretation. Metaparians seem actually to be most closely related to other Cambrian trilobites with functional opisthoparian sutures. Perhaps they are a group in which these sutures failed to develop or have disappeared without leaving a clear trace.

Hypoparians

The hypoparian division (Fig. 246) of the trilobites was originally set up to accom-modate blind species without dorsal facial sutures, and they were presumed to be very primitive. The theory was presented that the typical eyes of trilobites first appeared ventrally on ancestral organisms which were supposed to have been swimming planktonic creatures. The eyes then migrated to the upper surface of the cephalon and carried the sutures with them when these animals adapted to living on the sea bottom. Trilobites without eyes that migrated in this way, therefore, retained the primitively situated marginal suture. The eyes possessed by a few

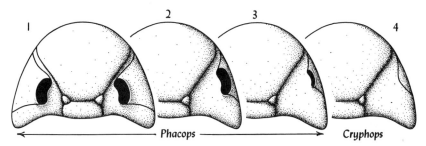

Fig. 253   A presumed evolutionary sequence observed in closely related Devonian trilobites shows progressive reduction of the eyes and free cheeks. Finally the facial sutures become marginal. (*After Harrington et al.,* 1959, *fig.* 47, *p.* 62.)

trilobites which have only marginal sutures are, according to this idea, different structures not homologous with the eyes of other trilobites.

This theory is no longer held. Hypoparians are now recognized to be a miscellaneous lot of relatively unrelated, specialized trilobites that had lost their eyes, rather than ones that never had them. Blind species are known which retain typical dorsal facial sutures (Fig. 253-4). The sutures of some descendants of trilobites of this kind migrated until they became marginal. Furthermore, there is an evolutionary series of eyed trilobites whose sutures before and behind the eyes (Fig. 253a) approached until they finally met and the functional part of the suture was restricted to a marginal position.

The most primitive trilobite eye is a long, narrow, curved organ which seems to be a continuation of one of the anterior glabellar segments (Fig. 247-1). Such eyes are typical of the metaparians and also occur on some Lower and Middle Cambrian opisthoparians. With evolutionary advancement, the eyes shortened but maintained their connection with the glabella by means of narrow raised ridges, the so-called eye lines (Fig. 253a). These characterize a variety of Cambrian and Ordovician trilobites but were lost in further evolution. The most primitive eyes are but little raised above the surface of the cephalon, and vision was mostly upward. From this condition, eyes developed which rise more or less abruptly above the nearby

Fig. 253a   The eyes of this Upper Cambrian trilobite are connected to the glabella by eye lines, or ridges, and the two branches of each facial suture are very near together. In other closely related trilobites the suture branches coalesced and disappeared, thus producing a hypoparian condition. (*After Harrington et al.,* 1959, *fig.* 46A, *p.* 61.)

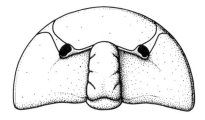

Loganopeltoides

surface, permitting lateral vision. Most such eyes are strongly curved, and it is evident that a wide range of visibility had been acquired. A few specialized trilobites, believed to have been burrowing forms, had their eyes raised at the ends of rigid upright stalks that probably projected above the sediment surface.

Eodiscids and Agnostids

The eodiscids and agnostids are two groups of small-to-tiny trilobites of Cambrian and Ordovician ages (Fig. 254). Most of them were blind, and none have more than three free thoracic segments. No other trilobites are known with less than five, and most have more than seven. The eodiscids are unmistakable trilobites, but the agnostids are so aberrant that their assignment to this group of arthropods has been questioned. This doubt, however, has found little support among paleontologists.

Formerly the eodiscids and agnostids were considered to be very primitive. Their small number of thoracic segments suggests comparison with the early growth stages of more conventional trilobites, in which these segments were added one by one. Applying the recapitulation theory, this was considered evidence that the eodiscids and agnostids exhibit a condition characteristic of the ancestors of other trilobites. The fallacy of this idea is evident when consideration is given to the probability that trilobites evolved from annelids, and the fact that the number of thoracic segments actually was progressively reduced in several well-known lineages of trilobites. The agnostids particularly are almost surely highly specialized, because they show the greatest deviation from the generalized trilobite form. Their ancestry is not known, but they probably evolved in Precambrian time from creatures basically similar to the eodiscids.

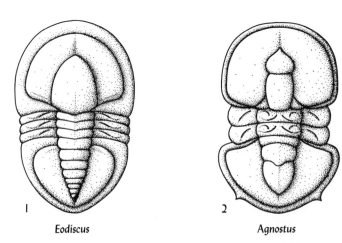

Eodiscus	*Agnostus*

Fig. 254   1, An eodiscid. 2, An agnostid. These peculiar early Paleozoic trilobites have been considered very primitive by some paleontologists, but they now are generally believed to be highly evolved and specialized (much enlarged). (*After Shimer and Shrock, 1944, "Index to Fossils of North America," pl. 252, fig. 1; pl. 251, fig. 1.*)

Table 255 Probable homology of appendages in the principal groups of arthropods

TRILOBITE	CHELICERATE	CRUSTACEAN	ONYCHOPHORE	MYRIAPOD	INSECT
Antennae	Absent	Antennules	Antennae	Antennae	Antennae
Legs	Chelicerae	Antennae	Jaws	Absent	Absent
Legs	Pedipalpi	Mandibles	Slime papillae	Mandibles	Mandibles
Legs	Legs	Maxillae	Legs	Maxillae	Maxillae
Legs	Legs	Maxillae	Legs	Maxillae	Labia
Legs	Legs	Maxillipeds	Legs	Legs	Legs
Legs	Legs	Maxillipeds	etc.	Legs	Legs
Legs		Maxillipeds		Legs	Legs
etc.		Maxillipeds		etc.	
		Legs			
		Legs			
		Legs			
		Legs			
		Legs			
		Pleopods			

The solid line indicates the termination of the fused segments of the head (dashed line) or cephalothorax of chelicerates and many crustaceans.

Because agnostids and metaparians do not conform to the structural plan of other trilobites in ways considered to be important, the opinion has been expressed that they evolved independently of other trilobites and of each other from different groups of annelids. Although this may be true, there is no substantiating evidence, and present knowledge does not seem to justify any such conclusion.

Trilobites have been extinct for well over 200 million years. Nevertheless, study of the structure of these fossils, coupled with determination of their stratigraphic ranges, has permitted the recognition of a variety of progressive evolutionary trends. Trilobites are better known in this respect than any other invertebrates, except possibly the ammonites.

Chelicerates

The chelicerates are characterized particularly by their legs and other ventral appendages. Antennae are lacking, and the first appendages are a pair with pincher claws which ordinarily are capable of grasping food and carrying it to the mouth. These are followed on the underside of the anterior fused segments, or cephalothorax, by five pairs of unbranched legs, some of which are specialized for different functions. Thus a chelicerate differs from a trilobite, whose cephalon bore ventrally a pair of antennae and four pairs of unspecialized similar branching legs which served at the same time for feeding, walking, swimming, and respiration. The presumed homologies of these appendages are shown in Table 255.

The abdomen of the earlier chelicerates typically consisted of 12 movable segments and a terminal spine-like telson. That number was reduced, however, and some of the segments coalesced in various lineages. The posterior part of the body commonly is divisible into a pre- and a postabdomen except in the advanced horseshoe crabs and such terrestrial forms as spiders and their allies. The general nature of the chelicerate body and its appendages shows that these animals were more advanced than the trilobites. Both groups probably evolved from related trilobitoid ancestors in Precambrian time.

Aglaspids

The aglaspids (Fig. 256) are the most primitive of the marine chelicerates. All known specimens are Cambrian except one discovered in Upper Ordovician strata.

The ventral structures of the aglaspids rarely are preserved. Some specimens show, however, that these animals were without antennae and that their first pair of appendages were clawed chelicerae. At least one species had legs on the anterior segments of the abdomen. These seem to be unbranched, like the walking legs of the cephalothorax. No gill structures have been observed, but they may have been too delicate to be preserved. Division of the abdomen is not clear, but the last two or three segments may be slightly different from the others, thus suggesting the postabdomen that is so evident in some of the other chelicerate groups.

Nothing can be determined regarding evolution of the aglaspids except that they appear to occupy a position intermediate between some of the trilobitoids and other more advanced chelicerates. They closely resemble a group of trilobitoids (Fig. 244-1) in body form and the possession of a telson spine, but the cephalothorax seems to consist of one or two more segments, and the appendages are different. Presumably aglaspids were derived from trilobitoids of this general type. Both the

Aglaspis

Fig. 256 Reconstructed drawing of an Upper Cambrian aglaspid. (Approx. × ½.) (*After Raasch, 1939, pl. 1.*)

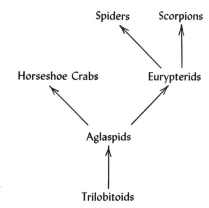

Fig. 257   Diagram showing probable evolutionary relations of the main groups of chelicerate arthropods.

horseshoe crabs and eurypterids probably evolved from a stock that is represented by the aglaspids (Fig. 257).

Horseshoe Crabs

The horseshoe crabs have left a long but meager geologic record. The oldest known
species was discovered in Upper Silurian strata, and three forms, closely related
to one another, are members of modern faunas. Although a few of these fossil
animals were marine, as are the living species, most of them seem to have inhabited
brackish or even freshwater environments.

The older fossils are much like the aglaspids, but they have fewer abdominal
segments, and a postabdomen of about three segments is differentiated in some
of them (Fig. 258-1). From the late Paleozoic onward there was a tendency for the
body to become shorter, and the abdominal segments were reduced from ten to
about six. At the same time these segments fused, generally from behind forward,

Table 257   Distinguishing characters of the main groups of chelicerates

Aglaspids: primitive marine chelicerates with articulated segments not clearly differentiated into pre- and
postabdomen. Legs seem to have been borne on abdominal segments of some forms. Cambrian

Eurypterids: aquatic chelicerates perhaps restricted to brackish environments. Articulated segments differentiated into pre- and postabdomen. Five pairs of legs, in addition to clawed chelicerae borne
by cephalothorax. First six of abdominal segments with flat overlapping plates that probably
were gills. Ordovician to Permian

Xiphosurans: horseshoe crabs, brackish water and marine chelicerates with five pairs of legs and fewer
abdominal segments than the foregoing. These segments, except for the final spine-like telson,
tend to fuse. Abdominal segments with flat overlapping gill plates. Silurian to Recent

Scorpions: terrestrial chelicerates with very slender postabdomen ending with a poison-bearing stinger.
Second pair of appendages with large pincher claws, followed by four pairs of legs. Respiration
accomplished by internal gill plates or tracheae or both. Silurian to Recent

Spiders: terrestrial chelicerates with four pairs of legs. Abdominal segments reduced, postabdomen not
differentiated. Respiration by internal gill plates or tracheae or both. Devonian to Recent

The chelicerates, originally marine, adapted successively to brackish water and terrestrial environments.

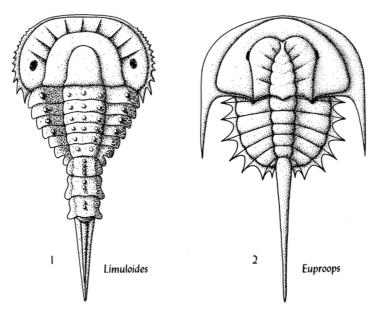

1   *Limuloïdes*   2   *Euproops*

Fig. 258 Two Paleozoic fossils related to modern horseshoe crabs. 1, A Silurian specimen, (enlarged.) 2, A Pennsylvanian specimen, (about natural size.) (*After Woodward,
1872, Brit. Foss. Crust., Palaeontography Society, fig. 64, p. 177; Størmer, 1955, fig.
13-4a, p. 19.*)

so that the posterior part of the body was covered by a rigid shield, much like the
pygidium of a trilobite but terminating with a movable telson (Fig. 258-2).

The appendages have not been observed on many fossil species. The earliest
seem to have had five pairs of clawless walking legs, which were not continued on
the abdomen. Later the four forward pairs developed pinching claws, but the last
pair terminated in spines, which is a primitive condition. The abdomen of modern
species bears six pairs of highly specialized appendages. These appear as broad,
flat, overlapping plates, and parts of them serve as gills. Traces of similar plates
have been noted on only a very few fossil specimens, but appendages of this kind
may have been possessed by all of them.

Eurypterids

The eurypterids are rare fossils ranging from the Lower Ordovician to the Permian.
Most of them seem to have been members of restricted brackish-water faunas. The
great majority were small creatures, but this group includes the largest arthropods
that ever lived.

All eurypterids had abdomens consisting of 12 freely movable segments (Fig.
259). The last five segments were different from the others and formed a postabdomen that tapered rapidly and was more slender and more flexible than the

remainder of the body. The first six segments of the preabdomen bore specialized appendages, whose most prominent parts seem to have been flat overlapping plates. Parts of them probably served as gills, but their exact structure is not known. The last pair of cephalothoracic appendages of most eurypterids was specialized as flat paddle-like legs. These commonly are assumed to have been used for swimming, but they may have been equally adapted to digging in soft bottom sediments.

The most generalized eurypterids were very similar to aglaspids (Fig. 259-1), and at least one species at different times has been assigned to first one and then the other of these groups. Progressive evolution cannot be traced in any detail among creatures of this kind, but several divergent types developed, probably in response to changes in their modes of life. The more conservative eurypterids seem to have been bottom dwellers (Fig. 259-1). Their heads were flat and shovel-like,

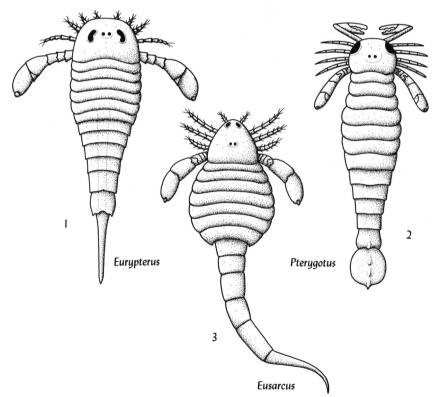

Fig. 259 Examples of three types of eurypterids. 1, The flattened form and upwardly directed eyes of this eurypterid suggest that it was a bottom-living creature. (× ⅓.) 2, This animal, with forwardly directed eyes and flattened telson, probably was more active and possibly was an able swimmer. (× ¹⁄₃₀.) 3, This form suggestively resembles a scorpion. (× ⅙.) (*After Clarke and Ruedemann, 1912, N.Y. State Museum Mem. 14, pls. 2, 27, 67.*)

and their eyes were directed upward. The telson retained its primitive spine-like form, and in some it became long and very slender. Perhaps these animals burrowed sluggishly in the mud.

Other eurypterids seem to have been more active creatures. One group had a form very similar to that of scorpions, with a long, slender postabdomen sharply set off from the rest of the body (Fig. 259-3). This ended in an acute curved telson, and there have been speculations that it may have contained a poison gland. The eyes were located on the anterior margin of the head and evidently provided much more effective forward vision. Some species had the forward pair of legs, or chelicerae, specialized as long claw-bearing organs that probably were useful in the capturing of prey.

Another group seems to have become particularly adapted to swimming (Fig. 259-2). This is suggested by the telson, which was transformed into a short, lobed structure with flattened lateral vanes. The eyes of these creatures also had migrated to anterior marginal positions. The chelicerae of some of them became long grasping organs, very different from the short ones possessed by most eurypterids.

Scorpions

The earliest known scorpions, which generally are regarded as the first air-breathing terrestrial animals of record, have been found in Silurian strata. That they actually were air breathers has been doubted, but insofar as preservation permits comparison, the respiratory structures do not seem to be significantly different from those of modern scorpions. Fossil representatives of this group are very rare.

Scorpions are so similar in body form to some eurypterids (compare Figs. 259-3 and 261) that their evolution from such ancestors has been confidently assumed. The possibility has been suggested that eurypterids were able to live temporarily out of water like some modern crabs. In the course of time, small animals of this kind may have become adapted to life on land by the invagination of their abdominal gill plates, which thus were converted to so-called book lungs. This theory is attractive, and no better interpretation has been offered. The embryology of modern scorpions, however, fails to provide evidence that book lungs developed in this way.

The Silurian scorpions have eight preabdominal segments (Fig. 261-1) which is one too many if typical eurypterids were their ancestors. The embryonic development of modern species shows, however, that the last two actually are subdivisions of a single original segment. Therefore, derivation from the eurypterids is not ruled out. Most post-Silurian species have what seem to be the expected number of seven preabdominal segments. Embryology also demonstrates that this evolutionary reduction resulted from the shortening and final disappearance of the first segment. Progressive evolution is most evident, however, in changes in the ventral structure of the cephalothorax, which reached its most advanced condition in Carboniferous species. Modern scorpions seem to be representatives of a less-advanced lineage that has altered very little since that time.

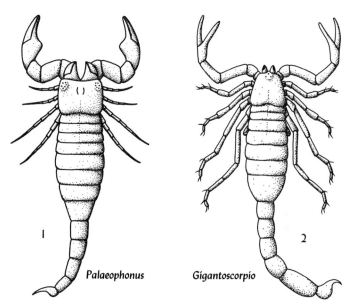

Fig. 261   Reconstructions of Paleozoic scorpions. 1, A Silurian form. (Approx. × ½.) 2, A Lower Carboniferous form. (Approx. × ¼.) (After Eastman-Zittel, 1913, "Text-book of Paleontology," 2d ed., vol. 1, fig. 1513A, p. 787; Stormer, 1963, Norske Videnskaps Akademi i Oslo Skrifter, Naturvidenskapelig Klasse, ny ser. no. 8, fig. 23, p. 61.)

Spiders

Although fossil spiders and their allies are very rare except at a few places, a considerable number of species has been discovered dating as far back as the Devonian Period. They have been studied by only a few specialists and are almost unknown to paleontologists.

The most conspicuous evolutionary trend in spiders has been the reduction of abdominal segments from an original number of 12 to less than 5. Segments were lost in different groups by (1) shortening of anterior ones whose remnants were incorporated in the cephalothorax, or (2) shrinkage and disappearance of segments at the posterior end. Much evolutionary diversification had been accomplished before the end of the Carboniferous, and some lineages show little change even to the present.

This group of animals respires by either book lungs or tracheae, or both. The tracheae presumably evolved from book lungs, but how this was accomplished is not known. The only abdominal structures that can be traced definitely to embryonic legs are the spinnerets. The ancestry of spiders and related creatures is not known. Probably they are an offshoot from the same stock that produced the scorpions after adaptation to terrestrial life had been accomplished. The form of

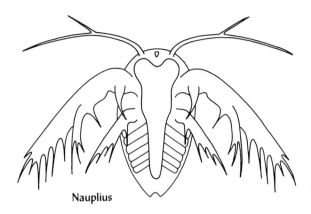

Nauplius

Fig. 262   The larva of *Apus*, a branchiopod crustacean, viewed from the ventral side. (Much enlarged.) (*After Parker and Haswell, 1940, "A Text-book of Zoology," 6th ed., vol. 2, fig. 363A, p. 396.*)

spiders suggests that these animals may be examples of arrested development because they retain some features that resemble embryonic stages in the ontogeny of other chelicerates.

Crustaceans

Crustaceans have been aptly termed the insects of the sea, because of their great diversity, incredible numbers, and the ways in which they have adapted to almost every marine environment. They also inhabit waters that are, on the one hand, fresh and, on the other, excessively saline, but only a few have acquired the ability to live out of water on the land. Most crustaceans are distinguished from other arthropods by possessing two pairs of antennae and three pairs of appendages specialized as jaw parts, all borne ventrally on the head. Many of them hatch from eggs as a

---

Table 262   Conspicuous characters differentiating the main groups of crustaceans known as fossils

Branchiopods: many-segmented crustaceans inhabiting freshwater and brackish water as well as marine environments; of greatly varied form, some shrimp-like. They may be provided with single- or double-valved shells, commonly horny, some not molted periodically. Appendages behind antennules generally leaf-like and similar. Cambrian to Recent

Ostracods: marine and freshwater crustaceans enclosed in a calcareous two-valved shell that is molted periodically. Paired appendages behind mouthparts few. Ordovician to Recent

Cirripeds: barnacles, marine crustaceans that become permanently attached after larval stage. Body enclosed by several roughly triangular, movable calcareous plates. Ordovician(?) to Recent

Various Mainly Shrimp-like Forms: uncommon fossils belonging to several variously classified not closely related groups, differing in number of segments, structure of cephalothorax, and type of appendages. Cambrian to Recent

Malacostricans: crabs, etc.; mostly aquatic crustaceans with body composed of 20 or 21 segments. Cephalothorax includes a variable number of thoracic segments with a calcified carapace. Some about as wide as long, with narrow, strongly turned-under abdomen, others lobster-like. First pair of appendages behind mouthparts may bear powerful pinching claws. Triassic to Recent

---

Other crustaceans, important in modern faunas, are unknown as fossils.

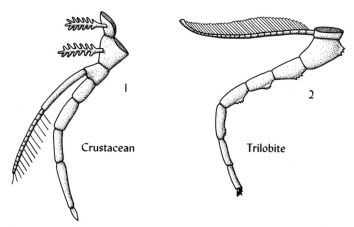

Crustacean   Trilobite

Fig. 263   Sketches illustrating supposedly characteristic differences in the structure
of the biramous legs of some advanced crustaceans and trilobites. (*After Størmer,
1939, Norsk. Geol. Tidsskr., vol.* 19, *fig.* 1, *p.* 155.)

larval form known as a nauplius (Fig. 262), which vaguely resembles a tiny trilobite
with three pairs of appendages.

The detailed structure of appendages is important in the classification of
crustaceans. Most appendages consist of two branches, one of which performs
respiratory functions, but otherwise they are variously specialized in adaptation to
swimming, walking, feeding, etc. Some of the appendages resemble those of trilobites, but the importance of certain differences has been both emphasized and
denied (Fig. 263).

Phylogenetic relations of the major crustacean groups are very doubtful (Fig.
263*a*). A few tiny modern creatures, known as cephalocarids, which inhabit protected
situations on the bottom of both the Atlantic and Pacific Coasts of North America,

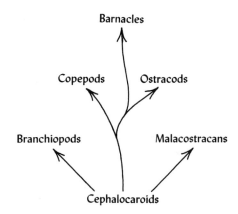

Fig. 263*a*   Diagram showing the possible evolutionary relations of the principal groups of
crustaceans. Partly because most of these groups
are poorly represented in the fossil record, their
relations are not well understood.

lack serial specialization of their segments and have generalized appendages of a type from which those of other crustaceans might have been derived. Perhaps they represent a primitive stock related to the trilobitoids that was ancestral to all crustaceans. The other better-known groups mostly differ so greatly from one another that evolutionary radiation is believed to have occurred very long ago among tiny creatures, followed in many lineages by increase in size and wide divergence.

Crustaceans undoubtedly were abundant in many ancient faunas. Some groups have long but scanty geologic records. Most of them were ill suited for preservation, and only the ostracods are common fossils.

Branchiopods

The branchiopods generally are considered rather primitive crustaceans, because of their many similar and relatively unspecialized appendages. Most of these animals are very small. Some modern species resemble certain Middle Cambrian trilobitoids, and the group probably evolved from creatures of that general type. Fossils have been reported from the Cambrian onward, but the identification of most older species is uncertain. Specimens of a few kinds are very abundant in some brackish water and nonmarine strata of late Paleozoic to Triassic age.

Several types of branchiopods have shells that generally are horny and little

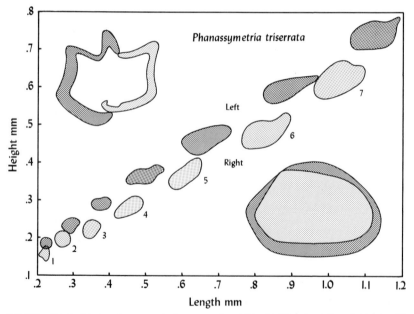

Fig. 264   Graph showing size ranges in the successive molts of a Devonian ostracod. Inserts, side view and transverse cross section. (× 30.) The left valve is considerably larger than the right. *(After Lundin and Scott, 1963, J. Paleontol., vol. 37, figs. 1d, 2, pp. 1276, 1277, pl. 179, fig. 1d.)*

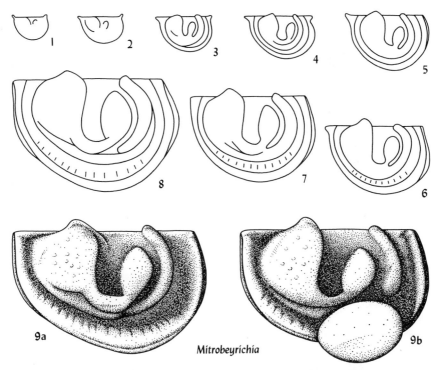

Fig. 265   Sequence of molt stages of a Silurian ostracod, showing increase in size and development of ornamentation. Sexual dimorphism is conspicuous in adults as the females are provided with brood pouches. (× 25.) (*After Martinsson, 1962, Bull. Geol. Inst. Univ. Uppsala, vol. 41, fig. 21, p. 80.*)

calcified. Some consist of a single more or less oval plate, but those most commonly found as fossils are bivalved like tiny pelecypods but lack the hinge teeth and sockets of these molluscs. Distinction may be difficult, particularly among early Paleozoic specimens. These branchiopods did not molt their shells, like ostracods, and the fossils are marked by concentric growth lines, each of which may represent a molting episode. This provides some information concerning their ontogeny but little evidence of their evolution.

Ostracods

Marine ostracods are more or less abundant in many formations of early Ordovician and all younger ages, and the record of nonmarine species begins in the Carboniferous. The fossils are paired bivalved shells composed of calcite. These were cast off five to eight times as the growing animals molted, and were rapidly replaced each time by a new and larger pair (Fig. 264). The shells of ostracods differ greatly in their shape, ornamentation, and hingement. Several reconstructed ontogenetic series show how these details changed progressively with growth (Fig. 265).

The ostracod body is not obviously segmented, and it has only five to seven pairs of appendages. The last one to three pairs are attached to the thorax, and there is none on the very short abdomen. Thus ostracods differ greatly from all other modern crustaceans, and they bear little resemblance to any known possibly ancestral form. Zoologists classify ostracods on the basis of the detailed structure of their appendages, and they generally have neglected shell characters. Paleontologists know them only by their shells. Ostracod ontogeny has been studied to some extent, but it has not been much utilized in an attempt to work out the phylogeny of this group of animals.

Some general evolutionary tendencies have been recognized in the hinge characters of ostracods (Fig. 266). Most primitively, the dorsal edge of one half of the shell fits into a simple groove in the other half. Irregularities that developed into

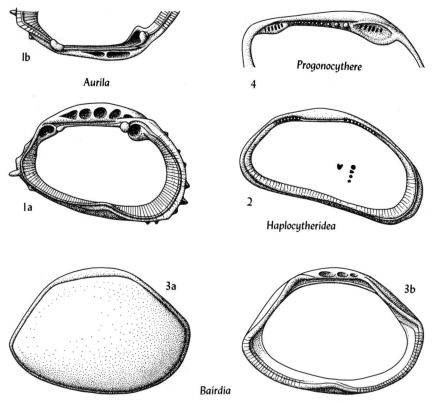

Fig. 266   Three modern and one Jurassic ostracod, showing details of hinge structure, muscle scars (2), and inequality of valves (3a). (*After Benson and Coleman, 1963, Univ. Kansas Paleontol. Contrib., Arthropoda, art. 2, figs. 6, 15, 21, pp. 17, 29, 35; Benson et al., 1961, fig. 248c, p. 323.*)

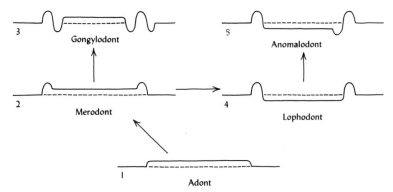

Fig. 267 Diagrammatic representation of some of the principal types of ostracod hinges as seen from above. A hinge ordinarily consists of a ridge and teeth in the right valve and a corresponding groove and sockets in the left. Either or both ridge and teeth may become crenulated, thus producing more complex hinges. The whole hinge pattern is reversed in some ostracods. (*See Scott, 1961, "Treatise on Invertebrate Paleontology," pt. Q, fig. 25, p. 35.*)

teeth and sockets appeared first at the anterior end of the hinge and later at the posterior end (Fig. 267). Both became progressively more elaborate. This condition may then have become simplified secondarily, so that return to a more primitive type of hinge resulted. Some evolutionary relations also might be recognized by comparing surface lobes (Fig. 267a), other ornamental features, and muscle scars, but very few studies of this kind have been attempted.

Barnacles

Barnalces are very aberrant crustaceans that do not resemble any other arthropods. For many years they were mistakenly considered to be molluscs. These animals, however, have a nauplius larva and then pass through a stage with a bivalved shell very similar to that of a tiny ostracod. When the larva becomes attached, this shell is molted and growth of the permanent plates that enclose the mature animal begins. The ontogeny suggests, therefore, that barnacles and ostracods may have evolved from similar ancestral stocks.

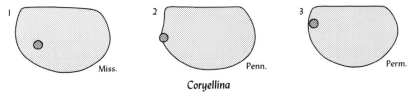

Coryellina

Fig. 267a Outline drawings of closely related ostracods, perhaps an evolutionary sequence, ranging from the Mississippian to the Permian, showing progressive backward and upward migration of a posterior spine. (*After Sohn, 1962, J. Paleontol., vol. 36, fig. 3, p. 1207.*)

Although barnacle-like fossils have been described from strata as old as the Cambrian, the identification of Paleozoic specimens is uncertain. Most of the fossils resemble two types living in the sea today that seem to have diverged at least by Mesozoic time (Fig. 268). The bodies of modern species attached by a fleshy stalk are enclosed by only a few large movable plates. The stalks of several fossil species also were enclosed by plates, which were reduced in number and size with passing time and finally were eliminated entirely. The conical barnacles that grow closely attached to rocks and other objects are surrounded at the base by a circle of relatively immovable plates, which seems primitively to have numbered eight. This number was reduced in various lineages by either the fusion of adjacent plates or by their shrinkage and final disappearance, or by both processes, until only four remained. The plates of fossil barnacles generally are found in a separated condition. They are nowhere common except at a few places where sediments accumulated very near a probably rocky shore.

Shrimp-like Crustaceans

A great variety of shrimp-like creatures is classified in at least half a dozen more or less well-differentiated groups. Some are closely related to the crabs. Their generally elongated bodies, many of which are laterally compressed, are adapted to swimming or darting backward by a quick and powerful downward thrust of the

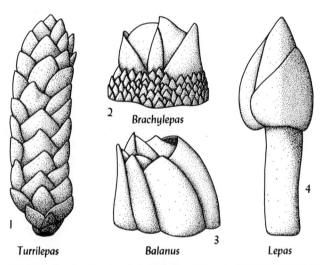

2 *Brachylepas*

1 *Turrilepas*   *Balanus* 3   *Lepas*   4

Fig. 268   Fossil and modern barnacles. The modern ones may have evolved from a possible Silurian ancestor by enlargement of the summit plates and either reduction of the pedicle or loss of lower plates that sheathed this part. 1, A Silurian form. 2, A Cretaceous form. 3, 4, Two modern forms. (*After Eastman-Zittel, 1913, "Text-book of Paleontology," 2d ed., vol. 1, figs. 1438a, 1445, pp. 744, 745; Parker and Haswell, 1940, "A Text-book of Zoology," 6th ed., vol. 1, figs. 454, 455a, pp. 578, 579.*)

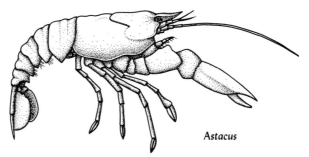

Astacus

Fig. 269   A common, modern crawfish. A great variety of crustaceans are similar in a general way but are not closely related to one another. (*After Parker and Haswell, 1910, "A Text-book of Zoology," 2d ed., vol. 1, fig. 430A, p. 540.*)

tail (Fig. 269). These creatures differ much among themselves in the number of thoracic segments joined to the head to form a cephalothorax, the nature of their dorsal skeletons, and the structure of their appendages. Many crustaceans of these kinds are very delicate, and consequently preservation is not likely to be good. Fossils are rare, but specimens are known as old as the Silurian, and there is a possibility that some Cambrian fossils may belong here also. Little can be inferred about evolutionary relations, but fossils clearly show that the radiation which produced most of the major modern groups was well advanced before the end of Paleozoic time.

Crabs
Crabs commonly are considered to be the most advanced crustaceans. The oldest known representatives of this group are Jurassic, and before the end of the Cretaceous Period a considerable number of modern families had differentiated. Fossils consist mainly of the dorsal part of the skeleton and the tips of large claws. Legs and other appendages which are important for the classification of modern forms are rarely preserved.

The ancestors of crabs and their near relatives probably were shrimp-like creatures with free thoracic segments and appendages on both thorax and abdomen that were branched, little differentiated, and used mainly for swimming. Progressive changes resulted in the gradual fusion of thoracic segments with the head region to produce a massive cephalothoracic shield, and differentiation of the appendages, bringing about (1) three thoracic pairs specialized to supplement the jaw structures of the head, (2) five pairs of walking legs on the posterior part of the thorax, and (3) gradual reduction of those on the abdomen. Animals so constituted became better and better adapted to living on the sea bottom. Eventually their swimming was restricted to rapid backward flight from enemies, accomplished by violent ventral retraction of the abdomen whose last segment and telson consisted of a set of fan-like plates. The first walking legs bore increasingly large pinching claws that were adapted to both offensive and defensive action. A lobster or a crawfish is a creature of this type.

The abdomen was shortened progressively, probably because it interfered somewhat with normal forward crawling. This disadvantage also was overcome by the turning under of the abdomen beneath the cephalothorax and reduction in its width (Fig. 270). The tendency for shortening of the body is evident in the shortening, widening, and general flattening of the dorsal skeleton. In the most typical crabs, the cephalothorax is about as wide as long and this is the only part of the body that is visible from above. The concealed abdomen is relatively short and very narrow, and it is not capable of much movement.

Crabs probably evolved from more than one lineage of shrimp-like ancestors. Relations of the fossils cannot be assessed on the basis of the structure of their appendages, because these parts rarely are available for study. Attempts have been made in a general way to reconstruct evolutionary lineages by tracing the changing patterns of supposedly homologous furrows in the dorsal shield. Most of these are related to areas of muscle insertion, and they provide some information concerning the mechanical relations of the appendages.

Onychophorans

The onychophores (Fig. 271-1) are a small group of obscure and peculiar terrestrial animals whose present distribution is restricted to the southern hemisphere and tropics. They are of great interest in zoology, however, because they seem to occupy a position intermediate between polychaete annelid worms and the arthropods. Several theories have been presented that identify them as the modern representatives of a very ancient group of animals from which either all arthropods or only

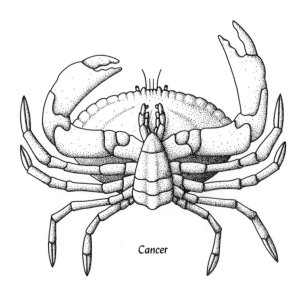

Cancer

Fig. 270   A modern crab viewed from below, showing the small, tightly recurved abdomen. (*After Parker and Haswell, 1910, "A Text-book of Zoology," 2d ed., vol. 1, fig. 468B, p. 590.*)

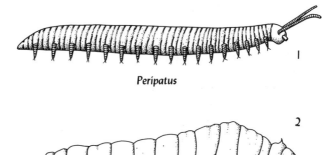

Peripatus

Papílio

Fig. 271   1. This peculiar modern creature, an onychophoran, superficially resembles a caterpillar but morphologically occupies a position inter- mediate between annelids and arthropods. 2, Caterpillar of a modern swallow-tailed butterfly. Entomologists are not agreed whether abdominal legs are homologous with the legs that persisted vestigially in some mature insects. (*After Parker and Haswell, 1940, "A Text-book of Zoology," 6th ed., vol. 1, fig. 479, p. 607; compare Tillyard, 1926, "The Insects of Australia and New Zealand," fig. Z60, p. 458.*)

certain branches of this phylum are presumed to have evolved. Fossils of similar animals are exceedingly rare. A few possible specimens have been collected from Middle Cambrian shale, and a single individual similarly identified was found in a quartzitic glacial boulder. The latter has been reported to be Precambrian, but actually its age is quite uncertain.

Onychophores resemble annelids in their worm-like form, with undifferentiated segments, soft integument, unstriated muscles, and simple eyes. On the other hand, numerous features of the head, their internal anatomy including tracheae, and their embryology are distinctly arthropodan. These animals differ from arthropods prin- cipally in lacking a stiffened horny exoskeleton and jointed appendages, in the presence of cilia in their genital tubes, and in the simpler organization of their heads. Obviously they are more advanced than annelids and more primitive than typical arthropods.

The proper position of onychophorans with respect to arthropod phylogeny has been much debated. Confusion evidently has resulted in the past because of unrec- ognized convergent evolution, without which all the similarities that have been noted in several major groups of arthropods cannot be explained. Probably modern ony- chophores are the remnants of an originally marine stock, some members of which emerged from an aqueous habitat in Silurian or earlier time and evolved to produce only the myriapods and insects.

Centipedes

Myriapods, or centipedes, and their allies are rare as fossils, but specimens have been found in strata as old as the Devonian or perhaps the late Silurian. Preservation generally is quite imperfect. Modern kinds are heterogeneous in form and perhaps represent more than one phylogenetic stock. All might have evolved, however, from onychophoran ancestors. This required the acquisition of a horny exoskeleton, the development of jointed legs, and increased cephalization resulting from the specialization and fusion of several anterior segments.

Insects probably evolved from a branch of the myriapods that is represented by one small modern group (Fig. 272-1). These animals resemble some of the primitive wingless insects, and in the past they have been classified as myriapods by some taxonomists but as insects by others. The heads of all myriapods are organized essentially like those of insects, although only a few have compound eyes. They differ from insects principally in the structure of their bodies, which consist of a series of almost identical segments. The myriapods that most closely resemble insects have 12 pairs of legs.

One of the most obvious evolutionary trends that can be traced from annelids to insects involves the progressive specialization of segmental appendages and their conversion to jaw structures. The annelids have a head bearing eyes, tentacles, and jaws, but none of these is derived from parts that correspond to segmental appendages. Onychophores do not have a well-defined head, but the feelers and single pair

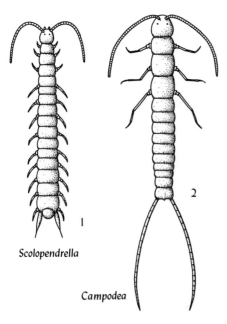

Scolopendrella

Campodea

Fig. 272   1, A modern myriapod of the general type from which insects may have evolved. The number of true segments is three less than the apparent number. 2, A modern wingless insect exhibiting very primitive characters, particularly in the strong segmentation of the thorax. (*After Tillyard, 1931, figs. 9, 10, pp. 28, 29.*)

of claw-like jaws are shown by their embryonic origin to be modified appendages (See Table 255). The distinct heads of myriapods have two or three additional pairs of appendages converted to jaw parts, and insects have a total of three pairs of similarly transformed appendages. This progression is accompanied by concentration of the nerves which control these parts and the development of a large ganglion below the esophagus which supplements the overlying brain.

Insects

If judged by their abundance, diversity, and adaptability, insects surely may be regarded as the dominant macroscopic creatures in the world today, and they have been studied more intensively than any other animals. They are not common fossils, although large numbers of specimens have been found at a few widely separated places where unusual conditions evidently were responsible for their preservation. Perhaps less-abundant and less-perfect specimens have been overlooked elsewhere, because many insects are represented only by their wings, and these are easily mistaken for leaves unless they are examined closely. Paleontologists generally have given little attention to insects, and these fossils are neglected in most textbooks. Much of what is known about them has been gained by entomologists whose interests were attracted to their study. Nevertheless, the insect fossil record, which probably begins in the Devonian, is on the whole impressive.

Although insects exhibit almost infinite variation with respect to many details, their basic organization has been remarkably stable. They seem to have achieved distinct individuality as a group in mid-Paleozoic time, and late Paleozoic evolutionary radiation produced some of the more important modern types before that era ended. Several very different theories of their evolution or relationships have been presented. Thus their supposed ancestors have been sought among the trilobites or crustaceans. They have been classed with the arachnids, or spiders, and their allies because both groups have tracheae, or with crustaceans because of similarities of their mouth parts. The insects, however, are most naturally grouped with the myriapods and onychophores. All these animals lack the large liver of the arachnids and crustaceans, and none of them has typically branched appendages. Furthermore, all have similar excretory organs and a so-called fat body which is lacking in the crustaceans. Insects probably evolved from myriapods with 14 postcephalic segments by reduction in the number of functional legs and specialization of some body segments to produce well-differentiated thorax and abdomen. Most ideas regarding their further evolution have been based on the characters of their wings.

Primitive Wingless Insects

A very primitive stage in insect evolution is, without much doubt, represented by a group of modern species which lack wings and in their general nature resemble the young stages of some more advanced forms (Fig. 272-2). Fossils of a similar kind are known, including the oldest certainly identified insects, which are middle Devonian in age. Wingless species which are specialized, rather than primitive, occur, however, in several of the insect groups. Some entomologists have postulated that all wingless forms evolved from normal winged ancestors, but this theory no longer finds much favor.

Differences between the primitive wingless insects and their probable myriapod ancestors are relatively minor and consist mainly in reduction of the walking legs to six. These are attached to the first three body segments and serve to distinguish a thorax from an abdomen. Otherwise the body segments differ little among themselves, and they are not segregated to form two body regions, as in many other insects. Some of the abdominal segments bear tiny paired appendages that may be the vestiges of legs. The primitive nature of these insects also is attested by other peculiarities possessed by some of them. Thus the tracheae generally are much less developed than in other insects, the mouth parts may be less specialized, and some species have simple eyes similar to those of most myriapods, rather than compound eyes.

Reduction in the number of functional legs in a terrestrial arthropod of limited length seems to have been an advantageous development. Studies of the gaits of myriapods have shown that if the legs are long, highly specialized coordination is required to prevent serious interference. Also the action of many legs on separate flexibly joined segments is likely to produce lateral undulations of the body that are not conducive to speedy locomotion. These disadvantages were overcome by leg reduction and close grouping beneath a relatively rigid thorax. Legs of the primitive wingless insects are not bunched together on a thorax (Fig. 272-2) as they are in many of the more familiar kinds. Reduction to three pairs, however, produced a structural system that could be perfected by further evolution.

Archaic Winged Insects

Most of the oldest winged insects yet discovered have been collected from strata of Pennsylvanian age. They are separated by a long interval of time from the earlier wingless Devonian ones, and their variety shows that much evolutionary radiation had been accomplished in addition to the primarily important development of wings.

Attempts to explain the acquisition of wings by insects have inspired much speculation by entomologists. One theory which was favored in the past postulates that insect wings evolved from the lateral parts of the thoracic segments of presumed trilobite ancestors. Another supposes that they evolved from the gill branches of crustacean legs. Neither accounts for the transformation necessary to convert a marine arthropod into a terrestrial air-breathing one. Also the conclusion is required that all wingless insects reflect the action of regressive processes. A more probable explanation is suggested by the presence of thin lateral outgrowths on the first thoracic segment and smaller ones on other segments of some late Paleozoic insects (Fig. 275). These are regarded as structures that developed on all postcephalic segments and permitted their possessors to glide through the air from plant to plant. The outgrowths on the second and third thoracic segments became functional wings when evolutionary processes supplied them with muscular articulation.

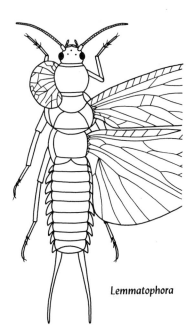

Fig. 275 Partial reconstruction of an archaic early Permian insect with the left wings and right expansion of the first thoracic segment removed. Lateral foliar expansions of the body segments probably provided the basis for wing development. Markings on the large expansion of the first thoracic segment are pigment bands, rather than true veins, but probably were underlain by tracheae. (× 7½.) (*After Tillyard, 1928, Am. J. Sci., ser. 5, vol. 16, fig. 1, p. 188.*)

Lemmatophora

Studies of wing venation indicate that a single fundamental pattern of the longitudinal veins can be recognized which is the basis for the many modifications exhibited by all insect wings (Fig. 275a). Therefore, all winged insects, and also those that are secondarily wingless, are believed to have evolved from a single, as yet unknown, ancestral stock. The most primitive wings are least modified from this pattern and have few or no cross veins. A primitive condition also is shown by

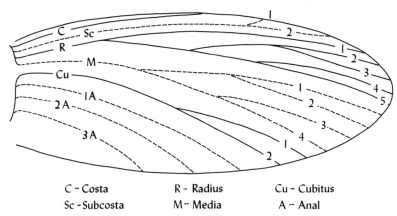

| C - Costa | R - Radius | Cu - Cubitus |
| Sc - Subcosta | M - Media | A - Anal |

Fig. 275a Generalized representation of the basic pattern of tracheation and venation of an insect wing from which all other patterns probably were derived. Some entomologists think there were primitively two rather than three anal veins. In this and the following illustrations alternate groups of veins are distinguished by solid and dashed lines. (*Modified after Snodgrass, 1909, Proc. U.S. Natl. Museum, vol. 36, fig. 5, p. 544.*)

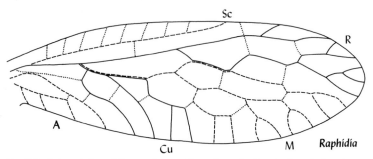

Fig. 276 Interpretation of the complex venation pattern of a Permian insect wing. Compare the veins indicated by solid and dashed lines with those of the preceding figure. Cross veins unrelated to tracheae are shown by dotted lines. Notice how the two main branches of the media have coalesced with branches of the radius and cubitus. *(After Tillyard, 1932, Am. J. Sci., ser. 5, vol. 23, fig. 3, p. 10.)*

similarity of the two wing pairs, both of which commonly are relatively long and oval, with wide bases of attachment. Wings of this kind are capable of up and down movement but cannot be turned back parallel to the body.

All Pennsylvanian insects were archaic in one or more respects. Among them, however, are several groups which bear sufficient resemblance to modern forms that they seem to be closely related to the stocks from which these later ones evolved. Most noteworthy are the relatively abundant roaches, which, although specialized in some respects, have altered less in post-Pennsylvanian time than almost any other insects. Two other modern groups that retain some conspicuously archaic characters are the dragonflies and mayflies. Insects of modern types first appear in the Permian faunas, and most of the archaic groups are not represented in post-Paleozoic collections.

Wings and Evolution

Modern insects have differentiated in so many ways that the phylogenetic relations of most major groups are obscure. Detailed comparisons reveal similarities and differences in such diverse combinations that both parallel and convergent types of evolution almost surely have occurred. Although no completely convincing pattern of evolution can be demonstrated, several important trends in insect evolution are indicated by the structural comparison of modern forms with the late Paleozoic fossils. Other trends also are suggested by the anatomy and ontogeny of modern insects that show various degrees of specialization from what was presumably a primitive condition.

The comparative morphology of insect wings has been studied intensively. Among the characters that are believed to be particularly significant in indicating genetic relations and evolutionary advancement are (1) wing venation, (2) position of wings when not in use, (3) differentiation of fore and hind wings, and (4) structures aiding in wing coordination.

(1) The main longitudinal veins of most insect wings originate as tracheae in the growing wing buds. They develop from a common generalized pattern (Fig. 275a) that is modified with growth. The changes effected in many groups of insects are so great, however, that the final patterns may seem to bear little or no relation to

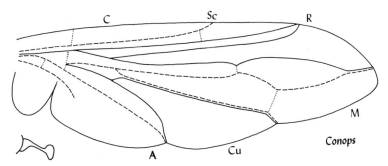

Fig. 277 The venation of this forewing of a fly is simplified by the obsolescence of some distal vein branches and the coalescence of others. The hind wings of flies are reduced to small knobs. (*After Macgillioray, 1923, "External Insect Anatomy," fig. 74, p. 331.*)

the original tracheation or to one another. The homologies of the primary veins in most modern groups, however, have been determined on the basis of ontogenetic studies. Similar studies of the development of fossil wings are not possible, but an understanding of the different types of modern venation aids in the interpretation of most fossil patterns (Fig. 276).

Starting with a presumed primitive wing whose venation corresponds with the generalized pattern that has been mentioned, evolution produced various changes, some of which resulted in simplification and others in complication of part or all of the venation. Simplification was effected by the coalescence of some of the primary veins, so that their number was reduced, or by their weakening and final obsolescence (Figs. 277; 277a). Some modern insects have wings almost lacking in

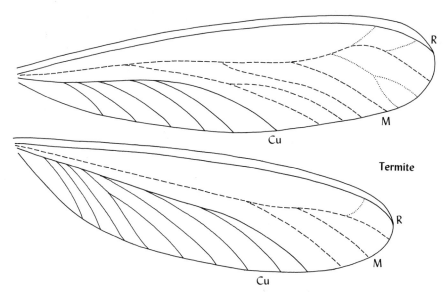

Fig. 277a Several of the main veins of *Hamitermes* weaken and tend to disappear. (*After Tillyard, 1926, "The Insects of Australia and New Zealand," fig. H1, p. 101.*)

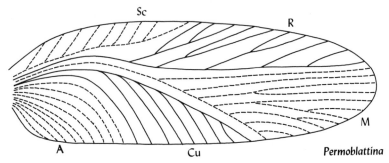

Fig. 278 This forewing of a Permian roach is an example of a type with repeated branching of the primary veins. (*After Tillyard, 1937, Am. J. Sci., ser. 5, vol. 34, fig. 10, p. 186.*)

venation. Complication resulted from (*a*) branching (Fig. 278), (*b*) anastomosis, (*c*) introduction of new longitudinal veins totally unrelated to any of the original tracheae (Fig. 278*a*), and (*d*) a few instances of crossing of primary veins so that their order is altered in the mature wing (Fig. 279). Ordinarily some features of the primary venation, no matter how much changed, are fairly stable within most of the major insect groups. Some of the secondary features, such as minor branching, may not be stabilized, and the venation of different individuals of the same species, or even in the two wings of a pair borne by a single individual, may not correspond in minor details.

Cross veins differ from the primary longitudinal veins because they do not develop from tracheae in the growing wings. The more generalized archaic insects had no true cross veins in their wings, although these are simulated in some by thickened lines. The venation of many insects, however, is complicated by cross veins which are very numerous in some groups (Fig. 279). The distinction of cross veins from irregular portions of some distorted longitudinal veins may be difficult.

(2) The horny wing bases of the most primitive insects are continuous with the

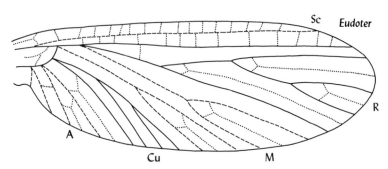

Fig. 278a This forewing of a Permian insect shows new longitudinal veins (dotted) between the primaries. (*After Tillyard, 1936, Am. J. Sci., ser. 5, vol. 32, fig. 3, p. 442.*)

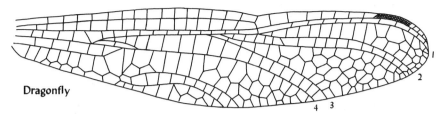

Fig. 279   This forewing of *Synthemis* has many cross veins but is relatively simple compared to the wings of some other dragonflies; branches of the medial vein are numbered; compare with Fig. 281-3. (*After Tillyard, 1926, "The Insects of Australia and New Zealand," fig. F16, p. 81.*)

thoracic integument. If a wing base of this kind is wide, only an up-and-down motion of the wing at this hinge is possible. Therefore when not in use, the wings are spread laterally at approximately right angles to the body or held vertically. Many archaic insects had wings of this kind, and the same general disposition is to be seen in modern dragonflies. Such insects were well adapted for flight but they could not hide from predatory enemies or search for food in restricted places. At the time that they developed, these early insects were the only flying creatures, and if they were agile, they had little to fear except from others of their kind. Perhaps this accounts for the large size that some of them attained. A few late Paleozoic species had wing spans of more than 2 ft and probably were the largest insects that ever lived.

Ability to fold the wings back along their bodies enabled insects to hide in cracks or under leaves and other objects and to seek food in similar places. Obviously this was advantageous for many of the early insects, which were mostly small and relatively defenseless. This ability was acquired very early, perhaps in Mississippian time and certainly well before the beginning of the Permian.

If wings are to be folded back, greater mobility at the hinge junction is required. A narrower wing base makes somewhat freer movement possible, as in the damselflies, but a more efficient structure is provided by a series of tiny horny plates separated by a flexible membrane, which intervenes between the rigid covering of the thorax and the wings (Fig. 280). Most modern insects have such hinges, and many ancient species undoubtedly were similarly constructed, although the details cannot be made out in most fossil specimens. The general similarity of these basal wing plates in many kinds of insects may indicate that they had a similar evolutionary origin, but this is far from certain. Several separate lineages may be indicated, however, by different dispositions of the retracted wings. Some lie flat upon the abdomen, overlapping each other, as in the true bugs. Others meet along the midline without overlap, as in the beetles. Almost all gradations occur between wings of these types and those which slope outward, roof-like, or lie alongside the body without actually meeting dorsally.

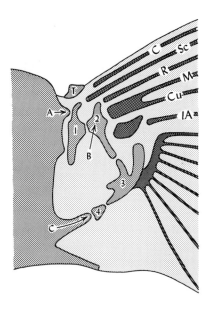

Fig. 280   Diagram showing the detailed hingement structure in the hind wing of an insect. *T*, Tegula, a horny pad. 1 to 4, Thickened articulating pieces in the wing base. Only 1 is present in mayflies and dragonflies. 4 is present only in bees and their near relatives. *A*, Principal point of articulation for up-and-down motion. *B*, Pivot that permits wings to be turned back along the body. *C*, Pivot where folding of anal part of wing is accomplished. (*After Snodgrass, 1909, Proc. U.S. Natl. Museum, vol. 36, fig. 1, p. 523.*)

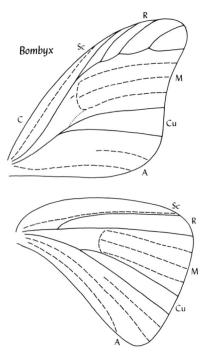

Fig. 280*a*   Wings of a moth. The media and first anal veins become obsolete proximally, and the subcostal and first radial veins of the hind wings coalesce. (*Compare Comstock, 1918, fig. 39, p. 58.*)

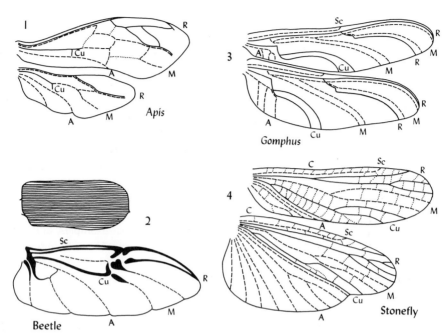

Fig. 281   1, Honeybee, much simplified venation and reduced hind wing. 2, Beetle, forewing transformed to horny wing case, hind wing enlarged; identification of veins uncertain because they develop ontogenetically before tracheation and are much modified by folding. 3, Dragonfly, cross veins omitted; notice curious crossing of branches of radial and media. 4, Stonefly, *Austroperla,* anal area of hind wing much expanded, folds up fan-like when wings are retracted. (*After Snodgrass, 1934, Smithsonian Inst. Sci. Ser., vol. 5, figs. 168C, E, p. 317; Tillyard, 1926, "The Insects of Australia and New Zealand," fig. K4, p. 117.*)

(3)   The most primitive flying insects almost certainly had two pairs of wings that were equal in size, shape, and structure. Probably they were elongate-oval (Fig. 277a). From this form evolution proceeded in two general directions: (*a*) Elongation resulted in relatively narrow, more or less parallel-sided wings, like those of dragonflies (Fig. 281-3). (*b*) Basal widening generally resulted in roughly triangular wings (Fig. 281-4). Changes of these kinds may have affected the wings of both pairs almost equally and, if so, the venation of both pairs is for the most part similar. More commonly, however, one pair became relatively larger than the other, and the venation may have altered accordingly. Insects with oval or parallel-sided wings ordinarily have the first pair longer than the second, although the difference in some groups is not great. Conspicuous widening commonly was restricted to the posterior wings; the butterflies and moths are the most noteworthy exceptions (Fig. 280a).

Many insects with widened hind wings have fore wings that are not comparably shaped or enlarged; in some of them the disparity in size is great (Fig. 281-4). Hind wings of this kind fold up to occupy less space and to fit beneath the fore wings when they are retracted. In the simplest method of folding, the enlarged posterior basal parts are merely turned back beneath the anterior parts. Wings of this type evolved among the roaches before the beginning of the Permian Period. A more advanced pattern of folding developed later and involved the accordian-like or fan-like pleating of the same wing parts. This permitted the accommodation of wings still more expanded basally. Wings of this kind are characteristic of modern grasshoppers and their allies.

Another type of wing differentiation produced fore wings that became protective coverings for the hind wings and abdomen (Fig. 281-2). An arrested stage in this development occurs in the true bugs. The basal parts of their fore wings are thickened and leathery, but the outer parts remain membranous, and these wings are used in flight. The fore wings of grasshoppers have been completely altered in this way. Other modern insects, of which the beetles are the best examples, and some ancient ones that became extinct, had their fore wings transformed entirely into hardened wing cases which fit snugly over the abdomen. These wing cases in the earliest species retained their wing venation, but this has been largely lost in modern beetles. Perhaps to compensate for loss of use of the fore wings in flight, insects of this kind developed enlarged and elongated hind wings. These are variously folded in the ways that have been mentioned, but because of their greater length, the wing tips also are folded back, so that they can be accommodated beneath the wing cases.

Evolutionary reduction of wing size in different groups of insects resulted in the elimination of one wing pair, as in the true flies (Fig. 277), or the loss of both pairs. Most of the major insect groups have some wingless representatives, and a few parasitic groups, like the lice and fleas, are entirely wingless. The flies have their hind wings reduced to small club-like processes, and the fore wings of another less-important group have been similarly transformed.

The variable development of insect wings seems to provide numerous examples of parallel and convergent evolution.

(4) Efficient flight requires perfect coordination in the movements of both pairs of wings. Improved coordination has been effected by evolutionary changes in (a) the nervous system, and (b) external structures associated with the wings. Primitively each thoracic segment of an insect contained a separate pair of nerve ganglia which controlled the movements of its legs and wings. Evolutionary advancement resulted in the progressive coalescence of these three pairs of ganglia, and also others, to produce a single large nerve center, which undoubtedly contributed to better coordination. This coalescence is recapitulated in the larval ontogeny of many modern insects. The nature of the nervous system, however, cannot be determined for fossil specimens.

When extended, as in flight, the fore wings of many insects overlap the hind wings, and their close contact aids in maintaining the synchronization of their movements. In others, special structures have evolved which serve to hold the wings together. The most noteworthy are tiny hooks situated on the forward edge of the hind wings of bees and their allies which grasp a fold on the margin of the fore wings. Other structures which have a similar function, although they probably are less efficient, are bristles, spines, or narrow projections of several kinds located near the bases of either pair of wings. These are recognizable on some well-preserved fossils.

Habits

Most insects are capable of flight, but they differ greatly among themselves in the extent to which this ability is utilized. Distinction can be made, for example, between those that depend primarily on this ability to move from place to place and others which take wing rarely except when they migrate, because they are disturbed, or for mating purposes. Among the strong fliers are dragonflies, wasps and bees, and many of the true flies. In contrast, ants and most beetles, as well as many others, are largely earthbound creatures.

Differences in flight habits may be reflected in adaptations of the legs. Undoubtedly swift movement was advantageous to the insects' wingless ancestors. Most insects have retained this primitive ability to run relatively rapidly, but some, like the termites, which live a very sheltered existence have lost some of this capability. On the other hand, some of the strongest fliers also lost the ability to run. The legs of dragonflies, for example, are adapted for seizing prey and perching and they are almost useless for any kind of locomotion.

An especially interesting evolutionary adaptation is illustrated by the jumping legs of grasshoppers and similar insects. For some, jumping has substituted for swift running or even flight. With others, jumping is a prelude to flight because their ungainly bodies cannot be lifted into the air by action of the wings alone.

Insects feed upon almost every kind of organic matter. Most of them are herbivorous, but many are carnivores, including both active predators and parasites, and some consume decaying substances. Insects feed mostly during their youthful stages, and some adults have nonfunctional mouthparts or are otherwise incapable of taking any food. Most adults, however, continue to feed to some extent throughout their lives. The feeding habits are reflected in the structure of their mouthparts and in some by adaptations of their legs. For example, the forelegs of mantids have been transformed to large raptorial appendages capable of capturing and holding prey, and they are not utilized in walking.

Insects evolved from ancestors with jaws adapted for chewing. Most of them have retained jaws of this general type, although these have been modified in their details in many ways. Insects of this kind include a great variety of herbivores, carnivores, and scavengers. Important modifications produced mouthparts adapted to sucking or to piercing and sucking. The structures of these parts show that evolution occurred independently in different major groups of insects to accomplish similar results. Most commonly an elongated tube consists of the greatly altered last pair of cephalic appendages, with which other parts also may be associated.

Piercing organs, if they are present, generally are the modified first or second pair of jaw appendages, or perhaps both. In the butterflies and moths, however, the sucking tube is formed by the second pair. Insects equipped with these structures feed on both plants and animals. Many of the sucking insects developed from young with chewing jaws.

Ontogeny

The early growth of animals was discussed briefly at the beginning of Chap. 3, and the metamorphosis which converts the larvae of many invertebrates to juvenile individuals was mentioned. The transformations which occur in insect growth also are termed metamorphosis, but this word is used here with a somewhat different meaning. With insects, all morphologic changes except increase in size, from the time a tiny creature emerges from its egg to sexual maturity, are included. Either true larval or juvenile stages are missing in the ontogeny of insects.

Like other arthropods, insects grow and develop by stages, between which they molt. The number of moltings varies between 1 and more than 20, although 4 to 6 is the common range. The most primitive wingless insects are considered to develop without metamorphosis. Almost all hatch in a condition essentially similar to the adults and only increase in size at each molt. They pass through a series of juvenile stages, but these cannot accurately be considered larval. Three different types of metamorphosis can be distinguished among other insects. These are (1) gradual metamorphosis, (2) incomplete metamorphosis, and (3) complete metamorphosis (see Table 284).

(1) Insects that develop by gradual metamorphosis hatch from the eggs as juveniles, which grow and pass by small changes at each molting to the mature form. The juveniles commonly occupy the same environment as adults and feed similarly to them. The young are wingless, but wings develop slowly as external buds and become functional only after the last molt. The term of larva has been applied to juveniles of this kind, but they are more properly known as nymphs. Insects that

Table 284   Three types of insect metamorphosis

Gradual metamorphosis: eggs hatch as nymphs which develop by small progressive changes at each molting to the final mature form. Nymphs commonly inhabit nearly the same environment as adults and feed similarly to them. Wings grow as external buds and become functional only after the last molt.

Incomplete metamorphosis: eggs hatch as aquatic naiads which inhabit an environment totally unlike that occupied by adults. Wings develop as external buds. Some morphologic changes occur at each molting, but the greatest transformation comes when the young emerge from the water, molt for the last time, and become air-breathing adults with functional wings.

Complete metamorphosis: eggs hatch as worm-like larvae, caterpillars, grubs, or maggots. These are quite unlike adults, occupy very different environments, and feed in different ways. The larvae do little more than eat and grow, with practically no morphologic change until they pass into a resting stage, or pupa, from which the adult emerges as a remarkably transformed individual. Wings develop as internal buds.

The evolutionary relations of these processes are doubtful, but advancement probably is reflected by this order.

develop in this way include the grasshoppers and true bugs. Fossil nymphs have been discovered in late Paleozoic faunas.

(2)   In incomplete metamorphosis the eggs hatch as aquatic young that live and grow in an environment totally different from that inhabited by adults. The wings of these insects also develop as external buds. Some morphologic changes occur at each molting, but a more important transformation is accomplished when the young emerge from the water and assume the air-breathing form of adults, with functional wings. These young have been termed both larvae and nymphs, but distinction is desirable and naiad is a preferable name. The best-known insects that develop in this way are dragonflies and mayflies.

(3)   The young of insects that undergo complete metamorphosis are worm-like creatures, known as caterpillars (Fig. 271-2), grubs, and maggots, that are very different from their parents. They occupy environments and feed in ways quite unlike those of the adults. These creatures grow with little morphologic change until they pass into a resting stage, or pupa, when remarkable transformation occurs, and adults are produced without the intervention of any juvenile stage. Wing development is not the same as in other insects, because these organs grow from internal buds. Young of these kinds are properly known as larvae. This type of development is characteristic of many of the most common insects, including the true flies, butterflies and moths, beetles, and wasps and bees.

The ontogenies of insects are exceedingly interesting from an evolutionary standpoint. The most primitive insects were terrestrial air-breathing creatures, and the earliest wing bearers almost certainly metamorphosed gradually. The naiads and larvae of some modern species resemble the primitive wingless forms, and they may be to some extent examples of recapitulation. This cannot be true, however, for other insects that develop by either incomplete or complete metamorphosis. Many young of both these kinds have become secondarily specialized in adaptation to ways of life very different from those of their ancestors. These are probably the best-known examples in the animal kingdom of evolutionary processes that affected the development of young and mature stages in different and unrelated ways. With them ontogeny certainly does not recapitulate phylogeny.

Both the incomplete and complete types of metamorphosis are modes of development in groups of insects which differ conspicuously and are not closely related to one another. This seems to indicate that evolution has operated independently in these groups and by convergence has produced some very similar results. This is most convincingly demonstrated by the several distinctively different types of gill-like structures possessed by various kinds of naiads and aquatic larvae.

Those insects whose young are adapted to an aqueous environment surely evolved from ancestors which were exclusively terrestrial. The immature development of naiads is gradual and similar to that of the nymphs of other insects except for the final stage, when abrupt transformation produces air-breathing adults. Thus incomplete metamorphosis is evolutionarily more advanced than gradual metamorphosis.

The processes of growth and morphologic development operate concurrently in gradual and incomplete metamorphosis. Complete metamorphosis, however, has largely accomplished the separation of these processes in a remarkable way, and each is restricted to a different part of the ontogenetic sequence. The larva does little more than eat and grow. Its lack of change in other ways is an example of arrested development and the perpetuation of a state that seems to correspond in some respects to a late embryonic stage in the development of other insects. Metamorphosis follows in a resting stage when development without any additional growth transforms the larva into an adult. Insects which undergo complete meta-morphosis also evolved from ancestors whose metamorphosis was gradual. Their development in this respect is evolutionarily more advanced than that of any other insects.

Social Organization
An extraordinary feature of insect evolution, not matched in any other animals, has been the perfection of social life in colonies consisting of morphologically differen-tiated members, or casts, which cooperate with one another instinctively in the performance of specific tasks. The whole colony almost has the quality of a single organism whose survival depends on the proper functioning of all its parts. The society so created has even been compared to human society, although, of course, there are obvious and fundamental differences. The development of such insect colonies involving the concurrent evolution of several types of individuals along divergent lines poses problems in genetics because adaptations of the different casts are advantageous to the colony as a whole, rather than to the survival of individuals in the ordinary way.

Colonial insects occur among the wasps, bees, and ants, which are closely related groups, and the termites, which are only very distantly related to the others. Surely the social habit has evolved at least twice, and some entomologists have be-lieved that it may have progressed independently in closely similar ways in more than 20 different lineages. The ants and termites are exclusively social, but the wasps and bees show all gradations from solitary to colonial types. Both sexes are included in all phases of the termite societies. In the other groups, males perform only a sex-ual function and the colonies are almost entirely highly organized female com-munities. In the most advanced colonies of both ants and termites, different com-munal tasks are assumed by separate worker and soldier castes.

The social evolution of these insects cannot be traced historically by means of fossils. Certainly it began very long ago, perhaps in pre-Mesozoic time for ter-mites and in the pre-Tertiary for ants. Nevertheless, a series of progressive stages can be recognized among modern insects that almost surely outlines the path that evolution followed. The stages as listed here must be considerably generalized, because the organizational details of insect colonies and the behavior of their members are so varied and complex that not all can be considered. The main stages, however, are: (1) Eggs are laid by the females at any convenient place. (2)

The eggs are laid near or on food suitable for the young. (3) The females store food in suitable places and lay their eggs on or in it. So far the behavior is that of solitary insects. (4) The mother brings food to the brood of growing young and may tend them in other ways. This stage demonstrates the existence of maternal instinct, and such behavior has been termed subsocial. (5) The first brood raised by the mother, which differ little from her morphologically, shares or completely takes over the duties of tending later broods. A social colony has developed, but it is not likely to outlast the favorable season of a single year. (6) After founding a colony the mother does nothing but lay eggs, and all labor is performed by morphologically differentiated workers. The colony has been perfected. It is likely to develop and grow for several years and may persist until some catastrophe strikes it.

Every colony is a family unit founded by a mother and inhabited by her off-spring. The latter part of the foregoing sequence is particularly characterized by the increasing differentiation of the mother, or queen, and one or more worker castes. The latter progressively take over more responsibilities in the increasingly complex and more highly organized colonies. Probably the most significant factor in this evolution has been the lengthening life-span of the queen. The female of most solitary insects soon dies after depositing her eggs. In the more advanced colonies the queen may live and produce eggs for as many as a dozen years.

A rough correlation exists, except among termites, between the stage of colony development and diet, with transition from carnivorous to herbivorous habits, and finally to the cultivation of fungus gardens. Likewise, advancement, particularly among wasps and bees, is exhibited by the nests, which range from simple burrows, to structures built of first mud and then chewed plant fibers cemented with either saliva or excreta, or both, and at last to honeycombs of wax. Internal construction also becomes more complicated, progressing from simple passages to special brood chambers and culminating in the intricate cell structure of a hive.

Colonial organization provides several advantages for an insect population. Most important probably are efficiency of reproduction and protection of the young. Curiously, however, most of the young do not develop into breeding individuals but become functionally sexless workers. Nevertheless, compensation is provided by the fecundity of the queen. In the more advanced colonies, breeding individuals seem to be produced only when they are needed for continued survival of the colony or at certain seasons before swarming and the establishment of new colonies.

Among wasps, bees, and ants, the differentiation of queens and workers, both produced from similar fertile eggs, results from differences in the quality or quantity of food provided for the growing grubs. The development of these two types of female individuals thus is controlled by an important part of their environment, not by difference in their genetic heritage. The males, which hatch from infertile eggs, exhibit no similar dimorphism.

The situation among termites is more complex and possibly somewhat differ-ent. The workers are juveniles of both sexes whose development has been arrested. The role of food probably is important with them also, but some deterministic differences in the eggs from which they hatch has been suspected. Some workers under certain circumstances may change their status and progress to another ontogenetic stage by molting.

All the peculiarities of social insects are, of course, basically determined by their genetics. Included in their inheritance is an extreme morphologic and behavioral reaction to differences in some environmental influences. This has evolved in the normal way by multiple mutations but is unique in the extent that it has worked to the advantage of the whole colonial community rather than to that of the individual.

Relative Evolutionary Status

Insects may be graded as evolutionarily more or less advanced, but the order which they take depends on the particular characters that are compared. Thus grasshoppers rank low on the basis of their gradual metamorphosis and chewing jaws, but high on the basis of their jumping legs and their wings, with the fore pair thickened as protective covers and the hind pair enlarged basally and folded in a fan-like manner. In another example, termites are relatively primitive if judged by their basic structure and gradual metamorphosis, but they are among the most advanced in their social organization and caste differentiation. Comparison in this way does not provide much aid in attempts to assess the relative evolutionary status of the different insect groups. Flies, however on the whole, seem to be among the most advanced. Their complete metamorphosis from legless larvae, sucking mouths, and reduction to one pair of wings are all highly evolved characters.

BIBLIOGRAPHY

R. H. Benson et al. (1961): Ostracoda, in "Treatise on Invertebrate Paleontology," R. C. Moore (ed.), pt. Q, Arthropoda 3, Geological Society of America, Boulder, Colo., and University of Kansas Press, Lawrence, Kans.
Phylogenetic diagrams appear on pp. 82, 83, and 87, but evolution is not discussed.
C. T. Brues, A. L. Melander, and **F. M. Carpenter (1952):** Classification of Insects, *Bull. Museum Comp. Zool.,* vol. 108.
See pp. 777–827 for remarks on relations of fossil insects.
J. H. Comstock (1918): "The Wings of Insects," Comstock Publishing Associates, Ithaca, N.Y.
This is a comprehensive account of the morphology and development of venation patterns.
————— **(1950):** "An Introduction to Entomology," 9th ed., rev., Comstock Publishing Associates, Ithaca, N.Y.
This is a standard textbook on the morphology, classification, and behavior of modern insects.
E. Dahl (1956): Some Crustacean Relationships, in "Bertil Hanström: Zoological Papers in Honour of His Sixty-fifth Birthday, November 20th, 1956," K. G. Wingstrand (ed.), pp. 138–147, Zoological Institute, Lund, Sweden.
The theoretical evolutionary relationships of crustacean subclasses are discussed.
W. T. M. Forbes (1943): The Origin of Wings and Venation Types in Insects, *Am. Midland Naturalist,* vol. 29, pp. 381–405.
The evolution of wings is considered, and the relations and homologies of wing veins are traced.
W. Garstang and **R. Gurney (1938):** The Descent of the Crustacea from Trilobites and Their Larval Relations, in "Evolution: Essays on Aspects of Evolutionary Biology, Presented to Professor E. S. Goodrich on His Seventieth Birthday," G. R. de Beer (ed.), pp. 271–286; Oxford University Press, Fair Lawn, N.J.
The authors' opinions do not conform to the ideas of paleontologists.
M. F. Glaessner (1957): Evolutionary Trends in Crustacea (Malacostraca), *Evolution,* vol. 11, pp. 178–184.
Most of the differentiation evident in modern forms had its beginning in the late Paleozoic.
H. J. Harrington et al. (1959): Trilobitomorpha, in "Treatise on Invertebrate Paleontology," R. C. Moore (ed.), pt. O, Arthropoda 1, Geological Society of America, Boulder, Colo., and University of Kansas Press, Lawrence, Kans.
This book is largely devoted to trilobites but also includes sections on onychophorans and a variety of arthropods, mainly from the Middle Cambrian Burgess Shale.
A. Petrunkevitch (1949): A Study of Paleozoic Arachnida, *Trans. Conn. Acad. Sci.,* vol. 37, pp. 69–315.
See pp. 102–112 for consideration of evolutionary trends and relationships.
V. Pokorný (1965): "Principles of Zoological Micropalaeontology," vol. 2, trans. by K. A. Allen, Oxford University Press, Fair Lawn, N.J.
Some casual remarks concerning evolutionary features are scattered through the section on ostracod morphology; see pp. 74–113.
G. O. Raasch (1939): "Cambrian Merostomata," *Geol. Soc. Am. Spec. Paper* 19.
See pt. III, pp. 69–84, for an interpretation of the relations and phylogeny of aglaspids and some other animals.
F. Raw (1955): The Malacostraca: Their Origin, Relationships and Phylogeny, *Ann. Mag. Nat. Hist.,* ser. 12, vol. 8, pp. 731–756.
An explanation is based on presumed evolutionary stages connecting trilobites and annelids; see bibliography for previous publications.
P. E. Raymond (1920): The Appendages, Anatomy, and Relationships of Trilobites, *Mem. Conn. Acad. Sci.,* vol. 7.

See pt. III, pp. 106–151, for comparisons of trilobites with other arthropods and conclusions regarding their evolutionary relations.

H. L. Sanders (1957): The Cephalocarida and Crustacean Phylogeny, *Syst. Zool.,* vol. 6, pp. 112–128.

The possible phylogenetic relations of crustacean subclasses are indicated by comparative morphology of appendages.

A. G. Sharov (1966): "Basic Arthropodan Stock with Special Reference to Insects," Oxford University Press, Fair Lawn, N.J.

The Russian author traces in much detail the supposed evolution to and among arthropods in a somewhat unorthodox way.

R. E. Snodgrass (1938): Evolution of the Annelida, Onychophora and Arthropoda, *Smithsonian Inst. Misc. Collections,* vol. 97, no. 6.

See pp. 132–149 for a detailed theoretical account of structural modifications, beginning with a planula and leading to all the main arthropod groups.

L. Størmer (1944): "On the Relationships and Phylogeny of Fossil and Recent Arachnomorpha," Norske Videnskaps Akademi i Oslo Skrifter, Naturvidenskapelig Klasse, ny ser. no. 8.

This is an important work, concerned mostly with marine arthropods other than crustaceans.

_____ **et al. (1955):** Chelicerata, in "Treatise on Invertebrate Paleontology," R. C. Moore (ed.), pt. P, Arthropoda 2, Geological Society of America, Boulder, Colo., and University of Kansas Press, Lawrence, Kans.

See this book for accounts of xiphosurans, eurypterids, and arachnids.

C. J. Stubblefield (1936): Cephalic Sutures and Their Bearing on the Current Classification of Trilobites, *Biol. Rev.,* vol. 11, pp. 407–440.

Hypoparians are polyphyletic, proparians are examples of arrested development, and metaparians are the most primitive of trilobites.

O. W. Tiegs and **S. M. Manton (1958):** The Evolution of the Arthropods, *Biol. Rev.,* vol. 33, pp. 255–337.

Theories of relations and systems of classification are reviewed, and conclusions based on morphology, ontogeny, biochemistry, and behavior are reached that arthropods are diphyletic; the suggestion is made that they may not have evolved from annelids.

R. J. Tillyard (1931): The Evolution of the Class Insecta, *Papers & Proc. Roy. Soc. Tasmania, 1930,* pp. 1–89.

The author reviews older theories and concludes that insects and myriapods evolved from a common terrestrial ancestor.

J. M. Weller (1937): Evolutionary Tendencies in American Carboniferous Trilobites, *J. Paleontol.,* vol. 11, pp. 337–346.

Most conspicuously, the number of thoracic segments decreased and the number of pygidial segments increased.

W. M. Wheeler (1928): "The Social Insects, Their Origin and Evolution," Kegan Paul, Trench, Trubner & Co., Ltd., London.

This somewhat out-of-date book contains much information on the morphology and behavior of wasps, bees, ants, and termites.

H. B. Whittington (1966): Phylogeny and Distribution of Ordovician Trilobites, *J. Paleontol.,* vol. 40, pp. 696–737.

Evolutionary relations are suggested in a series of pictorial diagrams.

_____ and **W. D. I. Rolfe** (eds.), **(1963):** "Phylogeny and Evolution of Crustacea," Museum of Comparative Zoology Special Publication. Harvard University, Cambridge, Mass.

The eight papers of this symposium present a great deal of information concerning modern crustaceans; much of the speculation on evolution is contained in the recorded discussions.

F. E. Zeuner (1940): Biology and Evolution of Fossil Insects, *Proc. Geol. Assoc.,* vol. 51, pp. 44–48.

This paper presents a very brief and generalized account of the occurrence and evolution of fossil insects.

CHAPTER SEVEN

ENTEROCOELATE INVERTEBRATES

ECHINODERMS
INVERTEBRATE CHORDATES

The coelomic cavities of enterocoelate animals develop as pinched-off invaginations of the entoderm. Two great phyla are recognized, (1) echinoderms, and (2) chordates, although numerous recent classifications list other phyla for the reception of the invertebrate chordates. For most zoologic and paleontologic purposes, a simple division between invertebrates and vertebrates is convenient, although the division falls within the generalized chordate phylum. Consideration of the vertebrates is reserved for the following chapters.

ECHINODERMS

The echinoderms are one of the best characterized of the invertebrate phyla, in spite of the great diversity of form which they exhibit. From at least the Ordovician Period onward, some of them have been highly successful animals, as shown by their abundance and variety in both fossil and modern faunas. Nevertheless, the echinoderms seem to be peculiarly primitive in some respects, especially in their failure to develop certain organs and structures common to other phyla. For example, they have no head, no coordinating nerve center comparable to a brain, no specialized excretory organs, no blood in any ordinary sense, and no organ like a heart to circulate blood-like fluid. They are, however, specially characterized by their complex so-called water vascular system, consisting of cilia-lined vessels, that opens to the exterior through a hydropore and contains modified seawater. Important parts of this system are the ambulacra, generally five in number, that radiate from the mouth region. Each ambulacrum is a simple to variously forked enclosed canal and related structures from which small tentacles branch off on either side. These tentacles, known as tube feet in some groups, serve feeding, respiratory, and locomotive functions in different degrees.

Most of the echinoderms possess a skeleton secreted by mesodermal cells, thus resembling in its origin the bones of vertebrates. The skeleton is composed of calcite and consists of plates, in each of which the mineral matter is arranged as in a single crystal. The plates of modern echinoderms have a delicate reticulated meshwork structure whose interspaces are occupied by living tissue. These spaces are filled with secondary calcite in most fossils so that the plates are solid and break along the characteristic cleavage planes of this mineral.

Five distinct groups of echinoderms, all exclusively marine and with long fossil records, are represented in modern faunas, and numerous other kinds occur in Paleozoic strata. (See Table 292.) The fossils provide little evidence of the interrelations of these groups. The major evolutionary patterns are especially obscure, because strange fossil forms share in various ways some of the characters seen in different modern groups. These fossils evidently represent a spectrum of evolutionary lineages that radiated from primitive common ancestors, many of which sooner or later became extinct. With few exceptions, however, the fossils are so specialized that their linking ancestral forms are difficult to visualize. Most of the evolutionary radiation which is so evident was completed in pre-Ordovician time, if not in the Precambrian.

Studies of the embryonic and larval development of living echinoderms furnish some hints concerning the origin and relationships of the surviving groups. The early ontogenies of all of them are generally similar and begin with a bilateral stage. This has led to the concept of the dipleurula as a common primitive bilateral ancestor, which is imagined to have had (1) an anterior nerve center, (2) a gut opening with an anus at the position of the blastopore and a mouth which broke through

---

Table 292   The main groups of echinoderms

Blastoids: stemmed, bud-like forms; bodies consist of 13 solidly joined plates in 3 circlets; with hydrospires. Silurian to Permian.

Crinoids, sea lilies: extremely diverse, mostly stemmed forms; bodies with variously modified pentamerous symmetry; with movable arms. See Table 312. Ordovician to Recent

Cystoids: a heterogeneous lot of stemmed forms; neither blastoids nor crinoids. See Table 294. Cambrian to Devonian

Echinoids, sea urchins: unattached, mostly solidly plated forms; without arms; with movable spines. Ordovician to Recent

Edrioasteroids: sac-like attached or adherent forms; bodies flexible; without arms. Cambrian to Devonian

Holothuroids, sea cucumbers: unattached elongated worm-like forms; bodies not plated; without arms. Mississippian to Recent

Stelleroids, starfishes: unattached, flexibly plated forms; with grasping or wriggling arms. See Table 339. Ordovician to Recent

---

These definitions are much generalized, and some members of several groups are exceptional in various ways.

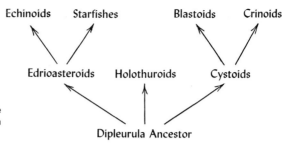

Fig. 293 Diagram showing possible evolutionary relations of the main echinoderm groups.

anteriorly as a new structure on the ventral surface, and (3) a coelom, produced by invagination of the entoderm, which consisted of three openings, each divided into right- and left-hand parts by a median longitudinal partition.

After a free-living existence, the dipleurula is supposed to have become attached by its anterior end and to have undergone changes that are recapitulated in an imperfect way by the metamorphosis of the echinoderm larvae. During this process, the mouth moved posteriorly. This distorted the creature's internal organization, and its bilateral symmetry was destroyed. Later developments resulted in the pentamerous symmetry acquired by most echinoderms. All living species which have been studied show the effects of these changes in their internal structure. All of them, therefore, are believed to have had a very ancient attached ancestor derived from the dipleurula.

The original typical echinoderm probably was a sessile animal with a looped gut and one or more tentacles located adjacent to the mouth. The tentacles were occupied by canals which were part of the water vascular system derived from the middle, left-hand coelomic cavity. This ancestor probably did not exhibit the radial symmetry which may have been acquired independently in more than one evolutionary lineage.

Ontogenetic developments suggest that the existing echinoderms represent two main evolutionary branches. Larval similarities seem to connect the crinoids, asteroids, and holothurians on one side and the echinoids and ophiuroids on the other. Holothurians possess some characters that seem to be relatively primitive, and they probably diverged very early in the evolutionary history of the phylum. Larval similarities suggest that the ophiuroids may be more closely related to the echinoids than to the asteroids, but this is not substantiated by the fossils. The evolutionary relations of most of the fossil groups are doubtful. Those shown in Fig. 294 are based on comparisons of some of the more obvious similarities, which may or may not be significant, and this pattern is highly speculative.

Early echinoderms probably secreted calcareous flesh spicules, somewhat like those of modern holothurians. From these, the plates later evolved that are so characteristic of most members of this phylum. Many of the earliest known fossil skeletons are massive, irregularly plated, and complex. They may have evolved as protection against predators. Later developments in several different lineages resulted either in more regularly organized plated structures or in more flexible skeletons. Elaborations of the water vascular system also were conspicuous and important.

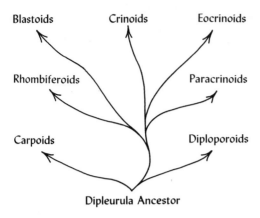

Fig. 294 Diagram showing possible evolutionary relations of the stemmed echinoderms.

Very little is known or can be inferred about evolution within several of the fossil groups which are represented by rare specimens. These belong to only a few recognized species which possess structures whose functions are not understood. The holothurians and ophiuroids among living forms have left very inadequate fossil records, because their skeletons disintegrated readily after death. Their presence in ancient faunas is indicated mainly by scattered spicules and disarticulated plates which provide little information about whole animals.

Carpoids

Carpoids are among the strangest of all extinct echinoderms. They exhibit no suggestion of the radial symmetry so characteristic of most members of this phylum, and the existence of a water vascular system is not evident. Were it not for their plated skeletons, these fossils could not be so identified. Carpoids are, in fact, so little understood that orientation of the body is not agreed upon and the functions of their appendages are differently interpreted. These animals seem, however, to have been very primitive echinoderms, and they may represent, in the most general way, a type of creature from which the other members of the phylum differentiated.

These fossils are known mainly from Cambrian and Ordovician strata, although a few younger examples have been reported. All had flattened bodies with one or more stem-like or arm-like appendages (Fig. 295). The structures, commonly

Table 294 Distinguishing features of cystidian echinoderms

Carpoids: flattened unsymmetrical forms with tail-like stems. Cambrian to Ordovician.

Diploporoids: mostly irregularly plated stemmed or otherwise attached forms; plates pierced by numerous pores ordinarily arranged in pairs; ambulacra extend over body surface without flooring plates, or raised on biserial arms. Ordovician to Devonian

Eocrinoids: stemmed forms, some resembling crinoids but with sutural pores and primitively biserial arms. Cambrian to Ordovician

Paracrinoids: more or less irregularly plated stemmed forms; with uniserial arms and pore rhombs which cross sutures between plates. Ordovician

Rhombiferoids: more or less irregularly plated to variously symmetrical stemmed forms; with pore rhombs and biserial arms. Ordovician to Devonian

Current classifications commonly unite the diploporoids and rhombiferoids as cystoids, and the other groups, formerly included with them, are given separate recognition.

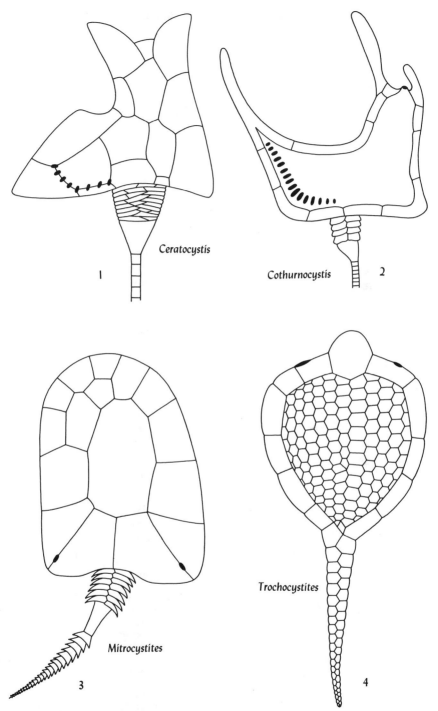

Fig. 295 Representative carpoids showing their flattened forms and tail-like appendages. The pores in some which suggest gill slits (1, 2) are particularly interesting. (*After Bather, 1925, Palaeo. Z., vol. 7, figs. 2, 10, pp. 2, 10, See Barrande, 1887, Syst. Sil., vol. 7, pt. 1, pls. 3, 4.*)

identified as stems but also interpreted as ambulacrum-bearing organs, generally taper to blunt or pointed terminations without obvious evidence of attachment. These animals probably lived in a horizontal position on the sea floor or, less probably, extending laterally from supporting seaweeds or other objects rising above the bottom. Some may have been capable of moving about in a clumsy way. All carpoids are laterally unsymmetrical, but some lineages developed increasing bilaterality of a superficial kind (Fig. 297).

The carpoids differ among themselves so remarkably in many ways that they cannot be considered a well-knit group (Fig. 298). The features of some which are of greatest interest are (1) the presence of what seem to have been slit-like openings that suggest comparison to the gill slits of invertebrate chordates (Fig. 295-1,2), although any homology with respect to them has been denied; and (2) a general resemblance to some ostracoderms among the most primitive fossil vertebrates. These similarities may have some possible significance in indicating remote relationships between creatures standing very low in the evolutionary sequences of diverging enterocoelate stocks. It is absurd, however, to accept them as evidence for a postulated derivation of the vertebrates directly from echinoderms.

Diploporoids

Most attached echinoderms that could not be classed with crinoids formerly were considered to be cystoids. These constituted an obviously very heterogeneous group, which gradually has been subdivided and restricted. The diploporoids are one of the two main groups still commonly identified as cystoids. They seem to be more primitive and only distantly related to the other group, the rhombiferoids. (See Table 294.)

Diploporoids are known to range from the Middle Ordovician to the Devonian. Most of them are constructed of relatively numerous, small, irregularly arranged plates (Fig. 299) pierced by tiny pores. These pores generally are associated in pairs (Fig. 299-7) and occur abundantly in all plates. Some primitive forms, however, have them more irregularly arranged (Fig. 299-5), and in certain specialized species they are fewer and may be restricted in their distribution (Fig. 299-4). These and other kinds of cystoid pores are supposed to have had functions connected in some way with respiration. The diplopores, or double pores, of typical diploporoids have been compared to the double tube foot pores of echinoids, but they cannot be homologous because the diplopores were not associated with the ambulacra, and they probably do not mark the sites of small tentacles. These pores may or may not have opened directly to the exterior. Many of them were covered by a thin layer of calcite continuous with the material of the plates, so that they are evident only on slightly weathered or abraded specimens.

The diploporoids exhibit what are believed to be early stages in the evolutionary development of plated stems. Some presumably primitive species were attached by a considerable basal area (Fig. 299-2). Reduction of this area and constriction of the lower part of the body (Fig. 299-1) produced an irregularly plated stem-like

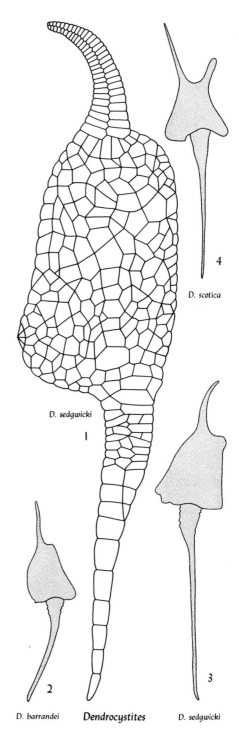

D. scotica

4

D. sedgwicki

1

D. barrandei   **Dendrocystites**   D. sedgwicki

2

3

Fig. 297 Evolution in a carpoid lineage. 1, Drawing showing the irregular plated structure, single biserial arm-like appendage, anus closed by a small pyramid of plates at the lower left, and the somewhat tail-like stem. 2 to 4, Outline drawings of three Ordovician species from progressively younger zones showing their gradual approach to symmetrical form. (See *Barrande, 1887, Syst. Sil., vol. 7, pt. 1, pl. 26.*)

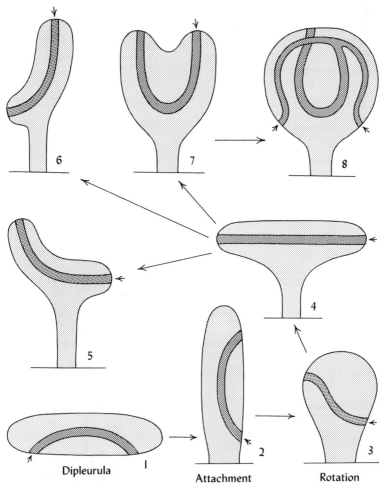

Fig. 298 Much-generalized diagrams showing possible morphologic transformations from a dipleurula to several types of carpoids. Location of mouth is indicated by small arrows. Compare No. 5 with Figs. 295-1, 2; No. 6 with Fig. 297-1; No. 7 with Fig. 295-4; No. 8 with Fig. 295-3.

structure. True stems, particularly as seen in crinoids, developed by the fusion of circlets of five plates surrounding a large central opening which became smaller and assumed several different forms.

The most primitive diploporoids reveal no evidence of ambulacra, but fleshy tentacles probably extended upward around the mouth (Fig. 299-1). Others had appendages supported by biserially arranged plates and protected by tiny covering plates. These arm-like structures arose around the mouth or were located at the ends of unfloored grooves (Fig. 300-1) which wander over the body plates in a generally aimless manner (Fig. 299-3).

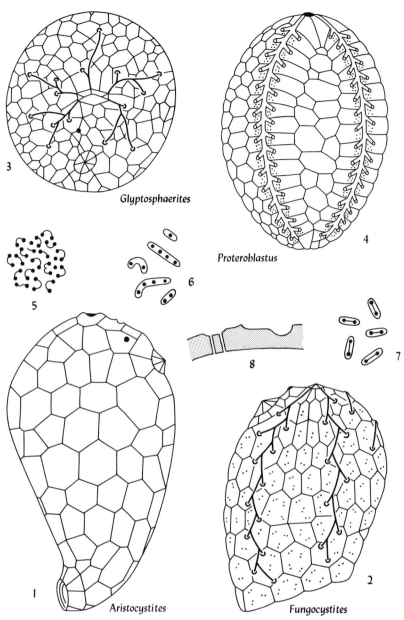

Glyptosphaerites

Proteroblastus

Aristocystites

Fungocystites

Fig. 299 Representative diploporoids. 1, This presumably very primitive form consists of a large number of irregular plates, was attached by its constricted base, and had no stem. The mouth is at the summit, without any indication of associated arms. The plates are pierced by very numerous small irregular pores (see 6). 2, Side view of a form attached by a broad basal area and showing ambulacral grooves wandering over body plates, which are pierced by a few typical diplopores (see 7). 3, Top view of another specimen showing the ambulacral grooves radiating from the mouth and ending in facets where arms were attached. 4, This form was stemmed. It is more regular in its plated structure, and diplopores are restricted to plates that bear the ambulacra. 5, Irregular pores, termed haplopores, of some primitive diploporoids. 6, Pores of a type that seem to be intermediate between haplopores and diplopores. 7, Typical diplopores. 8, Generalized section through body plates showing an ambulacral groove with a side branch to left leading to an arm facet and a longitudinal section through a diplopore. (See *Barrande, 1887, Syst. Sil., vol. 7, pt. 1, pls. 10, 17; after Volborth and Jaekel, see Bather, 1900, figs. 43, 46, pp. 44, 73, 75.*)

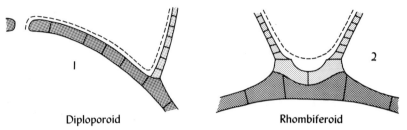

Diploporoid   Rhombiferoid

Fig. 300   1, Diagrammatic longitudinal section showing ambulacral groove in body plates not lined by flooring plates, and an arm rising from a body plate. 2, Cross section of an ambulacral groove with biserial flooring plates from which the arms arise.

Rhombiferoids

The rhombiferoids are characterized by pores or grooves which continue from one body plate to another transversely across the intervening suture (Fig. 301). Each system thus occupies portions of two adjacent plates and has a rhombic outline. These pore rhombs may be numerous and occur at every contact between two body plates (Fig. 301-1), or they may be few and restricted to only certain definite positions (Fig. 301-3). Fossils with these structures are known to range from the Middle Ordovician to the Devonian, and most of them bore arm-like appendages of typically cystidian biserially arranged plates.

The rhombiferoids are a rather heterogeneous lot, and not all the fossils included in this group may be closely related. Some consist of numerous irregular plates (Fig. 301-1), but others achieved a high degree of order and almost perfect symmetry (Fig. 301-4). Certain species exhibit resemblances to eocrinoids, paracrinoids, blastoids, and even to some crinoids. They seem to have evolved in a radiating manner from an ancestral group that may also have produced the blastoids and the crinoids.

Some paleontologists have assumed that the origins of pore rhombs and diplopores were related. Opinions have been expressed that each was derived from the other, but these two pore systems probably evolved entirely independently. The pore rhombs seem to have begun as simple pores located along suture lines between the body plates of very young individuals. As these individuals grew, the plates enlarged and the pores, displaced relatively from the suture, were transformed to grooves or canals. New pores appeared, successively flanking the earlier ones, and in their turn grew, thus producing the rhombic pattern. Diplopores may also have originated along the sutures, but if so, their subsequent development was very different.

The rhombiferoids as a group seem to be more advanced than the diploporoids in their general organization. Most of them had well-developed stems. Many consist of a relatively small number of large regularly arranged plates. The ambulacra, if they wander over the body surface, are floored by biserially alternating plates which

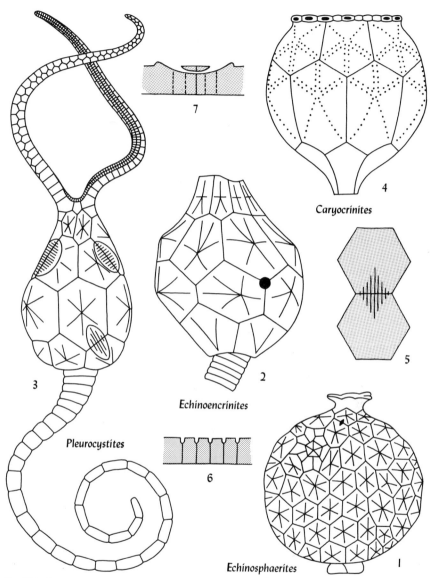

Fig. 301   Representative rhombiferoids. 1, Primitive type consisting of irregular plates joined by pore rhombs, attached at base apparently without a stem and having two to five appendages which arose around and above the mouth. 2, An advanced type with only three pore rhombs on side opposite the anus. 3, A highly specialized form with three pore rhombs, and an anus on the opposite side surrounded by many small plates. 4, A type with almost perfect sixfold symmetry and pore rhombs on all plates of two circlets. 5 to 7, Diagrammatic representations of pore rhombs as exposed on the surface (5), in transverse section parallel to a plate suture (6), and a highly specialized rhomb in longitudinal section (7). (1 to 4, *After Volborth and Jaekel, see Bather, 1900, figs. 14, 22, 34, pp. 53, 60, 65; Wood, 1909, U.S. Natl. Museum Bull. 64, pl. 2, fig. 3.*)

lie upon the body plates (Fig. 300-2) and seem to be homologous with the biserial plates of the free arm-like appendages. Finally, some of the pore rhombs became specialized in ways that probably indicate increased functional efficiency as their number was reduced (Fig. 301-3,7).

One group of the rhombiferoids is particularly interesting. Its members possess a constant number of body plates (Fig. 302), which, in spite of much distortion, can be identified individually in the various species. These show what seem to be one or more evolutionary series in which the number of pore rhombs was progressively reduced. The plate structure of the group can be manipulated theoretically, in ways suggesting that other cystoids, all blastoids, and some crinoids might have been derived from them. The conclusion that there actually are such evolutionary relations, however, is extremely doubtful.

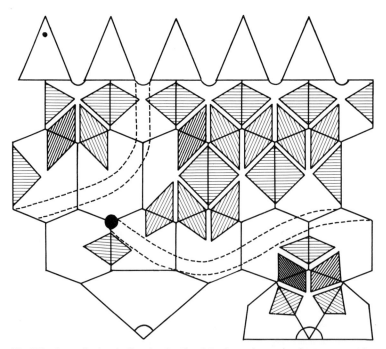

Fig. 302 Generalized projection showing the plate plan and pore rhomb arrangement in an interesting group of rhombiferoids which are basically similar but highly variable in detailed structure (see Fig. 301-2, -3). The absence of rhombs from a band leading to the anus is believed to have resulted from interference by the gut (dashed lines), which probably lay close against the inside of the plates. The number of pore rhombs was reduced progressively in this group. One, invariably present, is indicated by darkest shading. Three, next most commonly present, are shown by intermediate shading. Rhombs that may or may not occur are shaded the lightest. (*After Bather, 1914, Trans. Roy. Soc. Edinburgh, vol. 49, fig. 45, p. 439.*)

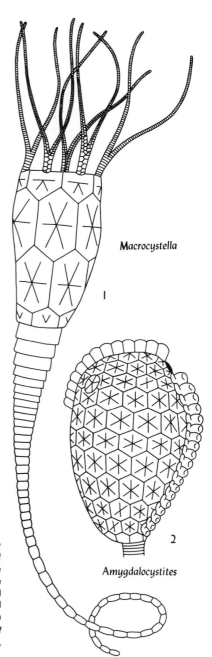

Macrocystella

1

Amygdalocystites

2

Fig. 303 1, An eocrinoid. This form is pentamerously symmetrical, like a crinoid, but has an extra circlet of body plates below the bases of the arms, which are biserial like those of cystoids. 2, A paracrinoid. The plates are connected by pore rhombs. The uniserial flooring plates of the ambulacral grooves wandered across the body plates and bore small arm branches on one side. (*After Eastman-Zittel, 1913, "Text-book of Paleontology," 2d ed., vol. 1, figs. 247, 249, pp. 156, 157.*)

Paracrinoids

The paracrinoids include a small number of Middle Ordovician fossils built of irregularly arranged body plates (Fig. 303-2). They resemble rhombiferoids in possessing pore rhombs, but the ambulacra were born on uniserial arms with small side branches, or pinnules, very different from the biserial nonpinnulated arms of cystoids. Their arms resemble those of crinoids, but whether they are homologous is an unsettled question.

The arms of crinoids are the upward extensions of the body plates, as shown by the continuity of their principal nerves. Their origin does not seem to have been related directly in any way to the ambulacral rays diverging from the mouth (Fig. 304). The association of these structures, developed as a result of evolution, is one of the most characteristic features of the crinoids. The biserial arms of crinoids evolved from those of the more primitive uniserial type, and the transition can be traced in detail (Fig. 318). In contrast, cystoid arms seem to be primitively biserial (Fig. 301-3), and there is no evidence of transformation to uniserial structure. Plated cystoid arms grew as supporting structures intimately associated with the ambulacra. They rise either almost directly from the margin of the mouth (Fig. 301-1) or at the ends of passages, most of which wander over the surface of the body plates (Fig. 299-2).

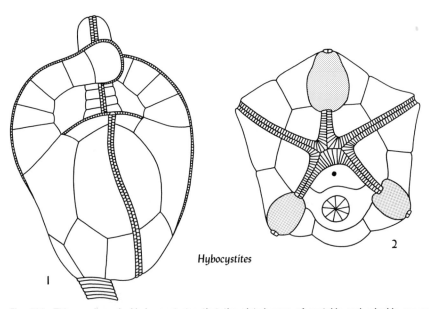

Hybocystites

Fig. 304 This peculiar crinoid demonstrates that the plated arms of cystoids and crinoids are not homologous. 1, View from the right side showing three stubby arms. 2, View from above with the arms removed. The arms grew upward as a single series of plates from three of the five radial body plates. The ambulacra, roofed by biserial covering plates, grew outward from the central mouth. Three of them extend up the insides, over the ends, and down the outsides of the arms onto the body plates. The others extend outward and down on the body plates. Specimens representing various growth stages show that the ambulacra lengthened progressively but the arms did not. The arms of cystoids are double series of plates flooring the ambulacra, and they grew as the ambulacra lengthened. (*Generalized after Springer, 1911, Geol. Surv. Canada, Mem. 15-P, pl. 2.*)

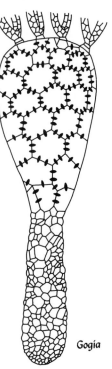

Gogia

Fig. 305 This Middle Cambrian eocrinoid has an irregularly plated stem-like appendage that, with its irregularly plated body, probably identifies it as being very primitive. There is nothing to indicate that the appendage was an organ of attachment. (*After Robison, 1965, J. Paleontol., vol. 39, fig. 3, p. 361.*)

Morphologically, the paracrinoids occupy a position intermediate between rhombiferoids and crinoids. Their generally irregular plating seems to be a primitive character, but the possibility of their being ancestral to either of these other groups is exceedingly remote. They probably are most closely related to primitive rhombiferoids, and perhaps these ancient animals sprang from common stock.

Eocrinoids

Fossils classified as eocrinoids constitute a small group known to range from the Middle Cambrian to the Middle Ordovician. They are characterized by having pores along the sutures between their body plates (Fig. 305) and by biserial arms like those of cystoids. Among them are the oldest discovered echinoderms with stem-like appendages. Two subgroups can be recognized, which may not be closely related to each other.

One subgroup, the older one that seems to be much more primitive than the other, has irregularly plated bodies and does not exhibit pentamerous symmetry. A stem-like extension consists of small irregular plates (Fig. 305) and encloses a central cavity that probably contained part of the viscera of these animals.

The other subgroup consists of specimens that resemble crinoids in some ways.

The body is composed of three or four circlets of fairly regularly arranged plates below the arms (Fig. 303-1), and some are perfectly pentamerous. One kind carried arms on two successive circlets. Stems are of the conventional type, made up of circular columnals pierced by a small central opening.

Conflicting suggestions have been made that eocrinoids were ancestral to or evolved from rhombiferoids. The former seems more likely. The possibility that these fossils represent a stock antecedent to some crinoids also is a more attractive speculation.

Blastoids

The blastoids, in contrast to all other main groups of the echinoderms, are re-markably constant in their structure. All consist of 13 body plates arranged in three circlets (Fig. 306). Those of the second circlet, which generally are the largest, are deeply notched for reception of the ambulacra. In spite of this basic morphologic uniformity, however, profound variations in shape, proportions of the plates, and ornamentation have produced forms of very different appearance. The fossils range from the Middle Silurian to the Upper Permian. Included among the earliest and latest representatives are those that deviate most widely from the few common types that are most characteristic of this group.

The evolutionary relations of the blastoids are uncertain, and their ancestors have been sought among both the diploporoids and rhombiferoids. On the one hand

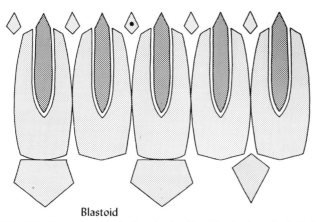

Blastoid

Fig. 306 Diagrammatic projection showing the plate structure of a typical blastoid. The small basal plate is in the right anterior position. The radials are forked, and each embraces a long narrow lancet plate that is more or less hidden beneath the tiny plates of the ambulacra. Small deltoid plates form the summit. The posterior one is pierced by the anal opening.

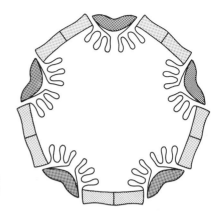

Fig. 307 Transverse section of a blastoid showing the hydrospire folds which underlie the lancet plates of the ambulacra.

some forms that seem to be assignable to the diploporoids resemble blastoids in a general way, and on the other the blastoid hydrospires (Fig. 307) which are the most characteristic features of these fossils cross suture lines between the plates and thus resemble pore rhombs. Most probably the blastoids and rhombiferoids evolved from the same ancestral stock.

Evolutionary trends among the blastoids are rather clearly shown by (1) nature and position of the hydrospires, (2) changes in the spiracles, (3) number and relations of small plates surrounding the anal opening, (4) plated structure of the ambulacra, and (5) modification of general body form and relative sizes of plates in the different circlets.

(1)   The hydrospires are delicate inward folds of some of the body plates, located beside or beneath the ambulacra (Fig. 308), which are believed to have served respiratory functions. Ciliary action probably drew water currents into them, which moved upward and then out through openings known as spiracles. Most primitively, several hydrospire slits parallel each side of an ambulacrum (Fig. 308-1). In more advanced forms, each ambulacrum is bordered by only a pair of slits, one on either side (Fig. 308-2). Finally these were partially closed by small plates of the ambu-lacrum, leaving a series of pores (Fig. 308-3) leading to the three to five hydrospire folds on each side.

Other evolutionary developments in some lineages resulted in either reduction or increase in the number of hydrospire folds associated with an ambulacrum. Also some of the hydrospires were transformed from simple parallel-sided folds (Fig. 308-1) to those which became dilated in their inner parts (Fig. 308-3). A few spe-cialized blastoids developed hydrospires in the posterior interambulacral position sharply differentiated from the others, or even lost them entirely. Some of these were without stems, except at a very early age, and lay loose but apparently im-movable on the sea bottom.

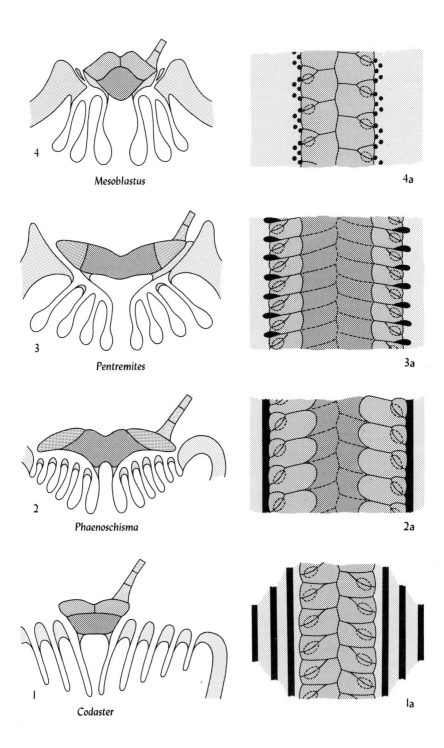

4

Mesoblastus

4a

3

Pentremites

3a

2

Phaenoschisma

2a

1

Codaster

1a

Fig. 308   (See facing page.) Diagrams showing relations of hydrospires to ambulacra; cross sections to left, surface views of ambulacra to right. 1, The most primitive type of hydrospires, with open slits paralleling the sides of the ambulacrum. 2, The hydrospire folds lie beneath the ambulacrum and open through a single continuous slit. 3, Ambulacral plates partly close the slits, leaving a series of pores on each side. 4, These highly specialized hydrospires open through a complex pore system. Paired double series of ambulacral plates conceal the lancet plate in 1a and 4a but have drawn apart and expose the median portion of the lancet plate in 2a and 3a. Facets of the delicate arms are shown along the edges of the ambulacra.

(2)   Most primitively, each hydrospire fold was an entirely separate structure, and its water escaped at the upper extremity of its slit (Fig. 309-1). As the hydrospires moved under the ambulacra, or as the ambulacra overlapped the hydrospires, entrance to a whole group of folds was gained through a single slit. Aside from this, the slits remained distinct, and each of them opened separately near the summit of the blastoid body. In the following evolutionary stage the folds on each side of an ambulacrum merged upward and led to a single spiracle (Fig. 309-3). Thus a pair of pores was present in each interambulacral position. Finally these merged to produce a single spiracle, which served the hydrospires of the adjacent sides of two neighboring ambulacra (Fig. 309-5).

Conditions in the posterior interambulacrum, where the anus was located, are

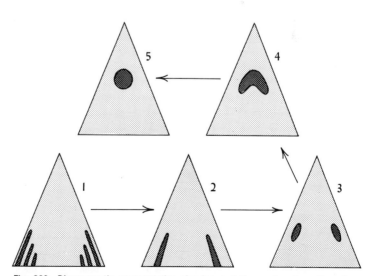

Fig. 309   Diagrammatic representation showing evolutionary transformations from hydrospire slits extending into deltoid plate (1) to a single opening or spiracle (5). The intermediate stages show a single hydrospire slit on each side (2), which is closed off behind to produce double spiracles (3), and these join across the deltoid plate (4).

somewhat different. There three openings merged into one similar to those in the other positions except that it is somewhat larger.

(3)   Single more or less triangular plates of the uppermost circlet lie between the ambulacra in all but the posterior position. Here there are most primitively six small plates surrounding the anal opening (Fig. 310-1), but this number was reduced progressively, probably by fusion of adjacent plates, until only one similar to those in the other positions remained (Fig. 310-5).

(4)   The ambulacra were borne on long narrow plates which occupied the notches in the second circlet of body plates (Fig. 306) and continued between the plates of the succeeding third circlet. These plates of the ambulacra carried two series of small plates on each side (Fig. 308-1*a* to 4*a*), from which rose biserial armlets that were very slender but otherwise like those of cystoids. The outermost plates are the ones that may extend across and partly close the bordering hydrospire slits. In the more primitive blastoids the side plates completely cover the larger plates (Fig. 308-1*a*). In more advanced forms, however, they have moved apart laterally, beginning at the top and continuing progressively downward, until the middle part of the underlying plate was exposed for its entire length (Fig. 308-3).

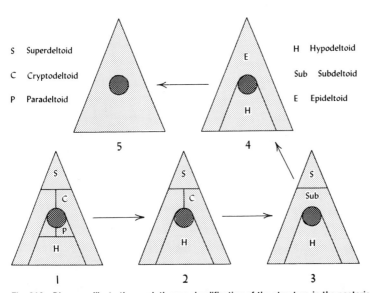

S   Superdeltoid

C   Cryptodeltoid

P   Paradeltoid

H   Hypodeltoid

Sub   Subdeltoid

E   Epideltoid

Fig. 310   Diagrams illustrating evolutionary simplification of the structure in the posterior deltoid region of blastoids. 1, The most primitive structure, as seen in *Polydeltoideus*, which possesses six small plates. 2, Fusion of the paradeltoids below anus to the hypodeltoid produces a larger hypodeltoid plate. 3, Fusion of the two cryptodeltoid plates produces a large subdeltoid. 4, Fusion of the subdeltoid plate and superdeltoid produces an epideltoid. 5, Fusion of the epideltoid and hypodeltoid produces the single anal deltoid. (*Data from Fay, 1961, Univ. Kansas Paleontol. Contr. Echin., Art 3.*)

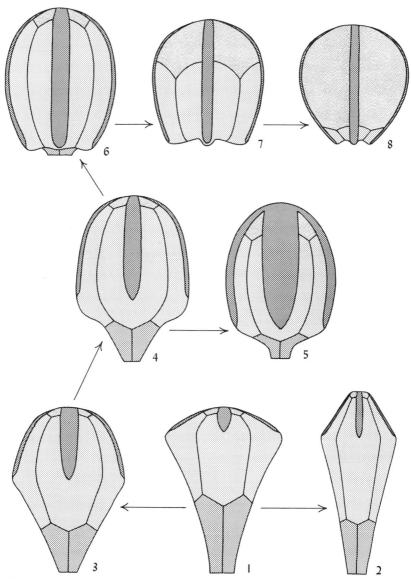

Fig. 311 Generalized sketches showing the probable evolutionary relations of some different types of blastoids. Changes from a form close to the most primitive type (1) involved lengthening of the body (2), lengthening of the ambulacra (3 to 6), decrease in the size of basals (1 to 8), widening of the ambulacra (5), and increase in size of deltoids (7, 8).

A final development in some lineages was reduction in the number of side plates until only two or three remained on each side of an ambulacrum.

(5) The most primitive form of a blastoid seems to have been steeply conical, with short ambulacra restricted to a more or less horizontal summit (Fig. 311-1). Longer ambulacra evolved notching deeper into plates of the second circlet. This

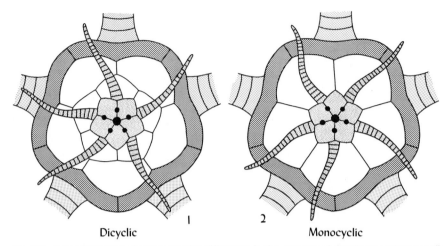

Dicyclic   Monocyclic

Fig. 312 Generalized basal diagrams showing differences in the structure of dicyclic (1) and monocyclic (2) crinoids. The basal and infrabasal plates are unshaded. Notice that the stem elements are inter-radial and the cirri, or appendages of the stem, are radial in the first but these relations are reversed in the second.

was accompanied by transformation of the summit to inverted pyramidal (Fig. 311-2) or dome-like form (Fig. 311-3). The latter development was continued by reduction in the relative sizes of plates in the two lower circlets and corresponding increase in size of those in the uppermost circlet (Fig. 311-6 to 8).

Crinoids

The crinoids have left a fossil record extending from the Ordovician to the present that is particularly full in the Paleozoic. The beauty of well-preserved specimens, their fairly common occurrence at many places, and the ease with which their very diverse plated structures can be observed have made them a favorite subject for paleontologic study. Aside from the conclusion that they undoubtedly descended from one or more kinds of cystoid-like creatures, their evolutionary origin is uncer-

Table 312 Distinguishing characters of main crinoid groups

Inadunates: mostly with solidly plated cups which ordinarily do not include brachials; arms free above radials; without completely plated dome. Ordovician to Triassic

Camerates: bodies solidly plated, consisting of cup below and dome above the level at which arms become free; cup ordinarily includes brachial plates. Ordovician to Permian

Flexibles: similar to dicyclic inadunates but plates loosely joined; upper limit of cup indefinite; arms mono-serial without pinnules. Ordovician to Permian

Articulates: similar to inadunates but without radianal or anal plates in cup; plates ordinarily loosely joined; arms monoserial without pinnules. Triassic to Recent

Monocyclic: with a single circlet of basal plates beneath the radials; both inadunates and camerates

Dicyclic: with two circlets of alternating plates beneath the radials; both inadunates and camerates

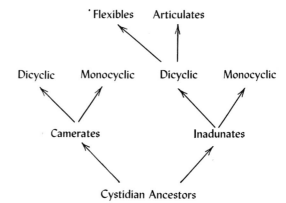

Fig. 313 Diagram showing the probable evolutionary relations of the main crinoid groups.

tain. Some species resemble in certain respects the eocrinoids, the rhombiferoids, and even some blastoids.

Most Paleozoic crinoids are assignable to two of four main groups: they are either camerates or inadunates, and likewise they are either monocyclic or dicyclic (Fig. 312 and Table 312). The evolutionary relations of these groups have been much disputed. One view held that monocyclic and dicyclic crinoids, with one or two circlets of basal plates, respectively, each evolved into camerate and inadunate branches or those with or without solidly plated upper vaults. It is now generally agreed, however, that monocyclic and dicyclic branches probably appeared independently in the primarily different camerate and inadunate stocks (Fig. 313). Opinions are not uniform respecting the relations of monocyclic and dicyclic forms in either of these lines. Each form has been supposed to have originated from the other, and the correct solution to this problem still is doubtful. It seems more likely, however, that infrabasal plates, or those of the lower or inner circlet, would be

Table 313   Names applied to crinoid plates

Infrabasals: five plates immediately above the stem in dicyclic crinoids; fusion may reduce their number
Basals: five plates immediately below and alternating with radials, overlying and alternating with infrabasals or, in monocyclic crinoids, occurring immediately above the stem, where fusion may reduce their number
Radials: five plates overlying and alternating with basals; the first and lowest plates of the arm series
Brachials: plates of arm series above radials; may be free or incorporated in the cup
Interbrachials: plates incorporated in the cup above basals or radials and between brachials
Radianal: a single plate in inadunate and flexible crinoids, ordinarily lying below and to the left of the right posterior radial
Anals: one or several plates in posterior interray, more or less different from plates that may occur in other interrays; leading upward toward the anus
Orals: five plates overlying the mouth; occupying interradial positions
Covering plates: small plates alternating in two series which roof the ambulacra
Axillaries: brachials which support two branches of the arms or arm series

eliminated during the course of evolution than that they would be additions to an older simpler structure. Several lineages of fossil dicyclic crinoids show progressive reduction in the size of infrabasals, to such an extent that they almost disappear (Fig. 315). Furthermore in the early ontogeny of the modern crinoid, *Antedon* (Fig. 317), infrabasal plates first are formed and then at a very early stage become fused with the first plate of the stem, so that in effect they disappear entirely.

A fifth group of Paleozoic crinoids, the flexibles, almost certainly evolved from an early dicyclic branch of the inadunates. One or more later branches of the inadunates probably were the source of the post-Paleozoic articulate crinoids.

Crinoids provide many examples of progressive evolutionary change which can be followed in considerable detail. Those evolutionary tendencies exhibited by many lineages of most main Paleozoic types include (1) fusion and reduction in number and size of plates in the lowest circlet, whether basals or infrabasals (Figs. 324; 325); (2) change in the shape of the cup below the arms from one that has steeply sloping sides or is globular to lower and wider forms which acquire a deeply concave base (Figs. 314; 315); (3) upward displacement of anal plates, which may be eliminated entirely from the cup (Figs. 316; 317); (4) transformation in arm structure from a relatively rigid, unbranched, uniserial condition to one in which the more flexible arms are regularly to irregularly branched and biserial (Fig. 318). These and other changes peculiar to certain restricted crinoid groups progressed independently at different rates in different lineages. Some of the more obvious or interesting evolutionary transformations are discussed in the following sections, but numerous others cannot be mentioned here.

A review of all crinoids indicates that the evolutionary kinships of many genera within groups most commonly considered to be families are reasonably clear but that the interrelations of the family groups themselves are obscure. It is not improb-

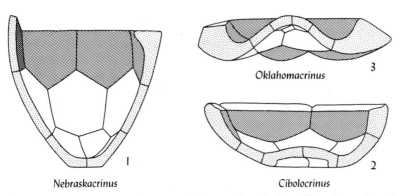

Fig. 314 Diagrams of dicyclic inadunate crinoids cut through the middle and seen from the inside, illustrating the general evolutionary tendency for the cups to become shallower and the base to become concave downward. (*After Moore, 1939, J. Sci. Lab. Denison Univ., vol. 34, figs. 2, 4a, 5b, 27, pp. 192, 200, 206, 259.*)

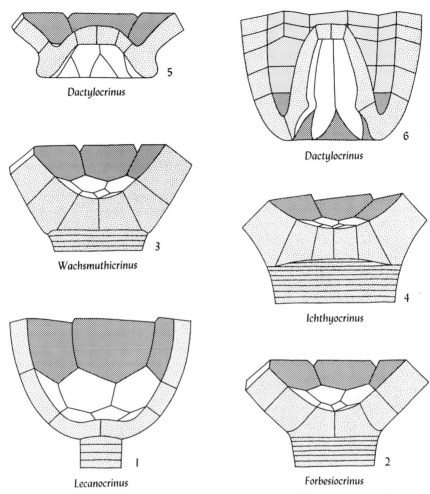

Fig. 315 Diagrams similar to those of Fig. 314 showing the same evolutionary changes in the shapes of flexible crinoid cups (1, 2, 5, 6) and also reduction in the size of infrabasal and basal plates (unshaded) so that they are withdrawn into the area of stem attachment (1 to 4). (*After Springer, 1920, Smithsonian Inst. Publ. 2501, figs. 9, 10, 11, p. 117; pl. 41, figs. 1d, 5d; pl. 43, fig. 10.*)

able that some different lineages of crinoids, even though they are classified together as camerates or inadunates, evolved from different cystidean ancestors.

Dicyclic Inadunates

The dicyclic inadunates range from the Lower Ordovician to the Triassic and are a more varied lot than any other major group of crinoids (Figs. 316; 319). Some of the earlier and more primitive examples are almost identical in their structure to contemporary monocyclic forms except for their extra circlet of infrabasal plates. It is very difficult to imagine how these could have had different evolutionary origins.

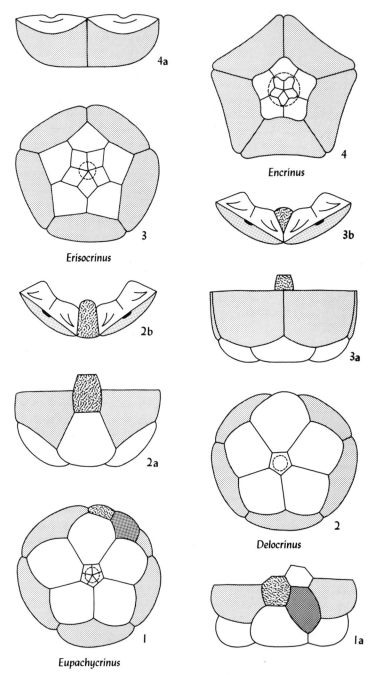

Fig. 316 Examples of dicyclic inadunate crinoids showing by bottom, side, and top views how anal plates were progressively eliminated from the cup. 1, A radianal (heavily shaded) and two succeeding anal plates separate the posterior radials (lightly shaded). 2, The single anal plate may or may not be a displaced radianal; perhaps the radianal has been eliminated. 3, The single anal plate has been crowded inward and does not interrupt the outer symmetry of this crinoid. 4, No anal plate remains in the cup. *Encrinus* is a Triassic genus; the others are of late Paleozoic age.

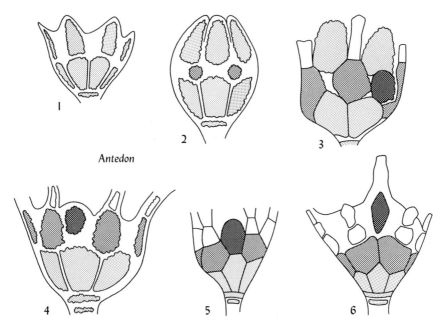

Fig. 317 Early ontogeny showing plate development in a modern crinoid that is stemless when mature. 1, Two circlets of plates, basals below and orals above, and a first stem segment with which tiny infrabasals are fused. 2, Radial plates appear between basal and oral circlets. 3, An anal plate (darkest shading) is introduced into the radial circlet, and arm plates have appeared above the radials. 4, Much like the last, but the anal plate begins to rise. 5, The cup plates are in contact, and the anal plate rests upon and truncates the posterior basal. 6, The anal plate has risen above the radials, and the two posterior plates of this circlet are in contact with each other. After a few months the young crinoid breaks loose from its stem and thereafter leads a free existence. (Enlarged but not drawn to uniform scale.) (*After Wilson, 1916, J. Geol., vol. 24, fig. 2, p. 497.*)

All the evolutionary trends mentioned in the last section occur in various lineages of the dicyclic inadunates. Parallel evolution of different features at different rates produced many structural types whose relations are very difficult to judge. Two developments, however, are particularly interesting and seem to be significant. Most crinoids have stems consisting of a column of circular plates. In contrast, the stems of several early dicyclic inadunates are composed of plates in five vertical columns surrounding a large central opening. This surely is a primitive condition and a structure probably inherited from a cyctidian ancestor with a many-plated stem. The common circular columnals evolved by the lateral fusion of circlets of these plates. Evolutionary advancement also resulted in reduction of the central opening, which acquired several different shapes and in a few camerate lineages almost disappeared.

Most inadunate crinoids have cups composed of plates with plane contact surfaces that adhere tenaciously. Among late Paleozoic dicyclic forms, however, are

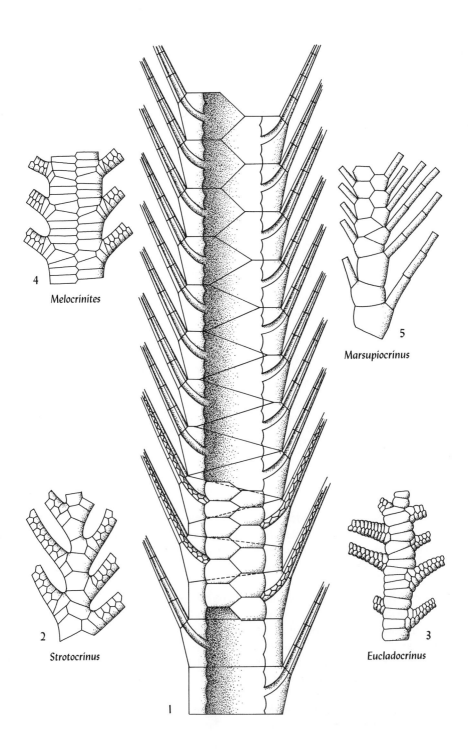

Melocrinites 4

Marsupiocrinus 5

Strotocrinus 2

Eucladocrinus 3

Fig. 318 (See facing page.) Details of crinoid arm structure. 1, Generalized representation of the inner or upper side of a crinoid arm, showing transition from monoserial to biserial condition, movable cover plates that roof the ambulacral groove, and pinnules which resemble small arms but are invariably uniserial. 2, 3, 4, Parts of arms with minor branches seen from the outer side. 5, Portion of arm with pinnules. (2–5 variously enlarged.) (*After Wachsmuth and Springer, 1897, Mem. Museum Comp. Zool., vols. 20, 21, pl. 23, fig. 3; pl. 65, fig. 1b; pl. 74, fig. 2; pl. 75, fig. 19b.*)

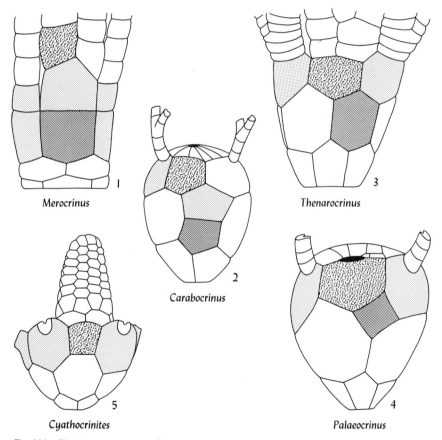

Merocrinus

Carabocrinus

Thenarocrinus

Cyathocrinites

Palaeocrinus

Fig. 319 Diagrammatic representations of the plate structures of several primitive dicyclic inadunate crinoids. Homologous plates are similarly shaded for identification; radials are shaded lightest. 1, Except for the dicyclic base, this structure is identical to that shown in 6 of Fig. 320. 2, The two small plates of the right posterior ray lie within the basal circlet. 3, Like 2 but the two small plates seem to have fused. 4, A radianal and first anal plate occur in the positions normal for many crinoids. The radianal may correspond to the two fused primitive plates. 5, A single plate between the posterior radials may be either a displaced radianal or, more probably, the first anal plate with the radianal missing. (*After Moore and Laudon, 1943, figs. 1, 4, pp. 14, 39.*)

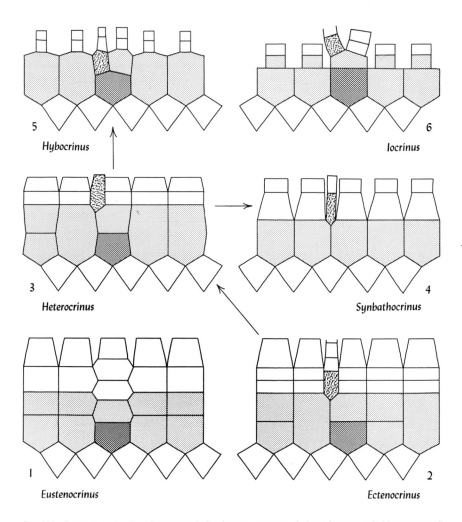

5 Hybocrinus

6 Iocrinus

3 Heterocrinus

4 Synbathocrinus

I Eustenocrinus

2 Ectenocrinus

Fig. 320 Projections showing diagrammatically the arrangement of plates in some primitive monocyclic inadunates. These suggest that the anal structures of the inadunate crinoids evolved as a specialization of the right posterior ray. Homologous plates are shaded similarly for identification; the radials are lightest. Arrows indicate possible evolutionary trends; although the genera are not necessarily related in this detailed way. 1, This is believed to be the most primitive structure known in any inadunate crinoid. The five rays are similar except that there is one more plate, in what seems to be the right posterior position, than in any other. No obvious anal structure is present. Homology of these plates is somewhat uncertain. 2, Two of the rays of this crinoid contain large plates that are believed to have resulted from the fusion of pairs of smaller plates similar to those of the other rays. A series of anal plates that seems to be a branch from the right posterior arm identifies the posterior position. 3, The two small plates of another ray are fused. 4, Either the small plates of all the rays are fused or, less probably, the upper unfused plates have become what seem to be the first brachials and the anal plate has moved downward to contact the lower plate in the right posterior position. 5, Small plates of the right posterior ray remain unfused. The arms become free above the plates identified as radials. The lower plate in the right posterior position seems to be the radianal of all more advanced inadunates. 6, None of the small plates in this crinoid seem to have fused. The upper ones in all rays except the right posterior seem to have become the first plates of the free arms. (*Modified after Moore and Laudon, 1943, fig. 3, p. 27.*)

many species whose cups disarticulated easily. The contact surfaces of their plates are hollowed out to produce spaces that seem to have been occupied by tendons or muscles. These cups probably were moderately flexible. A continuation of this evolutionary tendency in one or more unknown lineages almost surely led to the development of the post-Paleozoic articulates.

Monocyclic Inadunates

The monocyclic inadunates range from the Lower Ordovician to the Permian. They are a much smaller group of crinoids than the last, but they are especially interesting because they include some seemingly very primitive species whose evolution may explain the anal structures of all inadunates and the origin of several highly specialized and irregularly structured crinoids (Fig. 320).

The crinoid that is believed to be the most primitive of all inadunates has a slender cylindrical cup, little greater in diameter than its stem. Above the basals, four of its rays consist of three consecutive plates, and the other one has four (Fig. 320-1). Comparison with other somewhat similar crinoids indicates that in the course of evolution the two lower plates of each ray fused progressively in a definite order (Fig. 320-2 to 4). In derived forms among the inadunates, these fused plates are identified as radials in four rays and as the radianal in the fifth. Before its fusion to become part of the radianal, however, the second plate in the right posterior ray was transformed into an axillary (Fig. 320-3). It supports on the right side a plate that is the definitive radial in this ray and on the left the first plate of an anal series. After fusion the radianal shrank in size, shifted to the left, and assumed the form and position that are characteristic of this plate in many crinoids (Figs. 316-1a; 319-4). In the subsequent evolution of some lineages, the radianal either shrank still more and disappeared or moved further to the left and then upward until it was eliminated from the cup (Fig. 316).

The arms of almost all crinoids branch from pentagonal axillary brachials. Some specialized monocyclic inadunates are unique, however, in possessing radials bearing several to numerous arms directly without any intervening branching. The immediate ancestors of these crinoids are not known, but certain other species exhibit evolutionary developments that can account for this peculiar structure. In them the axillary brachials shrank in such a way that they were eliminated from the lower parts of certain arm branches (Fig. 322). Their disappearance and the loss of underlying brachials would finally produce multiple arm-bearing radials.

Dicyclic Camerates

Dicyclic crinoids constitute the smaller of the two camerate groups. They range from the Middle Ordovician to the Middle Mississippian but occur in greatest variety in Ordovician and Silurian strata. Some of the more primitive of these crinoids are structurally almost identical to early monocyclic species except for the presence of infrabasal plates. As with the inadunates, this seems to indicate that both camerate groups evolved from the same immediate ancestors. The fact that no camerate has a radianal plate strongly suggests that relationships with the inadunates are relatively remote.

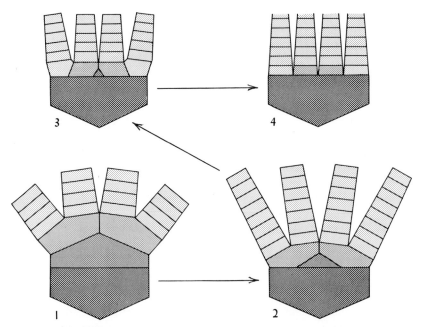

Fig. 322   Diagrams illustrating evolutionary stages leading to the development of radial plates which bear several arms directly. The axillary plates where branching is initiated shrink and disappear or are converted to the first brachial plates of the arms. 1, Stage showing first- and second-order branching on pentagonal axillary plates. 2, Theoretical intermediate stage with shrunken first axillary. 3, Further shrinking of axillaries. 4, Final stage characteristic of *Catilocrinus* and its near allies.

The dicyclic camerates exhibit no evolutionary trends that are not equally or better demonstrated by their monocyclic relatives. In the most primitive forms of both groups the radials are all separated from one another laterally by relatively small and irregular plates and in addition by a first anal and subsequent anal plates in the posterior interradius (Fig. 323-1). In progressively more advanced representatives, the interrays are occupied by regularly arranged larger plates (Fig. 323-2) that gradually moved upward. This brought the radials into lateral contact (Fig. 323-3). The anal plate was the last to be displaced from the radial circlet (Fig. 323-4).

Fig. 323   (See facing page.) Plate diagrams showing progressive evolutionary development in the posterior interray of dicyclic camerate crinoids. These genera, however, are not necessarily related in this detailed way. 1, Most primitive condition. The first anal plate rests directly on a basal, and the anal series resembles a sixth ray. Interbrachial plates are small, numerous, and irregular. 2, Structure similar to last, but the interbrachials are larger and regularly arranged. 3, The radial plates are in lateral contact with each other and the first anal plate. Interbrachials have been eliminated from the radial circlet. 4, The radials are in contact all around, and the first anal is above the radial circlet. This is the most advanced structure in this group of crinoids. 5, The anal plate is still in the radial circlet but is much shrunken. This condition is not known in dicyclic camerates, but it is characteristic of monocyclic *Pterotocrinus*.

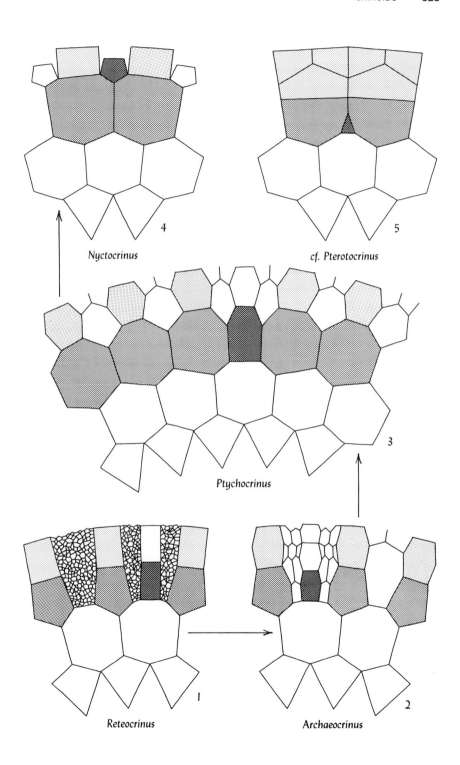

4
Nyctocrinus

5
cf. Pterotocrinus

3
Ptychocrinus

1
Reteocrinus

2
Archaeocrinus

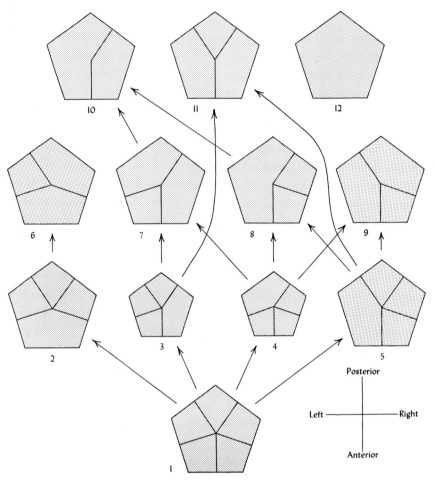

Fig. 324 Diagrams showing possible evolutionary modifications of basal plates in monocyclic camerate crinoids by progressive fusion of the five original plates. Structures shown in the large pentagons are known to exist; those in the small ones are theoretical intermediates. (*Adapted after Wilson, 1916, J. Geol., vol. 24, pp. 505, 507.*)

Monocyclic Camerates

The monocyclic camerates range from the Lower Ordovician to the Middle Permian and are a particularly conspicuous element in several mid-Mississippian faunas. In most parts of the world they almost disappeared at the end of the Mississippian Period. With very few exceptions, later representatives have been found only on the East Indian island of Timor.

Basal plates form the first circlet of the monocyclic crinoid cup. In combination, they produce a pentagon or hexagon, depending on whether the first anal plate

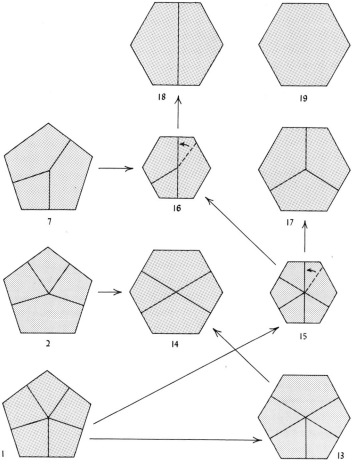

Fig. 325 Diagrams showing possible evolutionary modifications of structures in mono-cyclic camerate crinoids with hexagonal bases. The presence of an anal plate in the radial circlet necessitates a hexagonal base. An enlarged posterior basal may maintain its right suture in a diagonal position (13, 14). Enlargement of the right posterior basal causes the same suture to shift by rotation to the left and come to a posterior position (15, 16). (*Adapted after Wilson, 1916, J. Geol., vol. 24, figs. on pp. 666, 678, 680.*)

remains in the radial circlet (Fig. 326). The primitive number of basal plates is five, but this number was progressively reduced by the fusion of adjacent plates, until finally a base appeared consisting of only a single plate (Figs. 324; 325). Five basals are present in camerates occurring in the Ordovician and Silurian, four in those ranging from the Ordovician to Devonian, three in forms of Silurian to Permian age, and two in Mississippian to Permian specimens. A single-plated base may develop as a gerontic feature in several crinoid lineages, but it is characteristic of only a few specialized types found in Pennsylvanian strata. It is quite unlikely that plates

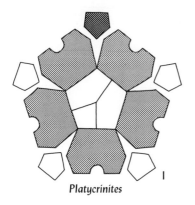

Platycrinites

1

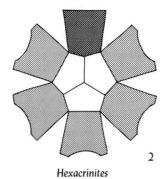

2

Hexacrinites

Fig. 326 Plate diagrams, as seen projected from below, of two very simple but highly evolved monocyclic camerate crinoids. 1, The base is pentagonal, and the anal plate is above the radial circlet. 2, The base is hexagonal, and the anal plate is included in the radial circlet.

once fused would later reseparate. Therefore, structure of the crinoid base may aid importantly in tracing evolutionary lineages. These structural variations have been studied for the monocyclic camerates but not for the other crinoid groups whose first circlet of plates, whether basals or infrabasals, fused in similar ways.

Progressive evolutionary modifications in the arms of many crinoids included reduction in the number of brachial plates between the radials, and successive branchings (Fig. 326a). As branching of the rays incorporated in the camerate cup increased, the number of arms becoming free was multiplied accordingly. This

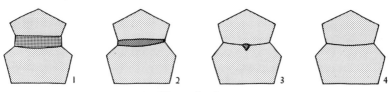

Alloprosallocrinus

Fig. 326a The loci of branching tend to move downward in the crinoid arm, probably by the progressive elimination of brachial plates. These stages of shrinkage and disappearance of a brachial have been observed in different specimens of a single species. (After Wilson, 1916, J. Geol., vol. 24, fig. 6, p. 534.)

resulted in crinoids with bunched arms (Figs. 327-1; 329-1). Continuation of this process or reduction in the number of interbrachial plates produced forms in which the bunches merged so that arms were given off continuously around the whole upper margin of the cup (Figs. 327-2; 329-2). This was likely to be accompanied

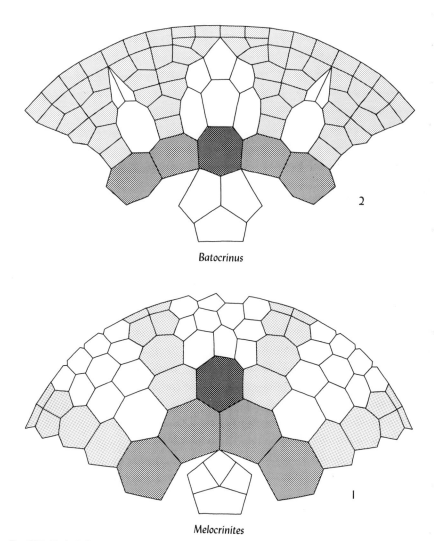

Batocrinus

Melocrinites

Fig. 327 Projected patterns of plate structure below the bases of the free arms in two monocyclic camerate crinoids. The fifth, or anterior, ray is omitted. 1, Bunches of arms (here two in each ray) are separated by interbrachial plates. 2, Bunches of arms (here four or five to a ray) are in contact all the way around the crinoid body.

Eucalyptocrinites

Fig. 328 Projected pattern of plate structure in the cup of a monocyclic camerate crinoid that is almost perfectly pentamerously symmetrical. No anal plates or other irregularities occur below the level where the arms become free except in the base, where two of the original plates have fused.

by the development of an increasingly prominent and long plated anal tube rising from the center of the crinoid's upper surface (Fig. 329-1,2).

Relatively little study has been devoted to the detailed structure of the upper surface of camerate crinoids because the plating is less regular there than in the cup below. Nevertheless evolution was effective in producing many noticeable modifications (Fig. 330). For example, various knobs, spines, and other kinds of outgrowths were developed (Fig. 329-5). One trend that seems to have been regular and progressive was related to the ambulacra. The food grooves of the arms pass beneath the crinoid's upper surface and into the interior of the camerate body (Fig. 330-1). At first, in converging toward the hidden mouth, they lay immediately below this surface. At this stage their courses are marked by double series of alternating

Fig. 329 (See facing page.) Representative camerate crinoids showing the solidly plated domes above the bases of the free arms. 1, 2, Examples with prominent anal tubes, which are characteristic of many camerates, with continuous or almost continuous circlets of arm bases. 3, 4, 5, Low-crowned forms, seen from the anal or posterior sides, showing variable development of dome plates above the arm bases ($\times$½–1½.) (*After Wachsmuth and Springer, 1897, Mem. Museum Comp. Zool., vols. 20, 21, pl. 28, fig. 9; pl. 31, fig. 2a; pl. 40, fig. 5a; pl. 45, fig. 13c; pl. 68, fig. 13b.*)

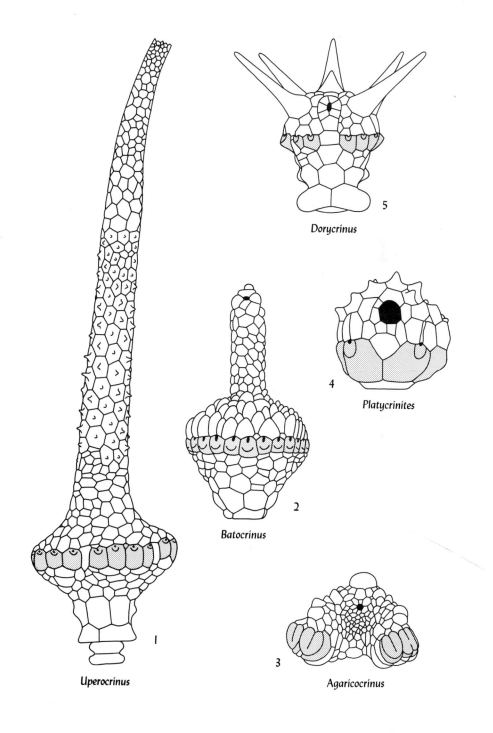

5
Dorycrinus

Platycrinites
4

2
Batocrinus

I
Uperocrinus

3
Agaricocrinus

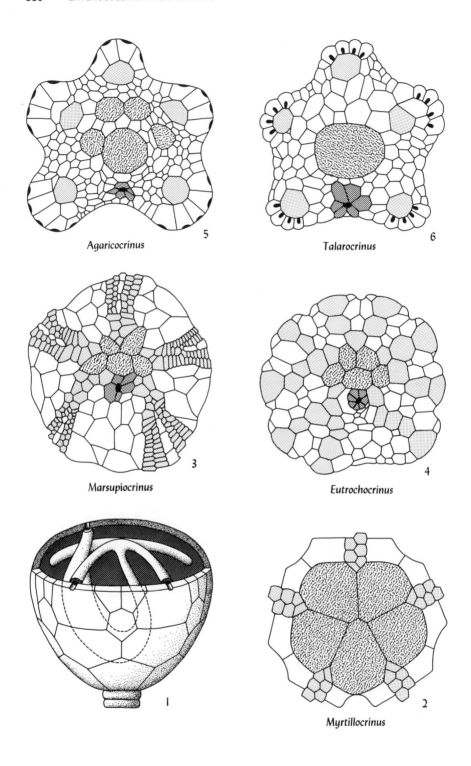

Agaricocrinus   5

Talarocrinus   6

Marsupiocrinus   3

Eutrochocrinus   4

1

Myrtillocrinus   2

Fig. 330   (See facing page.) Dome structures of camerate crinoids showing stages in progressive evolution. 1, Generalized sketch of a crinoid with half its dome removed to show the food passages leading from the arm bases to the central mouth. This represents an advanced condition, with the passages sunken below the dome. 2, This simply constructed crinoid has five large oral plates that overlie the mouth. The food passages, continuing from the arms and roofed by cover plates, pass inward between the outer angles of the orals. 3, This is a typical camerate. The food passages lie immediately below the dome. Their positions are shown by the double series of cover plates incorporated in the dome and leading to the five orals. The more central cover plates have begun to lose their characteristic appearance and are beginning to become indistinguishable from the irregular plates of the interray areas. 4, The food passages have sunk below the dome, and cover plates are absent or have been transformed to plates similar to the other dome plates. Courses of the food passages seem to be indicated, however, by series of plates leading from the arm bases to the orals. 5, Food passages are below the dome, and their positions are not clearly shown by the plate pattern. Orals are recognizable. Large plates located just inside the bunched arm bases are believed to show where the multiple food passages join as they converge toward the mouth. 6, Only the probable posterior oral can be identified as the large central dome plate. (2 to 6, *after Wachsmuth and Springer, 1897, Mem. Museum Comp. Zool., vols. 20, 21, pl. 3, figs. 13, 19, 21, 22; pl. 75, fig. 18.*)

plates that are continuations of the covering plates of the arms (Fig. 330-3). In somewhat more advanced forms, the ambulacral canals sank to lower interior positions, and single rows of plates overlie them (Fig. 330-4). Finally these plates disappeared or became indistinguishable from others, and nothing remained in the structure of the upper surface to indicate surely the positions of the ambulacra (Fig. 330-6).

Flexibles

The flexible crinoids constitute a relatively small, well-unified group ranging from the Middle Ordovician to the Upper Permian. They resemble dicyclic inadunates,

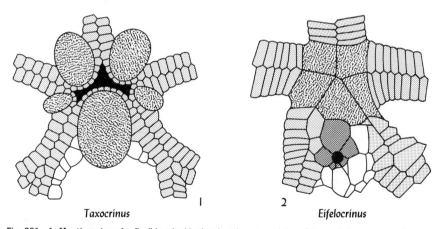

Taxocrinus	*Eifelocrinus*

Fig. 331   1, Mouth region of a flexible crinoid, showing the cover plates of the ambulacra converging and passing between the orals. This crinoid had an open mouth. 2, Plate structure above the mouth of a modern articulate crinoid. (*After Wachsmuth and Springer, 1897, Mem. Museum Comp. Zool., vols. 20, 21, pl. 53, fig. 1b; pl. 75, fig. 1c.*)

from which they are believed to have evolved, in possessing a radianal plate (Fig. 332), but they differ in many other ways. Particularly, brachials and interbrachials were incorporated in their cups, and all plates were loosely joined by ligaments so that some movement was possible in their flexibly plated bodies.

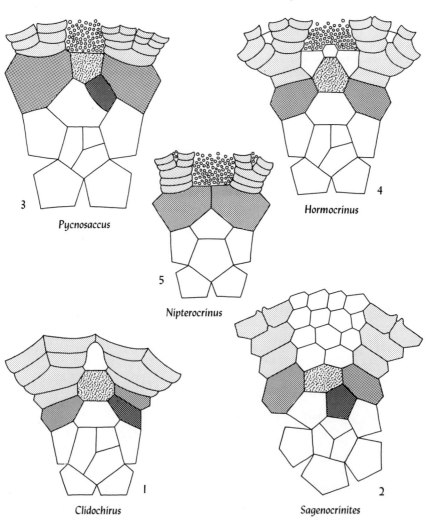

Fig. 332 Diagrams showing the dicyclic base (unshaded), posterior interray, and two adjacent rays of flexible crinoids. Progressive evolution involved the shifting and eventual elimination of the radianal plate (darkest shading). 1, The radianal is in its most primitive position in the right posterior ray. A first anal lies between the two posterior rays. 2, The radianal has descended into the basal circlet. 3, The radianal and first anal are in their most characteristic positions. 4, The radianal presumably has disappeared, but what seems to be the first radial remains between the rays. 5, All anal plates have been eliminated. *(Adapted from Springer, 1920, Smithsonian Inst. Publ. 2501, figs. 16, 19, 21, 24, 38, pp. 166, 180, 201, 215, 294.)*

Fig. 333 Part of the arm of a flexible crinoid. (*After Springer, 1920, Smith-sonian Inst. Publ. 2501, pl. 70, fig. 2a.*)

Onychocrinus

No flexible crinoids of late Ordovician or early Silurian ages have been discovered, and evolutionary trends that may have been established during this time are uncertain. According to some interpretations, advancement is shown by increase in the number of brachial plates below the first branching of the rays, and also in the number of interbrachials in comparable positions. If so, these developments run counter to those in all other crinoids. Later, simplification and elimination of brachial plates from the cup occurred in a more conventional manner.

A better-established evolutionary trend is indicated by displacement of the radianal plate. Most primitively, this lay directly below the right posterior radial (Fig. 332-1). First it moved downward into the basal circlet (Fig. 332-2), producing a structure known only in a very few other crinoids. Thence it moved upward and to the left, into a position where this plate commonly is seen in the inadunates (Fig. 332-3). In further developments it either disappeared or was pushed upward out of the cup, and perfect pentamerous symmetry was attained (Fig. 332-5).

Arms of the flexibles are conservative. They are all uniserial and do not bear pinnules. Regular equal division of the arms was succeeded by branching, which produced very unequal armlets (Fig. 333), more so than in any other crinoids.

Articulates

All modern crinoids and all but a few Triassic and later fossils are articulates. All of these actually are dicyclic, although the infrabasals generally are very small or have disappeared because of fusion with other plates or by resorption at early stages

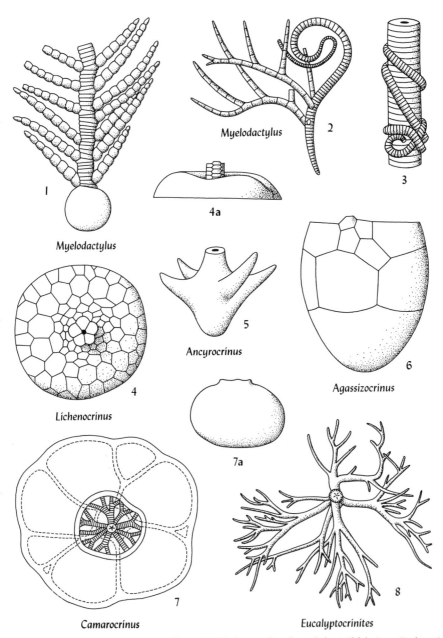

Fig. 334 The ends of crinoid stems. Many crinoids became free from their youthful stem attachments. 1, This stem, after breaking loose, developed a swollen terminal knob. The cirri branching from stem are peculiarly formed, perhaps for grasping. 2, This stem tapers to a blunt point. The upper part is slender and coiled. It and the crinoid body were protected by the stout branched cirri, which may also have functioned as grapnels. (1, 2, *after Springer, 1926, Proc. U.S. Natl. Museum, vol. 67, art. 9, pl. 4, fig. 2; pl. 5, figs. 3, 7.*)

of their growth. The arms are uniserial except for the tips in a very few species. This seems to be evidence of descent from inadunates that were primitive in this last respect.

Fossil articulates are abundant at very few places in North America. Little can be inferred concerning the evolutionary relations of these crinoids except that the stemless forms were derived from stemmed ancestors. All articulates are attached in a conventional way during the early stages of their development. About 85 percent of the known modern species soon free themselves by detachment from their stems. Thereafter they are able to move about, either swimming or pulling themselves along by the action of their arms and the cirri growing downward and outward from their bases. The numerical disparity between free and attached species probably reflects the fact that the stemmed crinoids of modern faunas live mostly in deep water and, therefore, are less commonly observed and less completely known than the stemless ones, many of which inhabit shallow environments.

The first stemless crinoid to appear was an inadunate of late Mississippian age (Fig. 334-6). This mode of life seems to have become steadily more common in post-Paleozoic time. Fossil crinoids of many kinds, however, broke free from their holdfasts but retained portions of their stems (Fig. 334-1,2). Some developed various types of anchoring structures (Fig. 334-5,8), and the stems of others seem to have been prehensile (Fig. 334-3).

Edrioasteroids

The extinct edrioasteroids are rare fossils ranging from the Lower Cambrian to the Pennsylvanian. They include the earliest known representatives of the echinoderms, and creatures of this general kind may have been ancestral to the starfishes and, somewhat less probably, the echinoids. Most edrioasteroids seem to have had more or less flexible sack-like bodies with upwardly directed mouths (Fig. 336-3b). They were attached by the lower surface, either permanently or, with a central adhesive organ, temporarily. Many starfishes pass through an ontogenetic stage when the larva settles to the bottom, becomes attached, and undergoes metamorphosis. The

3, Parts of two evidently prehensile crinoid stems are coiled around a larger stem. Drawn from specimen. 4, 4a, Top and side views of the attachment disk of a crinoid generally found adhering to other fossils. (*After Meek*, 1873, *Geol. Surv. Ohio, vol. 1, pt. 2, pl. 3, figs. 1, 1a, 1c, p. 00.*) 5, The grapnel-like growth at the end of a stem possibly belonging to *Arachnocrinus*. (*After Lowenstam*, 1942, *Bull. Buffalo Soc. Nat. Hist., vol. 17, diagram 3, p. 35.*) 6, A late Paleozoic stemless crinoid. The lower part is a solid mass composed of five fused infrabasal plates. (*After Springer*, 1926, *Proc. U.S. Natl. Museum, vol. 67, art. 9, pl. 15, fig. 7.*) 7, 7a, Top and side views of the bulbous growth at the end of stems of *Scyphocrinus*, which has long been known as *Camarocrinus*. These consist of a double layer of small irregularly arranged plates, are divided into compartments (shown by dashed lines), and attain sizes of 6 in. or more in diameter. They have been interpreted as floating organs but more probably reposed on the sea bottom. (*Generalized after Hall*, 1879, 28th *Ann. Rept. N.Y. State Museum, pls. 35, 36.*) 8, This is a root-like holdfast anchoring a crinoid in soft sediment. It consists of branching cirri. (*Simplified after Hall*, 1876, 28th *Ann. Rept. N.Y. State Museum, pl. 19, fig. 8.*)

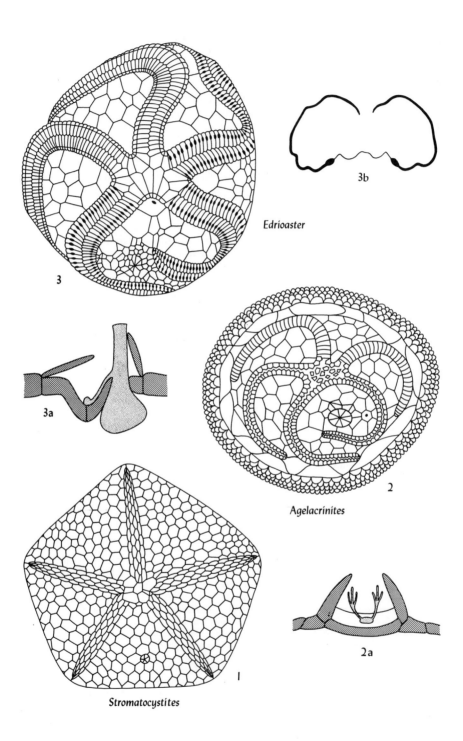

Edrioaster

3b

3

3a

Agelacrinites

2

2a

Stromatocystites

1

Fig. 336 (See facing page.) Representative edrioasteroids. These animals had flexible, flattened, sack-like bodies consisting mostly of irregularly arranged plates. They were attached by all or part of the lower surface, either permanently or by suction disks. 1, A Cambrian form believed to exhibit primitive features in its straight ambulacra. 2, Specimens of this kind, representing successive growth stages, show that the ambulacra lengthened without corresponding increase in body diameter, and the ambulacra were forced to curve. This Devonian form was attached by its entire lower surface. The cover plates of the ambulacra are removed from two of the rays to show the transverse flooring plates, which do not seem to reveal the openings of tube feet. 3, The ambulacra of this Ordovician form extend onto the sides and even part of the lower surface. It was attached by a restricted central area (see 3b, cross section as it probably was in life). Cover plates are removed from three of the ambulacra to show the alternating floor plates and pores believed to indicate the possession of tube feet and internal pressure sacs. (*After Bather, 1900, figs. 1, 6, pp. 206, 209; Hall, 1872, 24th Ann. Rept. N.Y. State Museum, pl. 6, fig. 14.*) 2a, 3a, Restored cross sections of ambulacra showing supposed relations of tentacles and a tube foot. (*2a, 3a, b, after Jaekel, 1899, Stammesgeschichte der Pelmatozoen, vol. 1, fig. 1, p. 19. See Regnéll, 1966, fig. 111, p. 139.*)

young starfish develops with its oral surface directed upward, then detaches itself, and begins its free existence. Edrioasteroids have been likened to the attached ancestor which this ontogeny suggests.

The ambulacra radiate from the central mouth (Fig. 336-1 to 3). A series of specimens of different sizes shows that in very young individuals the ambulacra are straight. With subsequent growth, however, they increased in length more rapidly than the body increased in radius. Consequently, the ambulacra were forced to curve in order to remain within the compass of the body. If they had not, a starfish-like creature would have been produced. The ontogeny of the ambulacra suggests that their relative length and curvature are measures of evolutionary advancement. This seems reasonable because longer ambulacra surely were more efficient feeding structures.

The ambulacra of the edrioasteroids seem to represent two types, one similar to those of starfishes (Fig. 336-3a), and the other more like those of crinoids (Fig. 336-2a). Both were provided with movable covering plates. Both also were specialized in such ways that the known fossils could not have been ancestral to any other echinoderms. A more primitive type of edrioasteroid, however, might have been.

The anal opening, closed by a pyramid of small plates, identifies the posterior interambulacrum. In this same area is located the perforated hydropore plate which was the orifice of the water vascular system. This opening seems to have moved first toward the right posterior ambulacrum, and then along the side of the ambulacrum nearer to the mouth in progressively more evolutionarily advanced forms (Fig. 336-2,3).

Starfishes

Starfishes are rare fossils. The skeletons of these animals consist of loosely bound calcareous plates which ordinarily become dissociated and scattered soon after the death of individuals. Most knowledge of the structure of ancient types has come from specimens found at a few favored places where burial evidently was rapid and

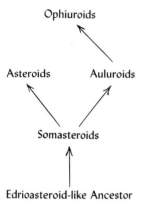

Ophiuroids

Asteroids   Auluroids

Somasteroids

Edrioasteroid-like Ancestor

Fig. 338 Diagram showing the probable evolutionary relations of the main groups of starfishes.

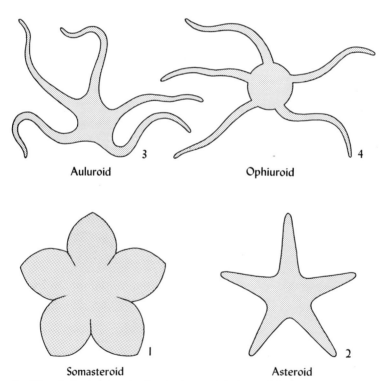

Auluroid   Ophiuroid

Somasteroid   Asteroid

Fig. 338a Outline drawings showing the characteristic shapes of the four main groups of starfishes.

Table 339   **Principal distinguishing characters of main starfish groups**

Somasteroids: primitive starfishes; water vascular canal enclosed between ambulacral plates or in open trough; without internal pressure sacs; ambulacrals alternating; arms wide with series of plates extending laterally from ambulacrals. Early Ordovician

Asteroids, ordinary starfishes: water vascular canal in open trough; pressure sacs internal; inner ambulacrals opposite, flanked by outer ambulacrals and marginals. Ordovician to Recent

Auluroids: water vascular canal enclosed between ambulacrals; without internal pressure sacs; arms slender, consisting of alternating or opposite inner ambulacrals flanked by outer ambulacrals. Ordovician to late Paleozoic

Ophiuroids, serpent stars: water vascular canal not enclosed by ambulacrals; without pressure sacs; ambulacrals opposite, typically fused to form vertebral ossicles; arms slender, enclosed by four series of covering plates, abruptly set off from circular body. Ordovician to Recent

The fossils include some forms intermediate between these groups that are difficult to place in any classification.

collecting has been intensive. Fossils are known from the Lower Ordovician to the most recent sediments, and many live in the sea today, but a disproportional number of fossils has come from strata of Ordovician and Devonian ages.

Two quite different types occur in modern faunas: (1) asteroids, or starfishes, as these are more commonly identified; and (2) ophiuroids, or serpent stars (Fig. 338a and Table 339). The general and obvious similarities of these animals invite the conclusion that they are branches of a single ancestral stock. Zoologists, however, who have studied the development of modern species have observed that the larvae of most asteroids and ophiuroids are different and that these young stages resemble those of holothurians and echinoids much more than they do each other. This has been presented as evidence that the two groups of starfishes evolved independently from different ancestral stocks and that the resemblances of mature individuals resulted from convergent evolution and adaptation to somewhat similar ways of life.

The evidence of fossils contradicts the idea that asteroids and ophiuroids are not closely related in their evolutionary origin. Fossils include specimens that are neither asteroids nor ophiuroids, and two other groups require recognition (Fig. 338a): (1) somasteroids, very primitive starfishes which probably include the direct ancestors of both modern groups; and (2) auluroids, whose characters are intermediate between those of the modern groups.

Somasteroids

Somasteroids, discovered in Lower Ordovician strata, are the oldest known starfish-like fossils. They consist of five broad petal-like arms radiating from a small central body. Each arm has a double column of ambulacral ossicles with hollow basins which housed tube feet. Series of rod-like plates extend laterally on both sides of the ambulacrals (Fig. 340). The animals represented by these fossils probably lived on or shallowly buried in the sea bottom, with the central mouth directed

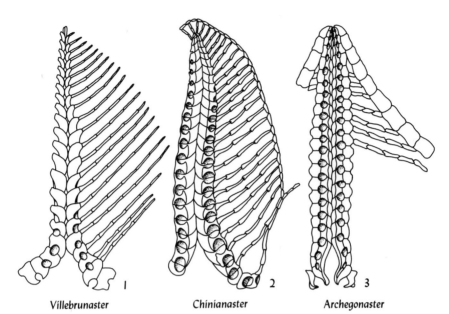

Villebrunaster   *Chinianaster*   *Archegonaster*

Fig. 340   Arm structure as seen from the oral surface of three primitive Ordovician starfishes, or somasteroids, which seem to show successive stages in evolutionary advancement. 1, The water vascular canal was enclosed between the ambulacral plates. Pressure sacs were external and lay in basins hollowed out between these plates (see Fig. 343-1). Tube feet were directed laterally, except near the mouth. The series of slender laterally radiating plates probably supported food-gathering areas; these animals may have lived with their mouths directed upward. 2, The water vascular canal lay in an open trough, as in the asteroids. The tube feet were directed outward (upward?), rather than laterally. The outermost lateral plates are differentiated as small marginals. 3, Much like the last, but the first plates of the radiating series have become outer ambulacrals, the marginals are large and stout, and the arms are pointed. (*After Spencer, 1951, figs. 1, 7, 9, pp. 92, 99, 102.*)

upward. Cilia-lined grooves between the lateral rods are believed to have moved food particles to the tube feet, which passed them inward to the mouth.

The skeletal structure of a somasteroid resembles that of pinnule bearing biserial crinoid arms, and the theory has been advanced that these animals evolved from crinoids. This is most unlikely, however, because somasteroids antedate all known echinoderms with this type of arm, the crinoid arm plates are not ambulacrals, and branches of the ambulacral system probably did not extend along the lateral rods of the somasteroids as they do in crinoid pinnules. The likelihood is much greater that a primitive edrioasteroid-like creature was ancestral to somasteroids.

The most primitive somasteroid skeleton consists only of ambulacrals and undifferentiated lateral rods (Fig. 340-1). The back surface was not plated, but small three-rayed spicules were present in the skin. From this beginning two successive stages of evolutionary advancement which seem to lead toward the asteroids are known. In the first, the outermost rod elements have been transformed to small marginal plates that outline the arms (Fig. 340-2). In the second, these plates have become much larger and the innermost rod elements have been transformed to adambulacral plates (Fig. 340-3). These are important parts of the axial skeleton in the arms which have lost their petaloid form, and such a structure persists in the arms of all asteroids. Specialization of some and elimination of other elements of the lateral rods, therefore, seems to have been particularly noteworthy in the early evolution of the starfishes. Vestiges of the rods remained, ordinarily in the interior of the arms, in several lineages of starfishes and are preserved in some modern forms. One living species of the eastern Pacific retains such primitive characters in this respect that it might be identified as a somasteroid.

Asteroids

The asteroid skeleton consists most typically of nine columns of plates, repeated in each arm, extending outward from a single plate occupying the center of the dorsal surface and the mouth at the center of the ventral surface (Fig. 341). These are (1) two columns of ambulacrals, ventral, along the middle of the ray, (2) two columns of adambulacrals, ventral, one on each side flanking the ambulacrals, (3) four columns of marginals, lateral, two on each side of a ray, and (4) one column of radials, dorsal, extending along the midline of a ray. The marginals originate at axillary plates that lie in the interradial angles of the skeleton. Ambulacrals, adam-

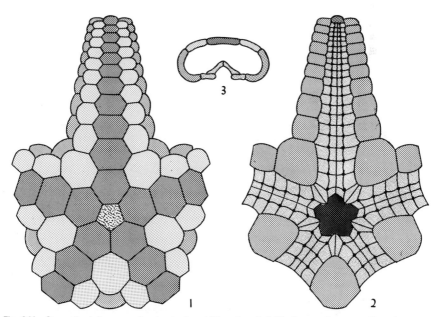

Fig. 341 Generalized diagrams showing in dorsal (1) and ventral (2) views and cross section of an arm (3) the theoretically primitive plate structure of an asteroid. (*After Schuchert, 1915, figs. 1, 2, p. 34.*)

bulacrals, and what are probably lower marginals are recognizable in the structure of advanced somasteroids.

Asteroids are known as fossils from the Middle Ordovician, and they are abundant in modern faunas. The earliest known species are so solidly and massively plated that they seem to have lacked much flexibility (Fig. 342-1). The heavy plating of the dorsal surface probably evolved as a protective covering when some early starfishes reversed the position of the somasteroids and acquired the habit of living with the mouth directed downward.

Primitively the ambulacral plates of the two columns in each arm alternate with each other. In somasteroids, as previously mentioned, each ambulacral bore a series of lateral rod-like plates so that the skeleton had an overall pattern of lateral continuity (Fig. 340). In the asteroids, the ambulacrals assumed opposite positions in the two columns (Fig. 344-2,3). Each was flanked by an adambulacral, but lateral continuity did not continue farther. In contrast a pattern of longitudinal continuity developed, because the ambulacrals and their associated adambulacrals did not correspond with the smaller number of marginals, and the plates of the marginal and radial columns all alternated with their neighbors. In more advanced forms, however, the plates in the two marginal columns on the side of an arm tended to occupy opposite positions. Relative increase in the number of ambulacrals, and consequently the more numerous tube feet, was an evolutionary advancement which contributed to more efficient feeding.

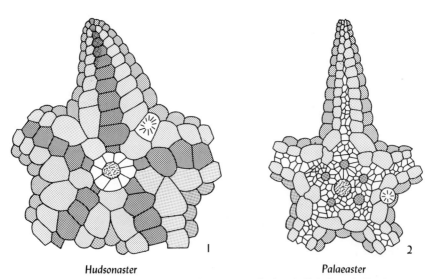

Hudsonaster   *Palaeaster*

Fig. 342   1, A primitive Ordovician asteroid whose structure is close to that shown in Fig. 341-1. Only one arm is illustrated completely. 2, A more advanced Silurian type which lacks the radial columns. *(After Schuchert, 1915, pl. 6, fig. 2; pl. 7, fig. 2.)*

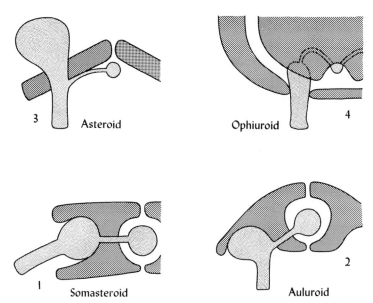

Fig. 343 Generalized diagrams showing the relations of water vascular canal, tube feet, and pressure sacs to the ambulacral plates in the four main starfish groups.

An open ventral groove between up-arched ambulacral plates extends the length of the asteroid arm (Fig. 343-3). It contains the radial water vascular canal with which the tube feet are connected. Similar grooves are evident in the arms of advanced somasteroids, but not in the most primitive type, whose radial canals were enclosed between the ambulacrals as in auluroids (Fig. 343-1,2). Each tube foot seems to have been provided with a pressure sac which controlled its extrusion. In somasteroids and auluroids these sacs were housed in basins excavated in two adjacent ambulacrals (Fig. 343-1,2). The sacs of modern asteroids are internal, and communication with the tube foot is maintained through a pore passing between adjacent ambulacrals (Fig. 343-3). When evolution accomplished this transformation is uncertain. Most Paleozoic asteroids are described as having had internal pressure sacs, but the supposed pores may be basins characteristic of the more primitive condition. Tube feet of asteroids commonly are arranged in two rows along each arm, but in some highly evolved modern forms with very short ambulacrals adjacent tube feet are offset to produce as many as eight rows.

Several evolutionary trends resulted in the development of increasingly flexible skeletons. Among them the most noteworthy were: (1) Plates of the radial and marginal columns decreased in size and lost contact with each other (Fig. 344-1a and b) or were transformed to star-shaped ossicles in contact only at their points. (2) Small accessory plates were introduced between the plates of all rows except

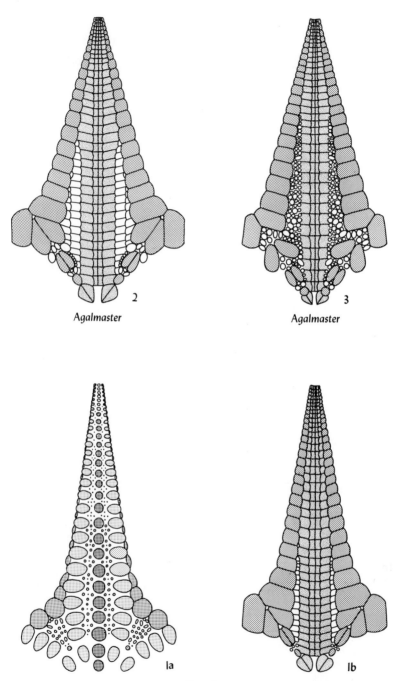

2
Agalmaster

3
Agalmaster

la

lb

Xenaster

Fig. 344   (See facing page.) The arm structures of some Devonian asteroids. 1a, Dorsal view showing tiny plates separating radials and upper marginals. 1b, Ventral view of same, showing columns of single plates separating lower marginals and outer ambulacrals. 2, A second column of plates intervenes on each side of the ambulacrals. 3, Space between marginals and ambulacrals is occupied by numerous small irregular plates. The conditions shown in these successive figures are interpreted as evolutionary advancements. Note also how the first few marginal plates have moved inward from the angles between the arms and are more or less separated from one another by small plates, which become more numerous. (*After Schöndorf, 1909, Palaeontographica, vol. 56, pl. 9, figs. 1, 3, 4, 5.*)

the ambulacrals and adambulacrals (Fig. 344). The arms grew at their tips by the addition of new plates at the ends of the primary columns, and the accessory plates made their appearance later. Therefore, a kind of ontogenetic series can be traced from the arm tips inward toward the body. The ambulacral and adambulacral columns were very stable and maintained their relations with each other. Plates of the other columns, however, especially the radials, tended to shrink, so that they became indistinguishable from accessories or disappeared entirely (Fig. 342-2). The marginals were relatively stable in some lineages, but the lower ones, which primitively outlined the arms, moved onto the ventral surface, so that the arms were bordered by the upper marginals. In some advanced forms, several of the marginals were drawn inside the angles between the arms onto the central body surface (Fig. 344-2,3).

Auluroids

The auluroids, as they have been distinguished in the past, are a miscellaneous group of starfishes ranging from the Middle Ordovician to at least the Mississippian. They are neither typical asteroids nor typical ophiuroids (Fig. 346). Although not all these fossils may be closely related, they possess certain unifying characters that suggest descent from primitive somasteroids. These characters are particularly (1) enclosure of the radial water vascular canal between the ambulacrals (Fig. 343-2), and (2) absence of a ventral ambulacral groove. Tube feet probably were equipped with pressure sacs contained in basins hollowed out between the ambulacrals (Fig. 346-2,3). Most of these fossils have slender arms with only four columns of plates, the dorsal integument contained spicules rather than plates, and the outline of the central body between the arms was concave (Fig. 338a-3), bordered by marginal plates in some (Fig. 346-1). The form of these animals suggests that they led more active lives than either somasteroids or asteroids.

Few distinct evolutionary trends are exhibited by the auluroids. In the one that is most evident, ambulacral plates of the two columns in each arm assumed opposite, rather than alternate, positions (Fig. 346-4,5). Almost all mid-Paleozoic species had this advanced structure. It seems fairly clear that auluroids and asteroids were related to each other only through their derivation from different types of somasteroids. Ophiuroids, however, may have evolved directly from specialized auluroids.

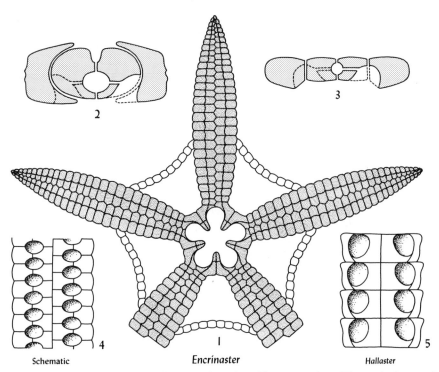

Schematic   *Encrinaster*   Hallaster

Fig. 346   Plate structure of an auluroid, as seen from above. The upper surface of these animals was not solidly plated like an asteroid but may have been provided with calcareous granules or very small disconnected plates. (*After Schöndorf, 1910, Palaeontographica, vol. 57, pl. 3, fig. 1.*) 2, 3, Cross section of the plates in an auluroid arm, showing central channel occupied by the water vascular canal and basins for reception of pressure sacs connected with the tube feet. (*After Spencer, 1914, fig. 22, p. 23; Schöndorf, 1910, Palaeontographica, vol. 57, fig. 4, p. 40.*) 4, Schematic diagram of a primitive auluroid arm, seen from below, showing alternating ambulacral plates and basins. 5, Similar diagram showing the opposite ambulacral plates of an advanced arm. (*After Schuchert, 1915, figs. 15, 31, pp. 214, 255.*)

Ophiuroids

The arms of modern ophiuroids contain a central column of ossicles that look much like tiny vertebrae. These articulate with ball-and-socket joints and are connected by muscles (Fig. 347-3). Each ossicle represents a pair of fused opposite ambulacral plates. The water vascular canal is located in a narrow groove incised in the lower surface of these ossicles (Fig. 343-4). Branches from the canal pass through the ossicles and emerge as tube feet, which lack pressure sacs. The arm is enclosed by four columns of cover plates—one dorsal, two lateral, and one ventral (Fig. 347-3). Because of the last, the ambulacral groove is not open, as it is in asteroids.

Ophiuroids are exceedingly rare as fossils, but the easily recognized vertebral ossicles occur in strata at least as old as the late Devonian. These animals probably evolved from an unknown type of advanced auluroid. If so, major structural modifications included (1) perfection of fused articulated ambulacrals, (2) change in position of the water vascular canals and their connections with the tube feet, and (3) development of more complete protective plating, both dorsally and ventrally. The lateral arm plates probably correspond to adambulacrals in other starfishes. A circular body disk is a distinguishing feature of the ophiuroids (Fig. 338a-4).

Sea Urchins

The echinoids, or sea urchins, constitute a particularly well-characterized group of animals that has been recognized since very ancient times. They are peculiar and differ from all other echinoderms except holothurians, because they completely lack what corresponds to a dorsal side as this is ordinarily identified in other groups within the phylum. The echinoid skeleton is like the ventral skeleton of a starfish with the arms arched upward and backward, joined tip to tip, and fused laterally along their margins.

The echinoids are readily divisible into two groups: (1) those with regular pentamerous symmetry, the anus opening opposite the mouth in the midst of a system

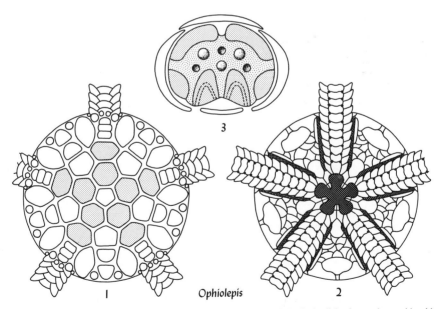

Fig. 347 Plate structure of the upper (1) and lower (2) surfaces of the body disk of a modern ophiuroid and a generalized cross section (3) of an arm. Structures vary greatly in different forms except in the arms. Some of the disk plates may be homologous with the radials and marginals of asteroids, but this is far from certain.

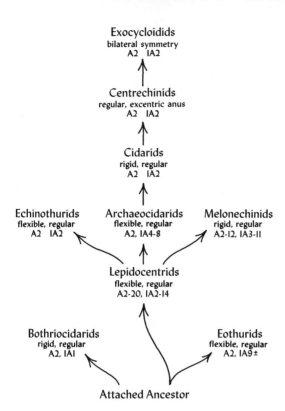

Fig. 348 Diagram showing the probable evolutionary relations of the main groups of echinoids. A descendant group, in most cases, is believed to have evolved from early unspecialized members of the ancestral group. Figures show the number of columns of ambulacral (*A*) and interambulacral (*IA*) plates.

of apical plates; and (2) those, commonly termed irregulars, which have developed bilateral symmetry, the anus having moved out of the apical system into the posterior interradius. The oldest known fossils are of mid-Ordovician age (Fig. 349-1), and these animals have continued to the present and are abundantly represented in many modern marine faunas. Well-preserved fossil specimens are rare in the

Fig. 349 (See facing page.) Plate structure of an interambulacral area and half of each adjacent ambulacrum of an early echinoid from the Ordovician. Very primitive features include the irregularly arranged imbricating interambulacral plates, ambulacral troughs occupied by the radial water vascular canals (see 3), and unpaired tube foot openings not entirely enclosed by the ambulacral plates. (*After MacBride and Spencer, 1939, fig. 8, p. 116.*) 2 to 5, Cross sections of half of the ambulacra of some echinoderms, showing relations of the water vascular canals, tube feet, and pressure sacs to skeletal plates, and illustrating stages leading to the enclosure of the water vascular canals within the skeleton, which is characteristic of almost all echinoids. 2, The canal lies in an external trough, as in the asteroids. 3, The ambulacral plates develop projections partially enclosing the troughs. 4, The troughs are completely closed, but parts of the original plates persist as inward projections. 5, The typical echinoid structure. (*After MacBride and Spencer, 1939, fig. 8, p. 116; and Mortensen 1940, vol. 3, pt. 1, fig. 181, p. 348.*) Pores through which tube feet and pressure sacs were connected originally passed between the ambulacral plates. This is the condition seen in the upper parts of echinoid ambulacra, where the plates begin to form. As they grow, the plates soon close around the stalks of the tube feet. This condition was not quite attained by *Aulechinus* (1). The single primitive pore was transformed to a group of pores in one Ordovician echinoid (4), but others developed two pores through which fluid flowed in opposite directions (5).

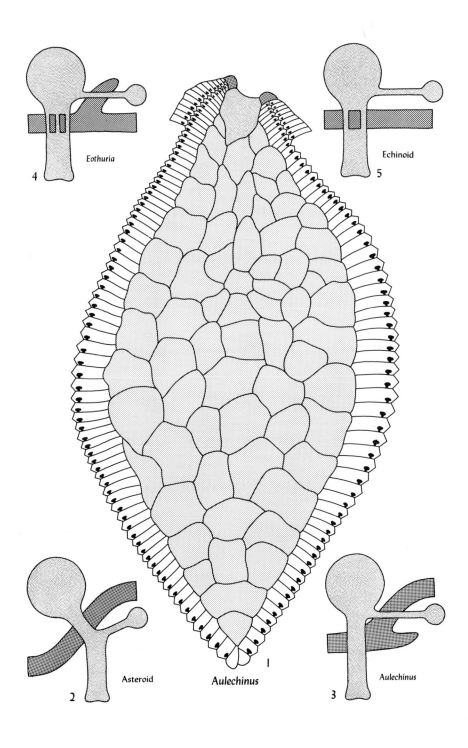

4   *Eothuria*

5   Echinoid

2   Asteroid

1   *Aulechinus*

3   *Aulechinus*

Paleozoic, although dismembered plates and spines are found at many places. In post-Paleozoic rocks, particularly from the Cretaceous onward, fossils may be numerous locally, but distribution is spotty and most faunas lack good specimens.

All echinoids normally have five alternating ambulacral and interambulacral areas extending from pole to pole in their generally spherical to hemispherical or biscuit-shaped, completely plated bodies. The water vascular canals lie internally beneath the ambulacral plates. Each ambulacral plate ordinarily is pierced by two pores, through which a single tube foot is connected with its internal pressure sac (Fig. 349-5); echinoids differ from all other echinoderms in this respect. Most echinoids have a complicated jaw structure, known as an Aristotle's lantern.

A circlet of ten plates located at the summit of the skeleton is basic to the structure of all but the earliest echinoids (Fig. 350). Known as the apical system, this consists of five ocular plates, occupying radial positions at the tips of the ambulacral areas, and ordinarily five alternating interradial genital plates. Where differentiation is possible, the right anterior genital is seen to be porous; it is the hydropore plate (Fig. 357). As echinoids grow, new plates are added at the upper

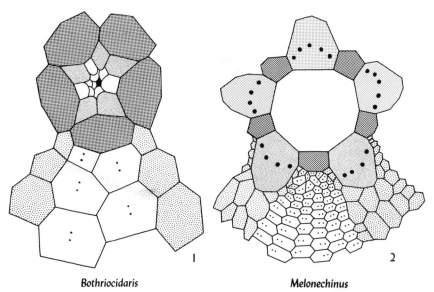

| Bothriocidaris | Melonechinus |

Fig. 350 Plate structure at the summit of two Paleozoic echinoids. The anal opening was located inside a ring of plates known as the apical system, consisting of oculars (heavy shading) and genitals (lighter shading). As an echinoid grew, all new plates, both ambulacrals and interambulacrals, appeared immediately adjacent to an ocular. 1, This structure of an Ordovician form is believed to be primitive. The ocular plates are in direct contact all around and enclose what seem to be the genitals between their inner angles. 2, This advanced structure of a Mississippian form shows the oculars separated all around by the genitals. Similar structure was not attained by post-Paleozoic echinoids until much later. (*After Jackson, 1912, pl. 1, fig. 2; Meek and Worthen, 1866, Geol. Surv. Ill., vol. 2, fig. 21, p. 228.*)

tips of both ambulacral and interambulacral areas immediately adjacent to the oculars (Fig. 357-2). This serves to push the previously formed plates downward in the skeleton, where they gradually increase in size and may change in shape. The oldest plates in both the ambulacral and interambulacral areas, therefore, are located on the lower surface of the mature skeleton near the mouth. Those echinoids whose structure changes progressively during growth preserve ontogenetic sequences in both types of areas. This suggests, in accordance with the recapitulation theory, the nature of ancestral forms and may indicate the direction of evolutionary development.

The evolutionary origin of echinoids has been sought in almost every other echinoderm group. Embryologists have supposed that they shared with ophiuroids a common ancestor different from another ancient creature that was antecedent to asteroids and holothurians. This idea now is generally rejected by paleontologists, but the problem of origin still defies a satisfactory solution. In the light of present knowledge either a primitive edrioasteroid or some cystoid-like animal seems to provide the most likely beginning for echinoid evolution.

Early Paleozoic Types

The few Ordovician echinoids that have been discovered are peculiar, and the interpretation of their structures and evolutionary significance is not agreed upon. None has an apical system comparable to those characteristic of later echinoids. Some late Ordovician species have a large porous hydropore plate at the top of one interambulacrum, but no other genitals can be surely recognized (Fig. 349-1). The water vascular canals were enclosed between the ambulacrals, or lay in only slightly open grooves (Fig. 349-3). This seems to be a primitive condition. The canals probably became internal after the atrophy of the inner portions of these plates and the closure of the overlying parts (Fig. 349-4). The tube-foot pores seem originally to have passed between the ambulacrals before they migrated upward and were divided by the development of median partitions. The interambulacral areas of these species are wide, with their plates arranged in irregular and somewhat discontinuous columns (Fig. 349-1).

An older mid-Ordovician species has a more simply constructed skeleton of more uniformly sized plates. Only one column is present in each interambulacral area (Fig. 352-2). Each ambulacral plate is pierced by two pores located centrally and arranged vertically, with one almost directly above the other (Fig. 350-1). The apical system includes five large plates, evidently oculars, in radial positions and in contact all around. They enclose several smaller plates that presumably surrounded the anal opening (Fig. 350-1). The largest of these, in interradial positions, may be genitals, although this is far from certain. Formerly this species was visualized as being very primitive, and efforts were made to derive all later echinoid structures by evolution from such a source. This theory no longer finds much favor among paleontologists, who recognize a combination of both relatively primitive and probably advanced characters in a fossil not near the main line of echinoid evolution.

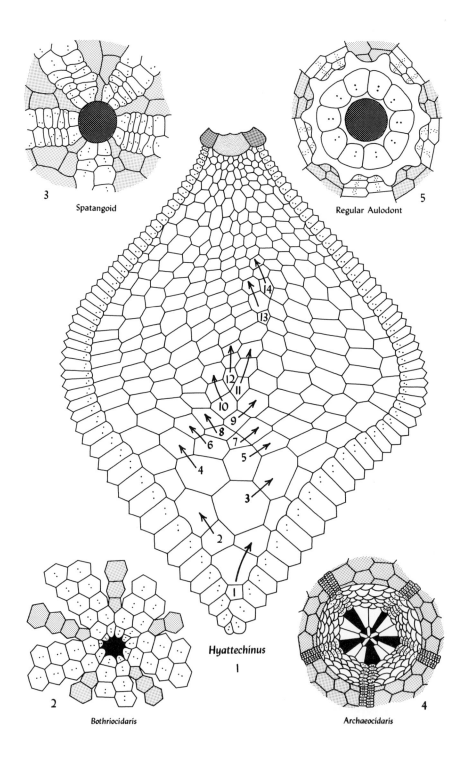

3 Spatangoid

5 Regular Aulodont

Hyattechinus
1

2 Bothriocidaris

4 Archaeocidaris

Fig. 352 (See facing page.) 1, Plate structure of an interambulacral area and half of each adjacent ambulacrum of a Mississippian echinoid. The plates are arranged in rather regular columns that are introduced successively. Some of these columns (1, 2, 10, 13) pinch out above, so that structure near the apex is somewhat simplified. (*After Jackson, 1912, pl. 25, figs. 1, 4.*) 2 to 5, Structures in the mouth regions in several types of echinoids. 2, This Ordovician form seems to be primitive in the equal size and regular alternation of ambulacral and interambulacral plates, and in the vertical arrangement and central positions of the paired pores. A peculiar but not unique feature (see Figs. 350-1 and 354-3) is the arrangement of interambulacral plates in single columns. 3, This type of structure is simple and relatively primitive. Each ambulacral area begins with two plates, and each interambulacral area with one. 4, As this Mississippian echinoid grew, the mouth area enlarged accordingly. The lowest and first-formed interambulacral plates were resorbed, and each interambulacral area begins with several columns of plates. The early ambulacral plates seem to have moved onto a flexible band of skin surrounding the mouth. The jaw parts and teeth of the so-called Aristotle's lantern are shown in the center. 5, The lowest plates of this modern echinoid also have been resorbed. Indentations between the ambulacral and interambulacral areas mark the positions of external gill-like organs. (*After Jackson, 1912, pl. 1, fig. 1; pl. 3, fig. 15; pl. 9, fig. 7; Mortensen, 1940, vol. 3, pt. 1, fig. 10, p. 16.*)

Paleozoic Regulars

All known post-Ordovician echinoids seem to have evolved from an original stock of flexibly plated, nearly spherical creatures with two columns in each ambulacral area and wider somewhat irregularly plated interambulacral areas. Pore pairs probably were central and may have been vertical in the ambulacral plates. Spines were short and were seated in pits or on very low rounded bases. The original form of the completed apical system is not understood. In most Paleozoic echinoids it consists of a circlet in which the genitals are all completely separated from one another by intervening oculars (Fig. 350-2). If these were post-Paleozoic fossils, their condition would be recognized as an advancement over a previous more primitive pattern of arrangement.

The results of evolution during the latter half of the Paleozoic Era include the following: (1) The flexible skeleton was altered to produce a rigid one composed of thicker plates. (2) Interambulacral plates promptly became arranged in regular columns that were introduced successively as the animals grew in size (Fig. 352-1). (3) The number of interambulacral columns decreased in some lineages. (4) The number of columns in ambulacral areas increased in some lineages (Fig. 354). (5) The pore pairs rotated, the upper pore ordinarily moving outward, so that the pairs became inclined or horizontal and approached the outer borders of the ambulacral areas (Fig. 354-1). (6) Some spines grew larger and articulated with prominent knobs rising from the body plates. These changes progressed at various rates in different combinations, and several distinctive types of echinoids evolved. Most of them became extinct shortly before the end of the Paleozoic Era.

The increase in ambulacral columns was particularly characteristic of one group of late Paleozoic echinoids (Fig. 355). This evidently was advantageous because it resulted in multiplying the number of tube feet. Their number could be increased in the original two columns only by reduction in height of the ambulacral plates so that more could be crowded into a restricted space. The structure of all but the

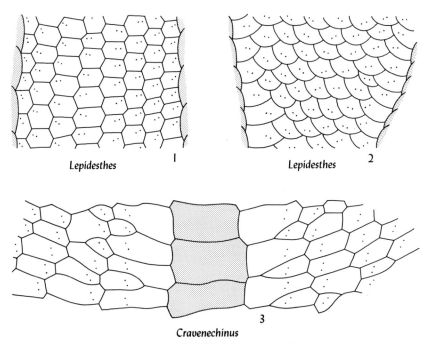

Lepidesthes 1   Lepidesthes 2

Cravenechinus 3

Fig. 354   1, 2, Ambulacral plates of two related Mississippian echinoids. In both, the plates imbricate downward but much more in 2 than in 1. The paired pores have rotated into more or less horizontal positions. 3, Another Mississippian form, peculiar in the single column of interambulacral plates and the location of paired pores farther from, rather than nearer to, the edges of the ambulacral areas. These pores seem to have rotated in a direction opposite to that characteristic of most echinoids. (*After Jackson, 1912, pl. 67, fig. 8; pl. 68, fig. 3; Hawkins, 1946, Geol. Mag., vol. 83, fig. 1, p. 193.*)

earliest echinoid shows that this process was in operation. When a practical limit had been reached, however, a change occurred. Some of the plates began to narrow and draw away from an edge of the column (Fig. 356). As the number of these plates increased, they rapidly became arranged above one another in new columns. Continuation of this process resulted in the development of up to 20 columns in the ambulacral areas of some species.

Fig. 355   (See facing page.) Ambulacral structures showing increasing complexity in an evolving lineage of late Paleozoic echinoids. 1, Each ambulacrum consists of two columns of plates; the most primitive condition. 2, Alternate plates have shortened so that they do not continue across their columns. 3, The shortened plates have produced two columns in each half ambulacrum. 4, Shortening of some plates results in the beginnings of two new columns. 5, Each half ambulacrum consists of three columns. (In this and the following figures, the middle of the ambulacrum is indicated by the heavier suture line.) 6, 7, 8, Continuation of this process results in the successive appearance of a fourth, a fifth, and a sixth column in each half ambulacrum. (*After Jackson, 1912, fig. 237, p. 231.*)

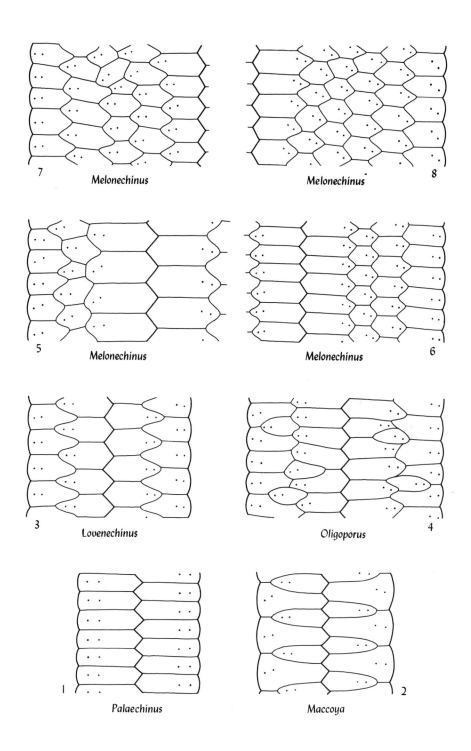

7   *Melonechinus*

Melonechinus   8

5   *Melonechinus*

Melonechinus   6

3   *Lovenechinus*

Oligoporus   4

1   *Palaechinus*

Maccoya   2

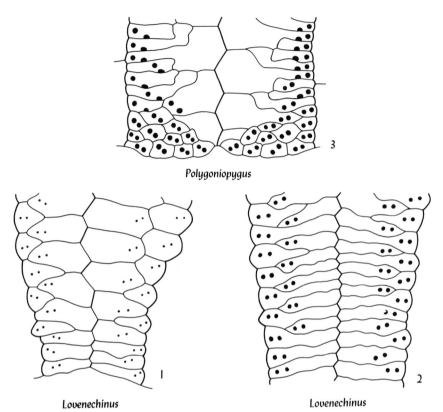

Polygoniopygus

Lovenechinus   *Lovenechinus*

Fig. 356   Modifications of ambulacral structures near the mouths of three echinoids. 1, 2, Transitions from single to multiple columns of plates in each half ambulacrum. 3, Transition from numerous small crowded plates to the compound ambulacrals possessed by many post-Paleozoic forms. (*After Jackson, 1912, pl. 42, fig. 1; pl. 45, fig. 2; Clark and Twitchell, 1915, U.S. Geol. Surv. Mono. 54, pl. 15, fig. 1h.*)

Every interambulacral area starts below with a single plate. This is followed above by the other plates of a column lying along one side of the area (Fig. 352-1). A second column begins with a plate above the original one just mentioned and extends upward along the other side of the interambulacral area. Further columns were introduced alternately to the right and left between the first two columns and those that appeared later. The full number characteristic of a species, which may reach 14, ordinarily occurs a short distance above the place where the skeleton attains its maximum diameter. Still higher, several of the columns commonly are pinched out between their neighbors as the interambulacral area becomes narrower (Fig. 352-1). These structural changes have been looked upon as an ontogenetic record suggesting a trend in evolution, first toward increase in the number of columns and then by reversal, toward decrease. The justification for such a conclusion is doubtful.

Post-Paleozoic Regulars

Most Mesozoic echinoids evolved from a single group that survived from the Paleozoic. They included both flexible and rigid types, but the latter became increasingly predominant. A few echinoids persisted in the Triassic with more than two columns of plates in the ambulacral or interambulacral areas, but their lineages soon died out. Essentially, therefore, the structure of the skeleton was stabilized early in Mesozoic time with two columns of plates in each of the 10 areas.

Evolutionary trends among the post-Paleozoic regular echinoids are especially noteworthy in four respects: (1) modification of the apical system (Fig. 362), (2) development of compound ambulacral plates (Fig. 357-2), (3) appearance of external gills (Fig. 352-5), and (4) growth of structures for the attachment of jaw muscles (Fig. 361).

(1) Both stratigraphic occurrence and ontogeny demonstrate that the apical system in its more primitive condition consists of a circlet of genital plates in contact with one another laterally, and oculars wedged into their outer angles. This is the structure of most Mesozoic species. Young specimens of subsequent epochs

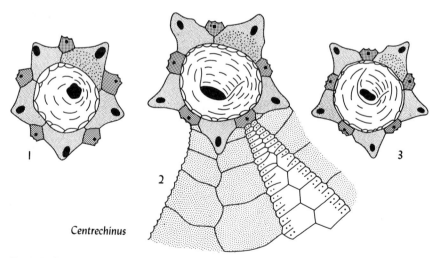

Centrechinus

Fig. 357 Ontogeny seems to recapitulate phylogeny in the apical structure of some echinoids. 1, The apical system of this modern form at a youthful growth stage shows the genital plates in contact all around, with the alternating ocular plates excluded from the central anal region. As individuals grew, the oculars moved inward to separate the genitals. 2, The structure commonly exhibited by mature specimens. Only one of the oculars is still excluded from contact with the anal region. 3, A progressive mature variant shows all the oculars inserted between the genitals. 2 also illustrates the fusion of ambulacrals to form compound plates. (Not drawn to uniform scale.) (*After Jackson, 1912, figs. 89, 94, 95, pp. 106, 107.*)

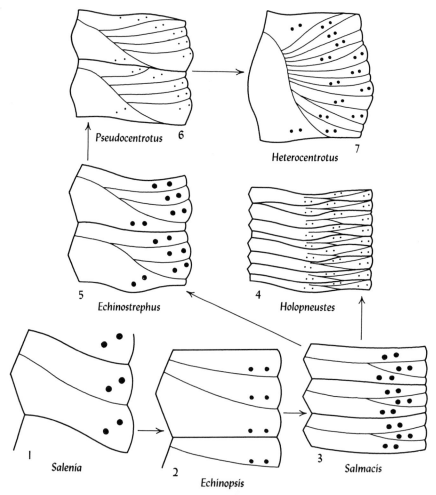

Fig. 358 Compound plates illustrating increasing complexity of ambulacrals. The patterns shown here all seem to be related because the lowermost of the fused platelets is largest or most prominent. Transition from the simplest type (1) involves the fusion of an increasing number of platelets (2, 5, etc.), the shrinking of one or many platelets (3, 4, 5, etc.), and the arrangement of pore pairs in arcs (3, 5, 6). The last (7) illustrates one of the most complex structures developed in plates of this type. Arrows indicate increasingly advanced evolutionary stages. They are not intended to imply that these genera are related in this detailed way. (*After Mortensen, 1935, 1943, vol. 2, fig. 196d, p. 377; vol. 3, pt. 2, figs. 35, 44, 67b, 240a, pp. 37, 46, 117, 384; vol. 3, pt. 3, figs. 132c, 145c, pp. 280, 307.*)

exhibit a similar arrangement (Fig. 357). As these later echinoids increased in size, however, their oculars grew somewhat more rapidly than the genitals. At maturity one or more oculars had forced its way between and separated the adjacent genitals. The insertion of oculars into the apical circlet in this way generally progressed in an order that was similar in groups of related species.

(2)   Ambulacral plates are introduced individually, one at a time, as an echinoid grows. Later, as these plates gradually mature in many post-Paleozoic regulars, they fuse in groups to produce compound plates (Fig. 357-2), which evolved as increasingly complex structures (Figs. 358; 359). Two general types of pattern developed that differ depending on which of the fused components is largest and most prominent in the compound plates. These developments made possible the crowding of an increased number of tube feet in an ambulacrum, which, like the multiple rows

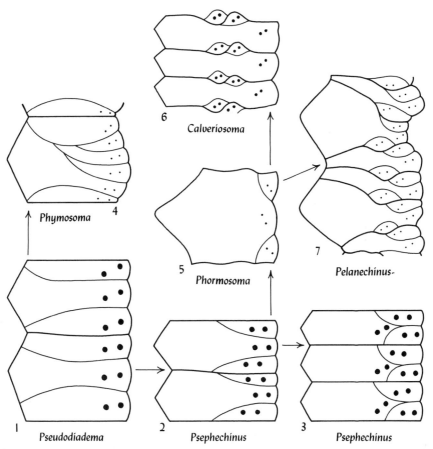

Fig. 359   Compound plates illustrating increasing complexity of ambulacrals. The patterns shown here all seem to be related because the middle, or next-to-lowest, platelet is largest or most prominent. Transitions from the simplest type (1) involve shrinkage of the outermost platelets (2, 3, 5, etc.), fusion of a larger number of platelets (4), or the development of superplates by the fusion of several compound plates (7). Arrows are not intended to suggest detailed evolutionary relations. (*After Mortensen, 1935, vol. 2, figs. 53a, 55c, 86a, 230a, 260c, 291c, pp. 83, 85, 143, 419, 458, 493.*)

of plates in some Paleozoic echinoids, probably was advantageous in effecting more efficient respiration.

(3)   Better respiration also was accomplished by external gills growing outward as fleshy structures around the mouth, which first appeared in some Triassic echinoids. To accommodate these gills, some of the first-formed plates in both ambulacral and interambulacral areas generally were resorbed (Fig. 352-5). This condition is recognizable, because the interambulacral columns begin with two plates instead of the original one.

(4)   The jaws of echinoids are anchored and activated by muscles attached to the inside of the surrounding body plates. A more efficient mechanism began to evolve in Mississippian time in some forms, with the appearance of a girdle of processes for muscle attachment which rose inward above the body plates (Fig. 361). This was continued in post-Paleozoic echinoids by the development of variant structures whose evolutionary sequence can be traced. Finally these structures degenerated and disappeared in several lineages of irregulars which adapted to a mud-eating habit and lost their teeth and jaws.

In addition to the foregoing, the teeth in some lineages of echinoids were strengthened by the evolution of a ridge or buttress upon the inner side which is not known in any Paleozoic form. Finally, considerable differences in the internal structure of the large movable spines have been observed; comparisons might indicate evolutionary trends, but comprehensive studies of this kind have not been made.

Irregulars

The first irregular echinoids appeared early in the Jurassic Period. Aside from the excentric anus which had migrated out of the circlet of apical plates, some of these fossils are almost indistinguishable from regulars that must represent the ancestral stock from which they sprang. Among the new forms are some with compound ambulacral plates and evidence showing that they had external gills, but others lacked these advanced characters. Evidently not all the irregulars had a common origin. Also it is clear that during later epochs similar displacement of the anus occurred in a few other stocks whose origins among the regulars are not so obvious. The irregular echinoids seem to provide several excellent examples of parallel, if not convergent, evolution.

Migration of the anus disturbed the symmetry of the apical system and was followed by a variety of distortions or rearrangements of its plates. Most of the resulting patterns can be ordered in a few morphologic sequences that seem to illustrate successive evolutionary stages of this process (Fig. 362). Loss of the posterior genital plate and enlargement of the right anterior genital plate are commonly recurring features (Fig. 362-4,5). In one group, the genitals fused to produce a single compound apical plate (Fig. 366-4).

An excentric anus impaired the perfection of radial symmetry, and steadily

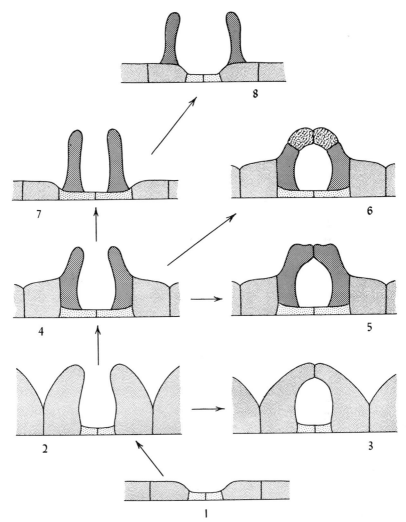

Fig. 361 Diagrammatic representation of stages in the morphologic evolution of the echinoid perignathic girdle which served for the attachment of jaw muscles. 1, These muscles, in early Paleozoic forms, were attached directly to the inner surfaces of interambulacral plates (intermediate shading). 2, In Mississippian time apophyses grew upward and inward from these plates. 3, Some of the apophyses joined to form an arch above the water vascular canal of the ambulacrum. 4, In the Triassic period new plates, the auricles (heavy shading), grew above the ambulacral plates (light shading) and between less-prominent apophyses. 5, In some forms these joined to produce an arch. 6, In others a similar arch resulted from the appearance of two new capping plates. This is the most complex type of girdle. 7, Apophyses shrank in size and disappeared. 8, The auricles of some echinoids shifted from the ambulacral to the interambulacral plates. Finally the girdle entirely disappeared in mature echinoids without jaws, although auricles may be present as temporary structures in young growth stages. (*Based on data from Jackson,* 1912.)

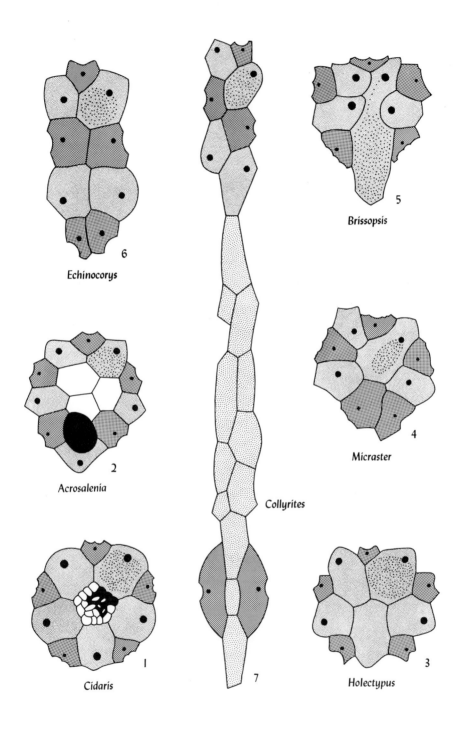

6
Echinocorys

Brissopsis
5

2
Acrosalenia

Collyrites

Micraster
4

1
Cidaris

7

Holectypus
3

Fig. 362   (See facing page.) Structures of the apical system in regular (1, 2) and irregular (3 to 7) post-Paleozoic echinoids. 1, This is a relatively primitive pattern. The genital plates are all in contact, and the oculars do not reach the central anal area, which is occupied by numerous small plates. 2, Evolutionary advancement is shown by the separated genitals, the excentric anus, and one large plate or a few large plates filling most of the anal area. 3, The anus has moved out of the apical system, and interference by the intestine has resulted in elimination of the sex gland beneath the posterior genital, which lacks a pore. 4, The right anterior genital of echinoids is pierced by many small pores which communicate with the water vascular system. This plate is enlarged and extends into the center of the apical system, from which the posterior genital has been eliminated. 5, Continued advancement is shown by further enlargement of the porous genital, which divides the apical system into two parts. 6, Evolution in another direction produced an elongated apical system. 7, In an extreme evolutionary development the posterior oculars have become separated from the remainder of the apical system by interambulacral plates (see Fig. 370a-2, 3). (*After Jackson, 1912, figs. 164, 168, 171, 173, 174, 175, p. 149; and Eastman-Zittel, 1913, "Text-book of Pale-ontology," 2d ed., vol. 1, fig. 412C, p. 291.*)

increasing bilaterality evolved in several lineages. The anus not only moved back-ward and downward, but in several groups it eventually passed forward on the ventral surface and approached the mouth (Fig. 364-6a). Some corresponding anterior migration of the mouth was much less conspicuous. These deviations from the almost perfect symmetry of less-specialized echinoids seem to indicate an important developing tendency for locomotion in a preferred direction. This probably was related to increasing adaptation to burrowing in bottom sediment and mud eating.

Several of the more characteristic features that evolved in the irregular echi-noids evidently were, in turn, related to such a way of life. These include (1) special-ization of the dorsal ambulacra, (2) specialization of the ambulacra ventrally around the mouth, (3) specialization of the ventral posterior interambulacrum, (4) the development of ciliated bands, and (5) loss of jaws.

(1)   Ambulacral plates in the central part of the dorsal surface became much shorter and wider (Fig. 366). The number of plates and tube feet thereby was increased, and the latter became flattened foliar expanses with enlarged surfaces adapted to more efficient respiration. The shape of these portions of the ambulacra somewhat resembles the petals of a flower. Primitively the petaloid areas graded outward into series of less-specialized ambulacral plates (Fig. 366-1). In more highly evolved forms, however, the petals close externally (Fig. 366-3) and some of the succeeding ambulacral plates seem to lack pores and, therefore, probably were without tube feet. Plates within the petals did not fuse, but differentiation of alter-nate plates developed in one group.

Many of these echinoids were shallow burrowers, and the extent of the petals probably indicates the part that ordinarily remained above the sediment surface. External gills around the mouth would be relatively useless for creatures of this kind; either they evolved from gill-less forms, or the gills were lost in their immediate ancestors.

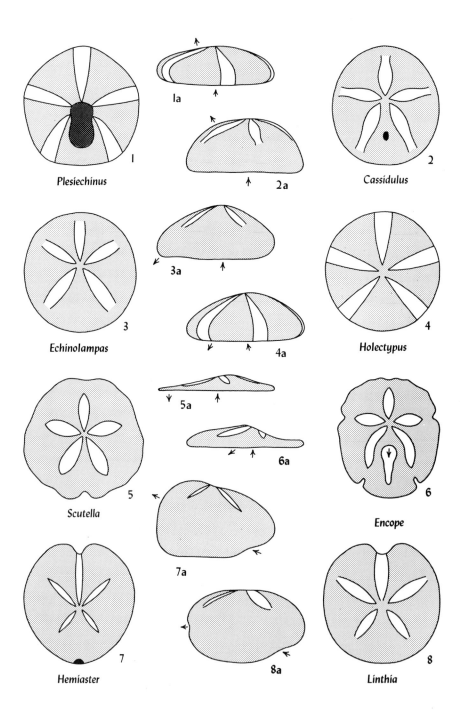

1 *Plesiechinus*

1a

2a

2 *Cassidulus*

3 *Echinolampas*

3a

4a

4 *Holectypus*

5 *Scutella*

5a

6a

6 *Encope*

7a

7 *Hemiaster*

8a

8 *Linthia*

Fig. 364 (See facing page.) Top and side views of a variety of irregular echinoids, showing positions of mouth and anus by small arrows. 1, The anus has migrated only a short distance from the apex. 2, The anal area is better defined and a little farther down the posterior slope. 3, The anus is located at the periphery. 4, It has moved onto the lower surface. Progressive migration of the anus in this way was not accompanied by comparable migration of the mouth. 5, 6, Much-flattened forms known as sand dollars, some of which show conspicuous indentations of the margin (see Fig. 370a-1) and perforations. 7, 8, Conspicuously bilateral forms, some of which are termed heart urchins, with posterior anus and anterior mouth. (1 and 4 after Mortensen, 1948, vol. 4, pt. 1, figs. 7a, 8a, 11a, pp. 20, 27. Others after Clark and Twitchell, 1915, U.S. Geol. Surv. Mono. 54, pl. 31, figs. 1a, c; pl. 48, figs. 3a, c; pl. 70, figs. 3a, c; pl. 81, figs. 1a, c; pl. 86, fig. 1a; pl. 87, fig. 1b; pl. 93, figs. 2a, b.

Differentiation of the ambulacra contributed to the bilaterality of many irregular echinoids. Extreme development in one group, the heart urchins, involved the evolution of an anterior ambulacrum much different from the rest (Fig. 367). It is not petaloid like the others and commonly occupies a distinct but shallow groove. Modern species show that the tube feet here are very long. They search the sediment surface for food particles, pick them up, and carry them to the mouth.

Widening of petals reduced the width of interambulacral areas. Evolution in one group of irregulars resulted in the substitution of an increasing number of single interambulacral plates for the double column of alternating plates that normally occupy these areas (Fig. 368).

(2) Somewhat similar but much shorter petaloid areas evolved in the ambulacra on the underside around the mouth (Fig. 366-1,2). In some groups these are recessed between projecting plates of the interambulacra. The tube feet evidently became differentiated for specialized feeding or locomotive functions. In the most advanced forms they communicated with the interior through a single pore, rather than double pores. The first pair of tube feet in each ambulacrum of some echinoids became different from all the others, and their positions are marked by especially large pores.

Progressive widening of the ambulacral plates in more than one group constricted the interambulacral areas and eventually wedged between and completely separated the first plate from those that follow it (Fig. 369). Increasing differentiation of the posterior interambulacrum together with the specialized dorsally nonpetaloid anterior ambulacrum gives some of these echinoids their strongly bilateral form.

(3) Each interambulacrum of most echinoids begins with a single plate, which is followed by two columns whose plates alternate with each other. Evolution resulted in the modification of this pattern in several ways. One of the most evident alterations is exhibited by a group of burrowing irregulars that developed more than ordinary bilateral symmetry. These show an evolutionary progression in the posterior interambulacrum between mouth and anus in which the alternating plates gradually assume opposite positions. At the same time there is a tendency for the

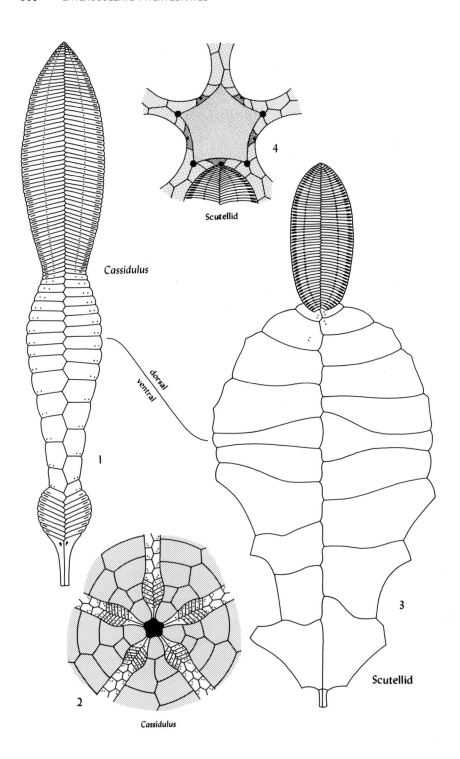

Cassidulus

dorsal

ventral

Scutellid

1

2

Cassidulus

3

Scutellid

4

Fig. 366 (See facing page.) Specialization of ambulacral areas in two kinds of irregular echinoids. 1, A complete ambulacrum extending from apical system to mouth area, showing both dorsal and ventral so-called semipetalloid structures. In these specialized areas the ambulacral plates are very short. The upper ones bore flattened leaf-like tube feet, whose main function is respiration. 2, Arrangement of semipetalloid structures around the mouth. These, as seen in an inverted specimen of a different species, are depressed below the intervening interambulacral plates. Their tube feet serve mainly feeding functions. (*After Clark and Twitchell, 1915, U.S. Geol. Surv. Mono. 54, pl. 30, fig. 2i; pl. 32, fig. 2b.*) 3, A complete ambulacrum showing a dorsal, fully petalloid structure drawn from a modern specimen. The difference is that this, unlike the other, seems to be closed below, and most of the lower ambulacral plates lack recognizable paired pores. 4, Apex of the same specimen showing how the genital plates are fused to form a single plate without a posterior pore. The oculars lying against the sides of this plate are very small.

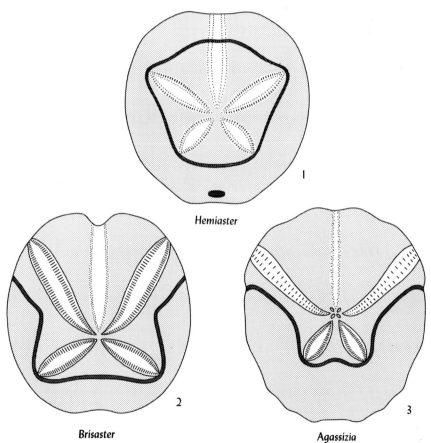

Fig. 367 Top views of three irregular urchins showing progressively increasing differentiation of the ambulacral areas and the disposition of a ciliated band or fasciole. Other forms have other or differently located fascioles. (*After Mortensen, 1950, vol. 5, pt. 1, fig. 272a, p. 380; Clark and Twitchell, 1915, U.S. Geol. Surv. Mono. 54, pl. 1, fig. 6; fig. 10, p. 115.*)

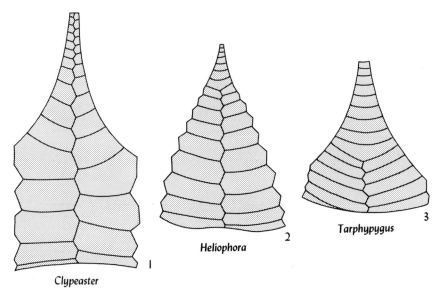

Clypeaster   1

Heliophora   2

Tarphypygus   3

Fig. 368   Diagrams showing progressive evolutionary stages in the transformation of the interambulacral structure from two columns to a single column of plates in clypeasteroid echinoids. (*After Durham, 1955, fig. 12, p. 91.*)

pair of plates following the first single plate to increase considerably in size (Fig. 369a). In the most extreme development, however, these paired plates become separated from the first one by the incrowding of adjacent ambulacrals.

(4)   Two groups of irregular echinoids have narrow and shallow surface grooves whose ornamentation is different from that elsewhere on the body plates. On the heart urchins, these grooves form closed loops of several different types whose development indicates evolutionary relations (Fig. 367). These echinoids lived wholly buried in the bottom sediments. The grooves bore innumerable tiny spines whose skin was ciliated. The currents that the cilia produced probably served to circulate water around the respiratory tube feet and removed sediment which settled on these creatures in their burrows.

Furrows of the sand dollars and their relatives are largely ventral and bear innumerable tiny tube feet. These furrows radiate from the mouth along the ambulacra and are simple and unbranched in the more primitive forms (Fig. 370-1). Branching evolved progressively, the grooves extending laterally outside the ambulacra and, on the most advanced forms, continuing upward for some distance on the dorsal surface. Water currents moving along these grooves probably carried food particles to the mouth.

(5)   Many of the burrowing irregular echinoids became mud eaters, subsisting on the organic matter it contained. Jaws are not required for this kind of diet, and

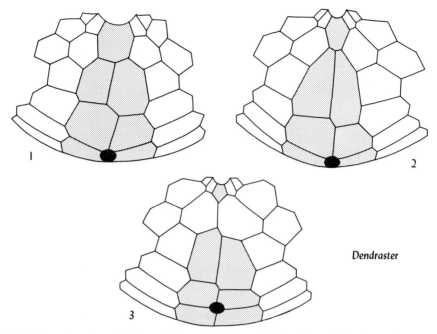

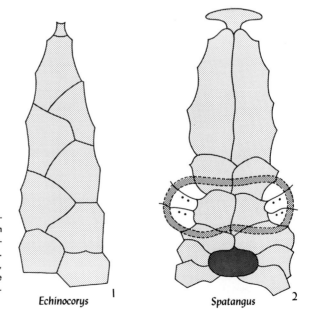

Fig. 369 Successive ontogenetic stages showing progressive modification of interambulacral areas between mouth and anus. These become interrupted by the relatively more rapid growth of ambulacral plates. *(After Durham, 1955, fig. 11, p. 89.)*

Fig. 369a Plate diagrams showing modification and specialization of structure in the posterior interambulacrum of the ventral surface. *(Compare Melville and Durham, 1966, "Treatise on Invertebrate Paleontology," pt. U, Echinodermata, vol. 3, fig. 181, p. 237.)*

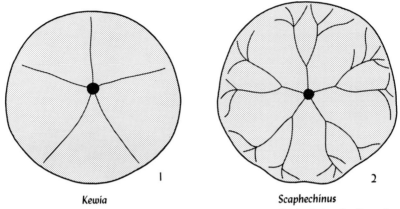

Kewia   Scaphechinus

Fig. 370 Narrow food grooves, provided with very numerous small tube feet, extending outward from the mouth along the ambulacral areas become increasingly complex in one group of echinoids. Some even continue onto interambulacral plates and the dorsal surface. (*After Durham, 1955, figs. 3A, 4E, pp. 78, 79.*)

these structures degenerated and disappeared in two of the main groups. Recapitulation is evident in the ontogeny of some of the more primitive forms. In their very young stages they possess jaws and specialized girdles for the attachment of jaw muscles or girdles only. With growth, however, these structures are resorbed and vanish.

Feeding among these echinoids is accomplished mainly by the tube feet, which stuff mud into the mouth. A somewhat different method evolved, however, among

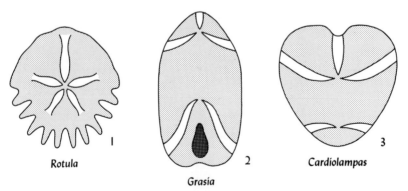

Rotula   Cardiolampas

Grasia

Fig. 370a Three irregular echinoids displaying distinct bilateral symmetry. 1, This is a flattened form with semipetalloid ambulacra and a scalloped posterior margin. 2, 3, Forms with disconnected apical systems (see Fig. 362-7) and posterior ambulacra widely separated from the others. The last is a typical heart urchin. (*After Mortensen, 1950, vol. 5, pt. 1, figs. 18a, 31a, 256a, pp. 18, 30, 461.*)

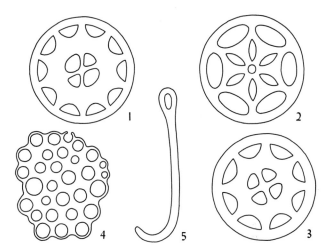

Fig. 371   Calcareous holothurian plates drawn from Pennsylvanian specimens. (Greatly enlarged but not at uniform scale.) 1 to 3, Wheel-like plates
exhibit a variety of forms. 4, A sieve-like plate. 5, A spicule shaped like a
fishhook. Somewhat similar plates occur in certain modern echinoids, but
they are much smaller, and fossil examples have not been recorded.

the bilateral heart urchins. Their mouths, although located on the undersurface,
became directed forward and, aided by the tube feet, acted as a kind of scoop as
these creatures moved slowly through the bottom sediment (Fig. 364-7a,8a).

Other noteworthy evolutionary features of some irregular echinoids include
internal buttresses, marginal indentations (Fig. 370a-1), and perforations (Fig.
364-6). All of these seem to have been functional developments which strengthened
the thin flat sand-dollar type of urchins. Buttressing began as internal pillars and
partitions connecting the dorsal and ventral body plates near the periphery, and
gradually became more complex and more extensive. The common circular shape
was modified by indentations of the margin which deepened to form slits. Finally
the projections between the slits joined externally, leaving perforations through the
body. Indentations and perforations evolved variously in both ambulacral and
interambulacral positions. They are likely to be most prominent and most common
posteriorly.

Sea Cucumbers

Holothurians, or sea cucumbers, lack a plated skeleton but are provided with tiny
characteristically shaped calcareous spicules embedded in the skin (Fig. 371). This
probably is an archaic condition that preceded the development of plates in other
echinoderms. A single plated species of Ordovician age has been described as a
holothurian, but almost certainly it is an echinoid. Impressions of holothurians in

fine-grained sediments have been found very rarely at a few places, but they reveal little about these animals. Otherwise the fossil record consists only of scattered spicules dating back to at least the mid-Mississippian. Consequently nothing is known about the actual evolution of holothurians, and speculation concerning their history is so uncertain that it has little value.

INVERTEBRATE CHORDATES

A small number of peculiar invertebrate animals included in the broadly defined chordate phylum seems to provide connecting links between the echinoderms and vertebrates. Such creatures have left no certainly recognizable fossil record, although some paleontologists believe that the graptolites were allied with them. These animals, however, are very interesting from an evolutionary standpoint.

The most distinctive feature of the chordates is a notochord, possessed by an animal during some stage of its existence. This is a structure developed dorsally from the entoderm; it directly underlies the dorsal nerve cord (Fig. 373). It consists of cells whose fluid contents are maintained under pressure, and thus provides a stiff but elastic axis for the body. In most vertebrates it is first enclosed by bone and then degenerates and is functionally replaced by the bony spinal column. Another feature peculiar to the chordates is the occurrence of gill slits, which may persist throughout life as in fishes, or which characterize only larval or certain embryonic stages, as in all four-footed land animals. Finally the dorsal nerve cord of chordates is pierced by a small central canal which is present in no other creatures.

The invertebrate chordates differ so greatly among themselves and bear so few obvious resemblances to either echinoderms or vertebrates that their probable

Table 372   **Classification of the chordates**

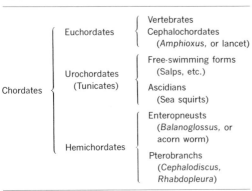

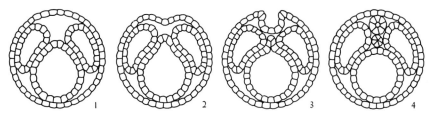

Fig. 373 Highly diagrammatic transverse sections, not drawn to uniform scale, showing embryonic origin of the hollow spinal nerve, derived from ectoderm, and the notochord, derived from entoderm.

relations long went unrecognized. The hemichordates reveal threefold segmentation of the coelom, like that of echinoderms, and possess exterior pores resembling the hydropores of these other animals. Anteriorly they have presumably vestigial structures that may represent the notochord, although this homology has been questioned. The larvae of *Balanoglossus*, the acorn worm, are very similar to those of some echinoderms (Fig. 381). This animal is worm-like and furnished with a double row of gill slits that function as a food-filtering device and for respiration. The pterobranchs are colonial and reproduce by budding. They bear tentacles that resemble somewhat the lophophore of bryozoans. *Cephalodiscus* has a single pair of gill slits, but *Rhabdopleura* has none. Graptolites have been compared to *Rhabdopleura* (Fig. 587).

Both an unmistakable notochord and numerous gill slits occur in the urochordates. The notochord is lost in mature ascidians, which are sessile, but it is present in the tails of their larvae and in the free-living salps and their kindred. The ascidian larva is termed a tadpole because of its resemblence to the larvae of amphibians (Fig. 382-2).

Amphioxus (Fig. 382-1) possesses all the essential characters of a chordate. It is similar in many ways to the lamprey larva, and its muscles are segmented as in the vertebrates, but it is not quite a fish. It differs from all vertebrates in lacking any semblance of a skeleton except for cartilage in its gill bars and fins. Also it does not have a brain, heart, kidneys, auditory organs, paired eyes, or hemoglobin in its blood.

The embryonic development of all invertebrate chordates resembles that of echinoderms. These chordates seem to be the somewhat specialized but otherwise conservative remnants of a group of animals that stood close to the ancestors of echinoderms and evolved to produce the vertebrates. Separately none of them provides much evolutionary information. Together, however, their evidence seems significant. The hemichordates show alliance to echinoderms, and their gills connect them with the tunicates, whose tadpoles resemble *Amphioxus*. This animal certainly is a chordate. It foreshadows the larva of the lamprey, which represents the most primitive group of fishes.

BIBLIOGRAPHY

F. A. Bather (1915): Studies in Edrioasteroidea: IX, The Genetic Relations to Other Echinoderms, *Geol. Mag. n.s.,* Dec. 6, vol. 2, pp. 393–403.
Theories are reviewed, and the conclusion is reached that starfishes evolved from primitive edrioasteroids.
_____ **et al. (1900):** "The Echinodermata: A Treatise on Zoology," E. R. Lankester (ed.), pt. 3, A. & C. Black, Ltd., London.
This book is old but still important, particularly for the consideration of pelmatozoan echinoderms.
J. W. Durham (1955): Classification of Clypeasteroid Echinoids, *Univ. Calif. Publ. Geol. Sci.,* vol. 31, pp. 73–198.
This mainly descriptive work includes incidental information on evolutionary trends.
_____ and **R. V. Melville (1957):** A Classification of Echinoids, *J. Paleontol.,* vol. 31, pp. 242–272.
A discussion of the main evolutionary trends of echinoids is contained in this article.
_____ **et al. (1966):** Echinoids, in "Treatise on Invertebrate Paleontology," R. C. Moore (ed.), pt. U, Echinodermata 3, vol. 1, pp. 211–366a; vol. 2, pp. 367–640, Geological Society of America, Boulder, Colo., and University of Kansas Press, Lawrence, Kans.
See pp. 266–270, 298–301, 312–313, 344, 493–498, and 544–548 for considerations of phylogeny and evolutionary trends.
R. O. Fay (1967): Evolution of the Blastoidea, in "R. C. Moore: Commemorative Volume," C. Teichert and E. L. Yochelson (eds.), pp. 242–286, University of Kansas Press, Lawrence, Kans.
Morphologic relations are traced in the two principal groups of blastoids.
H. B. Fell (1948): Echinoderm Embryology and the Origin of the Chordates, *Biol. Rev.,* vol. 23, pp. 81–107.
Echinoderm larvae are not recapitulatory, and they provide no evidence of relations to the chordates.
_____ **(1963):** The Evolution of the Echinoderms, *Smithsonian Inst. Ann. Rept., 1962,* pp. 457–490.
Starfishes are presumed to have evolved from crinoids.
D. F. Frizzell et al. (1966): Holothurians, in "Treatise on Invertebrate Paleontology," R. C. Moore (ed.), pt. U, Echinodermata 3, vol. 2, pp. 641–672; Geological Society of America, Boulder, Colo., and University of Kansas Press, Lawrence, Kans.
See pp. 641–645 and 656–657 for considerations of some evolutionary trends.
H. L. Hawkins (1920): The Morphology and Evolution of the Ambulacrum in *Echinoidea holectypoida, Phil. Trans. Roy. Soc. London Ser. B,* vol. 209, pp. 377–480.
The number of tube feet was increased by shortening of ambulacral plates, increase in the number of their columns, and combination in various ways to form compound plates.
_____ **(1931):** The First Echinoid, *Biol. Rev.,* vol. 6, pp. 443–458.
The earliest echinoids were not of the multicolumned flexible type, and they evolved from holothurian-like animals.
_____ **(1943):** Evolution and Habit among the Echinoidea, *Quart. J. Geol. Soc. London,* vol. 99, pp. lii–lxxv.
This article discusses the growth and evolution of echinoids as indicated by the fossil record.
L. H. Hyman (1955): Echinodermata, The Coelomate Bilateria, vol. 4 of "The Invertebrates," McGraw-Hill Book Company, New York.
See pp. 691–705 for consideration of evolution in this comprehensive zoologic treatise.
R. T. Jackson (1912): "Phylogeny of the Echini, with a Revision of Paleozoic Species," Memoirs of the Boston Society of Natural History, vol. 7.
This classical work includes a comprehensive consideration of echinoid evolution.
P. M. Kier (1962): Revision of the Cassiduloid Echinoids, *Smithsonian Inst. Misc. Collections,* vol. 144, no. 3.

Several morphologic evolutionary trends are described; see pp. 4–15.

———— **(1965):** Evolutionary Trends in Paleozoic Echinoids, *J. Paleontol.,* vol. 39, pp. 436–465. Increase in the number of plate columns and modifications of podial pores are the most conspicuous trends.

E. Kirk (1911): The Structure and Relationships of Certain Eleutherozoic Pelmatozoa, *Proc. U.S. Natl. Museum,* vol. 41, pp. 1–137. This article is concerned with the functions and evolution of pelmatozoan stems.

E. W. MacBride and **W. K. Spencer (1939):** Two New Echinoidea, *Aulechinus* and *Ectinechinus,* and an Adult Plated Holothurian, *Eothuria,* from the Upper Ordovician of Girvan, Scotland, *Phil. Trans. Roy. Soc. London, Ser. B,* vol. 229, pp. 91–137. These fossils probably represent the general stock from which all later echinoids evolved.

R. C. Moore (1952): Evolutionary Rates among Crinoids, *J. Paleontol.,* vol. 26, pp. 338–352. Evolutionary trends and relations among Paleozoic crinoids are described.

———— **(1954):** Echinodermata: Pelmatozoa, in Status of Invertebrate Paleontology, 1953, *Bull. Museum Comp. Zool.,* vol. 112, pp. 125–149. This paper points out possible homologies in the plate structure of various pelmatozoans which may have a bearing on their evolutionary relations.

———— and **L. R. Laudon (1943):** Evolution and Classification of Paleozoic Crinoids, *Geol. Soc. Am. Spec. Paper 46.* This mainly taxonomic work includes considerations of evolution within the main crinoid groups.

T. Mortensen (1928–1951): "Monograph of the Echinoidea," 5 vols. in 17 parts, C. A. Reitzel, Copenhagen. Many incidental references to evolution are contained in this comprehensive work, which is devoted mainly to modern echinoids; see particularly vol. 5, pt. 2, pp. 565–573.

G. M. Philip (1965): Classification of Echinoids, *J. Paleontol.,* vol. 39, pp. 45–62. This paper criticizes the evolutionary implications of recent echinoid classification.

———— **(1965):** Plate Homologies in Inadunate Crinoids, *J. Paleontol.,* vol. 39, pp. 146–149. Ideas regarding homologies of radial and anal plates have evolutionary implications and are criticized.

G. Regnéll (1966): Edrioasteroids, in "Treatise on Invertebrate Paleontology," R. C. Moore (ed.), pt. U, Echinodermata 3, vol. 1, pp. 136–173, Geological Society of America, Boulder, Colo., and University of Kansas Press, Lawrence, Kans. See pp. 156–157 for a consideration of evolutionary trends.

C. Schuchert (1915): Revision of Paleozoic Stelleroidea with Special Reference to North American Asteroidea, *U.S. Natl. Museum Bull. 88.* See particularly pp. 30–51 in this old but important monograph on fossil starfishes.

H. M. Smith (1960): "Evolution of Chordate Structure: An Introduction to Comparative Anatomy," Holt, Rinehart and Winston, Inc., New York. See chap. 2 for a consideration of invertebrate chordates and their phylogenetic relations.

W. K. Spencer (1913–1940): "A Monograph of the British Paleozoic Asterozoa," Palaeontographical Soc., London. In this unfinished work, see particularly pt. 1 (1914), pp. 3–33; pt. 2 (1916), pp. 61–68; and pt. 6 (1925), pp. 242–246.

———— **(1951):** Early Paleozoic Starfish, *Phil. Trans. Roy. Soc. London, Ser. B,* vol. 235, pp. 87–129. The primitive fossils described here have an important bearing on theoretical starfish evolution.

———— and **C. W. Wright (1966):** Asterozoans, in "Treatise on Invertebrate Paleontology." R. C. Moore (ed.), pt. U, Echinodermata 3, vol. 1, pp. 4–107, Geological Society of America, Boulder, Colo., and University of Kansas Press, Lawrence, Kans. See pp. 31–35 for a consideration of phylogeny and evolution.

CHAPTER EIGHT

VERTEBRATES 1

CARTILAGE AND BONE
ORIGIN OF VERTEBRATES
FISHES—AMPHIBIANS

The vertebrates are particularly interesting because (1) they include all familiar large animals, (2) they are economically important in many ways, and (3) man himself is a vertebrate. These animals have been studied more intensively from the standpoint of their evolutionary relations than any others, and their geologic history is relatively well known. Although vertebrates probably originated earlier, the oldest bones have been found in Ordovician strata. Very little is known, however, about the animals from which they came. Primitive fishes of middle and late Silurian age have been discovered, but the vertebrate record before the Devonian Period is rather scanty. From that time onward, however, the fossils are adequate to trace the evolutionary development and radiation of the vertebrates in considerable detail. Thus it is possible to reconstruct the general phylogeny of vertebrates, beginning with primitive fishes and progressing by stages through the amphibians and reptiles to birds and mammals, as shown in Fig. 378.

Cartilage and Bone
The descriptive term vertebrate seems to imply an animal provided with a bony spinal column, but not all creatures recognized as vertebrates have a structure of this kind. Actually most fossil vertebrates are identified by the presence of bone, no matter where within the body it was located. Some had no bony spinal column, and some modern vertebrates are entirely boneless, although they do have spinal supports composed of cartilage. This also was true for some known fossil animals.

There are several different types of cartilage and bone, and confusion may arise

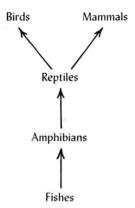

Fig. 378 Diagram showing evolutionary relations of the vertebrates.

because definitions of these materials differ. No phylogenetic sequence of the types has been established. The statement has been made repeatedly, however, that bone is more primitive than cartilage. This certainly is questionable. It may be a misconception deriving from (1) the presence of bone in the oldest known fossil vertebrates, (2) the fact that cartilage is not necessarily prerequisite to bone formation, and (3) the opinion that the cartilaginous condition of some vertebrates, such as lampreys and sharks, is not primitive. This last opinion also may be in part unjustified or mistaken. Lampreys and sharks certainly are primitive in some respects, although they may be much specialized in other ways. Their lack of internal bony skeletons may be truly primitive in some, even though in others it is an example of retarded development that has preserved characters in mature individuals which otherwise are indicative of youthful or embryonic stages.

Cartilage-like material almost surely was the first kind of primarily supporting connective tissue derived from the mesoderm to have evolved in chordates. Such material occurs in the gills and fins of the lancet *Amphioxus* and in some other invertebrate chordates, all of which are recognized to be more primitive than any vertebrate. Just how and in what kind of creature the internal vertebrate skeleton had its origin, however, is not known, but it probably first appeared as cartilage. If so, the first vertebrates had skeletons unsuited for fossil preservation, and their remains may never be discovered. The earliest ostracoderms and many of the older, more conventional fishes have left no trace of a central bony spinal column, but cartilage almost certainly was present in them.

Cartilage is a flexible, elastic, and translucent material produced by a special kind of cells (Fig. 379-1). It consists of a matrix bound together by a network of fibers which surrounds these cells. It has the potentiality of growth, both by internal expansion and by addition to its surfaces. In the ontogeny of vertebrates almost all bones that originate beneath the skin follow cartilaginous structures. Cartilage may be partly calcified and hardened, but it is replaced by bone, rather than transformed to bone.

On the other hand, dermal bones which originate within the skin are not preceded by cartilage. They are formed directly by a different kind of cells within the dermis or underlying layer of the skin, which is a tissue of mesodermal origin. Similar cells invade the cartilage of the immature internal skeleton. There, cartilage is resorbed and bone matrix is secreted. This consists of calcium phosphate with which various organic substances are associated. Bone is rigid and opaque, and this part of it is porous (Fig. 379-2). It encloses irregular star-shaped openings, in which the bone cells are located, and tiny canals connecting them. Unlike cartilage, it is traversed by nerves and blood vessels. There is no direct evidence that dermal bone is older phylogenetically than internal cartilage.

Bone can grow only outward from certain centers or by additions to its surface. Surficial bone is laid down in laminae that are more compact than cartilage-replacing bone. This also is the kind that forms within the skin. Despite its rigidity and partly inorganic nature, bone is living substance that continually changes. As bones grow, new layers are laid down on their outer surfaces and along their edges, and the older bone may be gradually resorbed and replaced by new. This process results in a spongy type of bone enclosing marrow in its openings or hollow centers. Thus both internal and dermal bones are altered. They retain their shape and strength without the addition of excessive weight.

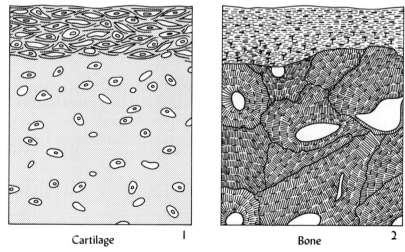

| Cartilage | 1 | Bone | 2 |

Fig. 379   1, Clear stiff cartilage below is transitional above into the membrane that surrounds it. 2, Replacement bone below is overlain by lamellar surficial bone. (Much enlarged.)

Origin of Vertebrates

The paleontologic record provides no direct information regarding the origin of vertebrates, and much speculation has been devoted to this problem. The oldest fossils are scraps of bony plates and scales from Ordovician sandstones, whose environmental origin has been disputed. These fossils are believed to indicate the existence of ostracoderms at that time, but the first reasonably satisfactory fossils are of middle to late Silurian and Devonian ages.

Vertebrates must have evolved from invertebrate animals, and their antecedents have been sought in several very diverse groups. Derivation from annelids, arthropods, or echinoderms has been most strongly argued, but even the coelenterates and molluscs have been viewed as possible ancestors. Although various theoretical morphologic transformations that might account for the origin of vertebrates can be imagined, the most pertinent evidence seems to be furnished by the comparative anatomy and ontogeny of modern animals.

Arthropod Theories

Theories seeking to explain the divergence of vertebrates from arthropods, or their annelid ancestors, formerly were much in vogue but no longer have much following. One of the most obvious difficulties that these theories must overcome is the difference in relative position, with reference to the gut, of the principal nerve cord, which is ventral in annelids and arthropods but dorsal in the vertebrates. One seemingly simple suggestion is that vertebrates evolved from creatures that swam upon their backs as some modern crustaceans do. Another theory supposes that the ventral arthropod nerves migrated upward and surrounded the gut which lost its digestive functions, shrank in size, and became the central canal of the vertebrate spinal cord. At the same time the ventral furrow between the arthropod appendages closed and became an entirely new gut. Some of the appendages were transformed to gill bars. The paired fins of fishes developed from arthropod pleural structures, in much the same way that these parts have been supposed to provide the evolutionary basis for insect wings.

Among the many serious objections that have been raised against theories of this kind, one of the most telling concerns the contrasting embryonic developments in these two groups of animals. From the first cleavage of the egg, arthropods and chordates follow two totally different paths. The arthropod egg cleaves spirally and the coelom originates schizocoelously, whereas in the invertebrate chordates the cleavage is radial and the coelom is enterocoelous. Embryonic development of the vertebrates, however, is more complex.

Echinoderm Theories

Few theories have attempted to show that vertebrates actually evolved directly from echinoderms, but a common ancestor generally is postulated. Much evidence seems to support this conclusion; e.g., (1) Embryonic development is similar in echinoderms and primitive chordates. (2) Larvae of the acorn worm *Balanoglossus,* which

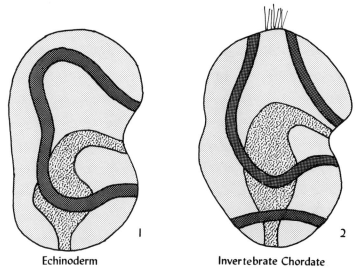

Echinoderm   Invertebrate Chordate

Fig. 381   Much-enlarged diagrams showing the fundamental similarity of, 1, an early stage in the development of an echinoderm larva and, 2, the larva of an acorn worm, an invertebrate chordate. Both resemble the theoretical dipleurula and are equipped with comparable ciliated bands. (*After de Beer, 1964, "Atlas of Evolution," figs. 102, 103, p. 40.*)

is a chordate, are very similar to those of some echinoderms (Fig. 381). (3) Skeletons in both groups originate mesodermally. (4) Serologic tests show that blood proteins are similar. (5) The same kind of phosphorous compound plays an essential part in the energy cycle of muscular action and differs from that in other animals.

Most zoologists believe that the echinoderms and chordates are closely related in their evolutionary origins and together constitute one of the two great branches of the animal kingdom. Certain general similarities in the forms of some peculiar echinoderms, known as carpoids, to some ostracoderms have received notice, but their possible significance as an indication of close connection is exceedingly remote. The chief phylogenetic uncertainty concerns the nature of the immediate ancestors of the vertebrates among the invertebrate chordates. Agreement is fairly general that the earliest vertebrates were swimming fish-like creatures. The nearest approach among modern animals to the theoretical vertebrate prototype is seen in either the tadpole-like larva of the tunicates (Fig. 382-2), of which the ascidian sea squirts are the most familiar, or in the lancet *Amphioxus* (Fig. 382-1).

One theory favors ascidians as the ancestors of vertebrates. It postulates that arrested development at the tadpole larval stage permitted evolution to proceed in a new direction. This seems to be based on (1) the idea that the sessile condition

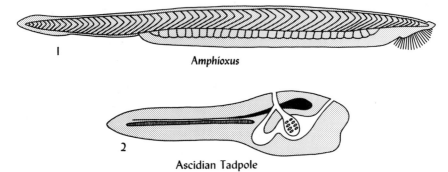

Amphioxus

Ascidian Tadpole

Fig. 382   1, The mouth is surrounded by cirri, and the gills, which are not visible, open into a long enclosed tube with a posterior aperture. Notice segmentation of the body muscles, and the ventrolateral fold that is duplicated by another on the opposite side. (Approx. ×2.) 2, Generalized diagram showing fundamental anatomy. Notice the notochord in tail, the overlying nerve cord extending backward from a "brain," and the gill slits associated with the alimentary tract. This creature is about 1 mm long and does not feed. (*After Moody, 1962, "Introduction to Evolution," 2d ed., figs. 4.17, 5.5, pp. 75, 101, and other sources.*)

of the adult ascidian is primitive, and (2) the fact that the growth of some tunicates does not continue beyond a sexually mature tadpole stage and, therefore, that they are an example of arrested development. No detailed analysis is necessary to show that this is a devious way of achieving a result that can be explained equally well in a more simple manner. Ascidians, like echinoderms, have bilateral free-swimming larvae that become attached anteriorly and then undergo metamorphosis. It is reasonable to suppose that both these groups evolved from a simply organized free-living ancestor of the general dipleurula type. Evolution beginning with such a creature might have produced a primitive chordate without passing through a sessile stage. Actually, several features of the ascidian tadpole, including its dorsally directed mouth and anus, seem to make a creature of this kind a very unlikely candidate for the ancestor of vertebrates.

Amphioxus comes much nearer to exemplifying the stock from which vertebrates probably evolved. This little animal has been considered by some zoologists and paleontologists to be degenerate, but few of the characters to which they point might not equally well be considered primitive. Moreover *Amphioxus* is remarkably similar in many ways to the lamprey larva. Lampreys and other jawless cyclostomes generally are regarded as the most primitive of modern fishes. *Amphioxus*, of course, cannot be the actual ancestor of vertebrates which evolved many millions of years ago. Even without this time discrepancy, some of its features would rule out such a possibility. This creature, however, probably represents a very ancient type of animal which has faithfully preserved many of the distinctive characters, including the segmental arrangement of muscles, possessed by the primitive ancestor of vertebrates.

Table 383   **The four great groups of fishes**

Agnaths: fishes without jaws or paired fins, fossil forms with heavy bony armor. Ostracoderms (Ordovician to Devonian), lampreys, and similar modern forms
Placoderms: a varied lot of mostly peculiar extinct bony fishes with a primitive type of jaws. Acanthodians (Silurian to Pennsylvanian), arthrodires (Silurian to Devonian), antiarchs (Devonian), stegoselachians (Devonian)
Chondrichthyes: jawed fishes with skeletons consisting of cartilage rather than bone, generally with dermal denticles instead of scales. Sharks and similar forms (Devonian to Recent)
Osteichthyes: relatively modernized jawed fishes with bony skeletons and scales (Devonian to Recent)

These groups are well characterized except for the placoderms, which are not all closely related to one another.

FISHES

All fishes look much alike to many people, and it may seem natural to divide the vertebrates into two equally important groups inhabiting water and the land, respectively. Fishes, however, have had a longer evolutionary history than land dwellers. When they are carefully observed, it is evident that they have varied structurally among themselves much more than have four-footed creatures. Four groups of fishes commonly are recognized on the basis of what appear to be their most important differences: (1) jawless fishes, (2) archaic jawed fishes, (3) shark-like fishes, and (4) advanced bony fishes (see Table 383). The general evolutionary relations of these groups are shown in Fig. 383. Amphibians evolved from one of the relatively minor side branches of the bony fishes.

The paleontologic study of fishes is rendered difficult by the nature of the fossils. These are rarely well preserved except at a few exceptional localities, and

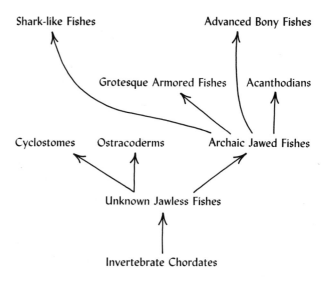

Fig. 383 Diagram showing the evolutionary relations of the principal groups of fishes.

even there a covering of scales or bony plates may hide internal structures. The skeletons of many fishes are not completely ossified, and some consist only of cartilage. Some fish bones contain much organic matter and are not likely to be preserved in good condition. Moreover, most fish skeletons are not firmly articulated, so that they are easily dismembered. All these factors are related to the medium in which fishes live. Being surrounded by water, these animals do not require bones for bodily support to the same extent as land animals; their bones function mainly for muscular attachment or protection. Finally the skeletal structures of fishes are so varied that the identity and relations of scattered parts are not easily determined.

Original Environment

Fishes undoubtedly had marine forebears, but there is much uncertainty whether the earliest vertebrates originated in a marine or in a freshwater environment. This problem has been argued vigorously from both points of view. Evidence has been sought in (1) the nature of fossil-bearing sediments, (2) the shapes of fishes, and (3) the physiology of fishes. Not uncommonly the same evidence has been presented to uphold opposite conclusions.

Evidence of Sediments
Most of the oldest remains of fishes of Ordovician, Silurian, and early Devonian ages have been found in strata which contain some marine invertebrates. The sediments commonly are sandy, and the invertebrates are not abundant. This suggests very shallow water and very near-shore marine environments. Such evidence has been presented as the basis for quite different conclusions, e.g., (1) that this is where the fishes lived, and (2) that fishes inhabited nearby streams, but their remains were transported to the sea, where they accumulated. The wide distribution of fish remains in the Middle Ordovician of Colorado and nearby states seems to favor the first conclusion, but their generally dismembered and fragmentary condition may be favorable for the second.

Evidence of Shape
The streamlined form of most fishes has been cited as evidence that these creatures evolved in rivers, where they had to contend with persistent currents. This may seem to be a reasonable guess, but it is by no means a necessary conclusion. Such a form would facilitate speedy movement in water anywhere. It certainly would have been advantageous and had survival value and might have evolved in fishes inhabiting a currentless environment.

Ostracoderms, which were the most primitive known early fossil fishes, probably were not accomplished swimmers. Their tail fins seem to have been less-efficient propelling organs than those of modern fishes, and most of them lacked stabilizing fins elsewhere on their bodies. Perhaps they swam as awkwardly as modern tadpoles. The best-known ostracoderms were depressed in form and had flattened ventral surfaces and upwardly directed eyes (Fig. 385). Such creatures evidently were

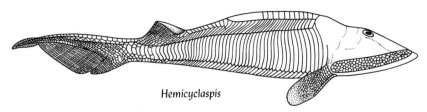

Hemicyclaspis

Fig. 385   This ostracoderm, of flattened form and with upwardly directed eyes, probably was a rather sluggish bottom-living creature. (Approx. ×¾.) (*After Stensiö, 1932, "The Cephalaspids of Great Britain," British Museum, fig. 15, p. 53.*)

adapted to living and moving along the bottom. This and their heavy armor, which must have weighed them down and restricted movements, do not suggest an environment dominated by strong currents. These observations, however, do not apply with equal force to all ostracoderms or to some of the other ancient fishes.

Evidence of Physiology

The vertebrate kidneys have been cited as evidence both for and against the origin of fishes in a freshwater environment. Kidneys, in addition to excreting nitrogenous wastes, function to maintain an ionic and osmotic equilibrium in the body fluids which is necessary to the existence of every animal. They control to a considerable extent salt content and concentration, and when necessary can remove large quantities of excess water from the body. If the body fluids of an aqueous creature are not in equilibrium with its surroundings, osmosis will result in the passage of water through any of its exposed permeable membranes. In salt water, fluids tend to be drawn out of a fish's body and in freshwater they tend to be drawn in. One argument is that the kidneys must have existed before fishes could have left the sea, in other words, their kidneys were preadaptive. The other is that kidneys developed in response to the need for acclimatization in the gradual adjustment to a freshwater environment.

Differences in the structure and function of the kidneys in modern sharks and marine bony fishes have been called upon to strengthen the contention for the nonmarine origin of all fishes. Sharks retain urea in their blood and thus maintain osmotic equilibrium with seawater. There is no necessity for their kidneys to conserve water and these organs eliminate excess salt taken in with food. The blood of marine bony fishes, however, is not in similar equilibrium. Their kidneys secrete very little water, and excess salt is eliminated in the gills. The point made here is that two kinds of freshwater fishes adapted to marine conditions in different ways. This difference has been considered unlikely if both kinds evolved in a marine environment.

None of the foregoing evidence or arguments is conclusive. At best an opinion can be formed only with respect to where the early fishes lived. This is quite different from an opinion as to the environment of their origin. Even if the explanation for the kidneys in sharks and other fishes is correct, marine origin cannot be eliminated because both groups may have evolved in freshwater from an ancestral stock of fish invaders from the sea. The preponderance of evidence might suggest that ostracoderms were at first marine and later adapted to freshwater and that all other known fossil fishes, except perhaps the sharks, evolved from nonmarine ancestors. The problem of the habitat of the first chordates to become vertebrates remains, however, an open question whose answer probably never will be known.

Ostracoderms

Ostracoderms are the most primitive of fossil fishes. They were small creatures, mostly not over a foot in length, which made their appearance in the Ordovician and became extinct before the end of the Devonian Period. Their most significant primitive feature was the lack of jaws. In this they were similar to modern lampreys and other cyclostomes. Probably some other features of ostracoderms that are not revealed by fossils can be learned by the study of these related modern creatures.

Ostracoderms differed conspicuously from the naked cyclostomes by their covering of bony scales and plates, which made their fossil preservation possible. An internal bony braincase has been observed in one group, but all lacked any bony structure in a spinal column. They varied greatly among themselves in the nature of protective coverings and also in body shapes. Some were much more fully armored than others. Certain species were broad and flat and seem to have been bottom dwellers (Fig. 385). Others had more familiar fish-like forms and probably were fairly active swimmers (Fig. 386). Lacking jaws, they fed mainly by sucking in material which was strained through the gills. There, food particles were extracted and passed into the esophagus. Ostracoderms could not have been predacious animals, although a few small, presumably movable plates associated with the mouths of some may have permitted nibbling.

The great diversity evident among ostracoderms indicates a long evolutionary history that is totally unknown. Some of these creatures seem to have been highly specialized in certain ways. For example, one group has been suspected of having possessed electric organs. There is little to indicate interrelations among the differ-

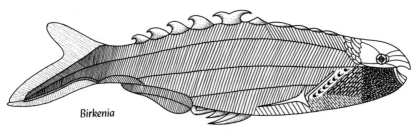

Fig. 386   This ostracoderm, of more conventional fish-like form and about 4 in. long, probably was a relatively active swimming creature. (*After Romer,* 1966, *fig.* 20, *p.* 19.)

ent types of ostracoderms and nothing to suggest that any of them was ancestral to other fishes, except possibly the cyclostomes. Most of them had a single nostril, as have the lampreys, opening upward centrally in the head. In one group, however, a pair of depressions in the bone above the corners of the mouth suggests the occurrence of paired nostrils more like those of other vertebrates. Perhaps these were more nearly related in their ancestry to other fishes.

Bony Armor
Bone seems to have developed first in the skin of small animals, where probably it functioned as protective armor. Another less-popular idea suggests that it reduced the danger of loss of body fluids by osmosis. Its appearance in ostracoderms during the Ordovician Period may have been related to the rise of cephalopods, which rapidly attained dominance in the marine realm as the largest and most active predatory animals of their time. In the Silurian Period the eurypterids occupied a similar position in brackish and perhaps freshwater habitats. Thus wherever they may have lived, dermal armor probably had survival value for relatively small defenseless creatures like the ostracoderms. Subsequently both these predatory groups declined greatly in importance, perhaps as the result of competition with jawed fishes. Competition with these more aggressive creatures also may have contributed to the extinction of ostracoderms, whose armor probably lost some of its advantages in the differently organized community.

Two kinds of dermal bone occur in the ostracoderms, some with and some without enclosed bone cells. The latter seems to be more primitive. Dermal and other bone also appeared in Silurian jawed fishes, which surely were not descended from any known ostracoderms. This seems to be good evidence that bone probably evolved independently in a relatively short span of time in several different groups of fishes as an adaptive response to similar stresses of their environments. It is unlikely that the history of bone can be traced among the early fishes without carefully considering and relating the lineages of the fossils that are compared. Thus the fact that the geologic record of sharks begins later than that of bony fishes is not acceptable as evidence that sharks had bony ancestors.

The dermal plates and scales of both ostracoderms and other early fishes generally consist of bone with a surface of enamel-like material. They commonly possess small projecting tubercles, or ridges, which resemble the isolated denticles occurring in the skin of sharks. The evolutionary relations of such denticles and the dermal plates and scales are not agreed upon. One view is that dermal bone first appeared in large plates and that subsequent fishes with smaller armor elements exemplify a regressive process. Thus sharks with isolated denticles and cyclostomes without any dermal bone have been considered to be degenerate in this respect. Conclusions based upon such a sequence in unrelated fishes are very insecure. Another view holds that dermal ossification probably began at many different centers and the elements so produced subsequently coalesced to produce larger structures. Therefore, creatures with heavy or continuous armor were not primitive in this way, but highly specialized. If this were so, more lightly armored possible ancestors might be found. Perhaps their existence is indicated by the separate denticle-like bone fragments reported from the Russian Lower Ordovician.

Archaic Jawed Fishes

The earliest jawed fishes have been recovered from Silurian strata, and a considerable diversity of archaic forms dominated the vertebrate fossil faunas of the Devonian Period. Only one or two groups are known to have persisted longer, and the last to disappear seems to have become extinct at the end of the Paleozoic Era. All these fishes must have evolved from jawless vertebrates, but their ancestors have not been recognized among the known ostracoderms. Many of the archaic jawed fishes, mostly small species, are believed to have inhabited freshwater. Others, however, certainly were marine, and these include the giants of their times, 30 ft or more in length. Most of these early fishes lacked bony vertebrae. Whether they evolved from unknown ancestors in freshwater or the sea and then migrated into the other habitat has not been established.

Several groups of Devonian fishes of varied and grotesque form bore heavy armor. Many of them had a completely plated head shield and another bony shield covering part or much of the thoracic region. A movable joint separating the two areas of armor in some of these fishes seems to indicate that opening of their mouths involved raising the head as well as dropping the lower jaws. A few specimens belonging to another group preserve traces of internal organs that have been interpreted as air sacs or lungs. This contributes to the opinion that a primitive lung evolved very early in these creatures and that the hydrostatic swim bladder of modern fishes is an evolutionary modification of a lung, rather than vice versa.

The most persistent group of archaic Devonian fishes, known as acanthodians, are particularly interesting. They were the fishes of this age whose appearance was most modern (Fig. 389). If the actual ancestors of the modern groups were known, they might be found very similar to some of these ancient creatures. The acanthodians were small, rarely a foot in length, and most of them were completely covered with small diamond-shaped bony scales. They possessed dermal spines that seem to have supported fin-like folds of skin located singly along the dorsal midline and ventrally in pairs. The gill structure of these fishes is especially instructive and seems to provide one of the keys to an understanding of the evolution of gills and jaws.

Paired Fins

Almost all fishes possess tail fins effective for propulsion. Most of them have median fins, both dorsal and ventral, which reduce the tendency for a fish to roll while swimming. Paired pectoral fins in front and pelvic fins behind serve similarly, but they also aid in steering and act as brakes; also, serving as movable planes, they are the principal means of directing a fish's upward or downward course. The development of these paired fins certainly was an important event in

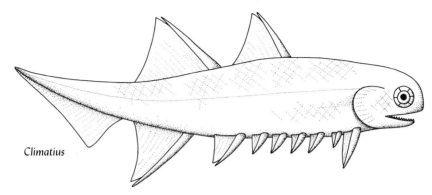

Climatius

Fig. 389 The body of this Lower Devonian acanthodian was covered with small rhombic scales, and the fins were skin folds supported by stout spines. ($\times 1\frac{1}{2}$.) (*After Romer, 1945, "Vertebrate Paleontology,"* fig. 25, p. 41.)

vertebrate evolution. They not only facilitated control in swimming and contributed greatly to the success of fishes, but they also provided basis for evolution of legs in all higher vertebrates.

Paddles, or primitive paired fins, of a sort located just behind the head shield were possessed by some ostracoderms. These were covered by bony scales, but the internal structure is unknown. Almost all of the jawed Devonian fishes seem to have had at least two pairs of fin-like structures which can be compared to the pectoral and pelvic fins of modern forms. Anterior paired structures possessed by some of the armored fishes were not true fins. They generally were covered by bony plates and look much like the paddles of eurypterids. Most of them were jointed, although some seem to have been immovable. One peculiar type of fish exhibits very large rounded pectoral fins containing radiating bones or rods of cartilage that give their owners much the appearance of modern rays. The fins of acanthodians seem to have been no more than folds of skin supported in front by stout bony spines (Fig. 389). Some of these fishes had several smaller pairs of spines between the larger anterior and posterior ones.

Homology of the main paired fins of all these fishes is not certain. Two modes of origin have been suggested. One theory is that two continuous laterally directed folds of skin developed ventrally on the fish's body. These separated into parts and acquired internal supporting structures, but only the first and last pairs of a primitive series were persistent. This is the way the median dorsal and anal fins commonly are supposed to have been formed. The other theory is that spines first appeared and were followed later by skin folds that became the fins.

Jaws

The advent of jaws was no less important than the development of paired fins to the success of vertebrates. In fact it probably was of more immediate importance to the fishes, because it opened up for them a new way of life and resulted in their

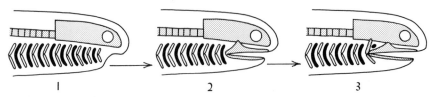

Fig. 390   Much-generalized and simplified diagrams showing how the jaws of fishes evolved from gill arch supports. (*Modified after Romer, 1962, fig. 149, p. 217.*)

rapidly becoming the dominant creatures of the ancient world. The transformations that led to the development of jaws in vertebrates is one of the best examples in all biology of the evolutionary adaptation of a structure or an organ to an entirely new function.

Gills probably originated in the invertebrate chordates or their immediate ancestors as a filter feeding mechanism, perhaps with respiration as a subsidiary function. Cartilaginous supports appeared in the tissues separating the external gill openings in some of these animals, as may be seen, for example, in modern *Amphioxus.* Transformation into the most primitive type of jaw may have been accomplished before these structures were replaced by bone, but no fossils have been found that surely represent such an early evolutionary stage. Homologous relations are believed to have been established, however, by the evidence of embryonic development and the courses of nerves traced in modern fishes.

Structures supporting gill arches in fishes consist of several parts forming units that are repeated serially on each side of a creature's throat. Only two of these parts need to be considered here. They are rods of cartilage or bone meeting at an angle pointing backward (Fig. 390). One of the angulated pairs enlarged, and the parts became freely movable against each other, in a hinge-like manner. They formed the lateral halves of upper and lower jaws (Fig. 390-2). It is generally believed that this pair of rods was the second in the original series of gill arch supports. The anterior pair either disappeared entirely or perhaps was incorporated in the braincase of the skull or is represented by a cartilaginous structure at the angle of a shark's mouth just in front of the jaw hinge.

At this theoretical primitive stage of its development, the upper jaw articulated with the overlying braincase, and a full gill slit presumably occurred behind it. The next following pair of angulated rods, which form what is known as the hyoid arch, was unmodified and continued to perform its original function as a gill support. Subsequent evolution resulted in further change. The upper element of the hyoid arch became a posterior support to the jaws. Its upper end was braced against the braincase, and the jaws were ligamentally attached to it below (Fig. 390-3). With this arrangement the gill slit in front of it was much restricted and reduced in size. It persisted as a small opening, the spiracle, located above the angle of the jaw of many fishes. In the most advanced bony fishes, however, even this remnant of the gill slit closed.

The armored fishes, in spite of their great diversities, may be similar in the primitive nature of their jaws. Specimens of some acanthodians have been supposed to show the details and reveal that the hyoid arch had not been greatly modified. The gill areas, however, commonly are covered by backward extending plates or scales or by part of the bony head shield in the heavily armored forms, so that the underlying structures are rarely seen. Nevertheless, the absence of any indication of a restricted spiracle, which is not obscured in later fishes, may be evidence that a full gill slit existed behind the jaws. If this is so, the gill cover is not homologous in acanthodians and the advanced bony fishes, because it must originate in these two groups in front of different gill slits of the original series. This casts some doubt upon the possibility that any known type of acanthodian was ancestral to later fishes.

Shark-like Fishes

The geologic record of the shark-like fishes extends from the late Devonian to the present. Through all this time they seem to have constituted a rather well-marked group, distinguished from other fishes by (1) the lack of any internal bony skeleton, although some of the cartilage may have been variously calcified, (2) no covering of true bony scales, (3) no gill covers like those of other fishes; the gills of almost all of them opened as individual slits in the throat region (Fig. 391), (4) no organ corresponding to a swim bladder or lung, as well as other less-conspicuous and less-important differences. Because of the absence of bones and bony scales, these creatures generally are known from less-complete and less-satisfactory fossils than other fishes. Except for a few fortunate discoveries, their presence in ancient faunas is indicated by little more than scattered teeth, skin denticles, and spines.

The sharks and their allies, unlike the bony fishes, have been almost exclusively marine, as far as can be determined. Only a few species ventured into freshwater

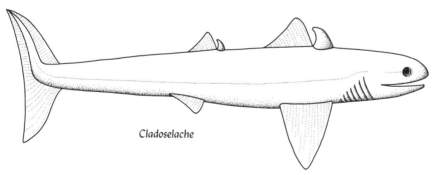

Cladoselache

Fig. 391 **This primitive Devonian shark grew to a length of about 4 ft. Notice the broadly based paired fins and the body extremity turned up sharply in the tail.** (*After Romer, 1962, fig. 22A, p. 42.*)

and were temporarily successful. Their evolutionary origins are uncertain. Although the ancestors of sharks probably had jaw structures similar to those of the ancient bony fishes, no specimens of these archaic creatures have been found that seem to represent the stock from which the sharks arose. There is very little evidence that any of the shark-like fishes had bony ancestors or that they were originally nonmarine. The supposition, therefore, that the cartilaginous skeleton of these fishes is evidence of degeneracy in this respect is not well founded. Probably they evolved from boneless marine ancestors, indications of whose existence are not likely to be discovered.

Most of the shark-like fishes have been active and accomplished swimmers of predacious habit. This is suggested strongly by their body form and teeth adapted to grasping and tearing prey. As in other groups of fishes, however, some altered to benthonic modes of life. They became flatter, had upwardly directed eyes, and developed pavement teeth suitable for crushing shellfish. Among these are the rays and similar creatures with enormously expanded pectoral fins and tails reduced to whiplash form.

On the whole the shark-like fishes have been conservative; little evolutionary advancement is evident since Mesozoic times. The early ones had paired fins with broad bases of attachment whose movement, other than in an up-and-down direction, must have been restricted. These fins in later ones are narrowed basally and evidently became much more flexible. As far as jaw structure is known, all sharks have had the jaws supported by the upper member of the hyoid arch. An early condition shows the upper cartilaginous jaw to be articulated to the braincase, but this connection did not persist in later forms. The open spiracle continued in most sharks. In the rays it is enlarged, and it, rather than the mouth, serves as the intake for water circulated through the gills which open ventrally.

Teeth

Teeth in vertebrates are similar structurally and in origin to the dermal denticles of sharks (Fig. 399-3). They seem to be no more than denticles that became specialized coincidentally with the development of jaws. Both consist internally of dentine which in some ways resembles dermal bone, and an outer covering of enamel-like material. One theory, as previously mentioned, holds that both denticles and teeth were derived from the surface features of dermal bony plates and became separated when, in the course of evolution, underlying connecting bone no longer formed. The other postulates that, phylogenetically, dermal ossification began at scattered centers and the resulting denticles later coalesced to produce continuous bony plates.

The enamel which coats the surface of most teeth is produced by ectodermal cells, whereas the underlying dentine, like bone, is of mesodermal origin. The enamel-like material of the shark's dermal denticles and at the surface of the dermal bones and scales of many ancient fishes differs from true enamel because it is not ectodermal. Teeth and dermal denticles undoubtedly are homologous, but their structures are not so completely similar as has commonly been supposed.

Teeth primitively were conical and adapted only to grasping prey or other food. Teeth of this type still occur in most classes of the vertebrates. Evolution, however, very early resulted in the appearance of other types. Some acquired multiple sharp cusps, some developed shearing edges, and others became flattened and may have combined with their neighbors to produce large dental plates. The earliest teeth, some with expanded bony bases, simply rested upon the jaws or other mouth parts to which they were loosely joined by connective tissue. In the next development firmer attachment was effected by various degrees of fusion with underlying bone (Fig. 453), as in more advanced bony fishes and most amphibians and reptiles.

Ray-finned Fishes

The advanced bony fishes are divisible into two groups (see Table 395) that have been distinct since their first appearance in the fossil record of early and middle Devonian time: (1) the ray-fins, and (2) the lobe-fins, including lungfishes. They differ from each other in many ways, most significantly in the form and bony structure of their paired fins and the nature of their nostrils, which open internally in the mouth roof of many lobe-fins but not in the ray-fins.

Nevertheless these fishes seem to be related. Both groups have jaws similarly supported by the upper bone of the hyoid arch, and both have air sacs or lungs. Their ancestry, however, is not known. The early ray-fins resemble acanthodians in the nature of their scales, and the Devonian representatives seem to have been more primitive in some skeletal features than the lobe-fins. Both groups are believed to have originated in freshwater. The lobe-fins continued mainly as inhabitants of that environment. The ray-fins expanded much more successfully as marine fishes.

The ray-fins are by far the dominant fishes of the modern world. In their evolutionary radiation they have adapted to almost every aqueous environment and constitute a confusing array of very differently appearing fishes (Fig. 394). Although they are so important now, they are not in the direct line of vertebrate evolution. For this reason, and also because their phylogenetic relations are not understood in any detail, only the main trends evident in their evolution are considered here.

The ray-finned fishes exhibit progressive evolutionary modifications with respect to (1) scales, (2) fins, (3) jaw structure, (4) spinal ossification, as well as other features. These changes progressed unequally, and more or less advanced features occur in various combinations in various groups of fishes. In a general way, however, five groups that represent successive grades of evolutionary advancement can be distinguished (Fig. 395): (1) palaeoniscoids, (2) subholosteans, (3) holosteans, (4) primitive teleosts, and (5) advanced teleosts.

Palaeoniscoids

These fishes had rhomboidal scales with a thick covering of enamel-like material consisting of many layers (Fig. 399-1). The posterior body extremity turned upward and extended far into the tail fin (Fig. 402-1). The mouth was long and opened far backward. These characters occurred together mainly in the late Paleozoic.

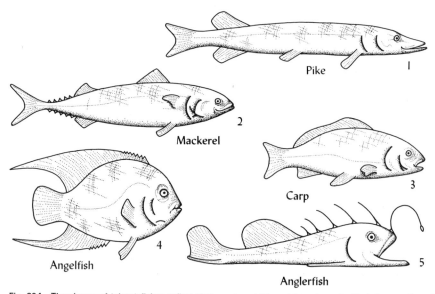

Fig. 394   The shapes of teleost fishes reflect their modes of life and have been duplicated many times in quite different and relatively unrelated groups. The carp is an example of the generalized form. It is not a very active fish. The mackerel is a pelagic fish, constantly swimming in the open sea. The pike is a swift predator. Its posteriorly located dorsal fin adds power to its tail in speedy movement. The angelfish is a typical basking kind, inhabiting shallow water. The flattened form of the anglerfish is related to its habit of lying on the sea bottom. The posterior position of the pelvic fins of pike and carp is relatively primitive. In advanced teleosts these fins have moved forward, as shown by the mackerel and angelfish.

Subholosteans

Scales were thinner and had lost their layer of dentine-like material. The tip of the body was shorter and extended less far into the tail fin (Fig. 402-2). The mouth opening was somewhat abbreviated. The number of bony rays supporting the fins was reduced, but they consisted of separate segments. Fishes of this type were mainly Triassic.

Holosteans

The body tip was very short, and the tail fin was nearly symmetrical (Fig. 402-3). The mouth was short, the jaw articulation had moved forward, and the upper jaw became free of the overlying skull bones posteriorly. The rays of the median dorsal fins were fewer, unjointed, and corresponded in number to the bony processes rising from the vertebrate. The spiracle had closed. Ossification continued into the axial part of the spinal column but generally was incomplete. This combination of characters occurred mainly in fishes of Jurassic and early Cretaceous ages.

Primitive Teleosts

Scales are thin and bony, without enamel-like surfaces, and they overlap posteriorly

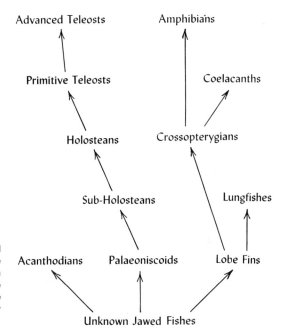

Fig. 395 Diagram showing in simplified fashion the evolutionary relations of the main types of bony fishes. Starting with the palaeoniscoids, there probably were several lineages that progressed to more advanced types entirely independently of one another.

Table 395 Principal distinguishing characters of the main groups of relatively modernized bony fishes.

Actinopterygians: ray-finned fishes, fins stiffened by bony rays derived from scales. Nasal passages do not open internally in mouth. Primitively with a single dorsal fin

    Palaeoniscoids: with thick rhomboidal scales. Posterior end of body turned strongly upward in tail. Mouth long. Devonian to Cretaceous

    Subholosteans: scales thinner. Termination of body in tail reduced. Mouth somewhat shortened. Triassic

    Holosteans: scales still thinner. Termination of body much reduced. Fin rays fewer, stouter, unjointed. Spiracle closed. Clavicle absent from shoulder girdle. Vertebrae more fully ossified. Mouth short. Triassic to Recent

    Primitive Teleosts: scales thin, commonly overlapping. Tail not quite symmetrical. Vertebrae almost completely ossified. Pelvic fins shifted forward. Air bladder entirely hydrostatic. Jurassic to Recent

    Advanced Teleosts: scales very thin, frayed out behind. Tail symmetrical. Dorsal fin rays stout spines. Pelvic fins located far forward. Dermal bones of skull sunk below skin, and head covered by scales. Cretaceous to Recent

Choanichthyes: lobe-finned fishes, paired fins with thickened bases, contain bones articulated with internal skeleton. Nasal passages opening into mouth. All early forms probably with functional lungs. Primitively with two dorsal fins

    Crossopterygians: skull structure foreshadowing that of terrestrial vertebrates. Teeth labyrinthodont with infolded enamel. Devonian to Recent

    Dipnoans: lungfishes with functional lungs. Lateral teeth reduced or lost, median teeth flat plate-like crushers. Scales progressively much reduced. Devonian to Recent

The actinopterygian subgroups are successive evolutionary stages which progressed concurrently in a number of different lineages. The choanichthyes subgroups have been distinct from their earliest fossil record.

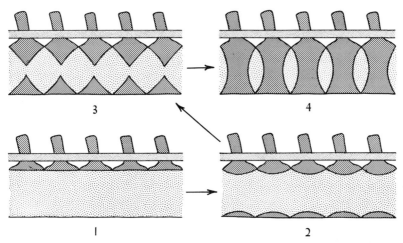

Fig. 396   Diagrams showing in longitudinal section stages in the evolution of a bony spinal column in fishes. 1, The first bony elements to appear were the neural arches overlying the notochord and enclosing the spinal nerve. 2, The notochord was encircled by a succession of bony rings. 3, These grew inward, constricting the notochord segmentally. 4, The notochordal perforation closed, and the vertebrae were separated by lenses of fibrous connective tissue.

(Fig. 400-2). The tail fin is almost perfectly symmetrical (Fig. 402-4). The lower jaw consists of a reduced number of bony elements. Minor changes appear in the skull structure. Vertebrae are completely ossified. Fishes of this type appeared in the Jurassic and were characteristic of the late Cretaceous; many persist in modern faunas.

Advanced Teleosts
The scales are frayed out or bear tiny spines behind (Fig. 400-3). Some of the dermal fin rays are stiff spines. Bones are without bone cells, the dermal skull bones have sunk below the skin, and the head is covered by scales. The pectoral fins have moved upward, and the pelvic fins forward to the thoracic region. These most advanced fishes are first noted in the late Cretaceous and dominate the Tertiary and modern faunas.

Spinal Column
Ossification of the fish's internal skeleton was a gradual evolutionary process. The fossil record shows that dermal bone was the first to appear, and dermal scales and plates were well developed before the advent of much internal bone. The bodies of early relatively boneless vertebrates were supported by a notochord enclosed in a fibrous constricting sheath. This evidently was adequate for small animals whose body density was nearly the same as that of the water in which they lived. Some ancient fishes grew even to very considerable size without any additional evident axial support.

Internal bones first appeared in the braincase at the anterior end of the notochord and posteriorly above the notochord in arches enclosing the spinal nerve. Later, ribs and dorsal spines developed, but the notochord remained unenclosed by any bony structure. The arrangement of all these bones reflects the longitudinal segmentation evident in the disposition of body muscles. In modern vertebrates the bones are encased by fibrous connective tissue that also separates the muscular segments, and doubtless similar relations were characteristic of ancient fishes. There is no positive evidence that these bones followed more primitive cartilage phylogenetically. The embryonic sequence of first cartilage and then bone, however, is considered by some persons to be presumptive evidence that this recapitulates an evolutionary sequence.

The central parts of vertebrae develop around and within the notochord sheath. Fishes show a progression in which thin bony rings enclosed the notochord (Fig. 396). These thickened inwardly, constricting the notochord by stages until it was almost pinched off in segments enclosed between the disks of successive vertebrae.

Structural details of the vertebrae vary greatly, particularly among the fishes, but do not require consideration here. Significantly, however, vertebrae differ in one way from the other internal bones associated with them. They do not correspond to the muscular segments but are in a sense intersegmental. Each develops from mesoderm belonging to the adjacent parts of two segments (Fig. 397). Because of

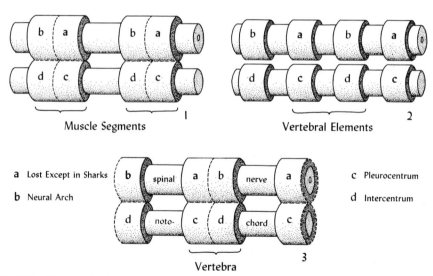

Fig. 397 Schematic drawings illustrating the relations of vertebrae to muscle segments as revealed by embryologic studies. 1, Four centers of cartilage-forming tissues are associated with each muscle segment and surround the spinal nerve and notochord. 2, The anterior and posterior centers of each segment separate. 3, The centers of adjacent segments join to form the basic elements of a single vertebra. *a* is lost in bony fishes and more advanced vertebrates. *b* becomes the neural arch. *c* becomes the centrum. *d* becomes the intercentrum, which disappears in reptiles and is not present in mammals. (*After Romer, 1949, "The Vertebrate Body," 1st ed., fig. 83, p. 152.*)

this relationship, a vertebra is bound to its anterior and posterior neighbors by the muscle fibers of two successive segments.

Lobe-finned Fishes

Remains of fossil lobe-finned fishes occur in rocks as old as the late early Devonian (Fig. 398-2). From this time onward to the end of the Paleozoic Era they are known to have been widely distributed, but thereafter they rapidly declined in relative importance and are now almost extinct. Only one deep-sea species, a coelacanth (Fig. 398-1), and three species of lungfishes are known to survive in modern faunas. The main division of this group, the crossopterygians or lobe-fins proper, are of great evolutionary interest, however, because they were ancestral to amphibians. The lungfishes either are an early offshoot from the crossopterygians or evolved from a common ancestor. They are especially interesting because the modern ones possess functional lungs and demonstrate a condition that must have existed among early crossopterygians and perhaps some other ancient fishes.

The crossopterygians differed from the ray-fins by the shape and bony structure of their paired fins and the nature of their nostrils, as already mentioned. They also

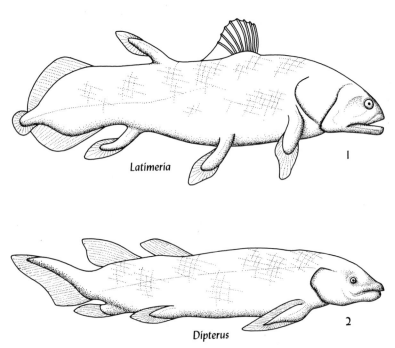

Fig. 398   Lobe-finned fishes. 1, The modern coelocanth inhabiting the Indian Ocean. 2, A Devonian lungfish. (*After Romer, 1945, "Vertebrate Paleontology," 2d ed., cover; 1955, fig. 25A, p. 48.*)

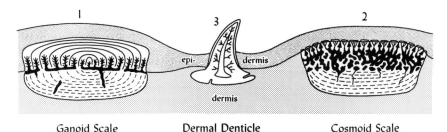

Ganoid Scale   Dermal Denticle   Cosmoid Scale

Fig. 399 Generalized drawings illustrating the two types of scales possessed by ancient fishes, and the skin denticle of a shark. They are all fundamentally similar, consisting of bone below, dentine in the middle, and enamel above, but the parts are differently developed.

had two dorsal fins instead of one, as in primitive ray-fins, and their scales were differently constructed (Fig. 399-2). Furthermore their skulls contained some bones that are homologous with those of land vertebrates (Fig. 408-1), and their larger teeth particularly have infolded enamel similar to that in the teeth of some amphibians (Fig. 414).

Several of the evolutionary modifications of the lobe-fins were comparable to those of the ray-fins. For example, their scales were simplified, the central parts of the vertebrae became ossified in some of them, and symmetrical tail fins developed from unsymmetrical ones (Fig. 398).

Scales

The scales of fishes are comparable in their origin to dermal bone. Primitively they consisted essentially of two parts, the lower one bone and the upper one shiny hard enamel-like material. The bony part was compact and laminated below but passed upward first into spongy bone and then into substance similar to the dentine of teeth, pierced by many tiny branching tubules. Two types of primitive scales are recognized which differ mainly in the prominence and relative thickness of these layers: (1) The scales of early crossopterygians had a thick layer of spongy bone beneath the relatively thinner upper layers (Fig. 399-2). These scales grew mainly about their margins and on their undersurfaces. (2) Primitive ray-finned fishes had scales with little or no spongy bone, thin dentine, and enamel-like material consisting of many layers (Fig. 399-1). Very similar scales were borne by the acanthodians. These also grew upward by the additions of surface layers. Which of the two types may have been derived from the other, if indeed they originated from a single kind, is not known.

The scales of both groups of advanced bony fishes were progressively modified in shape and simplified in structure. No living fish has scales exactly like either of the Devonian types that have been described. In the ray-finned fishes the dentine layer first disappeared, and then the enamel-like material thinned and finally also was eliminated. The scales of most modern species are thin and pliable (Fig. 400). They consist of compact bone-like material but do not contain living bone cells. Some

Cycloid Scale   Leptoid Scale

Fig. 400   1, The scales of fishes are situated within the skin, as shown in the cross section (anterior is to left). In teleosts they are reduced to thin bony plates. 2, Many of these fishes have cycloid scales. 3, Leptoid scales are characteristic of many advanced teleosts.

modern fishes have lost their scales entirely. The dentine and enamel-like material also disappeared from the scales of the lobe-finned fishes. The scales of modern lungfishes are reduced to a fibrous pliable bony substance.

Primitively the scales of both groups were rhomboidal and lay adjacent to one another in the skin. As they became thinner, they began to overlap posteriorly and downward on the fish's body. Obviously this produced a more flexible arrangement. At the same time the scales became progressively more circular.

As previously mentioned, the evolutionary relations of scales, more massive body armor, and the dermal denticles of sharks are not agreed upon. However this

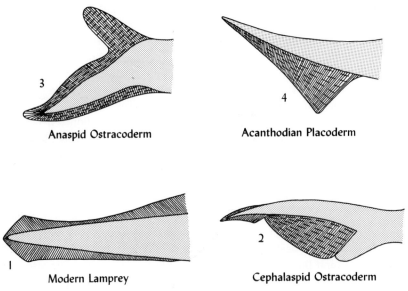

Anaspid Ostracoderm   Acanthodian Placoderm

Modern Lamprey   Cephalaspid Ostracoderm

Fig. 400a   The tails of primitive fishes showing relations of the posterior body extremity to fin webs.

may be, scales seem to have been modified to produce certain other dermal structures, particularly fin rays and spines. The bony fishes show transitions from fins
sheathed in scales, becoming smaller outward, to those supported by series of small
bony rods derived from scales which eventually united as continuous spines. The
larger cartilaginous rays in the fins of sharks are similar and may have originated
in a somewhat comparable way.

Tail Fins

The embryology of some modern bony fishes shows that, at a very early stage, a
web develops above and below the posterior part of the tiny creature's body to
produce a symmetrical tail fin. This condition, which is essentially similar to that
seen in an adult cyclostome and even in *Amphioxus* (Fig. 400*a*-1) is believed to
be very primitive. As the embryo grows, however, the end of the body turns upward and the tail fin that subsequently develops consists of little more than the
greatly enlarged lower portion of the embryonic web. The adult fish again has a
symmetrical tail fin, and there is little in its structure to indicate how this condition
was attained.

Almost without exception the earliest known fossil fishes had asymmetrical tail
fins. The body turned downward in one group of ostracoderms, and the fin was an

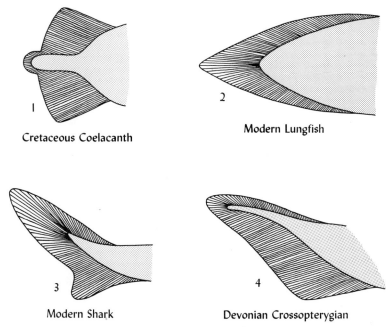

Fig. 401   The tails of primitive fishes showing relations of the posterior body extremity to
fin webs.

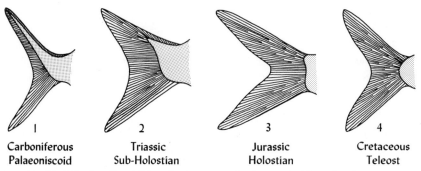

| Carboniferous | Triassic | Jurassic | Cretaceous |
| Palaeoniscoid | Sub-Holostian | Holostian | Teleost |

Fig. 402 The tails of ray-finned fishes showing progression from ancient to modern types.

enlarged upper web (Fig. 400a-3). This condition, incidentally, was duplicated by later evolution in the ichthyosaurs, a group of reptiles. All other very early fishes had the body end turned upward (Figs. 400a-4; 401-3,4). Starting with this condition, selected ray-finned fishes provide what appears to be a phylogenetic and evolutionary series that closely reproduces the ontogenetic sequence of their modern representatives (Fig. 402).

The crossopterygians, beginning with a similar asymmetrical tail, evolved a symmetrical fin but in a somewhat different way (Fig. 401-1,4). The early fishes of this group had a very narrow upper fin web and a much wider lower one. The symmetrical tail that finally developed here resulted from enlargement of the upper web and related reduction of the lower one, without shortening of the body as in other fishes (Fig. 401-1).

Only the sharks and a very few modern bony fishes like the sturgeons still possess asymmetrical tail fins of the primitive type.

AMPHIBIANS

An early amphibian was essentially a special kind of fish with legs. The presence of lungs in these animals is not diagnostic, because primitive organs of this kind probably were possessed by many ancient fishes, and lungfishes have them now. Modern amphibians can be distinguished from reptiles by their method of reproduction and certain features of their soft anatomy. The paleontologist, however, is faced with the problem of identifying ancient creatures as amphibians from their skeletons, which are likely to be very incomplete. For example, he may not know whether an animal represented by a skull had legs or fins. If the animal had legs, he must rely upon skeletal features only to differentiate it from a reptile. Because evolution was gradual in the transition of fishes to amphibians and of amphibians to reptiles, sharp divisions between these classes cannot be expected. Even if perfect skeletons were known, doubts and disagreements certainly would complicate the classification of some fossil animals which occupy transitional positions in the evolutionary sequence.

Origin of Amphibians

The idea has been attractive that certain fishes which were able to do so left the water in order to exploit superior opportunities provided by the land. This, however, is not how evolution ordinarily is accomplished. Every kind of creature must adapt to whatever environments are available to it or its extinction is likely to be swift and sure. Generally every kind of animal tends to remain in the environment to which it is well adapted, because this is where it is most successful.

During the Devonian Period many freshwater fishes seem to have inhabited environments that were changing in important ways. According to one theory some of them lived in more or less temporary streams and ponds that were subject to increasingly frequent drying up. Only fishes preadapted by the possession of a lung-like organ could survive under conditions of that kind. Even so, the dangers of desiccation were extreme, and the ability to wriggle or crawl from one shrinking puddle to a nearby larger one surely had survival value. Another theory postulates that aridity was not such a decisive feature. Some fishes may have made short overland excursions in order to find food or escape enemies, just as a few modern tropical species do, and this may have led to a new way of life.

In either case strong fins with which a fish could push itself along were advantageous, and natural selection favored individuals so equipped. Gradually evolution transformed fins into more effective organs for moving an animal when it was out of water. Thus legs finally developed as a result of the necessity for some fishes to migrate from one body of water to another to survive. Only after this transformation had been accomplished and animals were again preadapted was it possible for them to abandon water for life on land. Even then they were still bound to water because water was necessary for their breeding, and that was where food was most easily obtained.

Members of only one or two small groups of fishes accomplished the transition to a terrestrial existence. They were crossopterygians, which perhaps were uniquely favored by possessing unusually strong fins and relatively efficient lungs.

Legs

The paired fins of most fishes, both ancient and modern ones, have had mesodermal supporting skeletal structures of two basic types. The sharks and ray-finned fishes have structures that vary in many ways, but there generally are several cartilages or bones which articulate more or less directly with the shoulder or hip girdle. One or more of these bear a series of parallel or radiating rod-like elements. In the later sharks and in all ray-fins these elements occupy a minor part of the fin, whose main expanse is provided with dermal fin rays. This type of structure

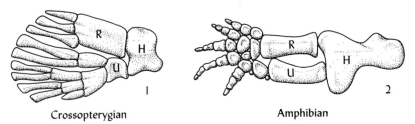

Crossopterygian   Amphibian

Fig. 404 Drawings showing homologies of large bones at the base of appendages of a Devonian crossopterygian (1) and a generalized early land vertebrate (2). *H*, humerus; *R*, radius; *U*, ulna. (*After Romer, 1966, fig. 117, p. 86.*)

is reflected by the small size or absence of a muscular fleshy lobe extending from the body into the fin.

The comparable bones of the lobe-finned fishes probably evolved from primitive structures of this kind. They consisted, however, of a single central jointed axis which extended relatively far into the fin. This bore a series of side branches, which also generally were jointed. Fins of this type had relatively large, scale-covered, fleshy lobes containing some of the muscles used to activate them (Fig. 398). Rather obviously they were much better suited structurally as the evolutionary prototypes of legs than were the weaker and less-mobile fins of other fishes.

Fossils have provided no examples of the transformation of fins to legs, but there can be few doubts concerning the relationships. Several of the bones present in the abbreviated fin supports of some crossopterygians find perfect counterparts in the limbs of terrestrial vertebrates (Fig. 404). For example, a single bone, the humerus of the upper arm, articulates directly with the girdle. This bears two bones, the radius and ulna of the forearm. Beyond these are numerous elongated bones that correspond to elements of the wrist and hand, although these cannot be identified individually. Furthermore, embryologic comparisons indicate the similarity of the upper and lower muscles of the fish's fin to those that operate the limbs of four-footed animals.

Lungs

Homology of the fish's swim bladder and the lung of land vertebrates has been postulated for many years. The belief that the lung evolved from an antecedent hydrostatic organ seemed quite reasonable because (1) the swim bladder is a characteristic organ of most fishes, and (2) fishes are recognized as the animals from which air-breathing land vertebrates evolved. Later reconsideration, however, based on a fuller knowledge of comparative anatomy, has resulted in the conclusion that this supposed evolutionary sequence is the reverse of the one which actually occurred. Most zoologists now believe that the fish's swim bladder is a structurally and functionally transformed lung.

The transition of a lung to a swim bladder cannot be traced in any series of closely related animals. All stages do occur, however, in different groups of modern fishes (Fig. 405). This sequence is so complete and so convincing that few doubts remain concerning the evolutionary relations of these organs.

The ability of some freshwater fishes to supplement their breathing by the intake of air probably evolved during the Devonian Period, if not earlier, in situations where stagnant water became deficient in oxygen. At first a limited amount of respiration may have been accomplished by the surface membranes in the mouths of fishes which gulped air. Then perhaps a pair of gill chambers became specially adapted for such a function. This is suggested by the embryology of some amphibians. What appears to be the most primitive stage in subsequent development preserved in any modern fish occurs in a peculiar African species. The relations of this lungfish to other fishes is not clear. It may be a ray-fin in the general holostean stage of evolutionary advancement, which, as the result of convergence, resembles crossopterygians in several ways. This fish possesses two simple elongated air sacs, perhaps derived from specialized gill chambers, connected by a tube that opens through the lower surface of the throat (Fig. 405-1). These sacs are mainly ventral in position, like the lungs of land vertebrates, but the right one is longer and is directed upward so that it partly overlies the digestive tract. The condition exhibited by this fish probably represents an evolutionary stage only a little in advance of the one that occurred in the crossopterygian ancestors of amphibians.

Ventral air sacs would disturb the center of gravity of a fish by raising it above the midline and thus creating an unstable balance. This was corrected by upward migration of the air sacs, whose beginning is seen in the African fish already mentioned. Further stages of this process are exemplified by the other lungfishes (Fig. 405-2). In these the air sac or lung, which may still be bilobed, is dorsal in

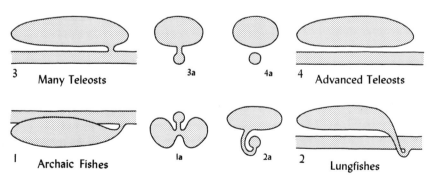

Fig. 405 Generalized diagrams showing in side view and transverse cross section how air sacs or primitive lungs possessed by some ancient fishes were transformed into the swim bladder of modern forms. (*Adapted from Romer, 1962, fig. 235, p. 329.*)

position but is connected by a tube opening ventrally in the throat, which curves around the right side of the digestive tract.

Almost all the ray-finned fishes became adapted to marine or adequately aerated freshwater environments, where a lung provided no advantage. In them the primitive lung was put to another use. It became a hydrostatic organ that, by expansion or contraction, altered a fish's density and aided it in ascending or descending in the water. Several successive evolutionary stages are exhibited by modern teleosts. In the first, the swim bladder is connected through a large duct with an opening in the dorsal surface of the throat (Fig. 405-3). The original ventral position of the organ is indicated, however, by the courses of the blood vessels which serve it. Perhaps air can pass into or out of such a bladder through the fish's mouth. The duct is narrower and longer in other fishes, and in some it has closed entirely and disappeared (Fig. 405-4). How the bladder operates in fishes of this last kind is not completely understood. Part of the control at least is muscular, but gas must be released into the organ from the fish's blood, and absorption of gas presumably is also possible. Some mature bottom-living fishes have no swim bladder, although one develops and is present as a temporary organ early in their ontogeny. Sharks do not have any comparable organ.

Amphibian Characters

The skeletal characters that serve to identify fossil amphibians are combinations of conservative features carried over from fish ancestors and progressive ones reflecting adaptation to a new way of life. In addition to the transformation of fins to legs already mentioned, these are evident in (1) the skull, (2) the spinal column, (3) the limb girdles, and (4) the ribs.

Skull

The skull is a complex structure consisting of several different kinds of bones or cartilages to which a confusing array of names has been applied. It is unnecessary here to become involved in their detailed terminology, but note should be made of the several kinds of bones: (1) Dermal bones, which form the outer skull, particularly in the lower vertebrates, and the roof of the mouth or palate. Dermal bones also are important elements of the jaws. (2) Internal bones, which form the braincase. This is continuous with the spinal column, and its posterior part and related structures may consist of transformed vertebrae. (3) Bones derived from the gill arch supports. The most important of these is the upper member of the hyoid arch.

The structure of fish skulls is particularly complex and variable. More than 150 different bones have been distinguished, occurring in various combinations and in various relations to each other. Although certain generalized patterns can be recognized in some groups of fishes, homologies in many cases are difficult if not impossible to establish. There have been many instances of fusion or elimination of these bones, and perhaps also of fragmentation. A persistent tendency is evident, however, for reduction in the number and extent of skull bones, not only in fishes but also in their terrestrial descendants. For example, 140 bones occur in the skulls of some fish species, but in mammals the number is reduced to no more than 27 or 28.

A characteristic pattern of skull bones evolved in the crossopterygians, many features of which were inherited by amphibians and later vertebrates (Fig. 408-1). The openings for nostrils which continue through the palate into the mouth, the eye sockets, and a pineal opening in the middle of the forehead mark important fixed areas, in relation to which various bones can be identified. The pineal opening is very ancient, as it occurs in the ostracoderms. It marks the location of a small third eye. This eye, or at least the opening in which it lay, was a persistent feature of most lobe-finned fishes, but it disappeared early in the ray-fins.

Although the primitive amphibian skull is very similar to that of progressive Devonian crossopterygians, some evolutionary changes had been effected (Fig. 408-2). These are most evident in skull proportions. The eyes are located more posteriorly, producing a longer snout and a shorter posterior cranium. Small, irregular bones at the front of the fish's skull increased in size, decreased in number, and fell into a pattern that has been carried on in later vertebrates. Perhaps these changes reflect development of a better sense of smell. At the same time, a few larger bones were eliminated from the posterior margin of the skull, and bones of the operculum which covered the fish's gills disappeared. This change probably was related to freeing of the shoulder girdle from the skull.

The crossopterygian braincase consists of two parts that were movably articulated. Flexibility extended to the dermal skull bones along a prominent transverse suture (Fig. 408-1a). This kind of structure was not duplicated in amphibians, although a line of weakness persisted at this position in some fossil amphibians and even in a few early reptiles, whose skulls may be found separated along this line.

Fishes are believed to possess a sixth sense that makes them aware of currents and vibrations in the water passing by their bodies. This is useful to them, because it provides a feeling of orientation, direction, and motion as they swim through water where no fixed reference objects are visible. Perhaps it also acts as a kind of sonar in locating nearby objects, avoiding enemies, or finding food. The receptors of this sense are located in or along what are known as lateral line canals. These extend the length of a fish's body on both sides and also in a rather definite pattern on its head. They are impressed on the outer skull as shallow grooves or a series of small pits, which also aid in identification of the bony elements. Lateral line canals persisted and are evident on the skulls of some primitive amphibians. They are not only decisive evidence of fish ancestry, but they also probably indicate that these early air breathers lived mostly in the water in a rather fish-like way. Such a life also is indicated by many amphibians' general lack of scales which might have protected their bodies against drying out. Ventral fish-like scales were retained by some amphibians where perhaps they dragged their bellies on the ground. A few primitive species had similar scales on other parts of their bodies. Most modern amphibians have slimy skins through which water is lost readily by evaporation, and

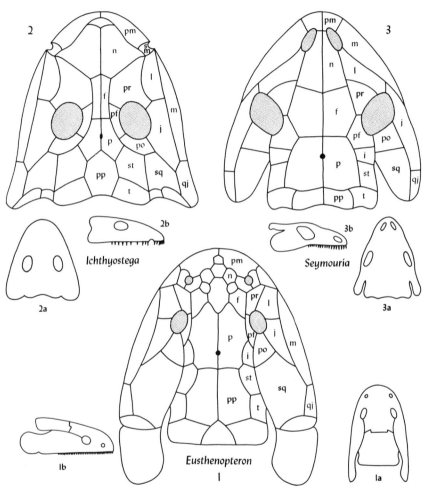

Fig. 408 Skull structures of three Paleozoic vertebrates. 1, Crossopterygian, late Devonian. 2, Very primitive amphibian, latest Devonian or earliest Mississippian. 3, Advanced amphibian (or very primitive reptile), early Permian. Several bones in the anterior and posterior parts of the crossopterygian skull are missing from the amphibian skulls. See Table 409 for identification of bones. (*After Romer and other sources.*)

Note: In these and all similar illustrations in the next chapter, skull patterns are projected onto a plane surface and consequently they are somewhat distorted, particularly around the front and lateral margins. The skulls as seen in top and side views are shown in proper proportions in the small outline figures.

Table 409   Bones identified in skull diagrams

f—frontal	pp—postparietal
it—intertemporal	pr—prefrontal
j—jugal	q—quadrate
l—lacrimal	qj—quadratojugal
m—maxilla	r—rostral
n—nasal	sm—supermaxilla
p—parietal	so—supraorbital
pf—postfrontal	sq—squamosal
pm—premaxilla	st—supratemporal
po—postorbital	t—tabular

These abbreviations are used in illustrations in this chapter and Chaps. 9, 10, and 11.

many ancient ones probably had skin of this same kind, requiring them to spend much of their lives in water.

Vertebrae

When they left their ponds and streams, the earliest amphibians moved into a medium much less dense than the one to which their fish ancestors had been adapted. Successful existence in this new medium of air necessitated strong skeletal structures, particularly in the spinal column and limb girdles, which were required to support the relatively much greater weight of the creatures' bodies. The Devonian crossopterygians had vertebrae that were well ossified in comparison with those of other contemporary fishes, and their spinal columns evidently were sufficiently strong to meet the demands made on them by small animals. As amphibians increased in size, however, sturdier supports were needed, and evolutionary modifications of the vertebrae were accomplished in several different ways.

Crossopterygian vertebrae (Fig. 410-1a,1b) consisted of four parts: (1) a dorsal piece, the neural arch, which enclosed the spinal cord, (2) a ventral piece, the intercentrum, which embraced the notochord, and (3) two small pieces, the pleurocentra, situated laterally between the other elements. The most primitive amphibians known as fossils had vertebrae built upon an identical plan (Fig. 410-2).

The bony spinal column of the crossopterygians was somewhat loosely formed, with the pieces of the vertebrae just mentioned bound together by fibrous connective tissue. Advancement in early amphibians is shown by some changes in the shape and proportions of these parts that produced a more compact and stronger type of structure. Despite these changes, homology of the different pieces is evident because the ribs, in addition to being articulated above with the neural arches, were attached below to the intercentra, rather than to the pleurocentra (Fig. 410-3 to 5).

Limb Girdles

The fin girdles of fishes are relatively simple structures. The anterior, or shoulder, girdle of bony fishes consists of both dermal and internal elements (Fig. 430), the

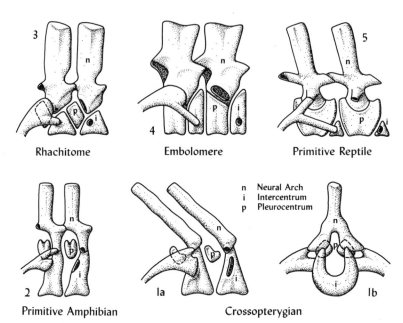

Fig. 410 Types of vertebrae illustrating stages in the main evolutionary sequence. 1, Starting with crossopterygian fishes, elements below the neural arch are a large crescentic intercentrum and two small pleurocentra. 2, Structure in the most primitive amphibians is essentially identical. 3, Beyond that stage the pleurocentra increase in relative size and importance. 4, In advanced amphibians leading toward reptiles, the pleurocentra fuse to form a disk. 5, In reptiles the intercentra shrink and disappear, and the pleurocentra become the centrum of more advanced reptiles and mammals. See also Fig. 397. (*Mostly adapted from Romer, 1962, fig. 104, p. 166.*)

latter variably ossified, and it is rather firmly attached above to the posterior portion of the skull. The posterior, or hip, girdle contains only a single pair of wedge-shaped ventral bones of internal origin (Fig. 431-1). These structures became the limb girdles of amphibians. They were called upon to bear all the weight of a creature's body and evolved rapidly to produce stronger structures. The internal bones became much more important than the dermal ones, because the limbs articulated directly with them and they furnished attachment areas for the principal limb muscles. In consequence the dermal bones were reduced in size, and the internal bones enlarged.

The shoulder girdle of amphibians was freed from its attachment to the skull. This was the beginning of the development of a neck differentiated from the remainder of the body. It permitted movement of the head independently of the body, which was not possible among ancestral fishes.

The rudimentary hip girdle of all fishes is simply embedded in the abdominal muscles and has no connection with the spinal column. Such a condition obviously is inadequate for a limbed land animal, and evolution among the earliest amphibians rapidly produced a better supporting structure. In the first stage, the small

Table 411   Types of amphibian vertebrae

Aspidospondylous: labyrinthodont types; each vertebra consists of a neural arch above, enclosing the spinal
nerve cord, and an anterior intercentrum and a posterior pleurocentrum below, enclosing
the notochord
Rachytomous: intercentrum crescentic, wedge-shaped in side view, median ventral. Pleurocentrum represented by two small diamond-shaped pieces between neural arch and intercentrum. Similar
to crossopterygian vertebrae. Pennsylvanian to Permian
Embolomerous: both pleurocentrum and intercentrum nearly circular, pierced for notochord. Mississippian
to Permian
Seymourian: pleurocentrum circular as in embolomeres; intercentra small wedges between their lower
parts. Similar to structure of primitive reptiles. Permian
Stereospondylous: pleurocentrum reduced or absent; nearly circular intercentrum notched at top for
notochord. Jurassic to Recent
Anuran: of frogs and toads. Vertebra lacks true centrum. Neural arch extends downward and encloses
notochord. Jurassic to Recent
Lepospondylous: of salamanders, etc. Each vertebra with spool-shaped centrum below neural arch. In
modern examples not preformed in cartilage. Mississippian to Recent

**The intercentrum disappeared in reptiles; their centrum, and that of mammals, corresponds to the pleuro-
centrum of amphibians.**

ventral bones of the fish's girdle were tilted upward laterally. Each of these parts
ossified from two centers, producing two intimately connected bones. A third connected spine-like bone also developed which projected upward and backward from
the others (Fig. 431-2). This presumably was attached to nearby ribs by ligaments.
In the next and final stage among amphibians, the third bone became attached
directly to a single expanded rib (Fig. 431-3). This is the structure, variously modified, that occurs in the pelvis of all later land vertebrates.

Ribs

Fishes swim by alternate lateral movements of their bodies produced by the contraction of powerful muscles. Very little of their musculature is attached to vertebrae,
but the ribs, situated between the muscle segments, contribute to the efficiency of
these motions. Two kinds of ribs are present in bony fishes (Fig. 411): (1) dorsal
ribs that extend more or less laterally from the spinal column, and (2) ventral ribs
that curve downward and more or less enclose the body cavity.

The legs of primitive land animals were advanced alternately in walking by body

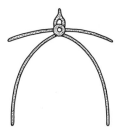

Fig. 411   Many fishes have both dorsal and ventral ribs. The ventral ribs are
represented only in the tails of other vertebrates. There they join below to
form ventral arches enclosing the main blood vessels of the tail, much like
the neural arches above which enclose the spinal nerve cord.

movements similar at least in part to those of fishes (Fig. 412). As legs were progressively perfected, however, this type of movement played a rapidly declining role, and the body musculature accordingly was profoundly altered. The fish's great muscles were much reduced except where their derivatives functioned to operate the legs. As a result of this and because of the need for more bodily support, the ribs acquired new functions. Most importantly they formed a flexible structure in the thoracic region whose muscular contraction and expansion facilitated breathing They also strengthened the body wall and provided some protection for vital inner organs.

Like the fishes, primitive land animals had ribs attached to every vertebra. The number later was reduced, by elimination of ribs first from the tail, then from the neck, and finally in the abdominal region. The ribs of all land vertebrates cor-

Fig. 412 In walking, the body motions of a primitive land vertebrate resemble the swimming motions of a fish. (See *Romer, 1962, fig. 120B, p. 181.*)

Eryops

Fig. 413 Restoration of a Permian amphibian that grew to be about 5 ft long. Most of these animals were squat, sprawling creatures with flattened heads. (*After Colbert,* 1955, *fig.* 31, *p.* 96.)

respond to the dorsal ribs of fishes. In early amphibians, they articulated with the vertebrae by two heads, one in contact with the neural arch and the other attached to the intercentrum (Fig. 410). As evolutionary changes altered structure of the vertebrae, however, these relations became different in several ways, both by the shifting of attachment areas and by modification of the heads. Ventral ribs are represented in terrestrial vertebrates only in their tails, where V-shaped arches, similar to those in the tails of fishes, are present below the spinal centra and enclose the principal blood vessels.

Amphibian Evolution

Amphibians are particularly interesting because they provide the evolutionary link between fishes and reptiles. Their oldest known fossils are of latest Devonian or earliest Mississippian age, and the reptiles gradually replaced them in the vertebrate faunas of the early Permian. Amphibians continued to prosper to some extent, however, through the Triassic Period as a generally minor element among the terrestrial animals. Thereafter they rapidly declined in numbers and importance, and only a few groups of small creatures such as frogs, toads, and salamanders have survived to modern times. During their period of dominance, however, they evolved into a number of radiating stocks and produced some animals of considerable size.

The early amphibians generally were small, rarely exceeding a few feet in length. The most primitive ones retained something of the form of fishes, with comparatively high rounded skulls and bodies somewhat laterally compressed. This shape soon altered, however, and most of the amphibians developed broad bodies and flattened heads (Fig. 413). Some of them, particularly in the Permian Period, had short bodies and stubby tails. These seem to have adapted rather successfully for a time to life away from water, except for breeding, but they eventually failed in their competition with the primitive reptiles. Most later amphibians probably rarely left the ponds and streams where they were more at home. A very few may have adapted to marine existence. A trend began among amphibians that led to delayed ossification of the bones, and skeletons reverted to a partly cartilaginous

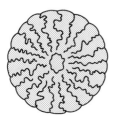

Fig. 414 Cross section of a labyrinthodont tooth showing the infolding of enamel. The outside is marked by closely spaced longitudinal grooves.

condition. The limbs of some also decreased in relative size and became less functional.

Labyrinthodonts

The labyrinthodonts receive their name from the teeth with infolded enamel which they inherited from their crossopterygian ancestors (Fig. 414). The distinctive structure of the vertebrae of some of these animals has been described (Fig. 410-2). Evolutionary advancement within this group, which became extinct at the end of the Triassic Period, can be traced in part through subsequent changes in this structure. From the primitive condition characterized by the presence of small pleurocentra, evolution proceeded in two directions: (1) The pleurocentra increased in size until they, as well as the intercentra, became complete disks (Fig. 410-4). In further evolution the intercentra shrank and were reduced to small ventral pieces,

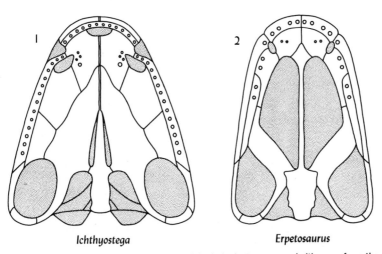

Ichthyostega   *Erpetosaurus*

Fig. 414a Ventral views of the skulls of two labyrinthodonts, a very primitive one from the latest Devonian or earliest Mississippian (1) and another from the Pennsylvanian (2), showing great enlargement of the openings between the bones that roof the mouth. (*After Romer, 1945, "Vertebrate Paleontology," 2d ed., figs. 105, 107, pp. 144, 146.*)

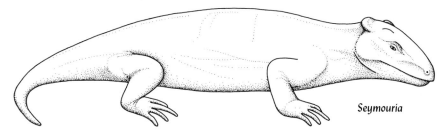

Seymouria

Fig. 415 This creature from the Lower Permian, restored about half natural size, is commonly classed with the amphibians, but it exhibits several distinctly reptilian characters. (*After Moody, 1962, "Introduction to Evolution," 2d ed., fig. 8.23, p. 168.*)

as in the primitive reptiles (Fig. 410-5). (2) The pleurocentra were eliminated, and only the intercentra remained as wide disks.

Evolutionary lineages, however, cannot be traced among the labyrinthodonts with much assurance. Most of the different types of vertebrae are found in Mississippian species in combination with features that are considered to be primitive. These include (1) the retention of some typically crossopterygian bones in the posterior part of the skull, (2) loose connection of the roofing bones of the mouth with the sides of the skull, and (3) the presence of small rather than large openings between some of these palate bones (Fig. 414a). At the same time, various relatively early species with different types of vertebrae closely resemble primitive reptiles in some ways. Actually the differentiation of certain skeletal features as either amphibian or reptilian is very difficult. One early Permian group, the seymourians, possessed so many reptilian characters that the identification of these animals as amphibians has been disputed (Fig. 415).

Frogs and Toads

Modern frogs and toads are not labyrinthodonts. They have small teeth only in their upper jaws, and their vertebrae are not of the proper type. Nevertheless they are believed to have evolved from labyrinthodonts, and traces of their ancestry may extend back to the Pennsylvanian Period. The skeletons of frogs and toads are peculiar and highly specialized in many ways. One striking feature is the nature of their vertebrae, which seem to consist only of neural arches which extend downward to embrace the notochord. A Pennsylvanian species, most of whose other characters seem to mark it as a labyrinthodont, has vertebrae of this type except that tiny intercentra and perhaps pleurocentra still remain. The skull of this little animal also resembles that of frogs, not in general form, but in the absence of several bones ordinarily present in other amphibians. Another Triassic species has a much more frog-like skull, but this is accompanied by a moderately long body, long tail, and unspecialized legs. Mostly fragmentary skeletons of very modern-appearing frogs have been discovered in Jurassic and more recent strata, but they are very rare.

Salamanders

Another group of amphibians, the lepospondyls, or salamanders and their relations, is distinguished because the central parts of their vertebrae consist of single pieces, not of pleurocentra and intercentra as in the labyrinthodonts. The vertebrae of modern species are not preformed in cartilage, but whether this was true for fossils cannot be known. None of these animals grew large. They are represented in living faunas by salamanders and some inconspicuous burrowing worm-like creatures of the tropics.

Lepospondyls first are known as fossils from the Mississippian. In the Pennsylvanian they inhabited the coal swamps in such variety that they must have had a long previous evolutionary history. Many of these creatures had small weak legs, and in some the legs had disappeared entirely, producing snake-like forms. A few seem to have been gill breathers. The skulls mostly resemble those of labyrinthodonts in general structure, except that some of the posterior bones are missing. Certain of the evolutionary tendencies exhibited by these animals parallel those of the labyrinthodonts. For example, large openings developed between the roofing bones of the mouth, and, as in frogs, the skeletons of salamanders have reverted to a partly cartilaginous condition.

BIBLIOGRAPHY

VERTEBRATES IN GENERAL

E. H. Colbert (1955): "Evolution of the Vertebrates: A History of the Backboned Animals through Time," John Wiley & Sons, Inc., New York.
Though the author has not written down to a popular level, this book is more readable than most others devoted to vertebrate paleontology.

G. R. de Beer (1954): *Archaeopteryx* and Evolution, *Advan. Sci.,* vol. 11, pp. 160–170.
This article briefly reviews the course of vertebrate evolution and lists the morphologic features characteristic of each class.

C. P. Hickman (1961): "Integrated Principles of Zoology," 2d ed., The C. V. Mosby Company, St. Louis.
This is a good general text containing much material related to evolution not found in most comparable books.

E. S. Goodrich (1930): "Studies on the Structure and Development of Vertebrates," The Macmillan Company, New York; reprinted, **1958,** paperback edition, 2 vols., Dover Publications, Inc., New York.
The thoroughness and detail of this work on comparative anatomy have rarely been equaled in later books.

W. K. Gregory (1951): "Evolution Emerging: A Survey of Changing Patterns from Primeval Life to Man," 2 vols., text and illustrations, The Macmillan Company, New York.
This is an outstanding work devoted mainly to vertebrate paleontology, with a wealth of illustrations and nearly 150 pages of bibliography.

T. J. Parker and **W. A. Haswell (1940):** "A Text-book of Zoology," 6th ed., vol. 2, The Macmillan Company, New York.
For many years this has been the standard text in the English language; particularly valuable for anatomical details.

A. S. Romer (1962): "The Vertebrate Body," 3d ed., W. B. Saunders Company, Philadelphia.
This book is an important source of descriptions and explanations of vertebrate structures that provide evidence for evolution.

_____ **(1966):** "Vertebrate Paleontology," 3d ed., The University of Chicago Press, Chicago.
This book is widely recognized as the foremost in its field.

H. M. Smith (1960): "Evolution of Chordate Structure: An Introduction to Comparative Anatomy," Holt, Rinehart and Winston, Inc., New York.
This book explains and illustrates morphologic and functional evolutionary changes fully and clearly.

H. H. Swinnerton (1947): "Outlines of Palaeontology," 3d ed., Edward Arnold (Publishers) Ltd., London.
The latter part of this book devoted to vertebrates is brief and general but good.

J. Z. Young (1962): "The life of Vertebrates," 2d ed., Oxford University Press, Fair Lawn, N.J.
The author is uncritical of theories but presents much information of interest to paleontologists.

FISHES AND AMPHIBIANS

N. J. Berrill (1955): "The Origin of the Vertebrates," Oxford University Press, Fair Lawn, N.J.
The ascidian tadpole, from which vertebrates evolved, developed as a specialized larva that does not recapitulate any ancestral form.

J. Brough (1936): On the Evolution of the Bony Fishes during the Triassic Period, *Biol. Rev.*, vol. 11, pp. 385–405.
The evolutionary changes that began the great radiation of bony fishes are described.
R. H. Denison (1956): A Review of the Habitat of the Earliest Vertebrates, *Fieldiana, Geol.*, vol. 11, pp. 357–457.
Fishes originated in the sea and then invaded freshwater, where much of their later evolution was accomplished.
————— **(1963):** The Early History of the Vertebrate Calcified Skeleton, *Clin. Orthopaed.*, no. 31, pp. 141–152.
Some kind of stiffened connective tissue, perhaps cartilage, antedated bone in vertebrates.
E. O. Dodson (1961): Neo-Lamarckism, Modern Darwinism, and the Origin of Vertebrates, *J. Paleontol.*, vol. 35, pp. 1065–1076.
The arthropod theory of vertebrate origin is criticized; echinoderm relationships are favored.
W. K. Gregory (1936): The Transformation of Organic Designs: A Review of the Origin and Development of the Early Vertebrates, *Biol. Rev.*, vol. 11, pp. 311–344.
The possibility is suggested that chordates evolved from early echinoderms resembling some known carpoids.
R. F. Inger (1957): Ecologic Aspects of the Origins of Tetrapods, *Evolution*, vol. 11, pp. 373–376.
The theory is questioned that aridity during the Devonian Period contributed to the origin of land vertebrates.
F. R. H. Jones (1953): The Structure and Functions of the Teleostian Swimbladder, *Biol. Rev.*, vol. 28, pp. 16–83.
The development, structure, function, and evolution of the swim bladder are discussed.
J. A. Moy-Thomas (1939): Early Evolution and Relationships of Elasmobranchs, *Biol. Rev.*, vol. 14, pp. 1–26.
Sharks were not closely related to other Devonian fishes, and all probably had jaws of advanced character.
————— **(1939):** "Palaeozoic Fishes," Chemical Publishing Company, Inc., New York.
See pp. 127–132 for review of early fish evolution.
G. K. Noble (1931): "Biology of the Amphibia," McGraw-Hill Book Company, New York.
In addition to biologic details, this book gives consideration to evolutionary relations.
J. R. Norman and **P. H. Greenwood (1963):** "A History of Fishes," 2d ed., Hill and Wang, Inc., New York.
This book is mainly devoted to modern fishes, but it contains much information of interest to paleontologists.
F. Raw (1960): Outline of a Theory of Origin of the Vertebrates, *J. Paleontol.*, vol. 34, pp. 497–539.
After reviewing theories, the author argues that vertebrates evolved from a primitive Precambrian arthropod before compound eyes developed.
J. D. Robertson (1957): The Habitat of the Early Vertebrates, *Biol. Rev.*, vol. 32, pp. 156–187.
The problem is reviewed and the conclusion reached that vertebrate evolution had its beginning in the sea.
A. S. Romer (1946): Early Evolution of Fishes, *Quart. Rev. Biol.*, vol. 21, pp. 33–69.
All fishes evolved from ostracoderms; reduction in dermal bone was a progressive and persistent process.
H. W. Smith (1936): The Retention and Physiological Role of Urea in the Elasmobranchii, *Biol. Rev.*, vol. 11, pp. 49–82.
Sharks absorb some water from their surroundings and excrete it as urine.
D. M. S. Watson (1951): "Paleontology and Modern Biology," Yale University Press, New Haven, Conn.

This book, based on lectures given in 1937, is devoted mainly to the evolutionary position of amphibians between fishes and reptiles.

P. B. Weisz (1966): "The Science of Zoology," McGraw-Hill Book Company, New York.
See pp. 185–191 for a brief but good account of the nature and formation of bone.

T. S. Westoll (1943): The Origin of Tetrapods, *Biol. Rev.,* vol. 18, pp. 78–98.
Terrestrial vertebrates evolved rapidly from crossopterygians during late Devonian time, probably from a single genus or even species.

E. I. White (1958): Original Environment of Craniates, pp. 212–234 in "Studies on Fossil Vertebrates," T. S. Westoll (ed.), Athlone Press, London.
The controversy is summarized and vertebrates are concluded to have evolved as unarmored creatures in a marine environment.

NINE

VERTEBRATES 2

REPTILES—BIRDS

REPTILES

The earliest known reptilian fossils have been collected from Pennsylvanian strata. By Permian time these animals had diversified considerably, and they rapidly supplanted amphibians as the dominant air-breathing vertebrates. All the principal groups of reptiles except snakes appeared before the end of the Triassic Period. Their evolutionary radiation produced a remarkably varied series of body types adapted to almost all the ways of life that subsequently were exploited by the mammals. In fact their success during the Mesozoic Era was quite comparable to that of the later mammals except that the cold-blooded reptiles were restricted to the warm and temperate regions of the earth. Few, however, of the many kinds of Mesozoic reptiles survived the Cretaceous Period. This great group of animals is represented in modern faunas only by lizards, snakes, turtles, and crocodilians, few of which attain large size, and a single archaic species living in New Zealand.

The Reptilian Egg

The spectacular evolution of the reptiles and their superiority over their ancestors among the amphibians resulted principally from the development of an advanced type of egg (Fig. 422). This is a relatively large air-breathing egg enclosed in a protective shell and richly supplied with yolk and other nutrient material. Such an egg is very different from the small fish-like eggs of most amphibians, which obtain oxygen and eliminate metabolic wastes by diffusion with surrounding water. The embryo of a reptilian egg grows within a liquid-filled sac, where it is protected

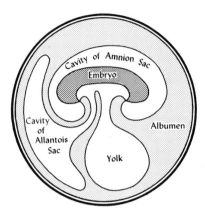

Fig. 422 Schematic diagram showing the functional parts that develop around the embryo in the reptilian egg.

against desiccation. Another sac develops which serves to exchange oxygen and carbon dioxide with the air, after the manner of a lung, and in which metabolic waste accumulates in the form of uric acid. The embryo receives its nourishment mainly from the yolk and likewise absorbs moisture from the white albumen, which does not occur in the eggs of any lower animals. Growth of the embryo continues within the egg until a self-sufficient young creature is produced that, from the moment of hatching, closely resembles its parents in both body form and way of life. This and other evolutionary developments freed reptiles from the dependence on water that has characterized most amphibians. Thus they were able to colonize the land and prosper in a wide variety of environments from which amphibians were excluded.

Origin of Reptiles

The partial escape of amphibians from water has led to the common belief that the origin of the reptilian egg was related to this progressing freedom. The most generally held theory postulates that, as some amphibians acquired more and more terrestrial habits, they became less and less dependent upon water for their breeding. Their better-protected and larger eggs produced young in an increasingly more advanced stage of development, until at last the reptilian condition was attained.

Another theory, however, reverses these relationships. It is based on the observation that a variety of modern tropical amphibians do not lay their eggs in water and thereby seem to have achieved an advantage over their relatives which still do. The suggestion is made that eggs laid in water and the tadpoles that developed from them were hunted out and eaten by a multitude of predatory creatures inhabiting the streams and ponds. Eggs laid by some early amphibians in damp places near the water, or otherwise removed from the environment where they were subject to destruction, had a better chance of surviving and hatching to produce less-vulnerable young. Perhaps the amphibian ancestors of reptiles continued to live in or near the water until the reptilian egg evolved and only then abandoned their habitual environment for a wholly terrestrial existence.

The immediate ancestors of the reptiles are not known. Morphologically transitional creatures like the seymourians (Fig. 415) do not stand in the direct evolutionary line, because a varied lot of reptiles was already in existence when these animals lived. The first reptiles probably evolved relatively rapidly, perhaps during the Mississippian Period, from primitive amphibians at the same time that evolutionary radiation was producing other kinds of more advanced amphibians.

Reptilian Characters

No single morphologic character can be selected that will surely serve to differentiate all reptiles from all amphibians. The most fundamental differences between these two groups of animals are provided by their methods of reproduction and the nature of their eggs. Complete transition must have occurred, however, in the course of evolution, and if all the connecting links were known, no sharp line between them could be drawn. The oldest known reptilian egg was discovered in the early Permian rocks of Texas, but nothing is known about the reproduction of the overwhelming majority of fossil species.

The identification of fossil reptiles must be based on skeletal characters. The comparative anatomy of modern species provides some useful information, but the structures of these animals are not necessarily similar to those of either the primitive Paleozoic reptiles or the specialized extinct Mesozoic forms. Much reliance is placed on skeletal features reflecting adaptation to kinds of terrestrial life that are believed to have been uncharacteristic of amphibians and to associated morphologic features that are different from those believed to identify the more primitive condition of amphibians. There can be no certainty, however, that some fossils classed as amphibians on the basis of their structures but possessing, for example, strong limbs indicating a dominantly terrestrial existence may not have developed from a reptilian type of egg, or that those with reptile-like structures had progressed beyond the amphibian type of reproduction.

Only a few skeletal features are ordinarily considered to be almost unfailing means for differentiating reptiles from amphibians. The more important are: (1) An otic notch (Figs. 408-1,3) which separates some of the bones in the posterolateral parts of the skulls of relatively unspecialized early amphibians is not duplicated exactly in any reptile. This notch was inherited from fishes, and originally it contained the opening of the spiracle. (2) The hip girdle of amphibians is connected with not more than one rib and vertebra. In reptiles it is joined to two or more. (3) Movement of the head in reptiles is facilitated by a structure consisting of two specialized vertebrae immediately behind the skull, instead of one as in amphibians. A great variety, however, of other features that are exhibited else-

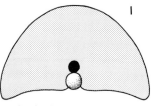
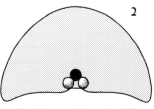

Single Occipital Condyle   Double Occipital Condyle

Fig. 424 Highly diagrammatic representations of the basal surfaces of two skulls. These articulated with the neck vertebrae by a single or a double ball-and-socket joint located adjacent to the opening occupied by the spinal nerve cord. 1, The simpler type is more primitive. It is characteristic of the earlier amphibians, most reptiles, and all birds. 2, The other type evolved twice. It appeared first in advanced amphibians and later in the mammal-like reptiles. It is characteristic of all mammals.

where in the skeleton, either singly or in combination, is more or less confidently considered to identify some fossils as reptiles rather than amphibians.

Skull

The skulls of many reptiles, in contrast to those of most amphibians, are relatively high, rather than low and flattened. They articulate with the neck vertebrae by a single ball-and-socket joint located below the opening occupied by the spinal cord (Fig. 424-1), instead of by two situated at the sides of this opening, as in most amphibians with depressed skulls. This type of articulation permitted freer movement of the head, which could be turned sideways, not only up and down as in many amphibians. Both these features are primitive. Because reptiles resemble the earliest amphibians in these ways, evidence is provided that reptiles evolved from an early amphibian stock which had not acquired some of the features characterizing most of the fossils belonging to this other group.

Evolutionary advancement beyond the amphibian grade of development is

Table 424   Five types of reptile skulls

Anapsid: without temporal openings in the skull behind the eye sockets. The most primitive type. Cotylosaurs
Euryapsid: one pair of temporal openings high on sides of skull bounded by parietal, postparietal, and supratemporal bones. Plesiosaurs
Parapsid: one pair of temporal openings high on sides of skull bounded by parietal, postparietal, and squamosal bones. Ichthyosaurs
Synapsid: one pair of temporal openings low on sides of skull, not bounded by parietal bones. Therapsids or mammal-like reptiles
Diapsid: two pairs of temporal openings, like euryapsid and synapsid types, separated by parts of postorbital and squamosal bones. Most Mesozoic land reptiles

These skull types are important in providing part of the basis for characterizing the primary divisions of the reptiles.

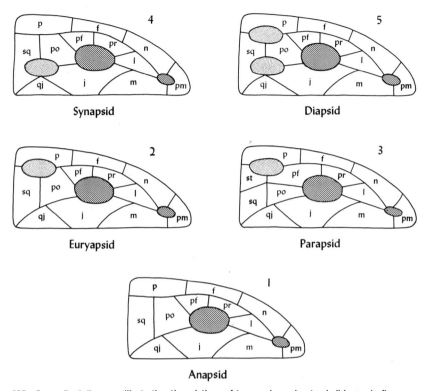

Fig. 425 Generalized diagrams illustrating the relations of temporal opening to skull bones in five morphologic groups of reptiles. (*See Romer, 1966, fig. 155, p. 108.*) See Table 409 for identification of bones.

indicated by decrease in the size, migration backward onto the basal area of the skull, or complete disappearance of some of the posterior bones of the amphibian skull. Closure of the otic notch also was an advancement which produced a stronger structural condition in the region where the jaw muscles were attached. An opening similar to the otic notch reappeared, however, in reptiles at a lower position nearer to the area of jaw articulation.

Other adjustments to more efficient jaw musculature seem to have resulted in the appearance of openings between the skull bones in the temporal region behind the eye socket (Fig. 425). The belief has been expressed that they accommodated bulging muscles when the jaws were closed. More probably, however, these openings were related to the functional areas of muscle attachment and the elimination of bone where it was not needed. The presence of temporal openings and their positions serve to differentiate five types of skulls (see Table 424): (1) Anapsid—without temporal openings (Fig. 425-1). This evidently is a primitive condition characteristic of the stock from which all other reptiles presumably evolved. (2) Synapsid—with a single opening low in the region of each cheek (Fig. 425-4). This is the condition existing in the group of reptiles from which mammals were derived. (3) Parapsid—with a single opening high on each side or near the top of the skull (Fig. 425-3). The ichthyosaurs had this type of opening. (4) Euryapsid—openings like the last but with slightly different relations to the skull bones (Fig. 425-2). Two groups, differing much in other ways, show this kind of structure. (5) Diapsid—with two openings on each side (Fig. 425-5). Most Mesozoic reptiles had skulls of this last

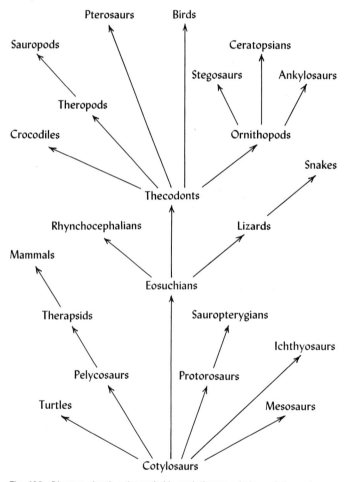

Fig. 426 Diagram showing the probable evolutionary relations of the main groups of reptiles.

Table 427   Some characters differentiating the main groups of reptiles exclusive of diapsids.

Cotylosaurs: anapsid. Diadectids and captorhinids respectively with and without otic notch indenting
    posterior skull border. Neural arches of vertebrae swollen. The group from which all other
    reptiles probably evolved. Pennsylvanian to Triassic
Chelonians: turtles, probably anapsid. Skull without closed temporal openings, but may be deeply indented
    behind. Body enclosed in a complete rigid shell of large dermal bones. Triassic to Recent
Pelycosaurs: synapsid. Neural arches less swollen and limbs more slender than in cotylosaurs. Some with
    long neural spines that probably supported a heat-regulating membrane above the back. Some
    with teeth differentiated to suggest canines. Intermediate evolutionary between cotylosaurs
    and therapsids. Pennsylvanian to Permian
Therapsids: mammal-like reptiles, mostly synapsid; an extremely varied lot. Temporal openings of some
    moved up to euryapsid position. Skull bones around eye socket smaller. Occipital condyle
    double. Internal nasal openings more posterior. Some bones of lower jaws smaller. Teeth com-
    monly well differentiated. Shoulder girdle simplified. Legs drawn under body. Permian to Triassic
Mesosaurs: synapsid. Small aquatic reptiles with long snout and tail. Pennsylvanian to Permian
Sauropterygians: plesiosaurs, nothosaurs, placodonts, euryapsid. Marine reptiles. Quadratojugal bone not
    present in skull. Ribs single-headed, attached to neural arch. Triassic to Cretaceous
Protorosaurs: parapsid or euryapsid. Small varied group. Ribs tend to be attached to vertebrae by a single
    head. Permian to Triassic
Ichthyosaurs: parapsid. Marine reptiles with very fish-like bodies. Triassic to Cretaceous

type, and the modern snakes and lizards are related in their structures. These differences provide the principal basis for reptilian classification, but they do not seem to be infallible guides to evolutionary patterns. Another pair of openings occurs in the skulls of many advanced reptiles between the eye sockets and nostrils (Fig. 451). The pineal opening in the skull roof disappeared in advanced forms.

Bones roofing the mouth bear teeth in many reptiles, but large fangs similar to those of some amphibians are not present in these positions. The number of bony elements in the lower jaws of reptiles generally is less than in amphibians. This is part of an evolutionary trend that was continued by further simplification in the mammals (Fig. 482).

Vertebrae

The vertebrae of primitive reptiles are similar to those of some labyrinthodonts (Fig. 410-2). The central parts consist of a large disk-like pleurocentrum pierced by an opening for the notochord and a very small intercentrum. Evolutionary advancement is shown by further shrinkage and disappearance of the intercentrum and closure of the notochordal perforation.

The principal areas of articulation in a fish's spinal column are between the central parts of the vertebrae. In terrestrial vertebrates, however, additional articu-lating structures develop on projections from the neural arches. These permit movement, both sideways and up and down, but they prevent rotation. The danger of vertebral displacement in fishes is small, because the spine is not called upon to bear much weight, but it increases in more and more active land animals, particu-larly those whose bodies are raised above the ground, like many reptiles. Conse-quently structures improving and strengthening articulation which appeared in the amphibians generally increased in their development and efficiency in reptiles.

Differentiation of neck vertebrae began in the amphibians, but only the first was structurally altered in this group of animals to permit some movement of the head. Additional evolutionary change in reptiles resulted in the specialization of the second vertebra as well, producing a composite structure capable of freer movement. This involved some rearrangement of the embryonic ossicles in the makeup of the definitive vertebrae known as atlas and axis in the necks of higher vertebrates. Most of the up-and-down movement of the head is accomplished at the articulation between the skull and atlas; most of the lateral movement occurs at the following junction of atlas and axis.

Ribs

Most primitively every vertebra presumably bore a pair of ribs corresponding to the dorsal ribs of fishes, but even in the tails of the earliest bony fishes the structure differed from that in other parts of the body. In the amphibians and early reptiles, ribs were borne on all vertebrae except possibly the first and second in the neck and those in the posterior part of the tail (Fig. 435). Originally all these ribs were similar except in length. Evolutionary modifications, however, beginning in the amphibians resulted in the regionalization and structural differentiation of several parts of the body that became especially conspicuous in some of the advanced reptiles.

Differentiation of parts of the reptilian body with respect to rib development was as follows: (1) Ribs in the neck shortened and eventually became parts of the vertebrae or disappeared as the mobility in this area increased. (2) Ribs in the thorax were long and joined to a ventral breastbone. The latter is not present in fishes but probably occurred in most amphibians and all reptiles. It is rarely recognized in fossils, however, because ordinarily this part of the skeleton was not ossified. (3) In the lumbar region behind the thorax the ribs were free below. They became progressively shorter posteriorly, and there was a tendency for the hindmost ones to disappear. Many reptiles also possessed so-called abdominal ribs. These, however, were very different from true ribs, because they were dermal rather than internal bones and they had no connection with the vertebrae. They had an evolutionary origin in the scales of fishes that persisted in the abdominal skin of some amphibians. (4) One or more sacral ribs was short and stout and served as a firm connection between the vertebrae above and the hip girdle below. One such rib occurs in most amphibians, but at least two are present in every recognized reptile. Some specialized reptiles had as many as eight. (5) Ribs in the tail were short, and limb muscles were attached to some of the anterior ones. Otherwise they served no useful function and tended to disappear.

The ribs of amphibians and early reptiles were two-headed and articulated with the intercentrum and neural arch (Fig. 410). As the intercentrum became smaller and finally disappeared in the more advanced reptiles, the heads tended to merge and lost their individual identity. At the same time the articulation shifted in various ways in different groups of reptiles. Thus ribs were attached only to the neural arches or to the single large centra, or they were partly wedged in between successive centra. Articulation with the vertebrae in most reptiles permitted some movement of all ribs.

Limb Girdles

Limb girdles originated in fishes in connection with their paired fins. The shoulder girdle developed more elaborately than the other (Fig. 430), probably because of the greater importance of the pectoral fins in guiding a fish's movements. The girdle consisted of both dermal and internal skeletal elements. The dermal elements predominated and joined the girdle to the fish's skull. The hip girdle consisted of only a pair of internal bones lying unattached in the wall of the abdomen (Fig. 431).

A fish's fins are not required to support its weight, and consequently no strong attachment of the girdles to the main axial skeleton is necessary. This situation, however, changed radically when amphibians crawled out upon the land. As the limbs of terrestrial vertebrates evolved, the supporting functions of the girdles became increasingly important. The internal bones of the shoulder girdle acquired strong muscular and ligamental attachments to the vertebrae and ribs. At the same time the development of a flexible neck resulted in severance of the connection of the dermal bones and skull. In the course of evolution leading from amphibians to the higher vertebrates, the dermal bones were progressively reduced, and finally they were lost entirely in some groups. The internal bones, on the other hand, became increasingly important and were structurally simplified (Fig. 430).

The forelegs of land vertebrates generally are less powerful than the hind legs and function mainly in supporting somewhat more than half the body's weight. The hind legs, which developed from the subordinate paired fins of fishes, not only share this supporting function but also provide much of the forward thrust when an animal moves swiftly. The added strength required of them would not be possible without extensive modification of the originally rudimentary and structurally weak hip girdle of the fishes. The evolutionary advancements that occurred in amphibians and reptiles involved enlargement of the pelvic bones and their increasingly firm attachment to ribs and vertebrae (Fig. 431). In most amphibians the attachment is to one pair of ribs, and in reptiles a stronger structure is attained by fusion to two or more pairs of modified ribs. Especially in the development of a bipedal habit in several groups of vertebrates, more ribs and vertebrae were incorporated in a still stronger structural complex.

The number of bones recognized in the limb girdles varies among different groups of vertebrates. Some of the reduction in the shoulder girdle resulted from the fusion of bones originating in different embryonic cartilages. On the other hand, increase in the number of bones of the hip girdle followed the development of three centers of ossification in each cartilage of the single embryonic pair.

Limb Structure

From their first appearance in amphibians the limbs of vertebrates have evolved

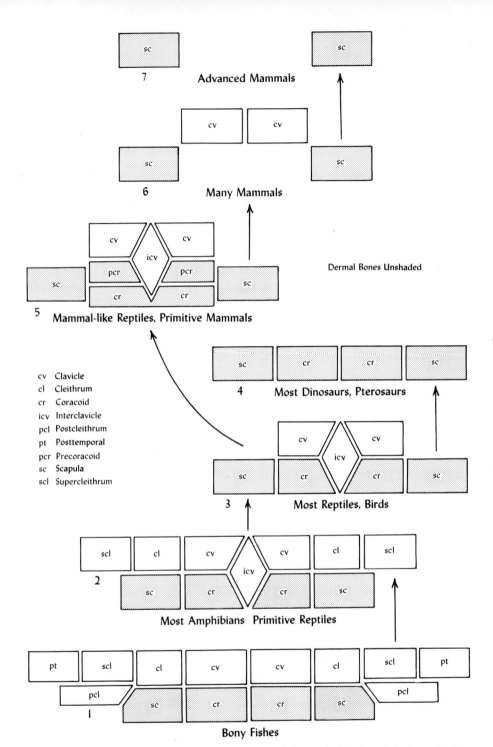

Fig. 430 Schematic diagram showing the progressive evolutionary simplification of the bony shoulder girdle structure in vertebrates. (*Modified from Smith, 1960, "Evolution of Chordate Structure," fig. 7.27, p. 189.*)

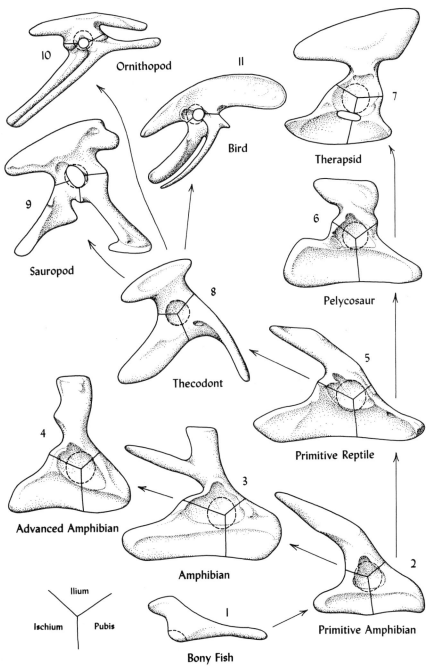

Fig. 431 Structure and evolution of the hip girdle in amphibians and reptiles. The primitive amphibian girdle (2) is not fused to sacral ribs. The more advanced girdle (3) is fused to one rib. The reptilian girdle (5) is fused to more than one rib. (*After Romer, 1962, "The Vertebrate Body," 3d ed., figs. 125–128, pp. 191–194.*)

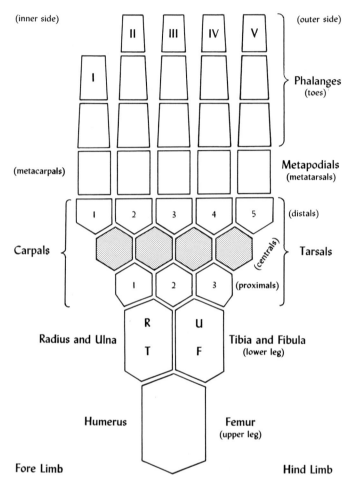

(inner side)

(outer side)

II   III   IV   V

I

Phalanges
(toes)

(metacarpals)

Metapodials
(metatarsals)

Carpals

1   2   3   4   5   (distals)

(centrals)

Tarsals

1   2   3   (proximals)

R   U

Radius and Ulna

T   F

Tibia and Fibula
(lower leg)

Humerus

Femur
(upper leg)

Fore Limb

Hind Limb

Fig. 432 Schematic diagram illustrating the basic bony structure of the vertebrate leg and foot.

in many different fashions in adaptation to many different ways of life. In spite of this, however, and in all but the most extreme cases, their essential structure has been preserved (Fig. 432). This is seen in the skeletal composition of the limb, which is divisible into four parts, (1) the upper leg containing a single long bone, (2) the lower leg with a pair of long bones, (3) the ankle consisting of a complex of small bones, and (4) the foot and toes, each formed by a succession of small bones. The relative proportions of these elements vary greatly among different groups of vertebrates, and evolution commonly has resulted in the elimination of some of the smaller bones. Much more rarely the number has increased (Fig. 445a).

The leg bones of reptiles generally are longer and more slender than those of amphibians, except in such highly specialized creatures as frogs and toads. The characteristic ankle structure also was somewhat simplified (Fig. 433-3,4). Thus the

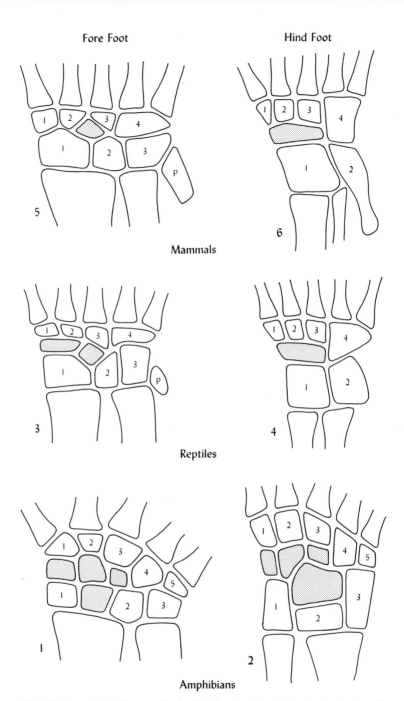

Fore Foot   Hind Foot

Mammals

Reptiles

Amphibians

Fig. 433 Generalized diagrams showing the characteristic wrist and ankle structures in amphibians, reptiles, and mammals. Compare with Fig. 432. These structures are modified and simplified in many ways, and the number of bones is reduced by fusion or elimination in connection with the loss of toes and other specializations. The bone marked P is the pisiform, which first appeared in reptiles. (*After Romer, 1962, "The Vertebrate Body," 3d ed., fig. 140, p. 206.*)

hind feet of reptiles have no more than two bones in the first series, and the forefeet have no more than two in the second series of small bones. There is some evidence that the most primitive amphibians had six or seven toes, but in many of them the number was reduced to four. By contrast the early reptiles had five toes on each foot, although reduction later occurred in several groups. Primitively the number of bones in the toes of reptiles, counting from the big toe outward, was two, three, four, five, and three in the little toe of the forefoot and four in the hind foot. Each toe had an additional metapodial bone at its base that primitively lay within the sole of the foot.

Leg Action

The earliest type of limb extended laterally from the body, and walking was accomplished by a kind of rowing action (Fig. 412), with a limited amount of backward

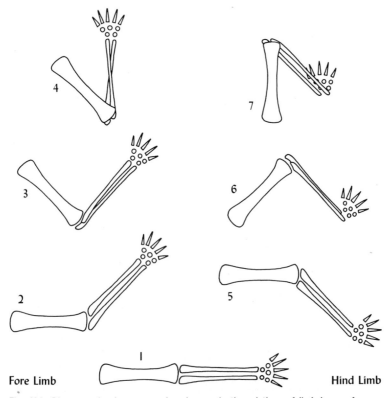

Fore Limb   **Hind Limb**

Fig. 434 Diagrams showing progressive changes in the relations of limb bones, from a primitive spraddle-legged condition to one with the legs drawn in under the body. Notice twisting of the radius and ulna in the foreleg. This contributed greatly to the flexibility of movement in the forefeet or hands of some animals.

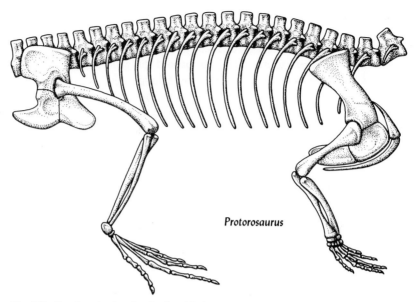

Protorosaurus

Fig. 435 Drawing showing the details of limb structure and attachment in an early Permian reptile. (*After Gregory, 1951, "Evolution Emerging," fig. 12.38A, p. 421.*)

and forward motion in a horizontal plane about a vertical axis at the hip or shoulder articulation of leg with girdle. This was an awkward way of walking that required the expenditure of much muscular strength to raise the body only a little above the ground. Improvement was attained by rapid evolution (Fig. 434) that modified the leg attitude in four ways: (1) The foot turned forward so that the toes became more efficient instruments for accomplishing a forward pull or thrust. (2) The elbow bent backward and the knee forward, and differentiation of the fore- and hind legs began. (3) The lower leg was directed downward. This raised the body above the ground. Its weight was borne partly by the leg bones, and some muscular strain was relieved. (4) The upper leg turned so that it approached parallelism with the body. Thus the legs were drawn more and more under the body, and muscular strain was reduced still further. In this position movement at the hip and shoulder joints took place in an approximately vertical plane about a horizontal axis.

These modifications were accomplished more or less concurrently, although they probably tended to progress most rapidly in the order mentioned. The perfected limbs became efficient spring-like structures, capable of powerful, rapid, and sustained motion (Fig. 435). The forelimb resembles a C-spring and is particularly adapted for exerting a forward pull. The hind limb became a Z-spring, especially adapted for delivering a forward thrust. The legs of most mammals are of these types, and similar legs were possessed by many advanced reptiles. Various degrees of more primitive legs, however, are characteristic of most modern amphibians and reptiles.

Transformation of the legs required some structural adjustments, mainly involving torsion necessary to maintain the forward direction of the feet. This is most conspicuous in the foreleg, where rotation through 180° is required (Fig. 434-4). It is accomplished mainly in the lower leg, where the two long bones permit twisting with little other disruption of structural relations. The hind leg does not present a similar problem.

Scales

The bony scales of fishes were mostly lost by the more advanced amphibians. Typically these animals possess thin, moist, and somewhat slimy skins. Modern toads have a modified kind of thickened skin that reduces the dangers of drying out, and some of the ancient amphibians probably were similarly protected. Few of these creatures, however, could exist for long outside of damp situations or away from places where they could return to the water at frequent intervals.

Development of the reptilian egg was only one factor that permitted some of the descendants of the early amphibians to become completely adapted to the dry environment of the land. The other was the acquisition of an impervious integument that was proof against desiccation. This was accomplished in the reptiles by the evolution of a new kind of scale. Unlike the bony scales of fishes, which are contained within the skin (Fig. 400) and lie beneath the epidermis, reptilian scales are external horny epidermal plates. Some reptiles, like the snakes, periodically shed their old epidermis with its scales, beneath which a new scaly epidermis has developed. In others, like most turtles, the scales are permanent and grow as the individual grows by the addition of new horny layers on the underside and around the margins.

Most reptiles probably had scales of this epidermal kind. A few impressions of scaly reptilian skin have been discovered in the rocks, but the scales themselves have not been found preserved with fossils. Bony dermal scales similar to those of fishes reappeared, however, within the skin of some reptiles, and in several groups these developed into thick and extensive bony armor. Evidently the deeper layers of the skin in some reptiles retained the potentiality to form dermal bone, and it was again produced when it had survival value. This potentiality persisted even in some mammals, e.g., armadillos.

Anapsids

Anapsids are reptiles with completely encasing bony skulls not pierced by openings in the temporal region behind the eye (Fig. 425-1). Because skulls of this type resemble those of the amphibians, anapsids are believed to represent a primitive reptilian stock from which all other members of this great group of animals evolved.

Cotylosaurs

The cotylosaurs (Fig. 437) are a group that undoubtedly consists of two or more

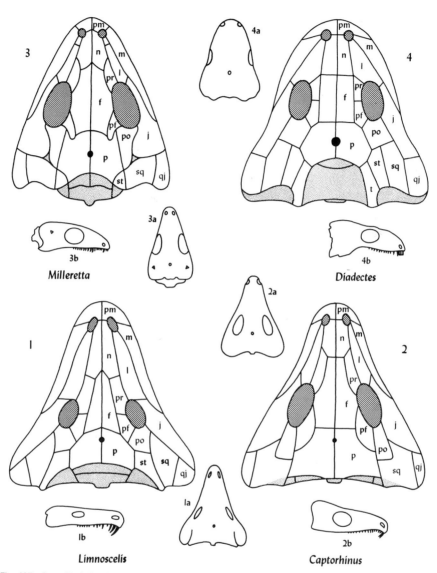

Fig. 437 Anapsid skulls of four cotylosaurs. 1, 2, Typical captorhinids. 3, A captorhinid with tiny openings between the bones in the temporal regions. Perhaps this is the beginning of synapsid structure. 4, A diadectid. (*See Romer, 1956, figs.* 35, 36, 41, 43, 54, 55, *pp.* 69, 70, 85, 89, 104, 105.) See also note following Fig. 408 and Table 409.

significantly different evolutionary lineages. The primitive nature of all of them, however, is attested by their anapsid skulls, amphibian-like shoulder girdles (Fig. 430-2), and generally short thick legs with unmodified feet. One peculiar character that seems to identify most cotylosaurs is the great lateral thickening of the neural arches of their vertebrae. These animals had reached the acme of their development by the beginning of the Permian Period, but very little of their previous history is known. Thereafter they steadily declined in both variety and abundance. Relatively few persisted into the late Permian, and none survived the Triassic Period.

The skulls of captorhinid cotylosaurs (Fig. 437-1 to 3) show no indication of the presence of an eardrum. The diadectid members of this group, however, have an indentation in the skull just above the jaw articulation (Fig. 437-4), where a drum possibly was located, directed either laterally or backward. Some of these beasts attained a length of 10 ft or more. The legs of the larger ones were drawn in beneath their bodies, probably in compensation for their great weight. The teeth of some do not seem to have been adapted to a carnivorous diet, and possibly these animals were the first herbivores to appear among the vertebrates. A few had protective bony dermal plates along their backs.

No other group of reptiles seems to have evolved from any of the known cotylosaurs. These creatures, which are a very important element of the early Permian terrestrial faunas, probably represent the latter part of an early reptilian radiation that failed in competition with the more progressive animals which evolved from them or from common ancestors.

Turtles

The turtles are generally considered to be anapsids, but their ancestry and possible connections with the cotylosaurs are not known. The earliest discovered turtle-like

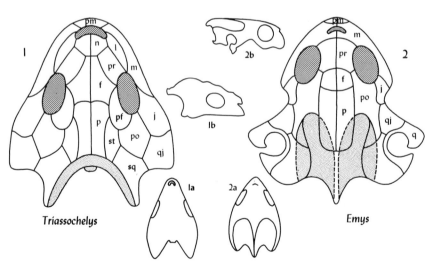

Fig. 438   Two turtle skulls. 1, From the Upper Triassic. 2, From the Paleocene. (See Romer, 1956, figs. 49, 50, 51, pp. 97, 98, 100.)

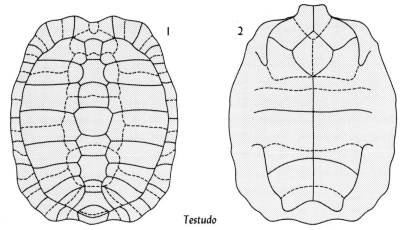

Testudo

Fig. 439   Shell of an early Cretaceous turtle showing its characteristic structure of dermal bones. The pattern of scales covering the shell is indicated by dashed lines. (*After Romer, 1966, fig. 163, p. 113.*)

fossil is from the Middle Permian, but the skull roof was not preserved and its nature is uncertain. Typical turtles date from the Triassic, and remarkably few progressive changes in their general skeletal structure have evolved in more recent time. Some of the oldest turtles, and also some modern ones, have solidly boned skulls without temporal openings (Fig. 438-1), so that they actually are anapsids. Many others, however, are characterized by deep posterior notches (Fig. 438-2) which have been compared to temporal openings except that commonly they are not closed behind by bars of bone. Such a condition has been explained as an accommodation for bulging neck muscles when the head was withdrawn within the shell, but this is far from certain.

The turtles have developed as perfect bony armor as any vertebrates. This is an extraordinary structure, consisting in most of them of solidly joined dermal bones (Fig. 439). Eight vertebrae and their ribs are fused to the inside of the dorsal armor, and this generally is covered by large horny scales that overlap the dermal bones in such a way that they may provide additional structural strength. The ventral dermal bones may have been derived more or less directly from the scales of fishes by way of the abdominal ribs of primitive reptilian ancestors.

Confinement of the turtle's body within a rigid shell seems to have preserved some primitive traits and at the same time induced some structural adaptive changes. For example, the spraddle-legged gait may be an inheritance from the awkward walking of very early land vertebrates. On the other hand, details of the girdle structures reveal modifications not duplicated in any other reptiles. The early turtles were not able to withdraw their heads into their shells. Evolution affecting vertebral articulations progressed, however, in two directions, producing one group with sideways-bending necks and another with necks retracting in a tight vertical S-shaped curve. Other advanced features of the turtles include reduction of some of the skull bones, development of horny beaks in place of teeth, the latter occurring only in the roof of the mouth in a few very early species, and a smaller number of toe bones than is common among reptiles.

Turtles may have evolved originally as fully terrestrial animals. Many of them, however, developed amphibious habits, and some adapted so completely to the aqueous environment that they returned to land only to lay their eggs. Some marine turtles grew to enormous size. Their feet were transformed to flippers supported by greatly lengthened toe bones. Some show much reduction in bony armor that was no longer needed for protection; thus lightened, they evidently became more proficient swimmers.

Synapsids

Synapsids are identified by the presence of a single temporal opening low on each side of the skull (Fig. 425-4). Two main groups require recognition: (1) The pelycosaurs and the therapsids, which developed many mammalian traits. They finally produced the stock from which mammals evolved in the latter part of the Triassic Period. (2) The mesosaurs, an early group of aquatic reptiles, which resembled ichthyosaurs in a general way. Their relations are doubtful, and they probably were not at all closely related to the other group.

Pelycosaurs

The known fossil record of the synapsids starts in the Pennsylvanian and thus is antedated by few other groups of reptiles. During the early Permian the pelycosaurs (Fig. 441-1,2) outnumbered all other terrestrial vertebrates, but in the late Permian they were superseded by the more advanced mammal-like therapsids (Fig. 441-3,4) and did not survive into the Triassic Period. These two groups present an almost complete morphologic transition from captorhinid cotylosaurs to mammals. This involved numerous specializations, but none of these animals departed from the conventional quadripedal form.

The early pelycosaurs had small temporal openings, but in many ways they resembled the cotylosaurs. The vertebrae included small intercentra, but the neural arches were less swollen. The shoulder girdle retained the primitive dermal elements but had added new internal bones (Fig. 430-5). The somewhat more slender legs were not drawn under the body and thus they were old-fashioned. One important advanced character which these animals acquired in varying degrees was differentiation of the teeth and beginning of the development of longer pointed canines (Fig. 441-2b).

The pelycosaurs seem to have included both carnivorous and herbivorous forms. Some species in both groups had enormously elongated neural spines rising

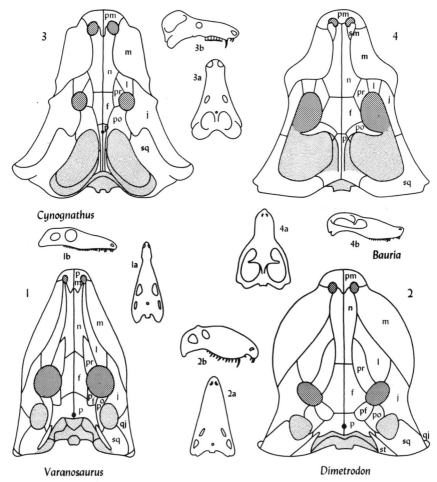

Fig. 441 Four synapsid skulls. 1, 2, Pelycosaurs from the Lower Permian. 3, 4, Therapsids from the Lower Triassic. (*See Romer and Price, 1940, pl. 4, p. 491; Romer, 1956, figs. 92, 93, 96, 102, pp. 175, 176, 180, 189; 1966, fig. 229, p. 283.*)

from the vertebrae of the back (Fig. 442). These probably supported a thin high sail-like expanse of skin whose function has aroused much speculation. The suggestion has been made it was a protective device or served to frighten potential enemies by making the animals appear much bigger than they really were. It has been supposed that this was a secondary sexual character of males, but such an explanation is not likely because sailless forms which might have been females do not occur in association with them. The most interesting idea is that these structures served as partial thermoregulators, absorbing or radiating heat under different circum-

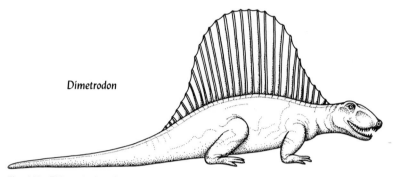

Dimetrodon

Fig. 442   This early Permian pelycosaur grew to a length of 10 ft or more. The function of its peculiar dorsal "sail" has provoked much speculation. (*After Colbert, 1955, "Evolution of the Vertebrates," figure on p. 121.*)

stances. If this were so, it might indicate that a trend toward the regulation of blood temperature had its origin far back among the relatively remote ancestors of the mammals.

Therapsids

Therapsids (Fig. 441-3,4) are believed to have evolved directly from carnivorous pelycosaurs. The detailed phylogenetic relations are not clear, however, and the evolutionary radiation that occurred probably had its origin in more than one branch of this ancestral group. The variable combinations in which distinctive mammalian characters appeared suggests that similarities resulted from parallel and convergent evolution. Possibly several of these evolutionary tendencies were related to the slow development of warm-bloodedness in many of these creatures. Increasing stability of body temperature almost surely would have been advantageous to animals, without regard to other environmental adaptations, in permitting continuous muscular activity under a variety of physical conditions.

Therapsids appeared in the mid-Permian, and they became the most abundant

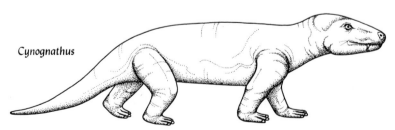

Cynognathus

Fig. 442a   Restoration of a mammal-like early Triassic therapsid that grew to a length of about 7½ ft. (*After Colbert, 1955, "Evolution of the Vertebrates," figure on p. 133.*)

reptiles in the late Permian and early Triassic. Among the mammalian characters that fairly rapidly evolved, the following were particularly important: (1) Enlargement of the temporal opening and its extension upward. In some advanced forms this was no longer separated from the eye socket by a continuous bar of bone (Fig. 441-4). (2) Closure of the pineal opening in the skull roof. (3) Reduction in the number of bones surrounding the eye socket. (4) Development of a double ball-and-socket articulation between skull and neck, and further specialization of the atlas and axis vertebrae. (5) Elimination of intercentra from the vertebrae. (6) Growth of a second bony roof of the mouth below the nasal passages, which opened posteriorly at the beginning of the throat (Fig. 475a). This permitted breathing at the same time that food was being chewed, an ability probably advantageous for warm-blooded animals. (7) Differentiation of the teeth to produce incisors, canines, and cheek teeth with cusps and crushing or grinding surfaces. These teeth became rooted in the jaw bones (Fig. 454-4), not attached to their surfaces, as in more primitive reptiles. Some were resorbed basally and replaced by others which grew up beneath them. (8) Lower jaws were simplified in structure, and the bones by which they articulated with the skull became very small (Fig. 482-5). (9) Limbs were drawn compactly under the body (Fig. 442a). This involved some change in structure of the girdles and shape of the upper leg bones in accordance with modified muscular requirements. (10) Reduction in toe bones to the number characteristic of mammals, represented by the formula 2,3,3,3,3.

Near the close of the Triassic Period most of the therapsids became extinct, apparently overwhelmed by the rapidly expanding and aggressive dinosaurians. Those which survived were small creatures that had evolved far in the direction of becoming mammals. Some rare fossils found in uppermost Triassic strata have such characteristics that it is very difficult to decide on which side of the reptilian-mammalian boundary they should be placed.

Mesosaurs

The mesosaurs were the earliest known reptiles to exhibit obvious aquatic adaptations, such as long laterally compressed tails, large hind feet that probably were webbed and included a few extra toe bones, and a long narrow snout with posteriorly located nostrils and bearing many sharp slender teeth like those of some other fish-eating animals (Fig. 444). They inhabited bodies of freshwater during Pennsylvanian or early Permian time. These creatures had vertebrae with swollen neural arches like those of the cotylosaurs, from which they probably were descended. The skull is not certainly pierced by a temporal opening similar to that of the pelycosaurs, and other resemblances to that group are common primitive reptilian characters. No close relationship is indicated.

The suggestion has been made that mesosaurs were ancestral to the ichthyosaurs, but this does not seem probable. Mesosaurs are interesting chiefly because they provide additional evidence of the very early evolutionary radiation of the most primitive reptiles.

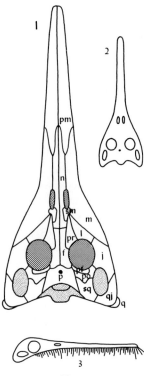

3

Mesosaurus

Fig. 444   Structure of a mesosaur skull from the Lower Permian.
(*See Romer, 1956, fig. 84, p. 159.*)

Parapsids

Two groups of reptiles are characterized by single temporal openings high on each side or in the roof of the skull (Fig. 425-3). The openings differ slightly in these groups with respect to the bones which bound them. Ichthyosaurs are the only typical parapsids. The other group, the euryapsids (Fig. 425-2), differs in so many ways that its members are believed to have evolved quite independently.

Ichthyosaurs

Ichthyosaurs were the most highly specialized of all marine reptiles. Although they probably evolved from cotylosaurs, their immediate terrestrial ancestors are unknown. These animals appear suddenly in the fossil record in strata of mid-Triassic age and seem to have become extinct well before the end of the Cretaceous Period.

Adaptive evolution of the ichthyosaurs produced a remarkably fish-like streamlined form (Fig. 445). The jaws were drawn out into a long thin snout, extending far in front of the nostrils (Fig. 446-1). The limbs were transformed to paddles that probably were used chiefly for steering. A dorsal fin, unsupported by bones, rose

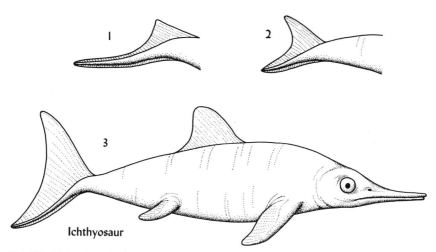

Fig. 445   Restoration of a Jurassic ichthyosaur (3). The two tails shown above it (1, 2) are less advanced and belonged to animals that lived earlier. (*After Romer, 1966, fig. 172, p. 118; Swinnerton, 1947, "Outlines of Palaeontology," 3d ed., fig. 306, p. 322.*)

above the back. A tail fin developed similar to that of sharks except that the vertebrae turned downward in its lower lobe (Fig. 445-3). The limb girdles, which were not required to function as supporting structures, were relatively small and weak, and the hip girdle was not attached to ribs or vertebrae. The limb bones were shortened. Bones of the feet were reduced to hexagonal or disk-like elements arranged in toe-like series (Fig. 445a-1). Their number increased both by multiplication in the individual series and by increase in the number of these series. The

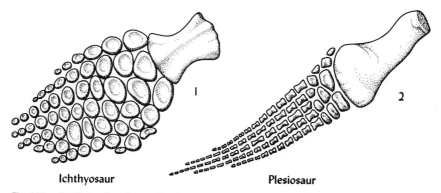

Fig. 445a   Forelimbs of marine reptiles that were converted to flippers. They contain many more bones than are present in the feet of more ordinary reptiles, and the leg shanks are relatively very short. (*After Romer, 1962, "The Vertebrate Body," 3d ed., fig. 129B, p. 211; 1966, fig. 147B, p. 187.*)

ichthyosaurs were so completely fish-like that they could not have left the water. Evidence has been discovered showing that they probably bore their young alive.

All ichthyosaurs were much alike. Progressive evolutionary trends from earlier to later forms are evident, however, in lengthening of the snout, increasing complexity of the bone structure in their paddles, and sharper downward bending of the vertebrae within the tail fin (Fig. 445).

Euryapsids

Evolutionary relationships of the several groups of reptiles with the euryapsid type of temporal opening (Fig. 425-2) are not clear. Two major groups can be distinguished: (1) protorosaurs, which were mostly small terrestrial animals, and (2) sauropterygians, which include large marine reptiles.

Protorosaurs

The creatures grouped as protorosaurs were a varied lot, and little can be determined about their evolution. They are sparingly represented in faunas ranging from the early Permian through the Triassic. Most of them were lizard-like. They retained some primitive characters suggesting cotylosauran ancestry, but also acquired various advanced features, some of which were quite peculiar.

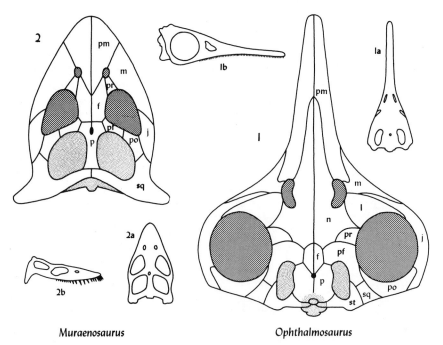

Muraenosaurus   *Ophthalmosaurus*

Fig. 446   1, A parapsid Jurassic ichthyosaur. 2, A euryapsid late Jurassic plesiosaur. (See *Romer*, 1956, fig. 85, p. 161; 1945, fig. 157, p. 195.)

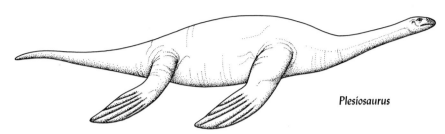

Plesiosaurus

Fig. 447   Restoration of an early Jurassic plesiosaur. (*After Fenton and Fenton*, 1958, *"The Fossil Book,"* figure on p. 321.)

Sauropterygians

Sauropterygians ranged through the entire Mesozoic Era. Possibly they evolved from protorosauran stock during the Permian Period. All were marine. Although they show increasing adaptations to aquatic life, no clear evolutionary sequence can be distinguished. Most of the more primitive known Triassic forms were specialized in such ways that they could not have been ancestral to the Jurassic and Cretaceous plesiosaurs. Some fossils of intermediate types, however, have been found.

Most sauropterygians had long flexible necks and relatively small heads, with nostrils located more or less posteriorly (Fig. 446-2). The teeth of some of the Triassic forms were large and of the pavement type, suitable for crushing shellfish, accompanied perhaps by anterior horny beaks. A few creatures of this kind bore dermal armor somewhat similar to a turtle's shell, except that the bones were much smaller and more numerous. The legs of Triassic species were little modified, but it is doubtful that they could walk easily on land. The shoulder girdles show a progressive tendency for reduction of the upper bones and enlargement of the lower ones. The hip girdle was not strongly attached to the axial skeleton, as in fully terrestrial creatures.

The best-known and most specialized sauropterygians were the post-Triassic plesiosaurs (Fig. 447). Their jaws were armed with sharp teeth set in sockets. Their feet were transformed into powerful swimming paddles, and these animals, like sea turtles, probably never left the water except possibly to lay their eggs. The bones of the feet, however, were not modified as greatly as in the ichthyosaurs (Fig. 445a-2). The toes remained as distinct series of somewhat elongated bones. The number of these bones is greater than in terrestrial reptiles, but the number of toe series was not increased.

Diapsids

The diapsid reptiles have two temporal openings on each side corresponding to those which occur separately in the synapsids and euryapsids (Fig. 425-5). The phylogenetic relations of the diapsids to these other groups, however, are not clear. They might have evolved by the appearance of a second temporal opening in a

Table 448 Some important characters of the main diapsid reptile groups

---

Lepidosaurs: relatively primitive diapsid reptiles

  Eosuchians: posterior part of skull with some bones that are lost in more advanced diapsids. Teeth in sockets. Permian to Eocene

  Rhynchocephalians: generally similar to eosuchians but with teeth fused to edge of jaw bones and upper jaw with a small overhanging beak. Triassic to Recent

  Lacertilians: lizards. Bar of bone missing below temporal opening. Triassic to Recent

  Ophidians: snakes, derived from lizards. Limbs almost or completely lost. Bar of bone below upper temporal opening also lost. Cretaceous to Recent

Archosaurs: relatively advanced diapsid reptiles

  Thecodonts: mostly small, active, lizard-like carnivores derived from eosuchians. Teeth in sockets. An opening in skull bones present in front of eye socket. Pineal opening in skull roof closed. No palatal teeth. Beginning of bipedalism, with hind legs lengthened. Many with dermal bones down back. Permian, the source of other archosaurs

  Phytosaurs: crocodile-like reptiles with nostrils far back between eyes in elongated skull. Triassic

  Crocodilians: crocodiles, alligators. Skull rather primitive, elongated. Nostrils at anterior tip, internal openings far back in throat. Dermal armor down back below scales. Triassic to Recent

  Dinosaurs: an extremely varied group, not all closely related to one another. Many bipedal. Others returned to four-legged walking but forelegs shorter than hind legs. Triassic to Cretaceous

  Pterosaurs: flying reptiles. Wings consisted of skin stretched along and behind a single much-lengthened fifth finger. Jurassic to Cretaceous

---

These were the most numerous and dominant land vertebrates of the Mesozoic Era. See Fig. 426.

stock derived from either of the other groups, or they might have arisen entirely independently from the cotylosaurs. The diapsids generally are considered to have radiated from a single closely related primitive group of reptiles, but this is speculative and the ancestral stock has not been identified.

Diapsids are present in the late Permian faunas, but a single Pennsylvanian species, if its doubtful structure has been properly interpreted, may be evidence of much earlier origin. The remarkable evolutionary radiation of the diapsids was in progress early in the Triassic, and before the end of that period these animals had attained a dominant position in all faunas of the land. Most Mesozoic reptiles except aquatic ones were diapsids. Near the end of the Cretaceous Period, however, many of these creatures became extinct. With only one exception, all that have survived to the present time are lizards, snakes, and crocodilians.

Eosuchians

The most typical eosuchians, or earliest diapsids, were small lizard-like reptiles with sharp teeth, borne in sockets, and skulls whose posterior bones suggest a primitive condition (Fig. 449-1). Also included as eosuchians is a variety of creatures that lack the identifying characters of the other diapsid groups. Among them is a species without a complete bony bar beneath the lower temporal opening. This is suggestive of the structure seen in lizards. Several of the later species were aquatic, and one superficially resembled a small crocodile. The youngest known fossils classed with the eosuchians are of early Tertiary age.

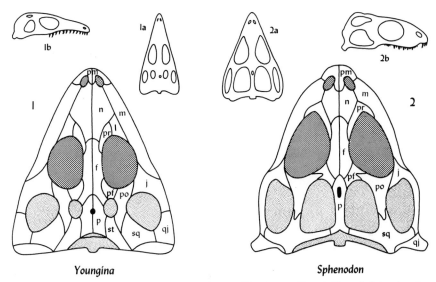

Fig. 449   Two relatively primitive diapsid skulls. 1, An early Triassic eosuchian. 2, The only living rhyncho-cephalian. (*See Romer*, 1956, *figs*. 56, 60, 62, *pp*. 106, 112, 114.)

Rhynchocephalians

The rhynchocephalians (Fig. 449-2) also were primitive diapsids. They differed from eosuchians in having teeth fused to the edges of the jaw bones, rather than con-tained in sockets. The anterior part of the upper jaw commonly dipped down in a kind of beak, which in some was encased in a horny sheath. These somewhat aberrant reptiles probably evolved from eosuchians. Known fossils are of Triassic and Jurassic ages, but one species still is living in New Zealand.

Lizards

Lizards are the most diverse and numerous modern reptiles. They evolved from eosuchian ancestors and were typically developed late in the Triassic Period. Their teeth are fused to the edges or inner sides of the jaw bones. Their skulls lack some of the bones present in primitive reptiles. More importantly, however, the lower temporal opening is not closed below by a bar of bone (Fig. 450-1). Freed in this way, the bone articulating with the lower jaw is somewhat movable. Still greater flexibility in the jaws results from a ligamental joint midway in the side of the lower jaw that developed in one group.

The legs of some lizards were reduced in size and eventually disappeared as external structures, thus producing creatures resembling snakes. Two groups of early Cretaceous lizards adapted to an aqueous environment. From one of these evolved the mosasaurs, or great marine lizards of the late Cretaceous. They had

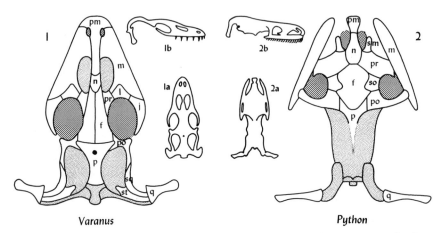

Fig. 450 Two highly specialized diapsid skulls. 1, A late Tertiary lizard. 2, A modern snake. (*See Gregory,* 1951, *"Evolution Emerging," figures on pp. 422, 424; Romer, 1956, fig. 67, p. 125.*)

long, laterally flattened tails that were powerful swimming organs. Their limbs became paddles. The girdles were reduced in size and strength. Extra bones appeared in their toes, but the number of toes was not increased. Certain lizards reacquired bony dermal plates that underlay some of their surface scales.

Snakes

Snakes evolved directly from one group of lizards which perhaps were burrowers and some of which may have been blind. Their record begins in the Cretaceous Period, but they are poorly known as fossils. Snakes were the last important group of reptiles to become differentiated.

Most snakes have lost all traces of limbs and girdles, although internal vestiges, some barely indicated externally, remain in a few primitive forms. Their vertebrae have multiplied to numbers that may reach several hundreds, and ribs are movably articulated with them. The jaw structure of the lizards was modified in the direction of still greater flexibility. Snakes have lost the second bony bar that separated the two temporal openings (Fig. 450-2). The skull bone articulating with the lower jaw thus became freely movable, and these animals are described as having double-jointed jaws. Moreover, lateral halves of the lower jaws are joined by elastic ligaments, so that the mouth opening can be much enlarged. This permits snakes to swallow prey whose size exceeds the normal diameter of their bodies. The teeth generally are sharp recurved hooks, and by alternate movements of the sides of the lower jaws food is gradually forced backward into the throat where it can be gulped down.

Snakes have adapted to many environments. Water snakes swim by sinuous movements similar to those by which they crawl along the ground. A few have developed somewhat flattened tails or bodies that increase the efficiency of these movements. Some are nearly helpless when out of water. The earliest poisonous snakes seem to have appeared in the mid-Tertiary.

Thecodonts

All the remaining groups of reptiles seem to be closely related in their origins. Some of their similar features probably resulted from parallel trends in evolution arising from the potentiality that existed in a common ancestral stock. The animals which constitute the base for this reptilian radiation, as well as others exhibiting a comparable stage of evolutionary development, are termed thecodonts (Fig. 451). Their known fossils are all of Triassic age.

One of the principal features of the thecodonts and their descendants was a tendency to rise up and walk on their hind legs. This was almost unique among the reptiles, although it was not shared equally by all thecodonts and reversionary tendencies toward four-footed walking were characteristic of several descendant groups. Bipedalism commonly involved important modifications in the hip girdles and legs that can be recognized in most of the thecodonts and their descendants. These modifications include the following: (1) Each half of the hip girdle, which was somewhat plate-like in other reptiles, became triradiate, with the lower bones ex-

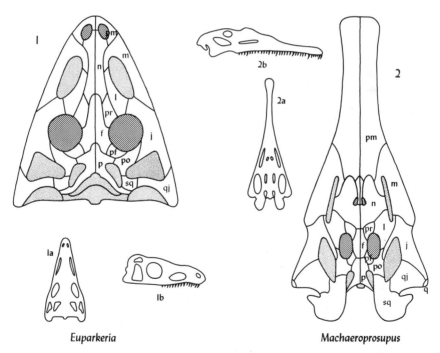

Euparkeria   *Machaeroprosupus*

Fig. 451   Two thecodont skulls. 1, A relatively primitive type from the Lower Jurassic. 2, A highly specialized late Triassic form adapted to aquatic life. (See *Romer*, 1956, *figs.* 69, 70, 74, *pp.* 131, 132, 138.)

tending forward, backward, and more or less downward. (2) The girdle became fused with one or more ribs in addition to the original two, thus producing a stronger structure. (3) The socket that received the head of the leg bone moved up nearer to the spinal column, and the leg bone pushed up into it. (4) The upper leg bone was directed forward and downward, rather than outward, and developed a head on the inner side of its upper end. (5) The ankle bones were reduced in number, some being fused with the leg bones above, others being fused below with the metapodials or first bones of the toe series. (6) The metapodials lengthened (Fig. 452-1). This and the foregoing modification were related to the tendency for these animals to raise themselves upon their toes for swifter running. Thus a leg was formed with the beginning of another sturdy joint in addition to that provided by the knee. (7) The toes were more strongly developed but retained the normal number of bones. The third toe became longest, the fifth commonly was lost, and the first may also have disappeared or was turned backward. (8) The forelegs became smaller and weaker. (9) Toes were eliminated from the forefeet in order from the outside inward.

By themselves, many of these modifications might be interpreted simply as similar and necessary responses induced by the development of bipedalism and thus

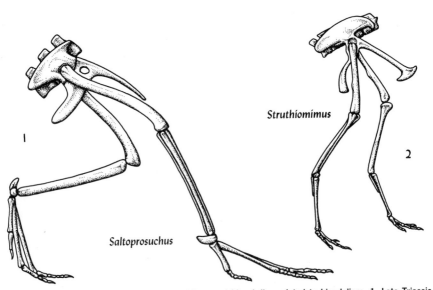

I

Struthiomimus

2

Saltoprosuchus

Fig. 452 Drawings illustrating structures of legs and hip girdles related to bipedalism. 1, Late Triassic thecodont with metapodials lengthened but not raised much above the ground. 2, Late Cretaceous ostrich-like theropod with metapodials lengthened, reduced in number, and constituting a well-marked third segment of the leg. (*After Romer, 1945, "Vertebrate Paleontology," 2d ed., figs. 178, 191, pp. 215, 235.*)

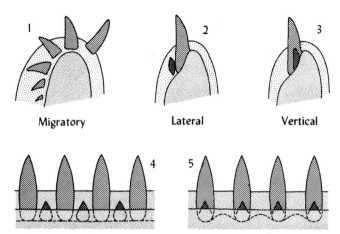

Fig. 453   1, The teeth of sharks grow in the skin inside the mouth and move outward to replace each other as the older ones are worn and drop out. These are believed to be the most primitive kind of teeth. 2, 3, Teeth of some other fishes, amphibians, and reptiles rest against the jaw bone and are replaced by new ones that grow behind or between them. 4, In many amphibians and reptiles the teeth are set in grooves in the jaw bone. 5, Pockets form in the grooves, and new teeth grow beneath the old ones. The pockets deepen to form sockets which contain rooted teeth, as in many advanced reptiles and all mammals.

be without significance in indicating close relationship. They are accompanied, however, by several unifying characters of the skull that seem to eliminate any such conclusion. These include the following: (1) Some skull bones disappeared, and others were arranged in a way somewhat different from that which is characteristic of most other reptiles. (2) The pineal opening was closed. (3) An opening generally was present between the eye socket and nostril (Fig. 451). (4) Teeth were rooted in separate sockets of the jaw bones (Fig. 454-4). (5) Teeth borne on the roofing bones of the mouth, common in many other reptiles, were absent except in a few primitive forms. (6) An opening generally occurred laterally at the junction of the bones which constitute the lower jaw.

The thecodonts probably evolved from eosuchians. The more generalized representatives were mostly small, active, carnivorous lizard-like animals. Their skulls had the characters that have been mentioned. The skeletons of most kinds show evidence of the beginning of bipedalism, but the modifications incidental to this mode of walking were not fully developed. Almost all had small dermal bones extending down their backs. Animals of this kind probably were ancestral to all the groups of reptiles that remain to be considered.

A variety of thecodonts was specialized in several different ways. Some developed a complete covering of heavy dorsal armor. Others returned to the water

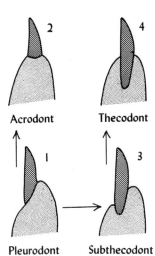

Fig. 454 Evolutionary progression in the types of tooth implantation is not clear. Either or both the pleurodont and subthecodont condition may be primitive in four-footed vertebrates. Rather clearly, however, the acrodont and thecodont conditions were derived from these.

and acquired a form very similar to modern crocodiles. Their skulls were different, however, as the nostrils had migrated far back and opened nearly between the eyes (Fig. 451-2). These animals had hind legs longer than their forelegs and evidently had returned to walking on all fours.

Crocodilians

Crocodiles and alligators are the largest and among the most aggressive modern reptiles. Their fossil record begins at about the opening of the Jurassic Period, but their development in most ways was conservative, and little evolutionary advancement is evident since Cretaceous time. All have been quadrupeds, but the large hind limbs, most obvious in Jurassic species, may indicate ancestors capable of walking on two legs. Ordinarily the legs of these creatures are rather sprawling, but when they wish to run rapidly these are drawn in beneath the body, which is thereby raised well above the ground.

These animals are characterized by certain peculiarities of their feet and hip girdles. The latter are attached to only two pairs of ribs, in a primitive way. The upper temporal opening in their flattened skulls generally grew smaller, and there was a tendency for the opening before the eye to close (Fig. 455-2). Bony plates of variable development occurred beneath the scales along the back. Most diagnostic was the progressive evolutionary development of a second bony roof of the mouth, very similar to that of mammals. The nostrils were situated at the end of the long snout, but nasal passages separated from the mouth cavity extended further and further backward until the inner openings attained a position well back within the throat.

The earliest crocodilians may have been terrestrial creatures, but the later ones all adapted to life in water. The upper temporal opening was large in some of the

Mesozoic forms (Fig. 455-1). From them evolved the specialized marine crocodiles of the late Jurassic. These beasts had no dermal bones that served as armor, their feet were converted into paddles, and a somewhat fish-like tail developed with the vertebrae turned downward in the lower lobe, as in the earlier ichthyosaurs. The largest of all crocodiles, reaching 50 ft in length, lived during late Cretaceous time. They were of modern form. An aberrant group of the early Tertiary had high narrow skulls and relatively short internal nasal passages.

Dinosaurs

Dinosaurs are famous because they include the most spectacular animals of the Mesozoic Era. They evolved from thecodonts but belong to two great groups that are not directly related to each other (Fig. 456 and Table 456). The key to their relations is believed to be revealed by the structures of their hip girdles. One group, the so-called lizard-hipped dinosaurs, had a girdle whose pubic bone extended forward and downward, as in thecodonts (Figs. 431-8; 452-2). This group evolved to produce first the theropods and then the sauropods. The other group, known as the bird-hipped dinosaurs because the structure of the girdle resembled that of birds, was

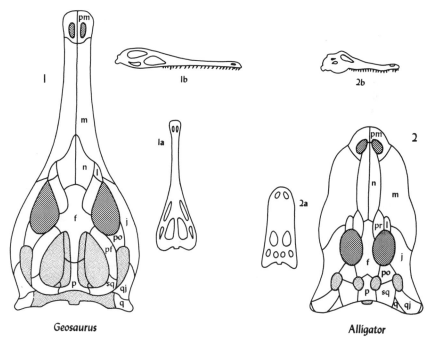

Geosaurus   Alligator

Fig. 455 Two crocodilians. 1, A late Jurassic or early Cretaceous crocodile. 2, An alligator, Oligocene to Recent. (*See Romer, 1956, fig. 76, p. 141; Gregory, 1951, "Evolution Emerging," figs. 14.3E, 14.6C, pp. 470, 472.*)

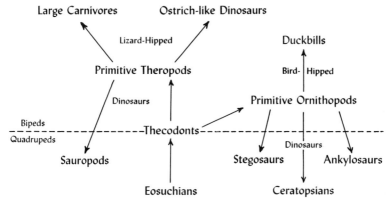

Fig. 456 Diagram showing the probable evolutionary relations of the dinosaurs. The tendency toward bipedalism so evident in the thecodonts was inherited and perfected to various degrees in many of their descendants. Later several groups of dinosaurs reverted independently to the four-legged type of walking.

characterized by a pubic bone which extended forward but also was drawn out posteriorly in contact with the other lower girdle bone (Fig. 431-11). Evolutionary radiation in this group resulted in the differentiation of ornithopods, stegosaurians, and the armored and horned dinosaurs.

Theropod Dinosaurs

The theropods include all the carnivorous dinosaurs. They were the animals that

Table 456   Conspicuous Differentiating Characters of the Main Groups of Dinosaurs

---

Saurischians: lizard-hipped dinosaurs; pubic bone of hip girdle directed downward

   Theropods: bipedal forms, mainly carnivorous. Skull much like that of thecodonts. Hip girdle perforated for reception of head of femur. Long hind legs, feet bird-like. Forelegs short. Shoulder girdle much reduced. Triassic to Cretaceous

   Sauropods: derived from theropods, mainly herbivores. Heads relatively very small. Most of these animals returned to four-legged walking; forelegs shorter than hind legs. Some beasts of enormous size. Jurassic to Cretaceous

Ornithischians: bird-hipped dinosaurs; pubic bone directed forward

   Ornithopods: herbivores, some amphibious. Bipedal or reverted to four-legged walking; forelegs shortened. Toes with blunt claws or hoof-like coverings. Teeth commonly missing from anterior part of jaws. Some animals seem to have had horny beaks. Triassic to Cretaceous

   Stegosaurs: quadrupeds but forelegs shortened. Back protected by large upstanding alternating dermal bones. Toes with small hoofs. Jurassic to Cretaceous

   Ankylosaurs: body broad, flattened, entirely covered by mosaic of small dermal bony plates. Temporal openings closed. Some animals toothless. Cretaceous

   Ceratopsians: quadrupeds; forelegs shortened. Heads large, skull with wide frill extending over neck and shoulders. Large horns commonly developed in various positions. With powerful beak. Toes with small hoofs. Cretaceous

---

The two large dinosaur groups probably evolved from different thecodont ancestors.

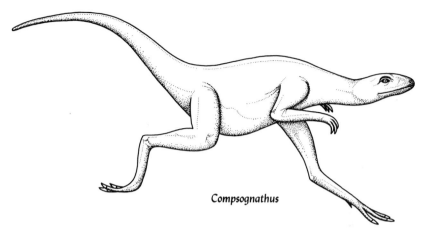

Compsognathus

Fig. 457   Restoration of a relatively primitive late Jurassic theropod. (*After Heilmann, 1927, fig. 119, p. 168.*)

supplanted most of the mammal-like reptiles of the Triassic Period. Primitive theropods of late Triassic age were mostly small active creatures (Fig. 457) with relatively small, lightly constructed skulls. All were bipeds. The hip girdle generally was attached to four pairs of ribs. Almost the only character which distinguishes these dinosaurs from thecodonts is a perforation in the socket of this girdle, into which the head of the thigh bone fits. Dermal bones had disappeared from the shoulder girdle, and the forelegs were relatively short. Some of these animals had hollow bones similar to those of birds, and the structure of their hind legs was very bird-like. Reduction in the number of toes on all four feet had begun, and all the toes were armed with sharp horny claws.

The results of evolutionary radiation are evident among the late Triassic theropods, and almost no small generalized species have been found in Cretaceous strata. A group of ostrich-like forms developed with long hind legs (Fig. 452-2), long necks, and small skulls whose teeth had been replaced by horny beaks. The feet were all three-toed, and the three metapodials of the hind feet were long and closely grouped and had become important elements of the lengthened legs.

Large carnivorous dinosaurs already existed in late Triassic time and continued through the Jurassic. They were heavily built beasts, whose relatively large heads had powerful jaws and long cutting teeth (Fig. 458-1). The metapodials of their hind feet were comparatively short for advanced bipeds. In one form these had fused into a single bone. Such animals, known mostly from fragmentary skeletons, may or may not have been ancestral to the great flesh eaters of the Cretaceous. The latter were the largest and most ferocious creatures of their kind that ever lived, some reaching a length of 50 ft and standing 20 ft tall (Fig. 458a). The forelegs of

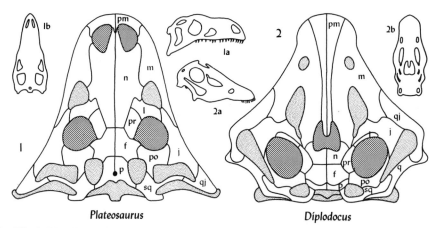

Plateosaurus   *Diplodocus*

Fig. 458 1, A late Triassic theropod dinosaur, probably herbivorous and of the type that was ancestral to the sauropods. 2, A late Jurassic sauropod dinosaur whose posteriorly located nasal openings (darker shading) indicate that it probably was a water dweller. This was a huge beast, growing to a length of 80 ft or more. (*See Romer, 1956, figs. 70, 78, pp. 132, 146.*)

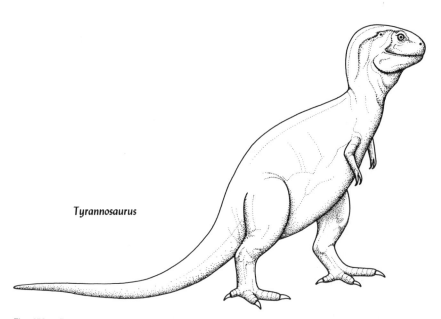

Tyrannosaurus

Fig. 458a Restoration of one of the huge carnivorous dinosaurs of the late Cretaceous. This beast stood nearly 20 ft tall. (*After Colbert, 1945, "The Dinosaur Book," figure on p. 69.*)

most of them were so small that it is difficult to understand how they could have had any usefulness.

Another evolutionary line, probably leading toward the herbivorous sauropods, also was in existence in late Triassic time. It was represented by rather clumsily built creatures with small heads and somewhat blunted teeth. They seem to have been reverting to walking on all fours. The structure of their legs and feet was somewhat primitive, and the forelegs were not so greatly reduced in length.

Sauropod Dinosaurs

The sauropods were herbivorous quadrupeds of the Jurassic and Cretaceous periods. They evolved from carnivorous bipedal theropods and provide one of several clear examples among the vertebrates of herbivores derived from carnivores. Evolution is not known ever to have trended in the reverse direction, at least not among land vertebrates.

These animals were large, and some were huge, up to nearly 90 ft in length and perhaps almost 50 tons in weight (Fig. 459), being exceeded among all animals only by the biggest whales. Their long necks mostly ended in remarkably tiny heads, with relatively weak jaws and small peg-like or laterally flattened teeth in continuous series around their mouths. The legs and girdles were very massive, as was necessary to support great weight, and the toes were short. The shorter forelegs of most forms and the claws on toes, which were reduced in number, seem to have been inherited from theropod ancestors.

The great size and weight of these animals, their small heads and apparently inefficient dentition pose problems concerning the sauropods' way of life. The most likely conclusion is that they ate succulent vegetation growing in or near swamps, ponds, lakes, and possibly shallow embayments of the sea. Probably they spent much of their lives more or less submerged in water, where their bodies were

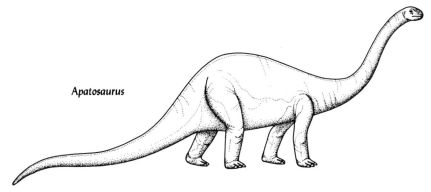

Apatosaurus

Fig. 459   Restoration of a large late Jurassic sauropod dinosaur that grew to a length of more than 60 ft. (*After Fenton and Fenton, 1958, "The Fossil Book," figure on p. 335.*)

partially buoyed up. Location of the nostrils of some forms high up on the skull (Fig. 458-2) seems to confirm this interpretation, because this permitted breathing when only part of the head was raised above the water surface.

Ornithopod Dinosaurs

The bird-hipped dinosaurs differ from the other group not only in the structure of the pelvic girdle (Fig. 431-10) but also in the absence of front teeth (Fig. 460-1) except in the upper jaws of some primitive forms. This lack of teeth generally was compensated by the presence of well-developed horny beaks, the lower one of which was partly supported by a bone not found in the jaws of any other reptiles. The numerous teeth of some forms, located more posteriorly, generally were small, laterally flattened, and closely spaced, so that several successional rows functioned

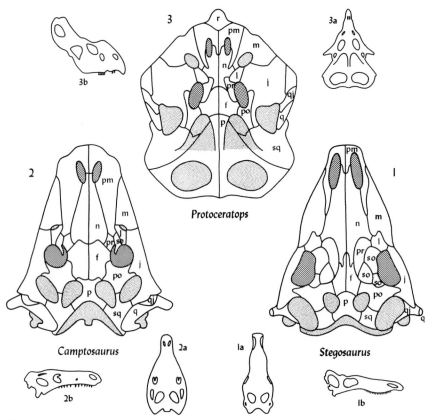

Fig. 460   Skulls of three bird-hipped dinosaurs. 1, A relatively primitive late Jurassic or early Cretaceous ornithopod. 2, A mid-Jurassic stegasaurid. 3, A late Cretaceous horned dinosaur. (See Romer, 1956, figs. 77, 79, 81, 82, pp. 143, 148, 153, 155.)

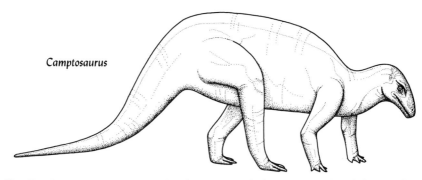

Camptosaurus

Fig. 461 Restoration of a primitive late Jurassic or early Cretaceous ornithopod dinosaur that attained a length of more than 15 ft. (*After Heilmann, 1927, fig. 108, p. 151.*)

together and provided grinding surfaces. These dinosaurs had hind legs not so well adapted to fast running as the theropods, and the front legs were not so much reduced in size. Many of the more specialized forms reverted to walking on all fours (Fig. 461).

No intermediate forms connecting the bird-hipped dinosaurs with their probable thecodont ancestors are known. The ornithopods were the most primitive members of this group. They probably walked as quadrupeds when they were feeding or moving about slowly, but rose up on their longer hind legs when it was necessary for them to run. A few had bony dermal plates extending along their backs. Ornithopods specialized in several different ways. Some possessed structures suggesting that they were arboreal. A few had strong pointed beaks like those of the horned dinosaurs of the late Cretaceous. Still others had skulls grotesquely thickened above by a large mass of solid bone.

Some of the best-known ornithopods are the so-called duck-billed dinosaurs of the late Cretaceous (Fig. 462). Their bodies were built much like those of their earlier relatives. The feet of some are known to have been webbed, and three toes on each foot ended in small hoofs instead of claws. The skulls of these animals exhibit several curious variations. Teeth were very numerous posteriorly, but the toothless anterior parts of the jaws were broad and flat and probably covered by a horny bill. The nostrils generally were located far back in the facial region, and several kinds of crests occurred containing hollows connected with the nasal passages. Evidently these creatures had acquired amphibious habits.

Stegosauran Dinosaurs

Evolutionary radiation of the ornithopods must have begun well back in the Triassic Period, because stegosauran fossils occur in Lower Jurassic strata. These animals were ponderous, probably slow-moving quadrupeds with long hind legs, short forelegs, and small heads (Fig. 460-2). Their principal peculiarity was the development of dermal plates and spines rising conspicuously above the spinal column in a sort

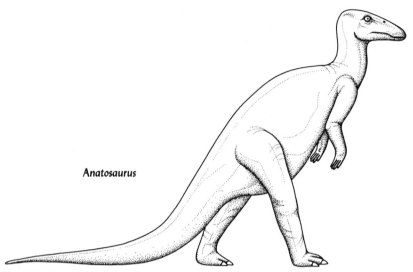

Anatosaurus

Fig. 462   Restoration of a late Cretaceous duck-billed dinosaur that grew to a height of about 15 ft. (*After Colbert, 1945, "The Dinosaur Book," figure on p. 78.*)

of serrated crest all the way from head to tip of tail. These were derived by modification of dermal bones similar to those of thecodonts and seem to have served a protective function. The stegosaurans increased in size as time passed during the Jurassic Period, and the dorsal structures seem to have become progressively more accentuated. This conforms to a common evolutionary tendency that is evident in a great many different groups of animals.

Armored Dinosaurs

The ankylosaurs were Cretaceous dinosaurs whose immediate antecedents are not known. They were broad, squat quadrupeds, more or less completely covered by dermal armor consisting of a mosaic of many plates. The tail was converted into a club-like organ bearing knobs or spikes of bone. The more primitive forms had typical diapsid skulls, but the temporal openings were closed in the later larger species. These creatures also developed some peculiarities of the hip girdle that partly obscure their probable relations to the ornithopods.

Horned Dinosaurs

The ceratopsians seem to have been the last group of dinosaurs to evolve. They were late Cretaceous quadrupeds, probably derived from certain specialized ornithopods. These animals had very large peculiar skulls, with a broad posterior plate-like extension covering the neck (Fig. 460-3). This probably developed originally as an area for the attachment of powerful neck and jaw muscles, but it surely acquired a secondary protective function. The jaws were tipped with pointed beaks.

Fig. 463   Restoration of a late Cretaceous horned dinosaur. Its skull is more than 5 ft long. (*After Fenton and Fenton, 1958, "The Fossil Book," figure on p. 358.*)

Another remarkable feature of these dinosaurs was a tendency for the development of horns indicated by bony spikes rising from the skull (Fig. 463). The most primitive species was hornless. In later forms a median horn grew out above the nostrils, a pair of horns projected above the eyes, and several spines or horns appeared along the posterior margin of the skull. These horns occur variously developed and in different combinations. In general they seem to have increased in prominence in successively evolving species. The famous dinosaur eggs found in central Asia belonged to creatures of this kind.

Flying Reptiles

The pterosaurs (Fig. 463a) constitute one of the most interesting and highly specialized groups of Jurassic and Cretaceous reptiles. They were small animals, some with

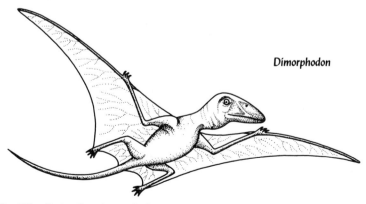

Fig. 463a   Restoration of an early Jurassic pterosaur. (*After Fenton and Fenton, 1958, "The Fossil Book," figure on p. 364.*)

Pterodactylus

Fig. 464 Wing structure of a late Jurassic or early Cretaceous pterodactyl, showing the greatly elongated metapodial and other bones of the outer finger. (*After Romer, 1966, fig. 225, p. 146.*)

long flexible necks, slender pointed skulls, peculiar hip girdles and weak hind legs, a shoulder girdle strengthened by attachment to the vertebrae, and forelegs greatly modified as wing bearers. The fourth finger of the hand was enormously elongated (Fig. 464) and served to stretch a thin membrane of skin outward from the body. Impressions of the skin in very fine-grained rocks show that these animals possessed neither scales nor feathers. Some, however, reveal a suggestion of hair-like outgrowths. There is a possibility that pterosaurs became warm-blooded, a condition advantageous for animals required to expend considerable muscular energy continuously, as in flight. Unlike most reptiles, they probably ceased to grow when they reached maturity. This is indicated by fusion of the skull bones, whose sutures are obscure in most specimens (Fig. 465-1). If so, pterosaurs were similar to birds in this respect.

These creatures probably evolved from a line of small unknown thecodonts that had adapted to arboreal habits. Flight is believed to have originated with climbing animals which had acquired structures permitting them to glide from tree to tree. With increasing facility for flight, the skeleton of these reptiles was modified similarly to that of birds in several ways. For example, the optical centers of the brain were much enlarged, a prominent keeled breastbone developed to accommodate the powerful flying muscles, the vertebrae above the shoulders became fused to produce a more rigid structure, and many of the bones were hollow and presumably filled with air. This is an excellent example of evolution following parallel or convergent trends in relation to similar functional requirements.

The most primitive pterosaurs may have been able to squat on the ground on all fours. It does not seem possible, however, that they or any of their successors could have taken off in flight from a level surface. The leg structure of later forms suggests that these creatures could not walk or hop in any ordinary way. Perhaps the legs were equally useless for perching. Probably pterosaurs clung to trees or cliffs by the short, clawed fingers of their forelimbs or upside down by the hind feet like bats. When they let go, they spread their wings and sailed away. Flight probably was accomplished more by gliding with the aid of air currents than by wing flapping like most birds. How the young developed and took to the air is a complete mystery.

The most primitive pterosaurs had long slender reptilian tails, ending, in one species at least, in a small expanse of membrane probably serving as a rudder, short bones at the bases of the fingers of the wing-bearing limbs, and sharp teeth. The more advanced Jurassic pterodactyls had very short stubby tails and progressively lengthening basal finger bones. Teeth disappeared in favor of horny beaks. Size increased after the Jurassic Period. The largest species, of late Cretaceous age, weighed perhaps 25 lb but had a wing span of more than 25 ft. The skull of this creature was drawn out behind into a very peculiar long thin crest, and its lower jaw seems to have been provided with a pelican-like pouch.

BIRDS

Birds are little more than warm-blooded reptiles clothed in feathers. Almost all their skeletal structures find parallels among the reptiles. If the earliest discovered fossil birds were not known to have borne feathers, there is little doubt that they would be identified as reptiles (Fig. 465-2).

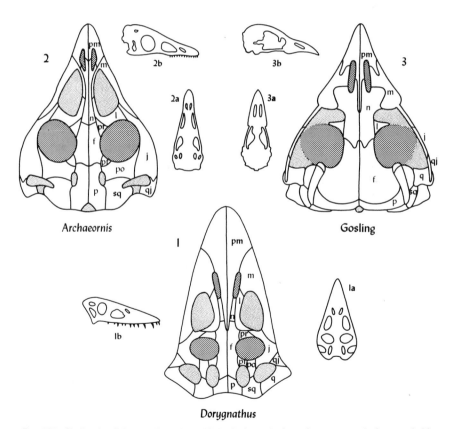

Fig. 465 Skulls of a flying reptile and two birds. 1, An early Jurassic pterosaur. 2, A very primitive late Jurassic bird exhibiting many reptilian characters. 3, A young modern goose. (See Romer, 1956, figs. 77, 81, pp. 143, 153; 1966, fig. 209, p. 261; Young, 1950, "The Life of Vertebrates," fig. 277, p. 442.)

Birds, however, are evolutionarily more advanced than any reptiles, but the differences are mostly such that they would not be recognized in fossils. Some advancement is evident in the soft anatomy, e.g., in the structure of the heart. Other advancements of modern birds are mainly behavioral and instinctive, such as nesting habits, care of young, migration, song, etc. The bird's egg is not significantly different from the reptile's, but it is relatively larger and in comparison to its size the amount of yolk is less. Brooding of the eggs by parents is a development related to warm-bloodedness and the necessity for maintaining approximately uniform temperature for the growing embryo. Care of the eggs, carried over to nurture of the nestlings, permits the young to continue growth to larger size before they are required to contend with the dangers of their environment alone. These behavioral characters have great survival value. Their evolutionary acquisition was a very important factor in the success of birds.

The fossil record of birds on the whole is very scanty. The evolutionary history is much less well known for these creatures than for many other vertebrates. The bones of birds are delicate and easily destroyed. Most birds have lived in such ways and places that preservation of their remains was exceedingly unlikely. Were it not for a very few fortunate discoveries, almost nothing would be known about Mesozoic birds. Tertiary fossils are more plentiful, but skeletons are mostly incomplete, and many of the bones are broken fragments. Great similarity in the skeletal structures of most birds, except for differences in the sizes and proportions of their parts, makes comparison and identification of fragmentary fossil specimens unusually difficult.

Origin of Birds

Modern birds have a single large temporal opening in the skull (Fig. 465-3). This is believed to have been derived from the double opening of diapsid reptilian ancestors by elimination of the intervening bony bar. Other skeletal features of the birds, such as the hip girdle and leg structure, are suggestively similar to those occurring in one or the other of the great dinosaur groups, and distant evolutionary relations seem probable. Birds, therefore, are believed to have evolved from a branch of the thecodonts somewhat similar to those ancestral to the dinosaurs. Perhaps this was the same branch which produced the pterosaurs.

Feathers and flight are the two most obvious attributes of birds. Flight could not have developed in this group of animals before feathers were acquired. Flight, therefore, is not a necessary attribute. Certainly the first reptile-like birds or the last bird-like reptiles could not fly. Numerous later birds, both fossil and modern, have been flightless. Some surely lost the ability to fly. Others, however, have been supposed to be primitive in this respect and to represent lineages that never acquired the ability to fly.

Feathers probably evolved as a concomitant to developing warm-bloodedness. They provided very effective insulation which, in this connection, would have had survival value. The perfection of warm-bloodedness and the attainment of an insulating coat of feathers probably evolved together rather rapidly in some of the small, very active, actually or incipiently bipedal thecodonts.

Feathers

Feathers consist of horny epidermal material that is produced within follicles in the skin at the contact between ectoderm and a core of mesodermal dermis. They grow at the base, like hairs, not at the tips like plants. Down and similar outgrowths which clothe fledglings are the simplest kinds of feathers. They probably are phylogenetically the most primitive, just as they are ontogenetically the earliest feathers to appear on young individuals. They serve primarily as an insulating coat.

The larger, more familiar quilled feathers evolved from down. In modern birds these larger feathers grow out and largely replace the early downy coat. By enclosing and entrapping air, they provide efficient insulation. They also furnish birds with a smooth streamlined surface. The enlargement of feathers of this type on the rear margins of the forelimbs and on the tail made flying possible. The evolutionary development, therefore, of these advanced and more complex feathers, whose original function was insulation, preadapted the very early birds for flight.

Flight

Two theories have been presented to account for the development of flight in birds. One supposes that feathers first appeared on small swiftly running bipeds. Some of these animals acquired the habit of beating the air with their small forelimbs to increase their speed. Gradual evolutionary improvements in the structure of these limbs and enlargement of their feathers produced wings, with which these creatures became able to fly for short distances. The advantage thus gained in escaping enemies was rapidly exploited, and proficiency in flight steadily improved.

At the same time other feathered animals grew larger. This also produced advantages in faster running and perhaps better defense against small predators. Further increase in size, however, resulted in creatures too heavy to adapt to flight. This was the beginning of the large birds whose forelimbs became useless, decreased in size, and finally, in some lineages, almost disappeared as external structures.

The other theory postulates that the first feathered creatures were small arboreal animals that climbed agilely in trees and shrubs. If they lost their footing or missed a hold in jumping from branch to branch, they fell, perhaps were stunned or injured and so became easy prey for enemies lurking below them on the ground. The enlargement of feathers provided these creatures with a sort of parachute that slowed their fall and lessened the impact when they struck. From this it was only one small step to the development of feathers that permitted some gliding from tree to tree, and another to the transformation of forelimbs to wings, useful first for more extensive gliding and then for actual flight.

Structure of the earliest birds discovered in Jurassic strata indicates that the second theory is much more likely to be correct. If this is so, the large flightless birds probably evolved from flying ancestors. As they increased in size their wings became progressively less useful. When the wings were no longer selectively advantageous, they tended to become vestigial.

Jurassic Birds

The three known specimens of Jurassic birds are among the most interesting fossils yet discovered. They are almost perfect connecting links between modern birds and thecodont reptiles. These creatures bore feathers whose impressions are preserved, but they had not yet acquired several of the bird's characteristic adaptations. Their reptilian features include the following: (1) General form of skull (Fig. 465-2). (2) Teeth set in sockets. (3) Relatively unmodified forelimbs ending in three, clawed fingers (Fig. 468-1). The feathered expanse of the wings was small, and there is no evidence of a large keeled breastbone. Therefore, these birds probably were weak fliers. (4) Vertebrae of the body not fused, except in the region of the hip girdle. (5) Bones not of the hollow type which in modern birds are connected with air sacs and the lungs. (6) Abdominal ribs present. (7) A tail longer than the body. This bore feathers spreading from both sides.

In addition to feathers, the principal bird-like features of these fossils are: (1) A somewhat expanded braincase. (2) Large eye sockets. These indicate keen sight and the development of improved nervous coordination, necessary for birds in their search for food and for balancing during flight. (3) A short body with a reduced number of vertebrae. (4) Hip girdle with pubic bone lacking a forward extension (Fig. 431-11).

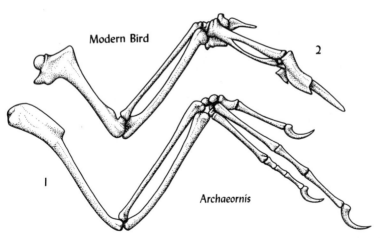

Fig. 468 Comparison of the bone structure in the wings of birds. 1, A late Jurassic bird that was very primitive in its other characters. 2, A modern bird. (*After Heilmann, 1927, fig. 17, p. 25.*)

Cretaceous Birds

The very few Cretaceous birds whose skeletons have been found demonstrate that much adaptational differentiation and evolutionary specialization had already been accomplished. The long reptilian tail had been suppressed, the large bones were hollow, and in other respects structures were distinctly modern. Only two types of birds are adequately known. One had well-developed wings whose bony structure differed little from that of living birds. The breastbone was large and strongly keeled. Obviously it was a proficient flier. The other was a swimming and diving bird, with vestigial wings, unkeeled breastbone, and large posteriorly directed legs. It could not fly and must have been a very awkward creature when it came ashore.

These birds are described as having teeth, except anteriorly in the upper jaws, which probably were furnished with short, sharp, horny beaks. Perhaps this is true, but possibly the jaw of a young ichthyosaur with teeth was mistakenly identified as part of one of the bird skeletons.

Large Flightless Birds

Among the best-known fossil birds of the late Tertiary and Pleistocene are a number of ostrich-like creatures, some of which grew very large. A few birds of this type, besides the ostrich, are still living, and some others became extinct in Recent time, probably exterminated by prehistoric human hunters. As previously mentioned, these birds have been considered primitive, in the sense that their ancestors never acquired the ability to fly, and closely related to one another. They are all generally similar in possessing long powerful running legs and small useless wings. Most of them bore plume-like feathers, rather than the more conventional type.

These characters do not necessarily indicate either primitiveness or close relationship, as they might have been acquired by birds of different lineage adapted to a similar way of life. Nevertheless, the roofing bones of the mouth are arranged in a relatively primitive manner, and similarity in this respect does suggest mutual relationship. The conclusion that flightlessness is primitive, however, is very doubtful. More probably the ability to fly was lost by birds inhabiting regions where predators were lacking or open country where escape from enemies was effected as easily by running as by taking flight. When legs replaced wings as the means of speedy movement, small size probably was no longer advantageous and natural selection favored increased stature. Perhaps these birds evolved in the very early Tertiary after the great reptiles had become extinct and while most predatory mammals were relatively small.

Modern Birds

The evolutionary radiation that produced the great variety of modern birds was well under way in early Tertiary time. This radiation involved adaptations to many different ways of life. The phylogenetic patterns are obscure, but it is easy to imagine progressive adaptational transitions such as those leading from a generalized type of flying bird to a shore wader, a swimmer, a diver, and finally to a bird like the penguin that has lost its ability to fly and uses its transformed wings for swimming under water. Several other types of birds, besides the ostrich-like forms already mentioned and penguins, sacrificed flight for greater abilities of other kinds.

Many birds developed structural specializations related to their dietary and feeding habits. These are seen principally in wing shape, beak form, and foot structure. The earliest birds probably were small, unselective carnivores. Some modernized forms specialized as birds of prey. They acquired short hooked tearing beaks and large sharp grasping claws. Others developed beaks strong enough to crack hard seeds and nuts, or long pointed jaws adapted to catching fish. Several other comparable examples of specialization might be cited. All these varieties radiated from some early generalized or common form. Perhaps this is best preserved today in the many birds that, instead of specializing, added variety to their diets, eating fruits, seeds, and other vegetable products, as well as insects, worms, etc. These are mostly small upland species. They include the so-called perching birds, although ornithologists consider these to be the most evolutionarily advanced of all. They have a specialized foot with a very large first toe, directed backward, that is adapted for grasping a twig or branch securely. Very few birds became exclusive herbivores.

Evolutionary Uncertainties

Attempts to trace evolutionary pathways among the birds are frought with more than ordinary uncertainty. Comparative fossil specimens are rare, particularly from the period of primary avian radiation, and the evidence that they might provide is lacking. Therefore, the similarity of many structures may indicate either close evolutionary relationship or adaptation to similar ways of life in quite different lineages. Among the characters employed in classification, (1) structure of the roofing bones of the mouth, (2) relative fusion of skull bones, (3) relative reduction or fusion of tarsal bones in the legs, (4) form of nostrils, and (5) disposition and number of specialized feathers may be evolutionarily significant. On the other hand, (6) form of beaks and (7) structure of the feet, including number of backwardly directed toes and degree of webbing, perhaps are not.

Some behavioral characteristics may be as revealing of relationships as any physical attributes. Among them are (1) habits of walking or hopping when on the ground, (2) disposition of the legs and neck while flying, (3) inclination to stand quietly after feeding, (4) quality of song, (5) flight patterns, and (6) nest-building abilities.

BIBLIOGRAPHY

A. d'A. Bellairs and **G. Underwood (1951):** The Origin of Snakes, *Biol. Rev.,* vol. 26, pp. 193–237.
Detailed morphologic comparisons suggest that snakes evolved from burrowing lizards.
E. H. Colbert (1951): Environment and Adaptation of Certain Dinosaurs, *Biol. Rev.,* vol. 26, pp. 265–284.
Contrasts are drawn between theropods, including early forest-dwelling types, ostrich-like forms, large carnivores, heavy herbivorous sauropods, and a variety of ornithopods, including duck-billed dinosaurs.
J. C. Ewart (1921): The Nestling Feathers of the Mallard with Observations on the Composition, Origin, and History of Feathers, *Proc. Zool. Soc. London,* pp. 609–642.
Feathers were not derived from scales but originated as hair-like outgrowths from beneath scales, as seen on the legs of many birds.
G. Heilmann (1927): "The Origin of Birds," Appleton-Century-Crofts, Inc., New York.
Detailed morphologic and ontogenetic comparisons of birds and reptiles demonstrate their close relationships.
F. R. Parrington (1958): The Problem of the Classification of Reptiles, *J. Linnaean Soc. London,* Botan. vol. 56, Zool. vol. 46, pp. 99–115.
Studies of the middle ear and other comparisons indicate that captorhinid cotylosaurs were ancestral to both synapsids and diapsids and that diadectids may have evolved from a different lineage of amphibians.
A. S. Romer (1956): "The Osteology of Reptiles," The University of Chicago Press, Chicago.
This is an encyclopedic account of the reptilian skeleton, with much information bearing on evolution and a long descriptive section on classification.
———— **(1957):** Origin of the Amniote Egg, *Sci. Monthly,* vol. 85, pp. 57–63.
Suggestion is made that the reptilian egg developed before quadrupeds abandoned an amphibious way of life.
———— **(1966):** "Vertebrate Paleontology," 3d ed., The University of Chicago Press, Chicago.
This is the foremost textbook in its field.
———— and **L. W. Price (1940):** Review of the Pelycosauria, *Geol. Soc. Am. Spec. Paper* 28.
See especially pp. 169–195 for a discussion of the nature and relationships of these mammalian ancestors.
D. B. O. Savile (1957): Adaptive Evolution of the Avian Wing, *Evolution,* vol. 11, pp. 212–224.
Aerodynamic considerations suggest that *Archaeopteryx* was not ancestral to other known birds.
B. W. Tucker (1938): Functional Evolutionary Morphology: The Origin of Birds, in "Evolution," G. R. de Beer (ed.), pp. 321–335, Oxford University Press, Fair Lawn, N.J.
Birds evolved from arboreal jumping reptiles; feathers originated as an insulating cover.
D. M. S. Watson (1954): On *Bolosaurus* and the Origin and Classification of Reptiles, *Bull. Museum Comp. Zool.,* vol. 111, pp. 297–450.
Very early divergence of reptiles, as shown by middle-ear structure, produced two lineages, the captorhinids (which led to mammals) and the diadectids (which led to archosaurs and birds).
L. W. Wing (1956): "Natural History of Birds: A Guide to Ornithology," The Ronald Press Company, New York.
See this book for much information having a bearing on bird evolution.

VERTEBRATES 3

ORIGIN OF MAMMALS
EVOLUTION OF THE EAR
MESOZOIC MAMMALS
EVOLUTION OF MAMMALIAN TEETH

Evolution of vertebrates reaches its culmination in the mammals. These animals on the land share with bony fishes in the water and birds in air a dominant position in the modern world. Evolution had, in a sense, long been trending toward the mammals; their general lineage diverged from the other vertebrates in the pelycosaurs of the late Paleozoic. The oldest fossils recognized as being mammals, however, are of latest Triassic age. This important group of animals, like the birds, branched from the reptiles when the latter were beginning their most spectacular evolutionary radiation.

Mammalian Characters
Modern mammals are easily distinguished from other vertebrates. They are warm-blooded and furnished with hair, they give birth to living young, except for a very few peculiar species that lay eggs, and the females suckle their young with milk produced by specialized skin glands. Several less-obvious features of the internal soft anatomy are equally characteristic. These include the structure of the heart and the existence of a muscular diaphragm that aids in breathing. Unfortunately features of these kinds are not of much service to paleontologists, whose knowledge of ancient animals ordinarily is restricted to what can be learned by observation and study of skeletal structures.

The mammalian skeleton is less variable than the reptilian. Although many differences are evident when the skeletons of different kinds of mammals are compared, these are concerned mainly with the relative proportions of their parts; the

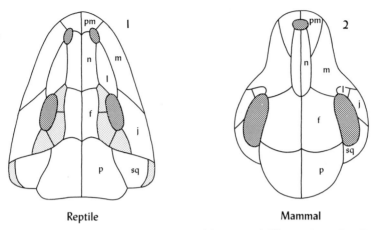

Fig. 474 Generalized diagrams showing some of the structural differences in reptilian (1) and mammalian (2) skulls. The shaded bones are not present in the skulls of mammals. (See Zangerl, 1948, "Evolution," vol. 2, fig. 2, p. 356.)

basic structure remains quite constant. The most noticeable differences distinguishing the skeletons of mammals and those of most reptiles are present in the skull. The number of bones has been reduced in mammals, especially above and behind the eye sockets, where several of the smaller reptilian bones have been eliminated (Fig. 474). The other bones of the skull roof behind the eyes are turned downward laterally, so that the brain cavity is completely enclosed, and the main jaw muscles are attached to the outside of the mammalian braincase. Less conspicuously, the long bones of the limbs grow in a somewhat different way. These bones in reptiles lengthen by the ossification of areas of cartilage at their ends. In mammals they have bony ends which provide surfaces of articulation, and growth occurs in zones of cartilage between these caps and the main bone shaft (Fig. 475). Structures of these advanced types, however, made their first appearances in the later therapsids.

A considerable number of additional noteworthy skeletal features are characteristic of most mammals: (1) the differentiation of teeth into simple single-rooted incisors and canines, and more complex multiple-rooted premolars and molars, (2) simplification of the lower jaw and the transfer of two small bones from the area of its articulation with the skull to the middle ear and the development of a new jaw joint (Fig. 482), (3) a single opening in the skull marking the position of the paired nostrils (Fig. 474), (4) a secondary bony palate roofing the mouth (Fig. 475a), (5) the skull and neck articulated at a double ball-and-socket joint (Fig. 424-2), (6) the upper bone of the hip girdle extending forward along the spine, (7) a heel developed in the hind feet, and (8) the toe formula of 2,3,3,3,3 in all feet if the number of toes is not reduced.

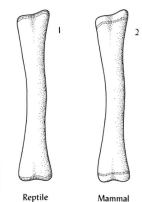

Fig. 475 The limb bones of mammals (2) grow in zones of cartilage (shaded) behind bony caps which are not present in the corresponding bones of reptiles (1). The weight of reptiles, including the huge Mesozoic beasts, was supported by cartilaginous articulating joints.

Reptile   Mammal

No skeletal character of these or other kinds, however, is by itself certainly diagnostic of the mammals. For example, turtles have a single external nasal opening (see Fig. 438), crocodilians have a well-developed secondary palate, and many amphibians had a double skull articulation. Other features appeared and were variably perfected in what seem to be several different lineages of mammal-like reptiles. Just when in their evolution these animals reached the point at which some of them became true mammals is not known. For the practical purposes of classification, however, a division must be made. Any solution of this problem is bound to be doubly arbitrary because (1) no abrupt transition in any evolutionary sequence is likely to occur and provide a natural division, and (2) skeletal structure cannot be equated certainly with the physiologic and reproductive characters that serve to identify modern mammals.

In any classification it is simplest to select a single character whose presence or absence determines whether an object or an organism is to be included or excluded from a group. The character that has commonly been relied upon by vertebrate paleontologists in the identification of fossil mammals is the nature of the jaw articulation and the fate of two small bones. If these bones occur at the

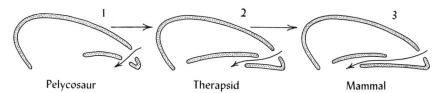

Pelycosaur   Therapsid   Mammal

Fig. 475a Schematic diagrams showing, in longitudinal section, the evolutionary development of a secondary bony palate in the mouth roof. The enclosed nasal passages lengthened, and the internal nasal openings migrated backward into the throat. (*Adapted from Romer, 1962, "The Vertebrate Body," 3d ed., fig. 168, p. 239.*)

point of articulation the fossil is accepted as a reptile. If, on the other hand, they have been displaced and enter into the structure of the middle ear, the fossil is regarded as a mammal. Unfortunately these relations ordinarily are difficult to observe because they involve tiny elements of small fossils that are uncommon and rarely well preserved. The recognition is general, however, that such a diagnosis may not accurately reflect the distinction between reptiles and mammals as this is based on the methods of reproduction and nurturing of young seen among modern animals. The attainment of true mammalian characters may have been accomplished either before or after this change in jaw and ear structure was completed.

Origin of Mammals

The ancestry of mammals can be traced back satisfactorily in a general way through the mammal-like therapsid reptiles to pelycosaurs and ultimately probably to captorhinid cotylosaurs (Fig. 426). Detailed study of therapsids has clearly revealed the appearance and progressive perfection of a variety of distinctly mammalian skeletal characters. These seem to be related to both structural and physiologic adaptations that conferred advantages upon the emerging mammals in two respects: (1) the ability to maintain vigorous and continuous physical activity under a variety of environmental conditions, and (2) the development of a greater awareness of the external conditions with which these animals contended.

The activity of cold-blooded reptiles is directly related to temperature conditions. These animals are well adapted to the relatively equable climates prevailing at low altitudes in tropical and warm temperate regions. Elsewhere, however, temperature fluctuations or continuous low temperatures severely restrict their activity. Under marginal conditions, lethergy induced by periods of low temperature confines their activity to certain times of day or certain seasons of the year. Under more rigorous temperature conditions they are unable to survive. The development of warm-bloodedness obviously was a very important factor in overcoming such restrictions. Warm-blooded animals not only acquired the capability of living successfully in regions unsuited to reptiles but also were provided with competitive advantages in marginal regions inhabited by these other creatures.

Extensive worldwide topographic changes that began in the late Paleozoic and continued in the early Mesozoic involved the uplifting of new mountains and resulted in the areal reduction of lowland regions which had provided the most suitable habitats for reptiles. At the same time related climatic zonation seems to have been accentuated. This combination probably stimulated the trend toward bodily thermal regulation in certain lineages of reptiles which may have had its distant origin much earlier among pelycosaurs. Cold-blooded reptiles continued to prosper and evolve in the warmer lowlands. On the margins of these areas, however, distinct advantages were gained by those advanced reptiles that were trending toward the prefection of warm-bloodedness.

Most of the evolutionary advances leading toward the mammalian condition seem to have been related to the development of warm-bloodedness. For example, more food was required to maintain body temperature. This necessitated teeth capable of more efficient mastication to prepare food for prompt digestion. Greater activity in the search for food and more rapid metabolism required a larger amount and more continuous supply of oxygen. The secondary palate, which displaced the internal openings of the nasal passages backward in the mouth (Fig. 477), permitted an animal to breathe while eating. A muscular diaphragm, probably indicated by the shortening and eventual disappearance of lumbar ribs, contributed to the efficiency of breathing. A hairy coat conserved body heat. The presence of hair may be indicated by shallow pitting on the snouts of a few fossil reptilian skulls where similar markings occur at the bases of the long "whiskers" of many modern mammals. Skin glands, producing sweat and oily secretions, which aid in regulating temperature and conditioning the skin, are suggested by small irregularities of the bone between eye and nostril, perhaps indicating that some ancient creatures had wet noses. The milk of mammals is produced by specialized glands of this general type. Growth of the embryo within the mother and live birth were advantageous, because better protection and nourishment and optimum conditions of uniform warmth were provided for development of the young. The nature of teeth in some fossil skulls, showing that they probably were replaced but once, has been regarded as suggestive that the young were much like those of modern mammals and possibly were born viviparously.

Greater activity by animals also required better physical coordination, which is controlled by the nervous system. In this connection improved sense reception and response thereto obviously are advantageous. Sight, smell, and hearing have compact reception centers located in the head, where they are intimately associated with

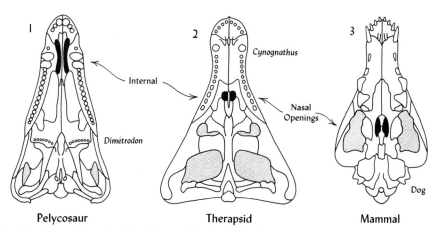

Fig. 477 Views of the undersides of three skulls showing their structure and the progressive backward migration of the internal nasal openings. (*After Romer, 1962, "The Vertebrate Body," 3d ed., fig. 167D, p. 238; 1956, "Osteology of the Reptiles," figs. 94A, 104C, pp. 177, 192.*)

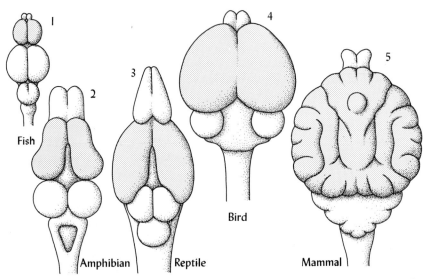

Fig. 478   The cerebrum (shaded) becomes relatively larger and progressively more complex in the succession of vertebrates from fishes to mammals. This is the part of the brain in which consciousness is centered and which controls voluntary actions. It is an evolutionary outgrowth of the primitive olfactory lobes, and its size generally is related to mental activity and intelligence. (*After Moody, 1962, "Introduction to Evolution" 2d ed., fig. 3.7, p. 32.*)

the brain. Enlargement of this organ, therefore, provides a good index to greater efficiency in the response to stimuli of the external environment. Slow but steady relative increase in the size and complexity of the brain is evident in passing from fishes through amphibians to reptiles, and this trend was further accelerated in the mammals (Fig. 478). Such development surely was one factor accounting for the superiority of mammals over the other vertebrates.

Sight was perfected to a relatively high degree in fishes. Although subsequent improvement undoubtedly occurred, no great accommodation was required to adapt this sense to the needs of terrestrial animals. Smell, likewise, generally improved from fishes to higher and higher vertebrates, as shown by enlargement of the nasal cavities in the forepart of the skull. A moist condition was maintained within these cavities, and again life in air required no great adaptive modifications. The sense of hearing, however, presented a more complex problem.

Hearing almost surely developed as a secondary function of an organ originally adapted to detecting bodily orientation and motion. In the relatively dense medium of water, sound waves striking a fish's body or received by its swim bladder are transmitted to the inner ear through the bones of the skull and vertebrae. Sound waves in air are too weak to register in a similar way. Without the development of a specialized receptor system, therefore, terrestrial animals would be deaf. Some amphibians and reptiles actually perceive only vibrations in the ground that are picked up by their jaws or legs and carried through their skeletons. The evolutionary acquisition of an eardrum and the adaptation of one or more small bones which transmit its vibrations directly to the inner ear made more efficient hearing possible for land animals.

EVOLUTION OF THE EAR

The ear of mammals is a very complex organ that has had a long and amazing evolutionary history. Some of the early developments, related to the perfection of jaws in fishes, which have a bearing on this subject, were explained in Chap. 8.

Inner Ear

The inner ear is a very ancient structure, buried posterolaterally in the bones or cartilages of the skull. Its origin is not known, although it may have been derived by specialization of part of the lateral line system possessed by even very primitive fishes. The inner ear existed in ostracoderms and occurs in modern cyclostomes in a condition somewhat simpler than that seen in other vertebrates. In bony fishes, however, it is essentially identical, except in rather minor details, to that of mammals.

The inner ear is membranous and is filled with a viscous fluid. It consists of a sac-like central chamber and three semicircular tubes, both ends of which open into this chamber (Fig. 479). Below the central chamber are two more extensions that vary in size and shape. One originally opened to the exterior, but its duct has been lost in all vertebrates except sharks, whose inner ear is filled with seawater. The other extension became long and coiled, especially in mammals. This whole structure is surrounded by lymphatic fluid and lies within a bony labyrinth that conforms to it in shape.

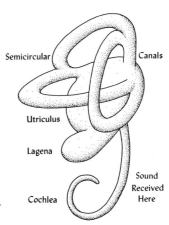

Fig. 479   Generalized drawing showing the main parts of the inner ear.

The three semicircular tubes, or canals, are disposed in three planes perpendicular to one another. Groups of sensory cells are located at specific places within the membranous inner ear and connect directly with the auditory nerve. When an animal moves, inertia of the fluid within the tubes causes it to press lightly upon these cells, and a sense of motion is transmitted to the brain. In a somewhat similar way orientation with respect to gravity is sensed. The inner ear undoubtedly was originally an organ that registered orientation and detected motion. Its function of hearing was a later acquisition, for which it evidently was preadapted.

Middle Ear

The middle ear is a cavity partly or almost wholly surrounded by bone adjacent to the inner ear. It is closed externally by a thin membranous eardrum. Internally it is continued by the eustachian tube, which opens into the throat. This tube is believed to have originated as the fish's spiracle, which in turn evolved from the gill chamber and opening that primitively lay behind the jaws and in front of the hyoid arch. It contains a bone known as the columella in amphibians and reptiles and as the stirrup in mammals. The middle ear is commonly supposed to have first appeared in amphibians.

Agreement is general that the columella, or stirrup, in all animals, except possibly frogs and similar amphibians, evolved from the upper element of the hyoid arch. In bony fishes, its precursor braces the jaws posteriorly against the braincase, with the inner end contacting the bone which encloses the inner ear (Fig. 481-1). Sound waves in water striking the skull and jaw are transmitted along this bone, through the bony covering of the inner ear, to the outer fluid surrounding this latter structure.

The evolutionary transformations and successive functions of the bony elements in the middle ear of higher vertebrates are uncertain. Contending theories postulate that (1) adaptive evolution resulted in the direct progressive perfection of hearing in terrestrial vertebrates, more or less influenced and directed by changes in the jaw structure, or (2) it produced several primitive types of hearing and the later independent development of a middle ear cavity and eardrum in more than one group of animals.

Orthodox Theory

The first, or orthodox, theory supposes that progress was made in three successive steps.

(1) In amphibians the hyoid bone shifted from its original position into the spiracle, or developed a projection within the spiracle, and became the columella (Fig. 481-2,3). It connected an eardrum, which closed the spiracle externally in the otic notch, and the inner ear. An opening appeared in the bone of the braincase opposite the inner ear, and this opening was occupied by the inner end of the columella. Vibrations were transmitted directly from the exterior through the columella to the outer fluid of the inner ear. This obviously was an improved mechanism especially adapted for the reception of airborne sound.

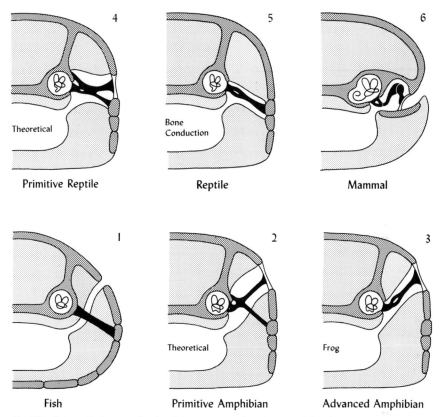

Fig. 481   Schematic diagrams showing, in transverse section, structure of the middle ear in several groups of vertebrates. (*Adapted in part from Romer, 1966, fig. 116, p. 85.*)

(2)   In the second step, the outer end of the columella moved downward to a position near the jaw articulation where the eardrum of reptiles is located (Fig. 481-4,5). Whether this drum is homologous to that of the amphibians is disputed. Some vertebrate paleontologists believe that an eardrum in this low position evolved independently twice. This belief is based on the fact that one branch of the cotylosaurs, the diadectids, has what seems to be a low otic notch but another, the captorhinids, from which pelycosaurs and therapsids presumably evolved, does not. However this may be, the reptilian columella bears the same relations to the drum and inner ear as in amphibians.

(3)   In the final step two other bones were added to the mechanism of the middle ear (Fig. 481-6).

The lower jaws of bony fishes are complex structures (Fig. 482-1) consisting of two lateral halves, each composed principally of dermal bones. Posteriorly, however, each half includes a bone or cartilage derived from the lower element of the

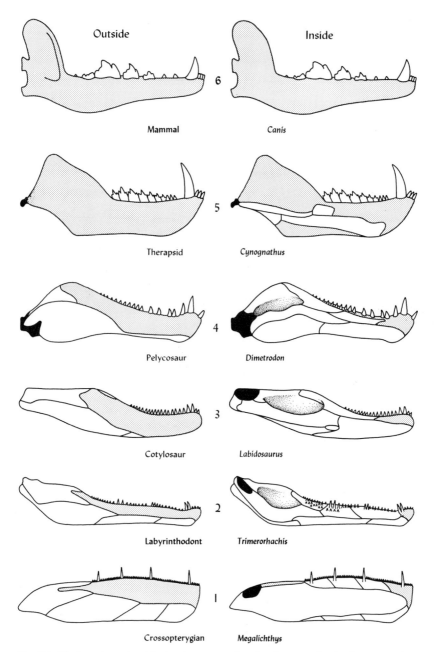

Outside

Inside

Mammal

Canis

6

Therapsid

Cynognathus

5

Pelycosaur

Dimetrodon

4

Cotylosaur

Labidosaurus

3

Labyrinthodont

Trimerorhachis

2

Crossopterygian

Megalichthys

1

Fig. 482 The lower jaws of vertebrates, from crossopterygians to mammals, showing progressive simplification in their structure. Articular bone, black; dentary bone, shaded. The articular passes into the middle ear, and all bones other than the dentary are eliminated. (*After Romer, 1962, "The Vertebrate Body," 3d ed., figs. 173, 174, pp. 246, 247.*)

primitive gill arch support. This bone is known as the articular, and it is the one upon which the jaw hinges. It is in contact with the quadrate bone of the skull, derived from the upper element of the primitive gill arch support, which constitutes the other part of the jaw hinge.

Reduction in the size and relative importance of the articular bone was a common evolutionary trend among some advanced terrestrial vertebrates. Forward migration of the jaw hinge contributed to this development in several groups of reptiles. Reduction is especially noteworthy among the mammal-like therapsids (Fig. 482-5). A few jaws of advanced therapsids have been found which seem to show that the hinge moved partly on the articular and quadrate and partly on bones located anterior or external to them—the dentary in the lower jaw and the squamosal in the skull.

The conventional division between reptiles and mammals is recognized to have been passed when the articular and quadrate were no longer included in the hinge (Fig. 482-6). Freed from their function as hinge parts, these bones became detached and entered into the structure of the middle ear. There the quadrate became the anvil, or incus, and the articular became the hammer, or malleus. This is the structure characteristic of the middle ear in mammals, where the sequence of three small bones, the hammer, anvil, and stirrup, carries sound vibrations inward from the drum. Perhaps this more complex structure is advantageous by amplifying these vibrations in some way.

Functional Theory

The other theory has been severely criticized, but it has some interesting aspects. It is based mainly on the consideration of two possibilities: (1) that the early amphibians and many reptiles did not possess an eardrum and middle ear cavity, and (2) that modern animals lacking these structures are primitive in this respect, rather than regressive. Both ideas have been strongly influenced by the opinion that hearing accomplished by the receipt of airborne vibrations through a drum and their transmission by a perfected middle ear is so superior to hearing by any other means that a lineage of animals possessing it would not have lost the advantage provided by this sensory ability except under extraordinary circumstances.

(1)   The otic notch in the skulls of labyrinthodonts does not demonstrate the existence of an eardrum, and no locus that might have been occupied by such a membrane has been recognized in some cotylosaurs. These animals may have been deaf to airborne sounds. Perhaps the columella was enclosed in a mass of muscle, as it is in snakes and certain modern lizards. The primitive reptiles and many of their descendants are considered to have perceived vibrations only by bone conduction.

A series of evolutionary stages of bone-conducted hearing can be reconstructed: (a) Sounds in water were received by the skull bones of early amphibians and passed through them to the middle ear. These animals were deaf when out of water. Some modern salamanders hear only in this way. (b) The early reptiles were sprawling creatures, and when at rest they lay prostrate on the ground. Vibrations picked up by their lower jaws were transmitted through the columella, which was in contact with the quadrate bone of the jaw articulation. These animals were deaf when they raised their heads above the ground. Some snakes seem to hear in this way today. (c) When the early terrestrial vertebrates rose up on their legs, a different kind of hearing was required. In one method, ground vibrations passed up the bones of the forelegs and then through connective tissues to the columella. Among modern animals only some amphibians hear in this way, but such a method may have been more general in the past. (d) A different method perhaps developed in the ancestors of some of the thecodont reptiles. These animals seem to have acquired the ability to perceive airborne vibrations picked up by membranes of the mouth and passed through the lower jaw bones to the columella. Such a method permitted the development of bipedal habits without sacrifice of hearing.

(2)   Up to this stage hearing was accomplished without an eardrum or middle ear cavity. If so, modern animals without these structures probably are the descendants of creatures that failed to evolve further in this respect. The conclusion also is necessary that the modernized and perfected middle ear evolved independently in several different groups of animals. There is some evidence for this. For example, ontogenetic development of the columella in frogs and toads suggests that this bone is not homologous with the columella in other animals, because it seems to be related to bones of the skull rather than to those of the hyoid arch.

The columella of early reptiles contacted the quadrate bone near the jaw articulation. Only a small shift in its position was required to free its outer end. At the same time part of the vestigial spiracle inherited from fishes enlarged to form a chamber around the columella. Where this chamber approached the outer surface of the head behind the jaws, an ear was developed. A modernized middle ear of this kind may have evolved more than once among the diapsid reptiles ancestral to many lizards, crocodilians, and birds.

The larger therapsid reptiles declined during the latter part of the Triassic Period and became extinct. Their failure may have resulted from the development of better hearing by diapsids, which contributed importantly to the competitive advantages of the latter group of reptiles. Evolution of the mammalian ear probably progressed rapidly among the small therapsids. In them, contact between the columella and quadrate bone was maintained. When the quadrate and articular bones were freed from the jaw articulation, they were drawn into an enlarged chamber of the spiracle by the columella. Thus the chain of small bones, stirrup or columella, anvil, and hammer, produced the structure characteristic of this type of middle ear (Fig. 481-6).

Outer Ear

The eardrum of all amphibians is flush with the body surface and continuous with its skin. In some reptiles the drum is sunken below the surface and lies at the bottom of a shallow depression or tube which is the beginning of the outer ear. This lengthens and is a narrow tube in a few reptiles and in all birds and mammals.

Crocodilians among the reptiles have a movable flap of skin that can close the opening of the outer ear. A somewhat similar structure, generally stiffened by cartilage, is the fleshy outer ear characteristic of most mammals. It forms a kind of funnel that gathers sound waves and deflects them into the passage to the eardrum, thus increasing the acuity of hearing.

MESOZOIC MAMMALS

Therapsid reptiles in various ways evolved far toward attaining the mammalian condition. Relatively little is known, however, about the details of their transition into mammals, because fossil specimens representing this critical evolutionary stage are incomplete and very rare. Although some therapsids grew to considerable size, many of them were relatively small animals and those from which the mammals actually evolved probably were no larger than rats and mice. The bones of such creatures were readily destroyed and, if they were preserved, are easily overlooked by collectors.

The fossil record of the most advanced therapsids and the most primitive mammals, spanning the time from the middle Triassic through the late Cretaceous, is very scanty, consisting mostly of scattered very tiny teeth and fragmentary jaws. The late therapsids and early mammals evidently were no match physically for their reptilian contemporaries. They probably managed to survive partly because they were small and inconspicuous. Perhaps the necessity for them to hide effectively contributed to the sharpening of their wits, which played an important part in their subsequent evolutionary advancement and success. Because the complete animals are not known, the remains of these questionable early mammals are difficult to evaluate, and evolutionary lineages cannot be traced with much assurance even in a general way. The relationships that are inferred are based almost exclusively on tooth structure.

Jurassic Fossils

The teeth of most Jurassic mammals indicate that these animals probably were carnivorous, but because of their small size they are believed to have preyed principally on insects. The teeth are differentiated into incisors, generally four on each side of both upper and lower jaws, one canine, four premolars, and five or more molars (Fig. 485). Such an array of teeth can be represented by the formula 4,1,4,5. Reduction in the number of teeth was general among more advanced mammals,

Fig. 485 Reconstruction of the lower jaw, 2½ cm long, of an early mammal, a Jurassic pantothere. Most Mesozoic mammals are known only from very tiny teeth. Many of these teeth have been recovered from material concentrated by ants in the hills built up around the entrances to their burrows. (*After Romer, 1966, fig. 305C, p. 198.*)

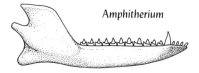

Amphitherium

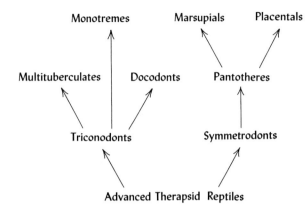

Fig. 486 Diagram showing the possible Mesozoic evolutionary radiation of early mammals. The relations of triconodonts and symmetrodonts are uncertain.

and this sequence is considered primitive, because both incisors and molars are more numerous than in most later mammals.

Five groups of Jurassic mammals are distinguished by their teeth: (1) triconodonts, (2) symmetrodonts, (3) pantotheres, (4) docodonts, and (5) multituberculates (Table 486 and Fig. 487).

Triconodonts

The molar teeth of triconodonts bear three or more sharp cusps disposed in a longitudinal row. Teeth of this form in the upper and lower jaws sheared against each other (Figs. 487; 495). Evolutionary advancement is believed to be indicated

---

Table 486   Types of Mesozoic mammalian molar teeth

Triconodont: teeth with three or more sharp cusps arranged in a longitudinal row. Late Triassic to early Cretaceous

Symmetrodont: teeth with three sharp cusps arranged in a nearly symmetrical triangle. Late Triassic to early Cretaceous

Pantothere: upper molars triangular but not symmetrical, with two main cusps arranged transversely. Lower molars triangular with an additional posterior shelf or heel that made contact with inner part of an upper tooth. Middle Jurassic to early Cretaceous

Docodont: teeth rectangular, upper ones nearly square with anterior and posterior sides pinched in somewhat. Cusps in two longitudinally parallel rows. Late Jurassic

Multituberculate: molars elongated with longitudinal rows of low cusps; two rows in lower teeth and more primitive uppers, and three in advanced uppers. Last lower premolar enlarged, with a high cutting edge. Late Jurassic to early Tertiary

Tribosphenic: upper molars unsymmetrical triangles with three main cusps. Lowers much like pantotheres, with posterior heel. Both kinds tend to become square and develop more cusps in several different lineages. Early Cretaceous to Recent

---

The evolutionary relations of the pre-Cretaceous kinds of teeth are uncertain. Tribosphenic teeth probably evolved from pantotheres, and the teeth of all modern mammals (except monotremes, which are toothless) seem to have had this origin.

by transition from teeth with a more prominent central cusp to those whose cusps are of approximately equal height (Fig. 497). Typical triconodonts are of latest Triassic to early Cretaceous ages. They are not known surely to have evolved into any other kind of mammal, although this is possible.

Symmetrodonts

Symmetrodont molar teeth also bear three main cusps, but these are arranged to form a nearly symmetrical triangle (Figs. 487; 495). Upper teeth have the bases of the triangles facing outward in the jaw. Orientation of the lower ones is reversed

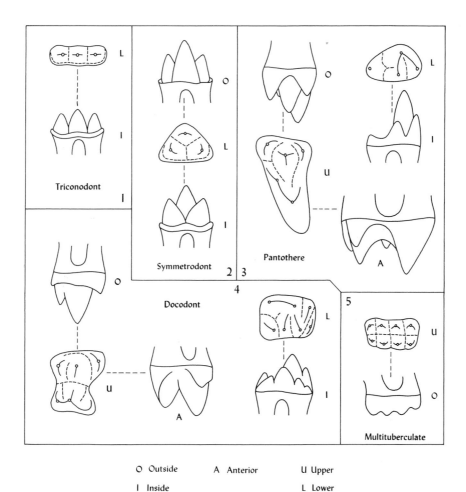

O Outside   A Anterior   U Upper

I Inside   L Lower

Fig. 487 Generalized sketches showing the characteristic features of various types of Mesozoic mammalian molar teeth. Solid lines in the crown views are ridges associated with the cusps. Dashed lines indicate troughs between the cusps. (*Mostly adapted from Simpson, 1929.*)

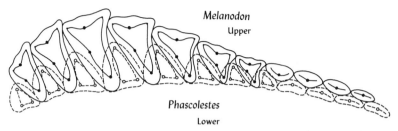

Melanodon

Upper

Phascolestes

Lower

Fig. 488 Semidiagrammatic sketches of the four premolars and seven or eight molars of a late Jurassic pantothere, showing the relations of upper teeth, in solid outlines, and lower ones, in dashed outlines. Notice the reversed orientations of the molar triangles, how the cusps of the lower teeth fit into the spaces between the uppers, and how the inner cusp of the uppers meets a heel posteriorly on the lowers. Although these upper and lower jaws are known by different names, they possibly belong to the same species. (*After Gregory, 1934, fig. 44C, p. 249.*)

so that their apexes point outward. Teeth of this type also sheared against each other when the jaws were closed. Symmetrodont fossils have been obtained from strata of latest Triassic to early Cretaceous ages. Pantotheres may have descended from animals with teeth of this kind.

Pantotheres

The upper molars of pantotheres bear two main cusps. These teeth are triangular and oriented as in symmetrodonts, but the triangles are not symmetrical (Figs. 487; 495). Each lower molar has in addition to its three cusps, similar to those of a symmetrodont, a posterior low flattened shelf or heel that contacted the innermost part of one of the upper teeth (Fig. 488). Pantothere teeth obviously are more specialized than those of symmetrodonts. They were capable of both shearing and a primitive kind of crushing or grinding action. Pantotheres are known from the uppermost Middle Jurassic to the Lower Cretaceous, where their remains are more abundant and more varied than either triconodonts or symmetrodonts. Animals with these teeth are believed to have been the ancestors of the more advanced later mammals.

Docodonts

Another type of tooth, the docodont (Fig. 487), is of uncertain relationships. These teeth commonly have been considered to represent specialized pantotheres, but possibly they belonged to an entirely different kind of animal. The teeth are not triangular. The upper ones in crown view are nearly square, with the anterior and posterior sides pinched in to give them a sort of hourglass form. The lowers are elongated and rectangular. Both kinds bear cusps in two longitudinally parallel rows. These teeth are less abundant than the other types. They are known mostly from the Upper Jurassic, but several specimens from the uppermost Triassic may be primitive teeth of this same kind.

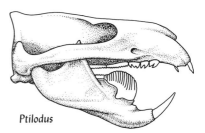

Fig. 489   Skull of a Paleocene multituberculate, two-thirds natural size, showing the extreme development of the last premolar in the lower jaw. (*After Romer, 1966, fig. 309B, p. 200.*)

Ptilodus

Multituberculates

Unlike most of the previously noted Jurassic mammals, other than possibly the docodonts, multituberculates probably were herbivores. They exhibited a tendency, duplicated in many other groups of vertebrates, for the herbivorous forms to grow larger than contemporary carnivores. The teeth of multituberculates are highly specialized, and the jaws are very rodent-like in general appearance (Fig. 489). A pair of relatively large chisel-like nipping or gnawing incisors was present in both upper and lower jaws. The canines were reduced in size or absent. The lower premolars, likewise, commonly were poorly developed, except for the last one which may be large with a high cutting edge. Thus a toothless space generally separated the incisors and cheek teeth. The molars are elongated and bear longitudinal rows of cusps (Figs. 487; 495). Two rows are present in the lower teeth and primitively in the upper ones as well, but a third row developed in the upper teeth of advanced forms (Fig. 489a). Teeth of this kind seem to have been adapted to grinding food. Remains of these animals first appear in the Upper Jurassic. Multituberculates lived on into the early Tertiary but then probably became extinct without leaving any descendants.

Cretaceous Fossils

Most teeth and bone fragments discovered in Lower Cretaceous deposits are much like those from the Jurassic. By late Cretaceous time, however, the Jurassic types, except multituberculates, seem to have become extinct. Their place is taken by fossils presumed to be both marsupials and placentals that first appeared in some-

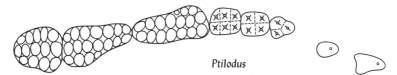

Ptilodus

Fig. 489a   Upper cheek tooth pattern of a Paleocene multituberculate with the dental formula 1, 1, 3, 3. (×4.) This is an advanced form with three rows of cusps on the molars. (*After Matthew, 1928, Proc. Zool. Soc. London, 1927, fig. 4, p. 953.*)

Table 490   The three great groups of modern mammals

Monotremes: a few species of peculiar egg-laying mammals native only to Australia and New Guinea. They are toothless. The females of one species have a pouch similar to that of marsupials. Fossils older than Pleistocene are not known.

Marsupials: young born in a very immature condition and suckled in a pouch on the abdomen of the females. These animals have a few skeletal peculiarities, particularly in the lower jaws and hip girdle. They have one less premolar tooth than placentals, and their milk teeth, except the last premolar, are not replaced. Upper molar teeth commonly have several outer cusps not present in most placental teeth. Abundant and varied in Australia; elsewhere represented by opossums. Cretaceous to Recent

Placentals: embryos nourished by an efficient placenta, and young born in a relatively advanced stage of development. These are the dominant modern mammals, native to all parts of the world except Australia. Cretaceous to Recent

These groups constitute a sequence of progressively more advanced mammals. Placentals must have passed through stages very similar to the others in their evolutionary development. It is possible, however, that the other groups, especially the monotremes, were derived from different lineages of mammal-like reptiles.

what older strata and probably had evolved from pantothere stock. These earliest representatives of modern mammals were small, probably arboreal animals, as shown by their toes, which were capable of grasping actions, and perhaps they were nocturnal.

Marsupials

Marsupial mammals represent a lower evolutionary grade than placentals. Their tiny young are born in a very immature state, and growth generally is continued in a pouch located on the abdomen of the mother. Few of these animals have developed an efficient placenta, which in the other group of modern mammals nourishes the embryo and permits it to mature to a more fully formed young animal. The skeletal features that serve to identify fossil marsupials include (1) a pair of marsupial bones articulated with the pelvis and extending forward in the abdominal wall (Fig. 490-2), (2) the dental formula of 4,1,3,4, which indicates

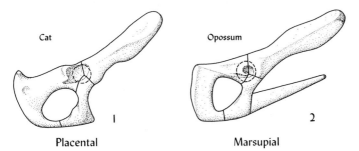

Cat   Opossum

1   2

Placental   Marsupial

Fig. 490   The pelvic structure of a marsupial mammal possesses a pair of forwardly projecting bones (2) that are lacking in placentals (1). (*After Romer, 1962, "The Vertebrate Body," 3d ed., fig. 128C, D, p. 194.*)

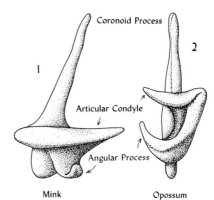

Fig. 491 Lower jaws of a placental and a marsupial mammal as seen from the rear. The incurved angular process is characteristic of marsupials. (*After Romer, 1966, fig. 312, p. 202.*)

the general presence of three premolars and four molars in each tooth series rather than four and three, as in primitive placentals, and (3) a characteristic incurved process at the posterior extremity of the lower jaw (Fig. 491). The most primitive marsupials had five incisors in each tooth series of the upper jaws, a greater number than is known in any other mammals. Most marsupials also are peculiar in that only the last premolar is a replacement tooth, all of the other first teeth being permanent.

In early Tertiary time the marsupials were eclipsed by the rapidly diversifying placentals in the northern hemisphere and almost disappeared from that region. Previously, however, primitive opossum-like carnivorous or omnivorous marsupials had gained access to Australia and South America. Changes in the distribution of land and sea cut off these regions from the northern continents before they were extensively colonized by placentals. The native modern mammalian fauna of Australia, which seems to have been completely isolated ever since the late Cretaceous, is almost exclusively marsupial. South America, whose land connection with the north was not reestablished until after the Miocene, lacked placental carnivores. This permitted marsupial evolution to progress in these two regions in ways that were not possible elsewhere in the world where competition with superior placentals was intense.

A variety of carnivorous marsupials evolved in both Australia and South America. Some of them were remarkably similar in their forms and habits to placentals that evolved concurrently in the northern continents. Herbivorous forms also evolved and differentiated. In Australia, particularly, marsupials adapted to almost all the ecologic niches occupied by placentals elsewhere. These creatures have persisted to the present in Australia, but most of the South American forms became extinct in the late Tertiary after the continents were reconnected and placentals invaded that region from the north. Some of the herbivores in both Australia and South America developed teeth so similar to multituberculates that a close genetic relationship has been suspected. These similarities, however, were not conservative

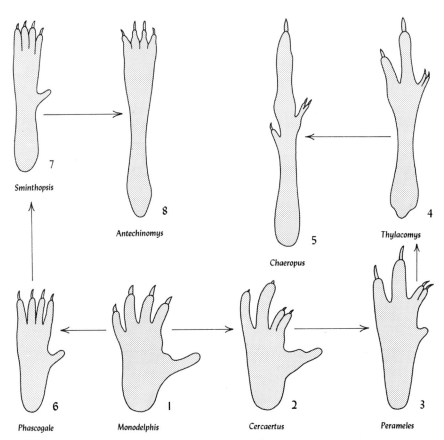

Fig. 492 The more or less specialized hind foot structure of marsupials shows two evolutionary trends, one characterized by the peculiar reduction of the second and third toes (2 to 5). The arrows indicate morphologic series but are not intended to suggest that these genera are related in such a detailed way. *(After Bensley, 1903, Trans. Linnaean Soc. London, ser. 2, Zool., vol. 9, pl. 7.)*

primitive characters, but specializations indicating evolutionary convergence toward a condition attained by an older stock at an earlier time. Parallel evolution in different marsupial lineages of Australia and South America is shown by the similarity of some species that developed in these two regions.

Placentals

Placental mammals, which dominate modern terrestrial faunas, are the most advanced vertebrates. Modern placentals are distinguished from marsupials primarily on the basis of the details of their reproduction. In them the embryo continues to develop and grow within the mother to a relatively advanced condition. Organic connection is maintained until birth by the placenta, which is the evolutionary derivative of one of the membranes occurring in the reptilian egg. Nutriment and oxygen diffuse through it from the mother's bloodstream, and metabolic wastes are similarly removed.

Placentals must have evolved from animals which reproduced in a more primitive manner. Probably they passed through a stage comparable to that exemplified by modern marsupials, but this does not necessarily mean that they evolved from creatures similar in other ways to animals of that kind. The reproductive methods of fossil mammals cannot be known, and distinction between the ancient representatives of these two groups is based on skeletal characters of the sort mentioned in the preceding section. Such differences can be traced back to late Cretaceous fossils. These provide no convincing evidence that marsupials evolved to produce placentals. Probably these two groups both originated by differentiation within a common ancestral stock derived from the pantotheres. Placentals, however, probably are of later origin than marsupials.

Placental fossils of Cretaceous age are very rare. Almost everything that is known about the evolutionary radiation of this great group of animals has been learned by the study of Tertiary fossils.

Monotremes

The duckbilled platypus and the spiny anteaters of Australia and New Guinea are the only representatives of a peculiar group of animals commonly regarded as the most primitive of mammals. There is no known fossil record of similar creatures older than the Pleistocene, but it is convenient to consider them briefly at this place. They possess most of the characteristic features of mammals such as milk glands, hair, a diaphragm, complex middle ear, and simplified lower jaw, but they lay eggs. When the eggs hatch, however, the young are suckled, and the anteaters have marsupial-like pouches. These animals also possess other more primitive characters, including a shoulder girdle of reptilian type.

Modern monotremes are highly specialized in certain ways, but they represent an evolutionary stage through which the more typical mammals must have passed. The evolutionary history of the monotremes is totally unknown. The adults are toothless, so that possible relationships cannot be traced by their teeth as for most other mammals. The general consensus is, however, that monotremes were not ancestral to other mammals but evolved independently from therapsids and since Triassic times have retained various primitive reptilian features with little change.

EVOLUTION OF MAMMALIAN TEETH

Most modern ideas regarding the evolutionary relationships of mammals and their classification have been based on the comparative morphology of teeth and limbs. The evolution of both certainly has been adaptive. The relations between morphology and way of life are somewhat more obvious in limb structure than in teeth (Fig. 536). Evolutionary change began, however, much earlier in the teeth than in the

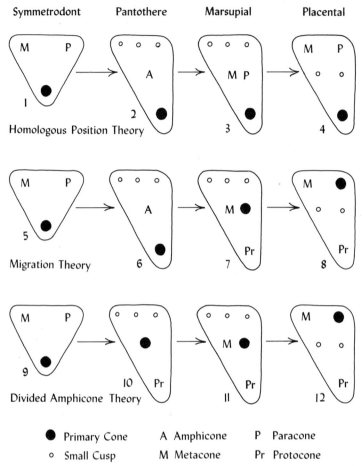

Primary Cone	A Amphicone	P Paracone
Small Cusp	M Metacone	Pr Protocone

Fig. 494   Diagrams contrasting three theories concerning the evolutionary transformation of mammalian upper molar teeth. Note that the cusps of all these teeth are named according to their positions in the triangles, and use of the same name in different types of teeth does not guarantee that these cusps are homologous. The two kinds of tribosphenic teeth are characteristic of advanced Cretaceous mammals and Tertiary marsupials, particularly carnivores (7) and the more primitive Tertiary placentals (8).

limbs of mammals, and dental structure commonly is accepted as the most reliable key to relationships among the principal mammalian groups. A few evolutionary sequences have been reconstructed in some detail, as among elephants and horses.

Mammalian teeth have provided an attractive field for study because of their presumed evolutionary significance. Teeth are relatively harder than bones, and commonly they are more perfectly preserved. The earliest mammals are known principally from their teeth. Studies devoted to these and later teeth have resulted in several contradictory theories regarding evolutionary trends and the homology of parts (Fig. 494), as well as numerous minor variations of these theories. Details of tooth structure have been expressed in a complex, confusing, and in some cases conflicting terminology and by systems of symbols that do not need to be considered here. Agreement is general, however, that most of the patterns of cusps, ridges, basins, etc., present in mammalian molars have evolved from a relatively simple triangular tooth whose summit bore three main cusps. The origin of this triangle, however, and the homologous relations of different triangles have been much disputed.

Reptilian Teeth

The teeth of most fossil reptiles are relatively undifferentiated simple pointed cones (Fig. 495). The cheek teeth are arranged so that the uppers and lowers alternate

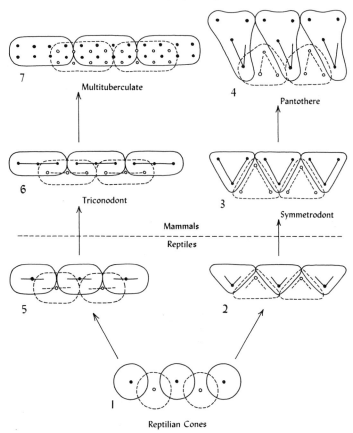

Fig. 495 Schematic diagrams showing the possible evolutionary relations of early mammalian molar teeth. Uppers are in solid outline, lowers are dashed. The relations of triconodonts and symmetrodonts are doubtful.

when the jaws are closed. They vary greatly in number in different groups of reptiles and commonly decrease in size backward. Most specializations derived from dentition of this kind, except in the therapsids. occurred in herbivores and in creatures which presumably fed by crushing molluscs and other shellfish. Tendencies leading toward mammalian dentition first appeared among the pelycosaurs, some of which possessed a few enlarged canine-like teeth (Fig. 496). These tendencies became strong and unmistakable among the mammal-like therapsids. In some of the more advanced members of the latter group the teeth are well differentiated into incisors, canines, and specialized cheek teeth. Multiple-rooted teeth appeared in some of them, and a pattern of single, rather than multiple, replacement may have developed.

Several types of complication or specialization of molar teeth originated and progressed in certain therapsid groups. They commonly involved some increase in the size of teeth and decrease in their number. One or more of these trends seems to lead to the types of molars possessed by early mammals. There is some evidence that different groups of animals classed as mammals may have evolved independently from different therapsid lineages. The most convincing trend is that which leads toward the triconodonts (Fig. 497). Here the reptilian teeth elongated and acquired anterior and posterior minor cusps, which increased in prominence. This produced a kind of shearing edge, which was a new functional development.

The theory has been presented that the multiple-rooted teeth of mammals are compound structures resulting from the fusion of two or more simple singly cusped and single-rooted teeth. Evidence does not bear this out. The earlier type of therapsid triconodont tooth has only a single root. Some later ones have roots with

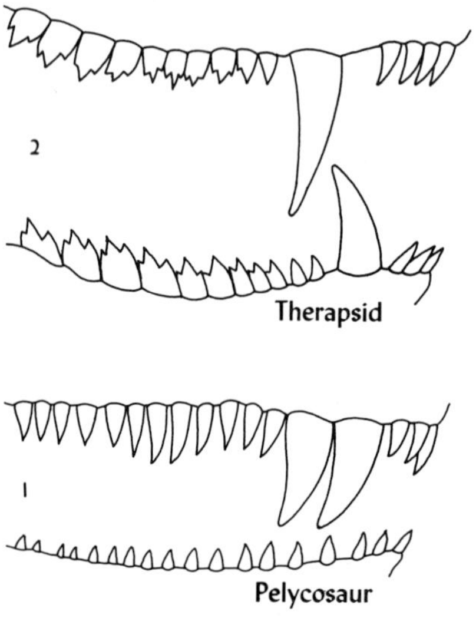

2

Therapsid

1

Pelycosaur

Fig. 496 Differentiation of teeth began with the appearance of canine-like fangs, first in the upper jaws of some Permian reptiles and later in the lower jaws. 1, *Eothyris*. 2, *Cynognathus*. (Romer, 1961, figs. 3, 4, 13, 15, pp. 16, 17, 39, 41.)

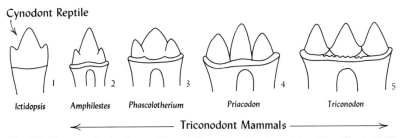

Fig. 497   Diagrams showing the possible evolutionary transition from a tricuspid reptilian tooth (1) to the mammalian triconodont molar (5). (*After Simpson, 1929, fig. 9, p. 32.*)

verticle grooves that begin to divide them into anterior and posterior parts. The simplest multiple-rooted teeth are not provided with a root beneath each cusp, of which the center one is largest. They were anchored by two equally strong roots located anteriorly and posteriorly, or more rarely by lateral pairs of roots in these same positions. Multiple roots also have developed on singly cusped premolars, on some canine teeth, and even on a few incisors. Teeth with three roots appeared later and certainly are more specialized than those with two.

Triangular Teeth

Triangular three-cusped teeth might have evolved in several different ways: (1) The two terminal cusps of the triconodont tooth may have migrated out of line with the central cusp to produce a triangle (Fig. 499-3,4). (2) Two cusps may have developed anew on the diagonal flanks or on a shelf at the base of the original reptilian cone and separated to form a triangle (Fig. 495-2,3). (3) One new cusp may have originated in this way on the upper molars and then divided to form two, thus completing the triangle (Fig. 494-2,3). Problems involved in the origin of triangular teeth in these or other ways concern not only the homology of parts in the teeth of different animals but also the homology of parts in upper and lower teeth and in the consecutive premolars and molars of the same animal. Some of the theories regarding the course of evolution among mammals and the phylogenetic relationships of various major groups depend very largely on which is the favored solution to these problems.

Triconodont Derivation

Derivation of the triangular tooth by migration of triconodont cusps (Fig. 499) is a simple and logical explanation that formerly was accepted widely. This theory found support in what seemed to be an evolutionary progression from very obtuse to more and more acute triangles among symmetrodont teeth (Fig. 498). According to this explanation, the cusp at the apex of the triangle corresponds to the original reptilian cone.

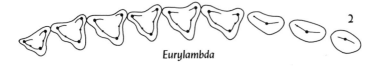

Eurylambda

Symmetrodonts

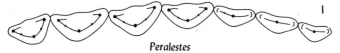

Peralestes

Fig. 498   Patterns of upper molars and premolars of two late Jurassic symmetrodonts. These might represent stages in the progressive evolution of triangles from a triconodont pattern. The apical cusps of the more obtuse triangles (1) are not conspicuously out of line with the cusps of the simpler premolar series. This is not so, however, for the cusps of the more acute triangles (2). (*After Gregory, 1934, fig. 44A, B, p. 249.*)

Diagonal Cusps

The appearance of new cusps diagonally on the flanks of an original central cone or rising from a basal shelf would produce a triangular tooth of exactly the same type. If the triangle originated in this way, however, triconodonts were not ancestral to symmetrodonts (Fig. 495). Appraisal of what little evidence there is has led some paleontologists to doubt that the symmetrodont triangle was an evolutionary modification of the triconodont pattern.

The apex cusps of all lower and some upper symmetrodont molars seem to stand in line with the main cusps of the simpler premolars which precede them. This has been interpreted as evidence that the apex cusps probably are homologous with the reptilian cone. The alignment of lower molar cusps in pantotheres suggests that these teeth originated in such a way. Finally, the molarization of some lower premolar teeth in more advanced mammals also produced triangles that seem to record a similar evolutionary development.

Pantothere Pattern

The upper pantothere molar is triangular but only roughly and unsymmetrically so. It has two prominent cusps instead of three, one near the inner apex and the other laterally near to or outward from the center of the tooth. In addition it has several smaller cusps that rise along its outer edge. The inner cusp, because of its similar position, seems to correspond to the apex cusp of the symmetrodont tooth (Fig. 494-1,2). These cusps, however, may not be homologous. When the jaws closed the pantothere inner cusp made contact with a low flattened posterior extension or heel of a lower molar that appears to be a new evolutionary addition to the symmetrodont type of tooth (Fig. 488). The complementary relations of these parts

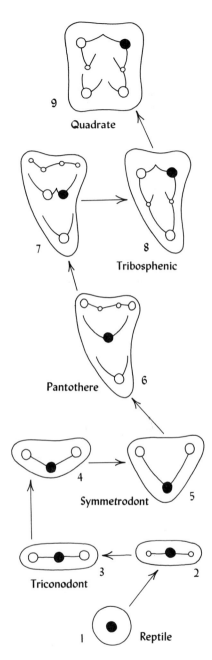

Fig. 499 Diagram illustrating the possible evolutionary transformation of upper molar teeth from the reptilian cone to the quadrate mammalian type, according to the divided amphicone theory. Cusps possibly homologous with the primary reptilian cone are blackened. In this sequence the protocone (apex cusp) of the pantothere tooth (6) supposedly was added to the symmetrodont triangle (5) at the same time that the posterior heel enlarged the lower molars. Appearance of a metacone next to the primary reptilian cone (7), perhaps by the division of the amphicone (black) of the pantothere tooth (6), produced the general type of molar characteristic of many carnivorous marsupials (7). Loss of the outer cusps and movement of the metacone and paracone nearer to the outer margin resulted in the typical placental tribosphenic molar (8). Outer cusps reappeared, however, in some highly specialized placental teeth. The appearance of a hypocone produced the quadrate type of molar (9).

suggests that possibly they evolved together. If this is so, the apex cusp of the pantothere upper molar also might be a new functional development.

Lower pantothere molars might easily have evolved from the corresponding teeth of symmetrodonts by no more than the addition of a heel (Fig. 500). The possibility of evolutionary transformation in the upper teeth is not so evident, and

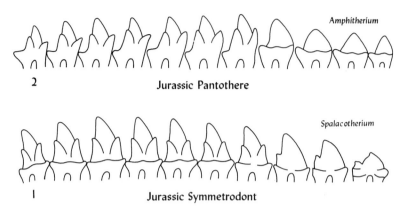

Amphitherium

2   **Jurassic Pantothere**

Spalacotherium

1   **Jurassic Symmetrodont**

Fig. 500   Lower molar and premolar teeth of Jurassic age, as seen from the inside. 1, A symmetrodont. 2, A pantothere. A conspicuous difference is provided by the low heels developed posteriorly on the pantothere molars. (*After Gregory, 1934, figs. 142B, 144B, pp. 245, 249.*)

an explanation is more difficult. Nevertheless the pantotheres may have had symmetrodont ancestors (Fig. 499).

The kinds of teeth previously discussed functioned primarily as piercing and shearing mechanisms. The cusps generally are high and sharp, and they may be connected by acute steeply sloping ridges, some of which bear minor cusps. The triangles of upper and lower teeth were oriented in the jaws in opposite directions. When the jaws closed the teeth did not meet directly but slid past each other, with the lower ones fitting into spaces separating the upper ones internally (Fig. 488). There seems to have been a tendency for these teeth to widen transversely in the jaws, and as the triangles became more acute, a longer and more effective cutting edge was formed.

The appearance of heels on the lower molars of pantotheres suggests adaptation to a new function. The wear evident on some fossil teeth shows that they moved back and forth against each other. This is the kind of motion required for grinding food. It increased the efficiency of mastication and may be evidence that some pantotheres became more omnivorous in their diet.

Tribosphenic Pattern

The more advanced mammals almost certainly evolved from pantotheres. The suggestion has been made that the main central or outer cusp of the pantothere upper molar divided and the two parts then separated to produce, with the apex cusp, the three-cusped tribosphenic pattern. The most prominent cusps of this new kind of tooth are disposed much like those of the symmetrodonts. The cusps in comparable positions, however, may not be homologous in these two kinds of teeth. The outer cusps in the new type of tooth are the ones that are aligned with those of the simpler premolars. In the tooth buds of the embryos of modern mammals, the beginning of the anterior member of the outer pair is the one that develops first. Both these facts suggest that the anterior outer cusp, not the apical one of the triangle, is homologous with the original reptilian cone.

Three main theories have sought to relate the types of upper molar teeth that seem to follow the symmetrodont stage. These are (1) the homologous position theory, (2) the migration theory, and (3) the divided amphicone theory (Fig. 494).

Homologous Position Theory

The homologous position theory, also termed the Cope-Osborn theory, was originated and developed by two of the most prominent American vertebrate paleontologists of a previous generation. Briefly it provides that the primary reptilian cone located at the inner apex of the symmetrodont triangle persisted in a comparable position in the specialized upper molars of later mammals. This is the simplest explanation of relations that can be devised, and formerly it achieved wide currency. Difficulties with respect to the comparison of premolars and molars and the early growth in tooth buds were later recognized. For example, complication of the premolars in the process of their molarization began with the appearance of a new inner cusp on the flank of the outer cusp, or primitive cone, thus producing a typical bicuspid tooth. This was followed by doubling of the outer cusp, either by the development of a new one in a posterior position or by the division of the old one to produce the fundamental upper triangle. Consequently the homologous position theory no longer seems acceptable and has lost most of its former following.

Migration Theory

The migration theory starts with the same assumptions as the homologous position theory, that the apex cusps of symmetrodont and pantothere teeth correspond to each other and to the primary cone. In the next step this cone is supposed to migrate outward to a position beside the second main pantothere cusp and a new cusp developed at the apex. The triangle so formed inside of several marginal cusps is characteristic of many marsupials, particularly carnivores, and also occurs in the teeth of the most primitive placentals. Finally the two outer cusps of the triangle moved outward and replaced the small cusps near the margin of the tooth. This pattern is seen repeatedly in the less-specialized placental upper molars. Migration in this way brings the presumed primary cone into the position suggested by the embryonic development of the tooth buds.

Divided Amphicone Theory

The divided amphicone theory differs from both the preceding theories by identifying the outer of the two principal pantothere cusps, or amphicone, instead of the inner one as the primary cone. This interpretation is based on the conclusion that the posterior heel of the pantothere lower molar and the apex cusp of the upper molar evolved at the same time as new features that performed a novel function. Thus the upper apex cusp probably is comparable to the inner cusp of a bicuspid premolar. From this stage onward the divided amphicone and migration theories are identical.

The evolutionary relations of the types of early mammalian molar teeth can be expressed at present more in terms of their possibility than their probability. In the light of current knowledge and some expert opinion, the divided amphicone theory seems to offer certain explanatory advantages and more favorable possibilities than either of the others.

Modernized Mammalian Teeth

The reasonably adequate fossil record of the mammals begins in the earliest Tertiary. This was a time of rapid evolution when the mammals were diversifying to occupy the various ecologic niches left vacant by the many kinds of reptiles that became extinct at the end of the Cretaceous Period. In a time that was relatively brief by geologic standards, differentiation produced the ancestors of all great later Tertiary and modern groups.

The complex molar teeth of most modern mammals are believed to have evolved from the basic tribosphenic pattern. Probably something in this kind of tooth conferred advantages upon the animals which possessed it and contributed to their success. The advantage may have been provided by the relations between the apical cusp of the upper molars and the enlarged and hollowed heel of the lower ones into which it fitted. These parts seem to have been adapted to crushing and grinding food, which was a new function in mastication. Teeth of this kind that could shear or crush or grind might evolve further in different ways advantageous to animals that became specialized in the use of any kind of food.

Primitive Placental Molars
The outer cusps of the pantothere upper molar generally decreased in size or disappeared in the teeth of the evolving primitive placentals. At the same time the outer cusps of the prominent tribosphenic triangle moved outward and occupied positions nearer to the corners of the tooth.

Primitive Marsupial Molars
The teeth of primitive marsupials differ so little from the early tribosphenic pattern

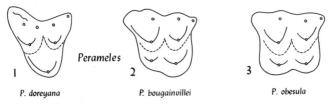

Perameles

1   *P. doreyana*   2   *P. bougainvillei*   3   *P. obesula*

Fig. 502 Upper molar patterns of three related species of marsupials that show transition from tribosphenic to quadrate form. Notice that these teeth have retained outer minor cusps corresponding to those possessed by pantotheres. Similar cusps commonly were eliminated in the early evolution of placental teeth. (*After Bensley, 1903, Trans. Linnaean Soc. London, ser. 2, Zool., vol. 9, fig. 4, p. 116.*)

Fig. 503   Upper cheek teeth of an early Tertiary creodont. The three premolars lack the third prominent cusp of the molars. The last two premolars, however, show a well-marked beginning of the evolutionary process termed molarization. (*After Gregory*, 1934, *fig.* 64, *p.* 286.)

Deltatherium

(Fig. 499-7) that some paleontologists have considered most of the Mesozoic mammals with tribosphenic teeth to be marsupials. The earliest placentals are identified by other features of their skeletons, but the similarity of teeth is commonly accepted as good evidence for the very close relationship of marsupials and placentals, which probably evolved from the same stock of pantotheres or their immediate descendants. Unlike the placentals, the marsupials did not so completely lose the minor outer cusps of the pantothere tooth. These cusps persisted (Fig. 502) and in some lineages became progressively accentuated. Evolution and specialization in teeth of some of the more advanced marsupials were remarkably similar to the developments in placental teeth.

Transformation of Premolars
Permanent premolars which lie in front of the molars in the jaws of placental mammals replace a first set of so-called milk teeth. Molars, on the other hand, are not replacement teeth. They probably are members of the same set that includes the first or deciduous premolars. The possibility that molars are members of the second set, however, and that a first set of milk teeth has been suppressed is unlikely but cannot be dismissed completely. Most mammals of late Cretaceous and earliest Tertiary ages had premolars and molars that were sharply differentiated from each other morphologically. The molars had become complicated in various ways, including those that have been described, but the premolars did not depart so greatly from the form of the original simple cones. In the Eocene, however, evolutionary transformation of premolars progressed rapidly in some groups of mammals (Fig. 503). This affected both the milk teeth and the permanent premolars. Similarity of form, however, was not attained in the same way, because premolars did not evolve through a pantothere stage and cusps similarly situated in molars and premolars are not necessarily homologous. In some lineages the premolars, or at least the most posterior ones, became more complex than any of the following molars. In others the permanent premolars are simpler than the milk teeth that they replace.

Cutting Teeth of Carnivores
The cutting teeth of many advanced carnivorous mammals are elongated narrow blades consisting mainly of two or three laterally compressed cusps (Fig. 504). They superficially resemble some triconodont teeth, but the tribosphenic nature of those

Last Premolars

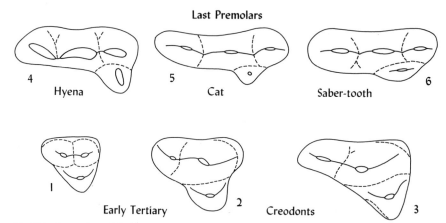

Fig. 504   Crown views of the last upper premolars of several carnivores. These teeth have been conspicuously molarized. The three early Tertiary teeth clearly show transitions to the long cutting teeth of modernized carnivores. The anterior cusp of the latter teeth is an addition to the older pattern. 1, Mesonyx. 2, Oxaena. 3, Vulpavus. 4, Hyaena. 5, Metailurus. 6, Smilodon. (*After Romer, 1966, fig. 334, p. 230.*)

in the upper jaws is indicated by the internal presence of a small low apical cusp. The lower teeth do not have a similar external cusp. In both kinds the highest cusp, which is either anterior or central, probably corresponds to the original cone. Intermediate stages connecting primitive with highly specialized teeth of this kind occur in several evolutionary sequences, or even in series, in the jaws of some fossil carnivores. Similar transformations have modified both molar and a variable number of premolar teeth.

Quadrate Teeth
Very early in their evolutionary development the teeth of mammals acquired cusps in addition to the three main ones that form the triangles. New cusps appeared, for example, upon the ridges that connect the main cusps, some arose from marginal shelves located near the base of the crown, and others may occur at the tips of ridges extending upward along the sides or at the corners of a tooth. Some of these supplementary cusps became as prominent as the other ones. Additions to some teeth, commonly bearing one or more cusps, altered shapes so that they are no longer triangular. Both the upper and lower molars of some mammals became rectangular or nearly square (Fig. 505). Teeth of this form did not shear past each other when the jaws closed but met in such a way that the cusps of one tooth were received between the cusps of two adjacent teeth in the other jaw (Fig. 505a). The cusps of such teeth commonly were reduced somewhat in height and the molars became better adapted first to crushing and later to grinding functions.

The triangular teeth were altered to quadrate form differently in the upper and

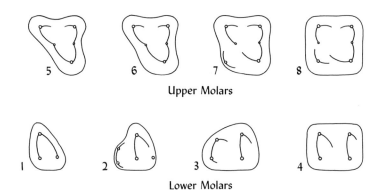

Upper Molars

Lower Molars

Fig. 505 Diagrammatic series illustrating the transitions from triangular to quadrate teeth in both the upper and lower jaws of placental mammals.

lower jaws (Fig. 505). The posterior heel that appeared in the lower molars of pantotheres, and is even better developed in tribosphenic teeth, was mentioned in a previous section. The basin of this heel in postpantotheres is margined by a ridge that bears two or three small cusps. As this heel became more prominent, two of its cusps moved to positions near the posterior corners of the resulting quadrate tooth. At the same time one of the original three main cusps, the anterior inner one, was reduced in size and disappeared. The remaining primary cusps are those that lie near the anterior corners of the transformed teeth.

A somewhat similar heel grew backward from the upper molar into the space that had separated the triangular teeth internally (Fig. 505). This was not basined like the lower heel, however, but bore a single central cusp. This heel squared up the upper molar, which now had four prominent cusps situated near the corners. Two smaller cusps generally appeared on the transverse ridges connecting the anterior and posterior pairs of larger cusps.

Quadrate teeth of these types or of kinds clearly related to them are charac-

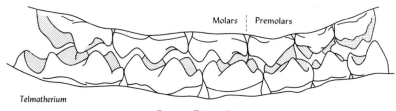

Eocene Titanothere

Fig. 505a Inside view of the upper and lower cheek teeth of an early Eocene titanothere, showing how the cusps of one series of teeth fit into the depressions of their mates. (*After Gregory, 1934, fig. 58A, p. 277.*)

teristic of most omnivorous modern mammals. The more complicated teeth of many herbivores also have been derived from quadrate molars and premolars of this type.

Grinding Teeth of Herbivores

Grinding teeth of hoofed mammals and other large herbivores evolved from quadrate teeth most of which had six cusps above and four below (Fig. 505). Some of these teeth developed very complex patterns (Fig. 506) that would be difficult to interpret if progressive changes could not be traced through series of morphologic stages. Factors variously contributing to the complications were (1) the appearance of new cusps or related crown elements that arose mainly on the tooth margins; (2) the transformation of cone-like cusps into crescentic structures; (3) the joining of cusps to produce prominent ridges; (4) the spacial rearrangement of these elements; (5) the development of folds and crenulations in the enamel layers on the flanks of ridges.

The early herbivores had low-crowned teeth with relatively sharp but low cusps and ridges (Fig. 553-1 to 3). Such teeth were adapted to the mastication of succulent vegetation. In Tertiary time, changing climate and the evolution of grasses resulted in the gradual transformation of large expanses of forested country to savanna. Some of the herbivores adjusted rapidly to these new conditions. This necessitated the development of legs constructed for swift running to escape enemies, from whom hiding was no longer possible, and the evolution of dentition capable of grinding the harsher grasses, which rapidly wore down the low-crowned teeth. The new types of teeth that were perfected in several different lineages of these mammals had high crowns and roots that did not mature at an early age (Fig. 553-5,6). These teeth continued to grow outward from the jaw as the tops were worn away and thus remained serviceable throughout the creatures' normal lives.

Growth by itself could not solve this problem of tooth abrasion because only soft dentine would remain when the top of the crown had worn away. Greater resistance was achieved and a better grinding surface was developed by the progressive

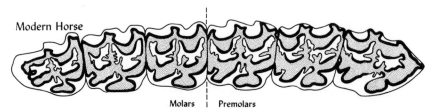

Modern Horse

Molars | Premolars

Fig. 506 Crown view of the upper cheek teeth of a modern horse, consisting of three molars and three completely molarized premolars. Black, enamel ridges thrown into folds. Shaded, dentine exposed in areas from which the originally overlying enamel was worn away. Unshaded, cement filling depressions and folds of the enamel. (*After Gregory, 1951, "Evolution Emerging," fig. 21.46J, p. 812.*)

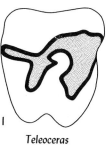
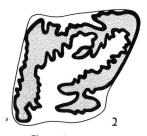

Fig. 507 Crown views of upper molars of two Tertiary rhinoceruses illustrating the increased complexity of enamel ridges that has characterized several evolving lineages of advanced herbivores. (*After Osborn, 1929, U.S. Geol. Surv., Monograph 55, vol. 2, fig. 718, p. 797.*)

I

Teleoceras
Miocene

2

Elasmotherium
Pleistocene

modification of cusps and ridges so that the enamel layers penetrated deep down within the teeth. The enamel also was thrown into increasingly complex folds, thus lengthening the grinding ridges (Fig. 507). Consequently, as wear continued, several edges of the hard enamel persisted in the grinding surfaces and provided an efficient cutting mechanism when the teeth were moved across each other. If enamel was the surface layer, as it is in most other kinds of teeth, deep hollows would remain between the cusps and ridges of such high-crowned teeth. These hollows became filled, however, in the teeth of many herbivores with bone-like cement that was continuous with the material anchoring the teeth in their sockets. Thus solid prismatic teeth were formed (Fig. 506).

As the surface of a newly errupted tooth began to wear, the tips of the cusps and then the crests of the ridges were the first parts to suffer. The enamel in these areas was worn through, and the underlying dentine was exposed. A tooth in this condition began to develop a pattern on the surface of its crown characteristic of its internal structure. Enamel rims surrounded basins worn down in the soft dentine, and if cement was present, this material, also worn to a lower level, occupied areas outside the enamel. Such a pattern gradually changed as wear continued, showing in effect a cross section of the tooth at lower and lower levels (Fig. 507a). The areas

I

2

3

Merychippus

Fig. 507a Three sections through an upper molar tooth of a Miocene horse showing changes in the pattern of enamel ridges that would develop in teeth worn down in older and older individuals. (*After Osborn, 1918, Mem. Am. Museum Nat. Hist., N.S., vol. 2, pt. 1, pl. 16.*)

of dentine became somewhat larger, the areas of cement somewhat smaller, and the folding of enamel somewhat simpler. These changes in the patterns of their teeth permit the ages of individual animals to be estimated.

Cusp-in-line Teeth

Several different kinds of teeth possess cusps that are aligned longitudinally. Some of these can be traced back to the tribosphenic patterns through intermediate types. Others seemingly cannot be connected in such a way. Among the latter the multi-tuberculates are the best examples. They evolved perhaps from a triconodont type of tooth by the development of cusps growing upward from marginal shelves on both the upper and lower teeth (Fig. 508).

The cusps of docodont teeth have been homologized with those of pantotheres, but the relations are uncertain. Some evidence suggests that docodont teeth did not evolve from a triangular pattern. If so, docodonts may have descended from triconodonts.

Rudimentary milk teeth of the duckbilled platypus are the only teeth possessed by any known monotreme. Their relations are very doubtful, but the opinion has been expressed that they were derived from some kind of cusp-in-line teeth.

The teeth of some herbivorous marsupials suggest cusp-in-line patterns (Fig. 509), and close ancestral relations of marsupials and multituberculates have been postulated. This does not seem possible. Less-specialized marsupial teeth are plainly tribosphenic, and even the most complex ones have patterns that can be related to tribosphenic teeth.

Tooth Number

The teeth of many different types of reptiles reveal no general plan of number or arrangement. An order began to emerge, however, among the therapsids, where differentiation into incisors, canines, and cheek teeth became evident. These mammal-like reptiles varied in many ways, and no more systematic dental pattern has been recognized. Cheek teeth ranged upward to a maximum of about 13 in a single

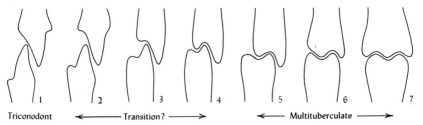

Triconodont ← Transition? → ← Multituberculate →

Fig. 508 Diagrams showing in transverse outline the possible transition from triconodont teeth (1) to an advanced type of multituberculate molar teeth (7). (*Partly after Simpson, 1929, figs. 3A to D, 12C, pp. 14, 35.*)

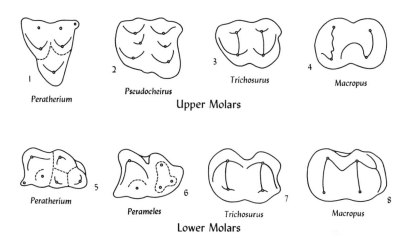

Upper Molars

Lower Molars

Fig. 509 Crown patterns of several kinds of marsupial molar teeth. Some of these teeth show little resemblance to a triangular arrangement of cusps. A cusp-in-line ancestral pattern has been suggested, but this seems unlikely. (*After Bensley, 1903, Trans. Linnaean Soc. London, ser. 2, Zool., vol. 9, fig. 1, p. 89.*)

series. This seems to be a number that was exceeded in few later animals, whereas progressive reduction was the rule.

Reduction in Mammals

The Mesozoic mammals were almost as variable as the therapsid reptiles in the number of their cheek teeth. The symmetrodonts and pantotheres possessed in each tooth series 4 incisors, 1 canine, and 7 to 12 cheek teeth, the last consisting of 3 or 4 premolars and 4 to 9 molars. Relative stability was attained, however, in the late Cretaceous and early Tertiary marsupials and placentals. The marsupials had 4 incisors, except for opossums with 5 in the upper jaw, 1 canine, 3 premolars, and 4 molars. The placentals likewise had 4 incisors and 1 canine, but 4 premolars and 3 molars. These primitive formulas were reduced progressively and regularly in various later lineages.

Reduction in number seems commonly to have been related to the increase in size of teeth. This is particularly evident in the premolars and molars of some carnivores and in the incisors of elephants and rodents. Shortening of the jaws in several lineages of herbivores also was accompanied by reduction in cheek tooth number. Canines on the whole have been the most stable of all mammalian teeth. Although they differ much in size, they are with few exceptions simple cones and have disappeared only in the rodent-like animals, some ungulates, and a few other variously aberrant creatures. Reduction of incisors and premolars commonly began with those next adjacent to the canines and continued forward with incisors and backward with premolars. The cud-chewing ungulates may be an exception, because camels have retained only the last of the three incisors counting backward in the

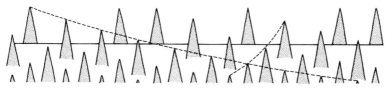

Fig. 510 Schematic illustration of reptilian teeth showing in a general way the pattern of replacement. Teeth are erupted and are simultaneously functional in about half the tooth positions. They seem to grow up in a wave beginning at the back of the mouth and moving forward, as indicated by the long dashed line sloping downward to the right. Either this is an illusion, however, or it was transformed into a series of shorter waves moving in the opposite direction, as indicated by the shorter dashed line sloping to the left. See Fig. 511.

upper jaw. Most other ruminants lack all upper incisors, and the lower ones are opposed by a calloused pad. This may in some way have been advantageous to certain herbivores but how is difficult to guess. Such a condition evolved perhaps more than once among the even-toed ungulates. It had been previously partly duplicated by some reptiles.

The elimination of one or more premolars commonly produced a gap in the tooth series behind the canine. A similar gap occurs, however, in some animals without the loss of any teeth. This is supposed to be related in part at least to some aspect of jaw musculature. Posterior molar teeth have been eliminated in regular order. This process was in operation in some of the therapsid reptiles and is very evident among the earliest mammals. The last human molar, or wisdom tooth, now exists in a reduced condition, never erupts in some individuals, and presumably will eventually be lost. Except for toothless creatures, elimination of molars has progressed furthest in cats, which retain only a single very small one in the upper jaw (Fig. 524-2).

Increase in Number

The general evolutionary trend toward tooth reduction in mammals is reversed in a few groups of highly specialized or otherwise peculiar animals that are dissimilar in ancestry, diet, and habits. They all have relatively simple undifferentiated teeth, more numerous than is normal for most mammals. Such creatures provide examples of convergent evolution that, unlike most others, seems to be entirely unrelated to convergence toward a similar way of life.

The largest and most important group with more numerous teeth occurs among the whales and includes many of the porpoises. These are actively swimming carnivorous inhabitants mostly of the open seas. Their peg-like or wedge-shaped teeth probably correspond to incisors, canines, and cheek teeth, but distinction between these types cannot be made. The number of teeth reaches about 75 in each jaw series of one species.

The manatees, or sea cows, of the American Atlantic Coast also are highly adapted to the marine environment, but they are rather sluggish herbivores, inhabit-

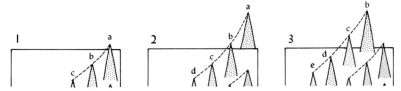

Fig. 511   Schematic diagrams showing three successive stages in the upgrowth of reptilian teeth and illustrating the short backward-moving waves that continued as long as an animal lived. This is the replacement pattern that was inherited by placental mammals.

ing mainly bays and estuaries. The teeth of these animals, numbering 20 or more in each series, are confined to the cheek regions. They are of the grinding type and less simple than the teeth of whales. These teeth grow up in succession at the back of the jaws and push their predecessors forward and out, so that only 5 or 6 are functional at the same time. Thus they resemble the teeth of elephants, and partly for this reason these animals are believed to be related.

The teeth of armadillos also are confined to the cheek regions. They range upward in number to about 25 in one species, are all of similar peg-like form, and lack an outer coating of enamel. These creatures feed mainly on insects and other invertebrates, but they will eat almost any kind of animal matter, including carrion.

Tooth Replacement

Most reptilian teeth are replaced continuously as long as an animal lives by new ones growing up at or near the bases of the older worn-out teeth. Replacement appears to progress along the jaw in forward-moving waves, affecting alternate teeth in such a way that about half of them are more or less mature and functional in every part of the jaw at any time (Fig. 510). The waves became fewer among the therapsids, and the cheek teeth of some advanced mammal-like reptiles seem to have been replaced only once in a single wave beginning at the front and extending backward, which affected each tooth in regular succession (Fig. 511). This pattern was inherited by the mammals. In them, however, the single wave of replacement is not complete and does not continue beyond the premolars (Fig. 511a).

The teeth of young mammals generally erupt regularly in order from in front

Fig. 511a Generalized pig-like dentition showing the pattern of replacement. The milk incisors have dropped out, and the permanent teeth are taking their places. Resorption of the root of the canine is almost complete. The permanent premolars are growing up in regular order. Molars are not replaced.

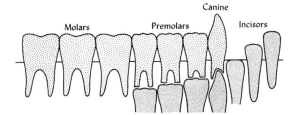

backward except for the canines, which may be somewhat delayed in their appearance. Before the first set of teeth is completely in place, however, the wave of replacement begins. The molars probably are members of the first set. They are not replaced, and the posterior ones may not erupt until after replacement of the premolars has been completed. The teeth of the second set are not exactly like those which they replace. The anterior teeth are larger and fill in the space in jaws that have grown in size. An adult animal with its molar teeth generally is less dependent than a young one on premolars for mastication. In conformity with their functional requirements, the milk premolars, except among some herbivores, generally are more molarized than the permanent teeth which replace them. Thus the morphologic distinctions between premolars and molars ordinarily are greater after the premolars have been replaced.

BIBLIOGRAPHY

A. S. Brink (1957): Speculations on Some Advanced Mammalian Characteristics in the Higher Mammal-like Reptiles, *Palaeontol. Africana,* vol. 4, pp. 77–96.
Some evidence suggests that certain reptiles acquired mammalian characters before evolution produced the mammalian type of jaw structure.
R. Broom (1932): "The Mammal-like Reptiles of South Africa and the Origin of Mammals," H, F, and G. Witherby, London.
See chap. 13, pp. 308–333, which traces the morphologic evolution of therapsid reptiles toward the mammalian condition.
P. M. Butler (1941): A Theory of the Evolution of Mammalian Molar Teeth, *Am. J. Sci.,* vol. 239, pp. 421–450.
This article traces in detail the theoretical evolution of cusp patterns in the molar teeth of early mammals.
A. W. Compton (1962): On the Dentition and Tooth Replacement in Two Bauriamorph Reptiles, *Ann. S. African Museum,* vol. 46, pp. 231–255.
The theoretical transition from reptilian to mammalian type of tooth replacement is discussed.
A. G. Edmund (1960): Tooth Replacement in the Lower Vertebrates, Roy. Ontario Museum, Life Sci. Div., Contrib. 52.
Tooth replacement is reviewed in all main reptile groups and early birds, and generalizations are reached regarding its wave-like occurrence.
W. K. Gregory (1934): A Half Century of Trituberculality: The Cope-Osborn Theory of Dental Evolution, with a Revised Summary of Molar Evolution from Fish to Man, *Proc. Am. Phil. Soc.,* vol. 73, pp. 169–317.
This work contrasts theories of dental evolution and presents the evidence in great detail.
J. A. Hopson (1967): Mammal-like Reptiles and the Origin of Mammals, *Discovery,* vol. 2, no. 2, pp. 25–33.
Evolution is traced in an easily understood, nontechnical way.
W. W. James (1960): "The Jaws and Teeth of Primates," Sir Isaac Pitman & Sons, Ltd., London.
This book is interesting for its many excellent photographs, but the text is particularly undistinguished.
E. C. Olson (1944): Origin of Mammals Based upon Cranial Morphology of the Therapsid Suborders, *Geol. Soc. Am. Spec. Paper 55.*
See sections Middle Ear, pp. 39–60, and Source of Mammals, pp. 120–126.
———— **(1959):** The Evolution of Mammalian Characters, *Evolution,* vol. 13, pp. 344–353.
The acquisition of mammalian characters probably was related to the development of warm-bloodedness rather than to external environment.
H. F. Osborn (1907): "Evolution of Mammalian Molar Teeth to and from the Triangular Type," The Macmillan Company, New York.
This book is mainly a compilation of material derived from the author's previous publications.
B. Patterson (1956): Early Cretaceous Mammals and the Evolution of Mammalian Molar Teeth, *Fieldiana, Geol.,* vol. 13, pp. 1–105.
The author presents a detailed discussion of the relationships and evolution of early mammals, based mainly on tooth structure.
A. S. Romer (1961): Synapsid Evolution and Dentition, in "International Colloquium on the Evolution of Lower and Non Specialized Mammals," 2 vols., text and plates, G. Vandebroek (ed.), Royal Flemish Academy of Science, Letters, & Fine Arts, Brussels, Belgium.
This article is devoted mainly to discussions of dental differentiation and replacement in the reptilian ancestors of mammals.

_____ **(1966):** "Vertebrate Paleontology," 3d ed., The University of Chicago Press, Chicago.
This outstanding book is the source of much evolutionary information.

G. G. Simpson (1929): "American Mesozoic Mammalia," Memoirs of the Peabody Museum,
vol. 3, Yale University Press, New Haven, Conn.
This is a monographic study; little can be added except with respect to later discovered early
Cretaceous fossils.

_____ **(1961):** Evolution of Mesozoic Mammals, in "International Colloquium on the Evolution
of Lower and Non Specialized Mammals," 2 vols., text and plates, G. Vandebroek (ed.), Royal
Flemish Academy of Science, Letters, & Fine Arts, Brussels, Belgium.
The many uncertainties in the phylogenetic relations of Mesozoic mammals are stressed.

J. M. Smith and **R. J. G. Savage (1959):** The Mechanics of Mammalian Jaws, *School Sci. Rev.,*
no. 141, pp. 289–301.
This article analyzes the functional relations of teeth and jaw muscles in carnivorous and
herbivorous mammals.

A. Tumarkin (1955): On the Evolution of the Auditory Conducting Apparatus: A New Theory
Based on Functional Considerations, *Evolution,* vol. 9, pp. 221–243.
Early vertebrates perceived sounds transmitted through their bones; eardrums and better
reception of airborne vibrations evolved independently in several groups of small animals.

D. M. S. Watson (1953): The Evolution of the Mammalian Ear, *Evolution,* vol. 7, pp. 159–177.
This article presents the orthodox theory and relates evolution of the middle ear to changing
jaw functions.

J. M. Weller (1968): The Evolution of Mammalian Teeth. *J. Paleontol.,* vol. 42, pp. 268–290.
Part of the preceding chapter is a condensation of this article.

CHAPTER
ELEVEN

VERTEBRATES 4

CENOZOIC PLACENTAL MAMMALS

Mammals evolved rapidly during the Tertiary epochs. The early branching of mar-supials, placentals, and probably monotremes had been accomplished before the end of the Cretaceous Period. The marsupials and monotremes were considered briefly in the preceding chapter. Subsequent evolutionary radiation of the placentals was the outstanding biologic event of the earliest Tertiary. In a relatively brief interval of time, at least seven great divisions had differentiated: (1) insectivores, (2) carnivores, (3) primates, (4) ungulates, (5) rodents, (6) edentates, (7) whales, and a few relatively more minor ones (Fig. 516). Each of these divisions had its own later complex evolutionary history. Consequently placental evolution requires the consideration of a multitude of animals. The facts that (1) Tertiary mammalian fossils occur at many places in considerable numbers and variety, (2) generally they are more favorably preserved than older vertebrates, (3) they have been intensively collected and studied, and (4) many mammalian lineages are represented by living animals available for comparison, all combine to provide a wealth of data to be organized and evaluated.

EARLY PLACENTAL RADIATION
The remains of animals believed to have evolved to the mammalian grade of devel-opment have been found in strata of latest Triassic or earliest Jurassic age. Mam-mals of primitive modern types, however, are not known to have existed before late Cretaceous time. Fossils of this age are rare and fragmentary, but evolutionary

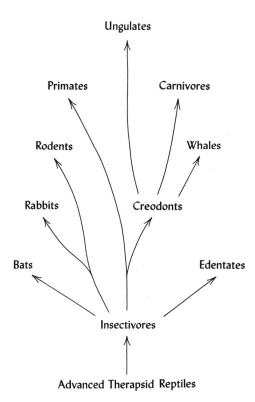

Ungulates

Primates   Carnivores

Rodents   Whales

Rabbits   Creodonts

Bats   Edentates

Insectivores

Advanced Therapsid Reptiles

Fig. 516 Diagram showing the probable evolutionary relations of the main groups of placental mammals. Ancestry of the rodents, rabbits, and edentates among insectivores is somewhat uncertain.

radiation seems already to have been well advanced, because both marsupial and placental types have been identified among them.

The reasonably adequate fossil record of the mammals begins in the early Tertiary. The dominant reptiles of the Mesozoic had but recently disappeared, and all the earth was suddenly opened to colonization by the mammals. These animals, particularly the placentals, rapidly took advantage of their opportunities. In a time that was relatively brief by geologic standards, evolutionary differentiation had produced the ancestors of all the great late Tertiary and modern groups.

This time in the early Tertiary, when the major part of placental mammalian evolutionary radiation was accomplished, is considered by most vertebrate paleontologists to be the first epoch of the Tertiary, termed Paleocene. Comparable evolutionary progress did not occur at this time among invertebrate animals or plants. No seemingly major stratigraphic discontinuity sets off the earliest Tertiary formations from the later ones. Consequently many geologists and many other paleontologists do not recognize a separate Paleocene Epoch but consider rocks and fossils of this age to be early Eocene. Thus confusion is possible because the expression early or Lower Eocene has different meanings for different persons. In this chapter, however, Paleocene is accepted as a useful division of the Tertiary.

Insectivores

The insectivores constitute the most primitive group of placental mammals, dating from the late Cretaceous (Fig. 517) and continuing to the present. These animals received their name because of the observation that many modern representatives of the group feed principally on insects. This generalization with respect to diet, however, certainly is not valid for all these creatures, either past or present.

On the whole the insectivores, which include such modern animals as shrews, moles, and hedgehogs, are a rather nondescript group of small creatures, not all of which are necessarily closely related to one another. They are classed together more because they lack the advanced characters of other mammalian groups than for any other reason. The upper molars generally are of the primitive triangular type (Fig. 517-2), although posterior heels may be developed. Some resemble the teeth of pantotheres rather closely, and the theory that an outer cusp divided to form two may have resulted from the study of such teeth. The incisors of one small section of this group resemble those of rodents. The limbs are unspecialized except in burrowing forms, and most insectivores probably walked flat-footed on the soles of their feet.

The Paleocene radiation of early insectivores, perhaps already begun in the late Cretaceous, may have produced the stems from which all other placental groups evolved (Fig. 516). Certain insectivores found in early Tertiary or later strata exhibit some primate-like or carnivore-like characters, but the diverging stages in these and other lineages generally are not known. Most of the insectivores were small animals. Many of them were soon supplanted by more advanced mammals. They are best represented in modern faunas on Madagascar, which lies off the African east coast. This large island seems to have been isolated from other land areas since early in Tertiary time. Its inhabitants escaped the pressures of competition with many types of mammals that evolved later in other regions.

Bats

Bats probably evolved from arboreal insectivores very early in the Tertiary. The skulls and teeth of some of them are so similar to those of typical insectivores that these animals would not be considered a different group had they not acquired the

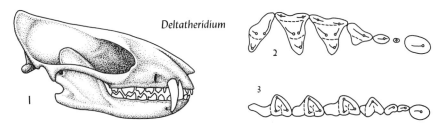

Deltatheridium

Fig. 517   Skull and cheek teeth of a late Cretaceous insectivore. Its dental formula is 4,1,3,3. (After Gregory and Simpson, 1926, Am. Museum Nat. Hist., Nov. 225, figs. 4, 8, pp. 5, 9.)

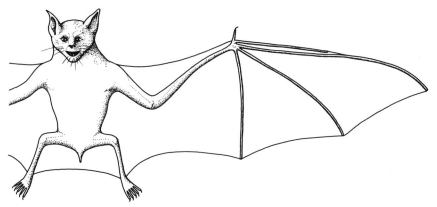

Fig. 518 The outer part of a bat's wing is supported by four elongated fingers. (*Drawn from photograph in Allen, 1939, fig. 9, p. 49.*)

ability to fly. The most ancient certainly identified fossil bats are of mid-Eocene age. Some of these fossils are as far advanced morphologically as modern species.

Wings are the most characteristic features of bats. These are thin expanses of skin much like those of the pterosaurs except that they are supported by four spreading fingers instead of one (Fig. 518). Progressive evolutionary modifications are shown by reduction or loss of some of the terminal finger bones. Claws are missing from the elongated fingers, except for one or two in some species. The shorter thumbs, however, remained free, clawed, grasping organs. The development of wings required strengthening of the shoulder girdle, but pelvic structures were reduced in size. The hind legs are weak and serve mainly as organs by which bats hang suspended upside down when they are at rest.

The teeth of some bats exhibit several conspicuous diversions from the primitive insectivore types, and a few attain the quadrate shape. These probably are adaptations to different feeding habits. Many bats have continued to prey on insects, which they catch while on the wing. Some, however, eat fruit, others flowers and leaves, and a few became blood lappers or capture fish. Probably the bats' most remarkable accomplishment was the perfection of a radar-like system of navigation. Reflections of high-pitched squeaks permit them to locate and avoid objects while flying in total darkness.

Other Relatives of Insectivores

If all later placental mammals evolved from primitive insectivores, intermediate forms leading toward the other groups might be discovered among the early Tertiary fossils. Although species that perhaps are actually the connecting links with other well-known mammalian groups are very rarely found, others that represent progressively advanced evolutionary grades serve to establish what seems to be satisfactory evidence of connections. This is so to some extent for the carnivores, primates, and ungulates, whose evolutionary developments are discussed in succeeding sections of this chapter. Comparable connecting forms leading to the rodents, edentates, whales, and possibly a few more minor groups, however, are not yet known. Present lack of evidence in this respect certainly is disappointing, but it is not regarded as being particularly significant in casting doubt on the evolutionary origin of these groups.

CARNIVORES

The large group of animals known collectively as carnivores includes most familiar predatory and carnivorous mammals but not insectivores, bats, whales, and some marsupials that also prey on other living animals. The question may be asked: Why did not carnivorous habits develop in many other groups of mammals? The answer is very simple. The primitive stock of the insectivores, presumably at the base of all placental evolution, was in a very real sense carnivorous because these animals preyed on other living creatures even if they were only insects or other invertebrates. The so-called carnivores simply continued this ancestral way of life while evolving into improved hunting-and-killing creatures (Fig. 519). Only the descendants of the more efficient ones have survived to the present day. Most of these are closely

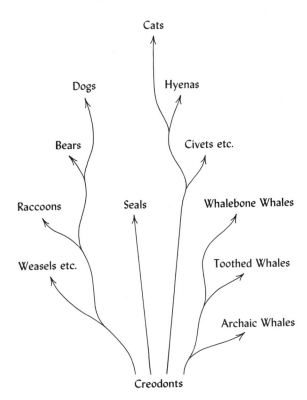

Fig. 519 Diagram showing the probable evolutionary relations of the main groups of carnivores and whales.

related in their origins, because they evolved from a single small group of Paleocene creatures equipped with a particular kind of teeth. The evolution of most other placental groups was directed toward other feeding habits. Some supplemented their originally carnivorous diets with fruit, carrion, succulent vegetation, and other readily digested foods and became omnivorous. Animals of this kind commonly developed crushing teeth like those of the primates. The acquisition of more specialized teeth and digestive organs led to the evolution of exclusive herbivores. Among these are the ungulates, or hoofed mammals, and most rodents that can eat harsher vege-table material and seeds.

The classification of mammals worked out by zoologists parallels rather closely the feeding habits of the major groups. Generalizations that are without exception, however, cannot be made. For example, among the carnivores most bears are omnivorous, and many of them eat more plant than animal food, and the giant panda, although technically a carnivore, is exclusively herbivorous.

Archaic Carnivores
The creodonts, or archaic carnivores (Fig. 520-1,2), evolved from small, probably arboreal insectivores. These animals are distinguished chiefly by the characters of

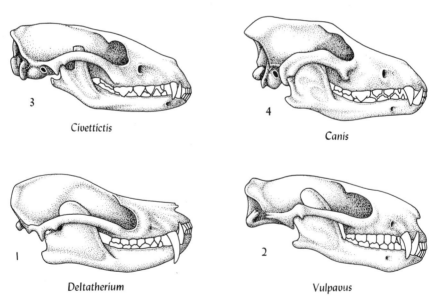

3

Civettictis

4

Canis

1

Deltatherium

2

Vulpavus

Fig. 520   Skulls of four carnivores. 1, A primitive Paleocene creodont. 2, A somewhat advanced Eocene creodont. 3, Modern civit, a viverrid. 4, A Pleistocene wolf. (*After Gregory, 1951, figs. 20.4B, 20.18C, 20.33A, pp. 731, 742, 757; Romer, 1966, fig. 295B, p. 190.*)

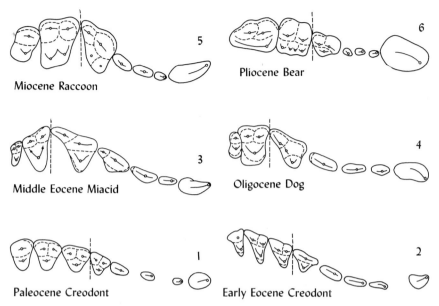

Fig. 521 Upper cheek teeth of a variety of carnivores. 1, *Mesonyx*. 2, *Sinopa*. 3, *Vulpavus*, belonging to the group from which all modern carnivores evolved. 4, *Temnocyon*. 5, *Phlaocyon*. 6, *Arctotherium*. Notice advance from the early triangular molars (left of dashed lines) to later square ones, loss of the third molar, tendency for the second molar to decrease in size, and molarization of the last premolar. (*After Romer, 1966, fig. 334, p. 230.*)

their teeth (Fig. 521-1,2). The early ones had a full complement according to the primitive placental formula of 3,1,4,3, but a few had lost the first premolar or last molar. The incisors became flattened nipping teeth, rather than pointed cones, and the canines were enlarged. Their molars were of primitive tribosphenic form. The collarbones or clavicles of their shoulder girdles, unlike those of most other carnivores, were not reduced or absent (Fig. 430-6). Evolution during the Paleocene and Eocene epochs produced a considerable variety of animals reflecting different adaptations. Most of them became extinct before the beginning of the Oligocene, although a few types persisted into the early Pliocene.

Many creodonts hunted on the ground. Some grew as large as bears. Most of them walked on flat feet whose number of toes was not reduced. A few had hoof-like nails, but sharp claws were common. The molar teeth of some became broad and flattened, suggesting omnivorous habits or other peculiarities of diet. The more advanced creodonts developed pairs of elongated shearing teeth, with sharp cusps arranged along their edges. In different lineages the transformed teeth were (1) the second molar above and the third molar below, (2) the first molar above and the second molar below, or (3) the last premolar above and the first molar below.

Modernized Carnivores

Primitive carnivores with shearing teeth in the last premolar and first molar positions were a minor element of the Paleocene and early Eocene faunas. In the late Eocene, however, evolution within this group began the radiation that produced all modern placental carnivores. Few of the older types survived the early Oligocene, and by the end of that epoch most of the modern families had differentiated. Skeletons of these newer beasts show few substantial structural alterations, the most noteworthy being (1) several small bones of the wrist were fused, (2) a bony capsule developed that surrounded and protected the middle ear, and (3) the third molars generally had disappeared (Fig. 521-3 to 6). A few of the older carnivores walked somewhat raised up on their toes. The metapodials of many of the newer ones became a little longer and were mostly raised above the ground, producing longer legs and resulting in a more springy gait. None of these changes seems adequate, however, to explain the rapid success of the newer carnivores. Probably this was a consequence of the evolution of a larger brain and the development of increased intelligence.

Miscellaneous Small Carnivores

Two groups of carnivores began a differentiation that was not completed until the Miocene Epoch. The modern representatives are mostly small, very aggressive flesh eaters. One group, the viverrids (Fig. 520-3), confined mainly to the Old World tropics, includes such animals as the civets and mongooses. These have lost the last molar from each tooth series, and the first premolar also may fail to develop. The other group, the mustelids, inhabiting the northern temperate zone, includes weasels, otters, minks, etc., and many of its members are more or less arboreal. These animals have only one molar remaining, and the anterior premolars commonly are reduced or lost. Some of this second group, such as badgers and skunks, are omnivorous, and their molars may bear four cusps.

Dogs

A group of animals with dog-like characters had evolved by late Eocene time (Fig. 520-4). Subsequently they differentiated in many minor ways. These beasts probably were derived from mustelid ancestors. They retained two molars above and may have had three below. The anterior molars were adapted to some crushing or grinding functions (Fig. 521-4). These animals walked on their toes, which were short and bore blunt nails. The first digits of all the feet were reduced and became nonfunctional. As a group, the wolves and dogs developed running habits in pursuit of prey, although the foxes hunt much more by stealth.

Raccoons

The raccoons, which are first recorded in the late Miocene, seem to have evolved from ancestors that were dog-like but arboreal. Raccoons have all toes functional and well developed, and these animals are good climbers. They became omnivorous, and their teeth largely lost shearing action. The molars are good grinders, and the

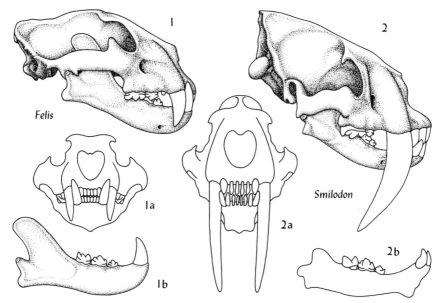

Fig. 523   Skulls and lower jaws of two Pleistocene cats. 1, A true cat. ($\times \frac{1}{8}$.) 2, A saber-toothed "tiger." ($\times \frac{1}{6}$.) *(After Merriam and Stock, 1932, Carnegie Inst. Wash., Publ. 422, pls. 1, 2, 26, 27.)*

first upper ones bear four cusps (see Fig. 521-5). The lesser pandas, in spite of their bear-like appearance, evolved from the same stock that produced raccoons. Giant pandas, as previously mentioned, are strict herbivores and may be more closely related to the bears.

Bears
Bear-like dogs appeared in the Oligocene, but the divergence of true bears did not occur until the Miocene. These relatively heavy animals retained all their toes and walked flat-footed. Two molars are present in each tooth series (Fig. 521-6). An omnivorous diet is reflected by the teeth, which lost their shear. They are good grinders with wrinkled surfaces bearing three or more low cusps.

Cats
The cats probably are the best-characterized group of carnivores. They are believed to have evolved from generalized viverrid ancestors. Typical cats existed in late Eocene time, and no noteworthy structural change has since appeared (Fig. 523-1). Most of these animals have retractile claws and are without the first digit of the forefeet. They have only one molar in each tooth series, and the upper ones are very small (Fig. 524-2). Two of the premolars generally function as shearing teeth, and the anterior premolars are small or absent. Unlike the dogs, most cats have

Fig. 524   Upper cheek teeth of two Pliocene carnivores. 1, *Hyaena*. 2, *Metailurus*. Notice that the number of premolars has been reduced from 3 to 2, the last one has become a shearing tooth, and only one small molar remains. (*After Romer, 1966, fig. 334, p. 230.*)

not been good runners. They hunt by stealthy stalking and suddenly leap upon their prey.

The early cats had very prominent canines. These teeth tended to become larger stabbing weapons in later species (Fig. 523-2), which are believed to have hunted large thick-skinned herbivores. Such animals persisted into the Pleistocene but are now extinct. Cats with short, more normal canines first are noted in the Pliocene. Their origin and possible relations to those with saber teeth have not been satisfactorily established.

Hyenas

The hyenas, whose earliest known fossils are of Miocene age, are believed to have evolved from cat-like ancestors. Their dentition, like that of cats, includes only a single tiny upper molar (Fig. 524-1). The other teeth are large, blunt, and well adapted to crushing bones. These beasts are mainly scavengers. Their general form is rather dog-like, and their toes bear blunt nails rather than sharp claws.

Fig. 524a   1, A true (earless) seal. 2, A sea lion. (*After Gregory, 1951, fig. 20.39A, C, p. 762.*)

Marine Carnivores

The seals and their allies have adapted so completely to an aqueous environment that their evolutionary origin is doubtful. Fossils are not known older than the Miocene. These animals exhibit some dog-like characters, but they may have descended from expert swimmers similar to the otters. The teeth do not shear and are rather uniformly peg-like, except for the canines, which are well developed. The incisors may be lost and the molars reduced in number. The body is streamlined, and the forelimbs are transformed to flippers without the loss of any digits. The hind legs are directed backward and are the main propulsive organs. They can be turned forward beneath the body by sea lions (Fig. 524a-2) and walruses but not by the true seals, which seem to have deviated furthest from their four-footed ancestors (Fig. 524a-1). All are very awkward creatures out of water.

PRIMATES

The primates, or the group which leads on to man, have in some respects attained the highest level of biologic evolution. Without qualification, however, this observation is an egocentric one, because in other respects the primates, including man, have retained several remarkably primitive characteristics. Most animals, and this includes the majority of primates, have become structurally specialized in various respects in adaptation to some particular way of life. This is less true of man, however, than of many other animals. The human lineage has evolved in a unique manner. Physically man's ancestors remained relatively unspecialized. Unaided by tools, machines, or other artificial devices, a human being is comparatively helpless. He cannot find food, run, swim, hide, or fight as well as many other animals, and of course he cannot fly. These obvious deficiencies, however, have been more than compensated by the increasing mental powers that evolved among the higher primates. Without this developing superiority, man's immediate nonhuman ancestors probably could not have competed successfully with many other animals and certainly man could not have attained his present dominance over all other creatures. By his intelligence, inventiveness, communication with others of his kind, and cooperative activities, man has met the challenges of every geographic and climatic environment and overcome them all. He is now in the process of learning how to adapt himself to conditions in outer space where no other earthly organism can exist without man's aid.

The evolutionary progress of the primates can be traced through a succession of advancing stages represented by (1) lemurs, (2) tarsiers, (3) monkeys, (4) apes, and leading finally to (5) men (Fig. 526).

Lemurs

The earliest mammals of modern types, as previously mentioned, are believed to have been arboreal and perhaps nocturnal. Although many insectivores adapted to

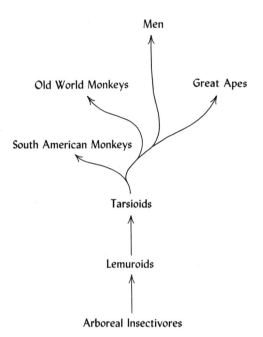

Men

Old World Monkeys   Great Apes

South American Monkeys

Tarsioids

Lemuroids

Arboreal Insectivores

Fig. 526 Diagram showing the probable evolutionary relations of the primates.

living on the ground and became variously specialized, some of them maintained these ancestral habits and retained many of the primitive characters of the group. The modern tree shrews of the oriental tropics are animals of this kind (Fig. 527-1). Morphologically they seem to bridge the gap between insectivores and primates. According to different classifications, they have been grouped with either the insectivores or the lemurs. Fragmentary fossils of similar animals have been found in strata of Paleocene and Eocene ages. These probably represent the stock from which the earliest lemuroids evolved.

The fossil remains of several varieties of Paleocene and Eocene lemuroids (Fig. 527-2) show evolutionary advancement beyond the insectivore stage by a general squaring up of the molar teeth which otherwise are primitive in structure. Incisors were reduced to two in each tooth series, one less than in insectivores. A bony bar separates the eye socket from the large temporal opening, a condition present in only a few of the most advanced insectivores. The first toes, especially those of the hind feet, were set off at an angle from the others (Fig. 528-2), suggesting grasping arboreal habits. Most of these animals had specialized in such ways that they could not be ancestral to later primates. With few exceptions, the skeletons are very incomplete, but some idea of the nature of ancient lemurs probably can be gained from study of the more primitive species now inhabiting Madagascar. These are mostly small nocturnal and omnivorous creatures resembling squirrels in general form and behavior. They sit up on their haunches and use their forefeet to some extent in manipulating food.

Tarsiers

Tarsiers are known mainly from Paleocene and Eocene fossils (Fig. 527-3) and a single modern species. Advanced lemurs show some shortening of the snout and enlargement of the eyes, and in the tarsiers, these evolutionary trends were carried considerably further. In addition, the tarsiers' eyes were directed forward, rather than laterally as in most animals, so that stereoscopic vision became possible. Good eyesight is especially important for active arboreal and nocturnal animals. Advantages gained in seeing, however, seem to have been obtained at some sacrifice in the sense of smell. Large eyes set close together in the front part of the skull encroached upon the nasal passages and produced a flatter face. Reduction in the field of vision in this way was compensated by the development of a slender neck and increased mobility of the head. Gradual enlargement of the brain had been in

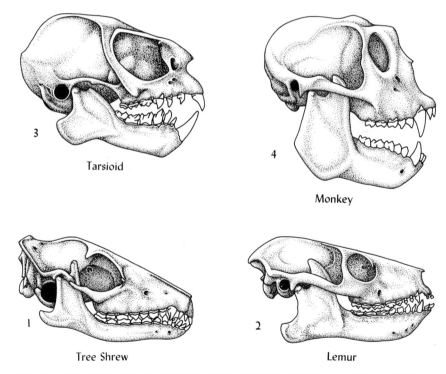

3

Tarsioid

4

Monkey

1

Tree Shrew

2

Lemur

Fig. 527 Four primate skulls. 1, Oligocene, *Anagale*. 2, Eocene, *Notharctus*. 3, Eocene, *Tetonius*. 4, Pliocene, *Mesopithecus*. (After Romer, 1966, figs. 329, 330, pp. 218, 222.)

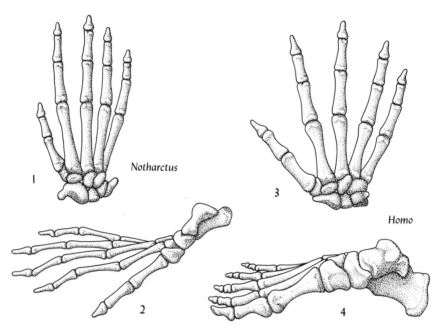

Notharctus

Homo

Fig. 528   Bones of the primate hand and foot. 1, 2, An Eocene lemur. 3, 4, Modern man. Resemblance is very close in the hand, but the foot has changed considerably in adaptation to walking on the ground rather than climbing in trees and bushes. (*After Gregory, 1920, Mem. Am. Museum Nat. Hist., N.S., vol. 3, pt. 2, fig. 83, p. 238.*)

progress among the early primates. In tarsioids this produced a higher and wider skull. Such change in shape seems partly responsible for shifting the area of neck articulation from the posterior surface of the skull downward toward its base. The teeth of tarsioids remained in a relatively primitive condition.

South American Monkeys

South and Central American monkeys are not on the main line of primate evolution. They appear much like other monkeys, but their ancestors migrated from northern regions in the early Tertiary and subsequently they have been isolated and have developed entirely independently. Nevertheless these animals represent an evolutionary stage more primitive than that of the Old World monkeys. This is shown particularly by their tooth formula of 2,1,3,3, which is similar to that of the lemurs and tarsioids. Dentition of the other primate branch, with only two premolars (Fig. 527-4), must have passed through a similar stage. Other characteristic features of the American monkeys, such as the prehensile tail and nostrils that are located far apart and face outward, are peculiar to them. These are believed to be evidence of descent from a common probably simian ancestor that had evolved among the tarsioids.

Old World Monkeys

The Old and New World monkeys share many common features. The brain is conspicuously larger than in the lower primates. The eye sockets have a bony partition behind them extending inward from the postorbital bars. The molar teeth of both upper and lower jaws are quadrate with four low cusps, and a fifth one begins to be evident on the posterior margins of the lower teeth. The Old World monkeys do not have prehensile tails, and their nostrils are closer together and face forward or downward. They have one less premolar in each tooth series, so that their dental formula is 2,1,2,3, as in the great apes and man. Most monkeys are gregarious, and many of the Old World species have developed a somewhat more advanced type of social organization. Under some circumstances they cooperate in intelligent ways and help one another. Rather obviously, these animals have evolved a little further toward the human condition.

The similarities and differences of these two groups of monkeys have been variously interpreted with respect to their evolutionary origin and the time of their divergence from an ancestral stock and from each other. The earliest known Old World fossils come from the Middle Oligocene, but monkeys probably were fully developed during the Eocene Epoch. They might have evolved from either advanced lemuroids or early tarsioids. If the monkeys' immediate ancestors were known, they probably would be classified as tarsioids.

Most monkeys have been arboreal and omnivorous, but some, adapted to living on the ground, evolved from the Old World group and became almost exclusively herbivorous. These are the baboons, which are rather dog-like in appearance. They run on all fours, with the soles of the feet and palms of the hands flat on the ground, and their toes and fingers have been reduced in length. Tails, no longer useful for balancing, became short. The muzzle grew longer to accommodate larger teeth that probably evolved as more formidable defensive weapons.

Great Apes

The great apes include the modern gibbons, orangoutangs, chimpanzees, and gorillas. These beasts are much larger than most monkeys. They are mainly herbivorous and arboreal, but their size has prevented them from running along branches like smaller monkeys. They progress mostly by swinging through the trees suspended by their arms, which are very powerful. In contrast, the hind legs are less developed and relatively short. Both these features are adaptations to their way of life. These animals are able to stand and walk erect, but customarily they move about in a four-footed way. They are without tails, and their brains approach human conformation.

The ape's dentition is much like that of human beings. The molars have somewhat wrinkled crowns, and the lowers possess a fifth low posterior cusp (Fig. 530-1).

The canines, particularly in the males, are large, as in the monkeys, and the teeth are arranged in a broad U-shaped pattern with premolars and molars of the two sides approximately parallel to each other. Ape-like fossil jaws have been discovered in Upper Eocene and Lower Miocene strata; animals of this type branched from the other primates at an early date. They are believed to have evolved from arboreal monkeys of the Old World kind.

Men

One or more types of creatures sharing some ape-like and human-like characters have been found in Europe, Africa, and Asia in Tertiary strata of either late Miocene or early Pliocene age. Fossils of still more human type are widely distributed but rare in deposits of early and mid-Pleistocene ages. They seem to represent ground-living creatures that, outside the tropics, commonly sought shelter in caves. Among them are a few fragmentary jaws and teeth of relatively enormous size discovered in southern Asia. They are of primitive human type, with five cusps on the lower molars, but nothing more is known about the creatures to whom these scraps belonged. Somewhat more complete fossils from southeast Africa provide evidence of a race with bodies like men but heads resembling those of apes (Fig. 532-1). They are of a type that might have been ancestral to human beings. Their age, however, seems to be too recent for them to be considered likely ancestors. Java man (Fig. 532-2) and Peking man, which may represent a single species, also are mid-Pleistocene and are regarded as definitely human. These men made crude stone implements and are believed to have been accustomed to the use of fire. Although their brains were smaller than in modern man, they were relatively much larger than in any ape.

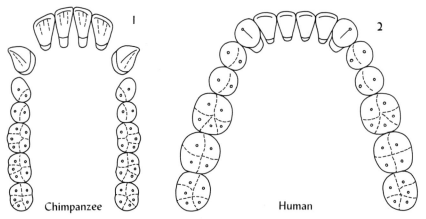

Fig. 530 Lower dentitions compared. The chief differences are in the shape of the jaw and relative prominence of the canine teeth. (After Gregory, 1951, fig. 23.80A, C, p. 990.)

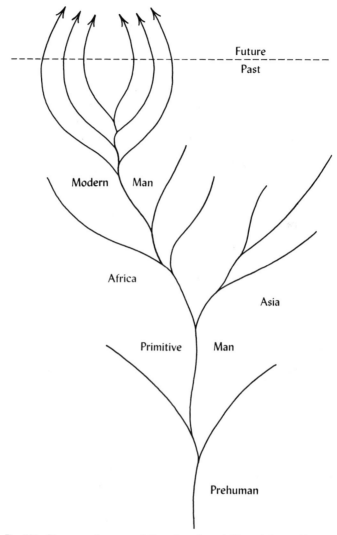

Fig. 531 Diagrammatic representation of man's probable evolutionary history.
(Adapted after de Chardin, 1956, in "Mans' Role in Changing the Face of the Earth," W. L. Thomas (ed.), fig. 50, p. 104.)

The fairly consecutive history of man in Europe begins at the time of the last interglacial stage, although a few scraps of older fossils have been found. Complete skeletons of Neanderthal man (Fig. 532-3) show all the characteristic human features. These differ from those of apes only in relatively minor ways, including the following: (1) A brain equal in size to that of modern man. (2) Completely bipedal walking, with comparatively long legs and feet modified from the hand-like feet of apes (Fig. 528-4). (3) Relatively broad hands with thumbs perfectly opposable to all the fingers. (4) Teeth that are more rounded in outline, with small canines, and with premolars and molars converging toward the front in a parabolic type of curve (Fig. 530-2). Neanderthal man retained certain features, however, suggesting ape-like ancestors. These include (1) a skull with heavy ridges above the eyes, (2) a receding chin, and (3) legs that were slightly bent, perhaps indicating a walking posture that was somewhat stooped, although recent studies suggest that this is an exaggeration. The Neanderthals might be the direct descendants of Java or Peking man or an offshoot of a more modern type of man.

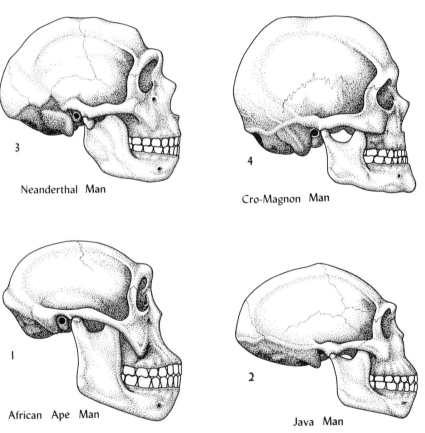

3

Neanderthal Man

4

Cro-Magnon Man

I

African Ape Man

2

Java Man

Fig. 532 Skulls of primitive and modern men. 1, *Australopithecus.* 2, *Pithecauthropus.* 3, 4, *Homo.* Notice increase in brain capacity, straightening of face, and development of chin. (*After Gregory, 1951, fig. 23.66B, p. 977; Romer, 1966, fig. 332, p. 224.*)

Table 533 The fossil record of human evolution

Modern Man	Postglacial Epoch
	Fourth Glacial Epoch
Neanderthal Man	Early Fourth Glacial
	Epoch
Primitive *Homo sapiens*	Third Interglacial Epoch
	Third Glacial Epoch
Java Man	Second Glacial Epoch
	First Interglacial Epoch
African Ape Man	First Glacial Epoch
	Late Preglacial Epoch

This succession probably is not a simple straight-forwardly connected genealogical sequence. Data from Clark (1955), "The Fossil Evidence for Human Evolution," fig. 3, p. 50.

Toward the close of the last glacial stage a new type of man invaded Europe from some unknown region. This was Cro-Magnon man (Fig. 532-4), the ancestor of modern Europeans. He was modern in every way and in a relatively brief interval of time completely supplanted the less-progressive Neanderthals. Cro-Magnon walked erect, with straight back and legs. His skull had a high forehead without protruding brow ridges. The face was short, with a well-developed nose and projecting chin. The teeth were somewhat smaller, with a fifth cusp imperfectly developed. Aside from the large braincase of his skull, his erect posture, and the nature of his teeth and feet, modern man exhibits few specializations in skeletal structure that are importantly different from those of the early primates.

UNGULATES

The ungulates cannot be easily defined, although they are commonly thought of as the large hoofed mammalian herbivores. All modern animals of this sort are included in the group, and they are the ones most familiar to mankind. Because of close relationships or other similarities, however, a variety of animals of different types are joined with them. Among these are small creatures as well as large, animals with claws or nails instead of hoofs, some that are carnivorous rather than herbivorous, and even a few marine mammals whose legs have been so transformed that they cannot leave the water. On the other hand, some relatively large herbivores are not included, such as the bigger marsupial kangaroos, the giant panda which is technically a carnivore, and gorillas among the primates. Even some herbivorous rodents equaled small bears in size.

The evolutionary sequence beginning with relatively primitive carnivores and proceeding to omnivores and finally to comparatively advanced herbivores has occurred many times in different lineages of animals, but at least among land vertebrates it has never been reversed completely. Advantage evidently was gained by the ability of an animal to subsist on a variety of different kinds of food. The final herbivorous adaptation also was advantageous because plants provide easily obtainable and nutritious food that exists almost everywhere and does not need to be hunted down and killed. Adaptation to a diet of leaves and grass, however, required specialized digestive organs capable of reducing cellulose-like material to more easily assimilable substances.

The ungulates are not so unified a group as either the carnivores or primates (Figs. 534; 542; 548; 559). The evolutionary origins of ungulates are tangled and uncertain, and the nature of the immediate ancestors of many of them is totally unknown. In all probability, however, several stocks differentiated very early in the evolutionary radiation of the primitive placentals. Adaptive evolution within these groups then progressed in a roughly parallel manner, and some remarkable examples of convergence can be recognized.

The evolutionary histories of the ungulates and carnivores are interrelated and very similar. In both groups the initial radiation, that may have begun in the late Cretaceous, produced a considerable variety of archaic early Tertiary animals. A second radiation, that was well under way in the late Eocene, introduced more modernized forms. These rather rapidly supplanted their less-progressive predecessors, and they became the ancestors of many animals that populate the world today. This parallelism is natural because of the close predator-prey relations of the carnivores and ungulates. As animals became more adept at either hunting or escaping their pursuers, compensating improvements in members of the other group were necessary for survival. The major impetus of the second radiation probably originated with the carnivores. Their evolutionary progress seems to have anticipated somewhat that of the ungulates, and more of the archaic ungulates than

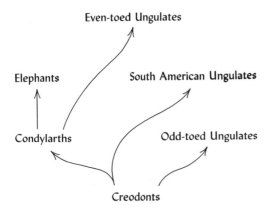

Fig. 534 Diagram showing the probable evolutionary relations of the main groups of ungulates.

of the comparable carnivores persisted among the newer animals for considerable periods of time.

Evolution of Legs and Feet

Some ungulates developed tusks or horns as weapons of defense. Other adaptations, however, also served to guard these animals against predatory enemies. Two that were successful in different groups were (1) increase in running speed so that attack by slower enemies was more easily avoided, and (2) increase in size so that attack by smaller enemies was in part at least discouraged. The first was accomplished by, and the second necessitated, evolutionary modifications in the structure of legs and feet.

The legs of primitive mammals were relatively short, with the upper and lower parts of approximately equal length. The feet contained short bones, and the soles were planted flatly on the ground in walking. The first toe commonly was directed inward at an angle to the others and was a grasping organ, useful to arboreal creatures.

Many small mammals developed feet slightly altered from this primitive condition. They rose up on their toes so that the metapodials, or first bones of the toe series (Fig. 432), and wrist and ankle bones were not in contact with the ground. This was an improvement, because it produced a structure delivering more effective forward thrust in running. In this position, the divergent first toe may have lost contact with the ground; there was a tendency for it to become smaller and eventually to disappear. This process of alteration seems generally to have progressed more rapidly in the hind feet than in the forefeet. As larger animals evolved in many lineages, foot structure of this type was maintained, but greater weight produced increasing strains that influenced the course of further evolutionary development in some of them.

Running Ungulates

The ability to run more swiftly was advanced by lengthening and slimming of the legs. The first stage in this process involved lengthening of the metapodials. This provided the legs with a third segment that articulated at a second joint corresponding to the primitive wrist or ankle (Fig. 536-3). Next, animals rose up on the tips of their toes, and the toe bones became parts of the new segment of the transformed leg (Figs. 542a; 552). Because the toes differed in length, some lost contact with the ground and became nonfunctional. Differences in the inequality of the toes resulted in the development of two kinds of structure. In the horses and their allies, the third toe was longest. It bore most of the animal's weight and became progressively stronger, while the others decreased in size (Fig. 552). First the inner toe was lost, then the outer one, and later the other two also disappeared and a foot consisting of a single toe eventually resulted.

The third and fourth toes were similarly long in cattle and their relatives whose legs evolved almost identically. These toes persisted, and the others disappeared, producing a two-toed type of foot (Fig. 542a). Differences in foot structure in the odd- and even-toed ungulates date back to a very early evolutionary stage. They serve to distinguish two groups of animals that diverged at the very beginning of ungulate evolutionary radiation.

The running leg perfected in this way has an upper segment that is relatively short (Fig. 536-3). The longer middle segment contains two parallel bones, but one of them carries most of the animal's weight, and the other, which generally is more slender, may fuse with it or perhaps disappear. The lengthened metapodials are the principal bones of the third segment. As toes lost their function or were eliminated, their metapodials became vestigial and tended to disappear. The remaining metapodials of some two-toed animals fused to form a single bone.

Hoofs are transformed claws (Fig. 537). The primitive narrow claws widened to form flattened nails, such as developed in the early primates. This trend continued in the ungulates, and the nails enlarged still more until they nearly enclosed the last toe bone. At the same time a spongy pad developed on the

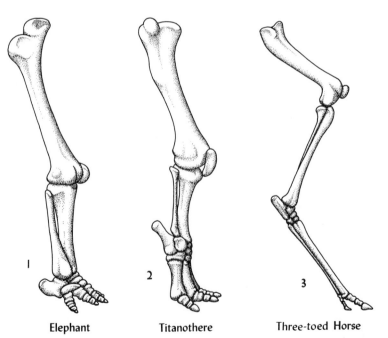

Elephant	Titanothere	Three-toed Horse

Fig. 536 Structure of the hind leg in heavy-bodied and running mammals. Notice the relative length and orientation of the metapodials. (*After Gregory, 1912, figs. 3.2, 5, pp. 280, 283; Osborn, 1929, fig. 606, p. 670.*)

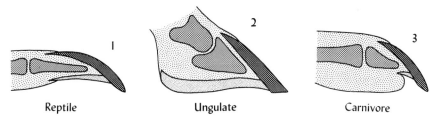

Fig. 537   Claws and hoofs are similar to the scales of reptiles in their epidermal origin. Very few amphibians have claws.

undersurface of the toe. Thus a very effective protective cover was provided for the
toe of an animal adapted to running on firm ground.

Heavy-bodied Ungulates

As animals grow larger, their weight tends to increase as the cubed power of their
dimensions but, without changed proportions, the strength of bone and muscle
increases only as the squared power. Obviously the slender, rather delicate legs of
running ungulates are not well adapted to support ponderous bodies. These require
much stouter legs. Consequently it is not surprising that the legs of elephants and
other even larger animals that existed in the past are built on a somewhat different
pattern. Not only are the bones more massive, but the legs have almost the form
of erect pillars, so that the weight is supported more directly by the bones. In
animals of this kind, the upper leg is characteristically longer than the lower and
the metapodials are not lengthened to form a well-defined third segment.

Evolution that took the direction of greatly increasing size and weight seems
to have begun in animals whose foot development had progressed to different
stages. If walking on the toes had not been well established, flatfeet resulted (Fig.
536-1). If the metapodials had been raised well above the ground, these bones
became directed almost vertically upward but did not lengthen greatly (Fig. 536-2).
In both cases broad rounded feet developed, with short toes. The first toe was lost
in some, but the number was not reduced further. The toes in feet of this kind bore
thick flattened nails, and hoofs were not produced.

Archaic Ungulates

Condylarths

A rather miscellaneous lot of fossil mammals found in Paleocene and Eocene strata
are grouped together as the condylarths (Fig. 538-1). Included among them are
hoofed mammals with teeth not yet evolved to a form distinctive of the herbivores,
and clawed animals whose teeth had reached that stage. Some had the slender
bodies, short legs, and long tails characteristic of many early carnivores, others were
differently formed, and some may have been arboreal. Most seem to have lacked

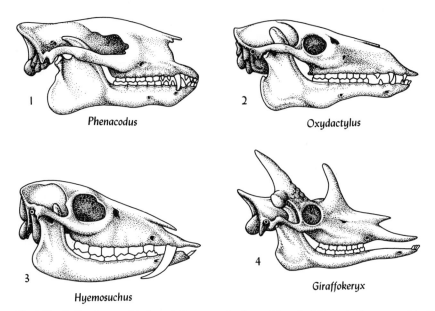

Fig. 538   Skulls of a condylarth and three even-toed ungulates. 1, This early Eocene condylarth resembled a carnivore in appearance. 2, An early Miocene camel, 3, A Pleistocene animal closely related to the modern mouse deer. 4, A Pliocene two-horned giraffe. (*After Gregory, 1951, figs. 21.69A, 21.102B, 21.107A, 21.111A, pp. 838, 872, 877.*)

clavicles, or collarbones, in the shoulder girdle (Fig. 430-7), an advanced feature, but the structure of their feet was primitive, with all toes present. These animals varied among themselves so much that their relationships are dubious, and they undoubtedly represent many different lineages. All show some features that seem to ally them with the ungulates. They lack, however, other features that would identify them as members of any group characterized more restrictively.

Animals of these kinds might have evolved from early creodonts or directly from insectivores, or both. Some of them might have been ancestral to more advanced ungulates. Presumably many of the later ungulates had ancestors that would be classed as condylarths if they were known. The possibility should be recognized, however, that the divergence of different ungulate lineages may have been a confused and complex radiation accomplished at a very early stage of placental evolution. Many relations are uncertain because distinctions among some fossil insectivores, creodonts, lemurs, and condylarths are neither clear nor sharp.

South American Ungulates

The Tertiary ungulate faunas of South America were unique. This continent was colonized in late Cretaceous or very early Tertiary time by a variety of primitive mammals. Breaking of the isthmian connection with North America isolated the region, and its inhabitants thereafter pursued a course of independent evolutionary radiation (Fig. 539). Placental carnivores did not reach South America at that early time, and the ungulates that developed there had to contend only with the less-progressive carnivorous marsupials. When the continental connection was reestablished in the late Tertiary, more advanced mammals, both carnivores and herbivores, invaded from the north. The indigenous ungulates suffered greatly from the resulting predation and competition, and before the end of the Pleistocene, all were driven to extinction.

The main group of these South American animals, known as notoungulates, diversified adaptively in many different environments, producing creatures ranging in form and size from rats and rabbits to large rhinoceroses. They possessed certain unifying features, including (1) peculiar bony structure in the region of the ear, and (2) teeth generally unreduced in number and grading transitionally from small canines to molars without any intervening gap. Teeth range from relatively primitive triangles to advanced square prismatic types that grew continuously and had enamel patterns resembling those of rhinoceroses and horses and an outer coating of cement. There were forms with rodent-like incisors. Some species primitively retained clavicles in the shoulder girdles, but many more did not. The skull structure of several different types suggests the presence of large mobile snouts that may have attained trunk-like form. None had a bony bar behind the eye·socket. Some of these beasts had metapodials raised above the ground, but none progressed to walking on the tips of toes. Both clawed and hoofed forms occurred, and the toes of some were reduced to three. Similar animals are almost unknown in other parts of the world. Certain features indicate that these creatures evolved entirely inde-

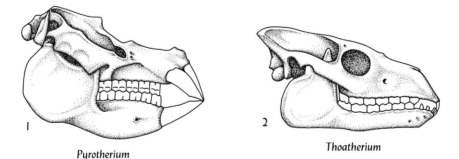

Pyrotherium   *Thoatherium*

Fig. 539 Skulls of two South American ungulates. 1, This beast from the Oligocene has nasal openings high in the skull between the eyes, suggesting the existence of an elephant-like trunk, and its incisors somewhat resemble tusks. 2, This Miocene litoptern represents a group of animals that evolved in much the same way as the horses of the northern hemisphere. (*After Romer, 1966, figs. 358, 381A, pp. 246, 260.*)

pendently of the northern condylarths. Understanding of them is imperfect, partly because there are no living animals with which they can be compared.

Other South American Tertiary ungulates probably evolved from different ancestral stocks. Some of them are especially interesting because they developed features remarkably similar to those of certain modernized ungulates of other regions. One group, the litopterns (Fig. 539-2), evolved in a way that closely paralleled the horses and produced both three-toed and one-toed species (Fig. 540). Members of another group, the pyrotheres, grew to resemble elephants in size and had tusk-like incisors (Fig. 539-1), large molars similar to those of primitive elephants, and probably long trunk-like noses. In spite of such similarities, it is very doubtful that these animals are closely related to others in distant regions. Some beasts were quite unlike any animals known elsewhere.

Other Archaic Ungulates

A few other types of relatively primitive ungulates have been discovered in Paleocene to Oligocene strata of the northern continents. Many of them, like the amblypods, were heavy beasts that grew to large size and had stout legs and feet with short, spreading toes ending in small hoofs. One of them, however, was equipped with claws, suggesting that it may have fed on roots which it dug up. Multiple horns of peculiar form evolved in another group, the dinocerates (Fig. 541-2). Their molars varied from relatively simple triangles to quadrate teeth with diagonal ridges unlike those of any other animals. Some of the males possessed large canines. The primitive ancestors of these creatures have not been found, and their relations to other mammals are not known.

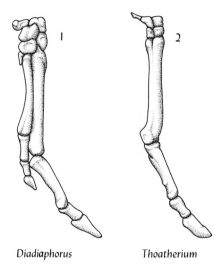

Diadiaphorus   *Thoatherium*

Fig. 540 Forefeet of three-toed and one-toed litopterns from South America. Their structure closely resembles that of North American horses. These animals provide an excellent example of convergent evolution. (After Howell, 1944, fig. 30D, T, p. 193.)

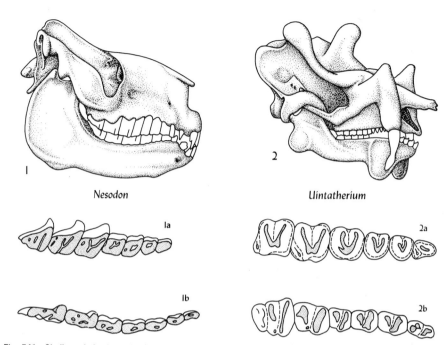

Nesodon

Uintatherium

Fig. 541 Skulls and cheek teeth of two peculiar archaic ungulates. 1, From the Miocene of South America. 2, From the Lower Eocene of the western United States. The tooth homologies of these animals are uncertain. (*After Romer, 1966, figs. 353B, 354C, E, 369C, 371A, B, pp. 244, 245, 256, 257.*)

Even-toed Ungulates

The even-toed artiodactyls are the most diversified and abundant ungulates in the world today (Fig. 542). Their fossil record begins in the early Eocene, and by the end of that epoch they had differentiated into the three great groups typified by (1) swine, (2) camels, (3) deer and cattle, and the more or less close relatives of these animals. Numerous now extinct minor groups also evolved during the Tertiary, some of which are difficult to classify.

All the ungulates belonging to the even-toed group are characterized by peculiarities in the articulation of the small ankle bones not shared by any other animals. Otherwise some of the more primitive species have features suggestive of insectivores, carnivores, or primates. The teeth in particular are somewhat like those of early carnivores. There can be no doubt that these ungulates differentiated very early from the primitive placental stock, and perhaps they were most closely related in their evolutionary origin to the carnivores.

Several parallel evolutionary trends are evident in different divisions of the even-toed ungulates. Perfection of the running leg as described in a previous section is one example. None of these animals except the hippopotamuses evolved in the

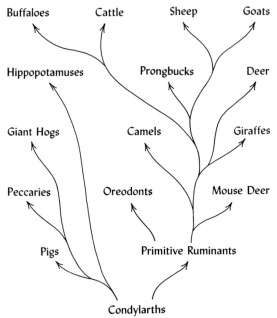

Fig. 542   Diagram showing the probable evolutionary relations of the main groups of even-toed ungulates.

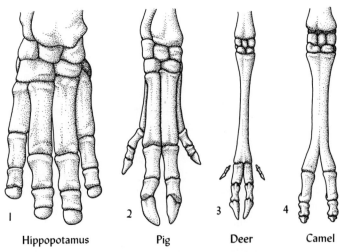

Fig. 542a   The bone structure in the feet of these modern animals illustrates stages in the evolution of the two-toed condition. (*After Howell*, 1944, fig. 20, p. 157.)

direction of ponderous form and the other type of leg. The first toe had been lost even in some of the earliest representatives, and four-toed species were common in the Eocene. Later the second and fifth toes were reduced and disappeared, and the metapodials of the third and fourth toes fused (Fig. 542a). Primitively the eye socket was fully open behind, but bony processes appeared and lengthened until they joined to form a postorbital bar. The teeth series originally were complete, with large canines and tribosphenic molars. The upper incisors and canines subsequently were reduced and vanished, the lower canines assumed forms similar to incisors, and the first lower premolars commonly were lost, leaving a gap in the tooth series. The molars, originally low-crowned, squared up in the Eocene and generally bore five cusps, two or more of them crescentic (Fig. 543). Higher crowns developed, the fifth cusp was lost, and cement coated the outer surface. The premolars never were molarized to the extent that was typical of horses.

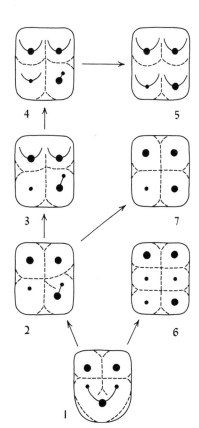

Fig. 543 Diagrammatic representation of the probable evolution of upper molar teeth in even-toed ungulates. Development seems to have started from a tribosphenic tooth that squared up without the appearance of a heel and its fourth major cusp (1). A minor cusp seems to have moved back to the position generally occupied by the fourth cusp (2, 3). Cusps in the four quarters then successively acquired crescentic form (3, 4, 5). Teeth of the last type (5) are characteristic of most ruminants, as this term is used broadly to include oreodonts, giraffes, and camels. A crescentless tooth with four main cusps (6) is characteristic of most pigs and their near relatives. An early side development (7), perhaps including a heel and fourth major cusp, occurs among some primitive forms.

Primitive Types

Several small, incompletely preserved Eocene and early Oligocene species, the palaeodonts, have the ankle structure of the even-toed ungulates but are so primitive that they cannot be included in any of the better-known groups. These animals generally were four-toed, but some had a small first toe on the front feet. Most Eocene forms had rather blunt cusped tribosphenic molars, but in the Oligocene six-cusped teeth developed which are different from those in the more advanced groups. The canines of some were large, but others were small and little different from adjacent teeth. Certain of the early Eocene species might have been ancestral to later ungulates, but the teeth of others seems to rule out any such possibility.

Swine

The pigs and their relatives constitute the most primitive surviving division of the even-toed ungulates. Their legs are relatively short, with metapodials not much lengthened. The teeth, which are low-crowned, have blunt cusps, and do not develop crescentic patterns, are consistent with their omnivorous feeding habits (Fig. 543-6). Pig-like animals appeared in the Eocene, and in the Oligocene had differentiated into three groups. All were four-toed. The true pigs seem to be the most primitive. Their leg bones have not fused. The lateral toes are small but functional on soft ground. The teeth have wrinkled crowns. The canines grow continuously, and in some forms the upper ones become large tusks, curving outward and upward. The peccaries, which also still survive, are more advanced. The bones of the lower leg have fused, and fusion of the metapodials has begun. The lateral toes may be much reduced. The molars are smooth and the canines normal. Giant pigs, the entelodontoids, existed in the Oligocene and persisted into the early Miocene. Their lower leg bones were fused, but not the metapodials. Their lateral toes were vestigial. These were the only pigs with a bony bar behind the eye socket.

Hippopotamuses

The hippopotamus is the only amphibious even-toed ungulate. It also is the only one whose ponderous body necessitated the development of short thick legs which end in broad four-toed feet (Fig. 542a-1). The molars are somewhat pig-like, and the outer incisors may be lost. A bony bar has formed behind the eye socket. Fossil specimens are not known before the Pliocene, but possible ancestors existed in a group of late Eocene to Miocene pig-like animals, the anthracotheres. These were relatively large beasts with four functional toes, and the first toe was still retained by some. Their teeth resembled those of pigs, but the molar cusps show the beginning of crescentic form. These creatures may also have been amphibious.

Primitive Ruminants

All the remaining even-toed ungulates are classed as ruminants, i.e., the cud-chewing animals which possessed a complicated stomach. This is an advanced character, but it cannot be determined for fossils, and paleontologists look to the bones in the ear region of the skull for features that seem to differentiate these beasts from the pigs and their near relatives. The ruminants also are unified, in all but the most primitive members, by evolution of the long running type of leg with fused bones and reduction of the lateral toes, the possession of molar teeth with crescentic cusps, and a tendency for the modification of anterior teeth and reduction of canines.

The earliest ruminants of the Eocene resemble the contemporary pig-like animals, with short legs and five-cusped teeth. The teeth, however, show crescentic patterns, and the cusps were soon reduced to four. These were primitive grinding teeth and probably indicate that these animals had become herbivores. One group resembled rabbits in size and some structural features. Others were larger and differed in the details of their teeth, and some of them had claws instead of hoofs. Most had four toes in the hind feet, but five persisted in the forefeet of some of them. None of these animals survived the early Miocene.

Oreodonts

Evolutionary differentiation among the early ruminants produced two important groups that preserved several primitive characters. One, the oreodonts, is known from the most abundant vertebrate fossils of the Oligocene and early Miocene in North America, and it persisted into the Pliocene. These animals were rather pig-like, with long bodies and short primitive legs. All had four toes, and the fifth one also was retained in the forefeet of some species. A few had claws instead of hoofs. The tooth series was complete. The upper canines were stout teeth, but the lowers resembled incisors, and the lower first premolars assumed the form of canines. The molars generally were low-crowned but had well-developed crescentic cusps. Bony bars closed off the eye sockets. The evolutionary origin of the oreodonts is obscure.

Camels

As the ancestry of the camels is traced back it seems to merge with the oreodonts and other more primitive ruminants of the Eocene. Many side branches evolved during the Tertiary, but gradational changes can be recognized leading from short-legged animals about the size of sheep, with four toes on the forefeet, to the modern camels and llamas. Besides increase in size, lengthening of legs, and loss of lateral toes, these changes include disappearance of two pairs of incisors in the upper jaws leaving only the outer pair, development of a wide space between molars and reduced premolars, and completion of a bar behind the eye socket (Fig. 538-2). The metapodials of cameloids are fused, but they did not lose their identity in a single bone (Fig. 542a-4). Spreading of the two remaining toes occurred in the late Tertiary, and the toe bones are contained within large circular padded feet particularly adapted for walking in loose sand. The toes bear flattened nails, rather than hoofs. No cameloid had horns.

Extinct Advanced Ruminants
More advanced ruminants are characterized by the further fusion of small wrist and ankle bones. Several groups of this sort, the traguloids, appeared in the Eocene (Fig. 538-3). They may have evolved from early cameloids or from even more primitive animals, and probably they provided the evolutionary base from which all familiar horned ruminants were derived. These were mostly very small creatures, some not over a foot tall. Only two genera, the so-called mouse deer, have survived to the present time. These animals were without horns, but the males commonly had long pointed upper canines.

Types of Horns
Horned ungulates appeared in the Oligocene. The early ones had bony prominences rising at one or more positions on the nose, above the eyes, or from the top of the skull. Such horns were permanent structures and probably were covered with persistent skin, as in modern giraffes (Fig. 546-1). From them two types of horns are believed to have evolved. The antlers of deer are solid branching bone (Fig. 546-5), which grows rapidly beneath a covering of skin. When growth is complete after a few months, the skin dies and is rubbed off. Annual resorption of bone near its base causes the antler to fall away, and it is replaced by another the next year that is larger and more branched.

The true horns of cattle and their allies are very different and are not shed (Fig. 546-2). They have solid unbranched bony cores that grow slowly during the whole lifetime of an individual. The cores are enclosed in a sheath of horn that grows by addition to its inner surface and around its lower edge, in much the same

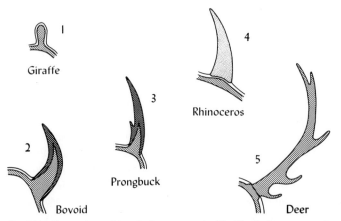

Fig. 546 Horns of ungulates. 1, Bone covered with skin. 2, Bone core and permanent horny cover. 3, Bone core, horny cover shed annually. 4, Horn consisting of elements homologous with stiffened and cemented hairs. 5, Bone, shed annually.

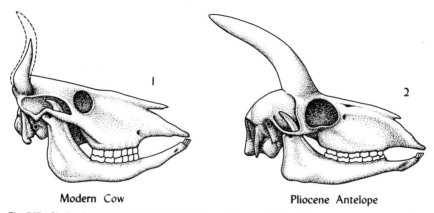

Modern Cow   Pliocene Antelope

Fig. 547 Skulls of advanced even-toed ungulates. 1, *Bos.* 2, *Gazella,* related to sheep and goats. The paired permanent horns of both animals had large bone cores. Incisors are missing from the upper jaws. (*After Gregory, 1951, fig. 20.2, p. 729; Romer, 1966, fig. 417H, p. 286.*)

way as the scales of reptiles. The American so-called antelope, or prongbuck, whose ancestors date back to the Miocene, has horns of a unique kind (Fig. 546-3) which probably are evidence of the long evolutionary separation of these animals from other ruminants. The bony cores are permanent, but the outer horny sheath, which is slightly forked like an antler, is shed annually and renewed.

Antlers or horns may or may not be borne by both sexes. Generally only males bear antlers. True horns commonly are possessed by females also, but in this sex they are likely to be smaller than those of males.

Deer

Deer, including moose, elk, caribou, etc., have a fossil record that begins in the early Miocene. The earliest forms had relatively simple antlers and large canines. These teeth later were reduced, but the antlers increased in size as larger species evolved. Very complex antlers had developed by the mid-Pliocene. Animals of this group commonly retain very small lateral toes (Fig. 542a-3). They have low-crowned teeth, and most of them are browsers, inhabiting mainly forested areas. Their ancestry seems to merge with that of the giraffes among the primitive Oligocene horn bearers.

Cattle and Their Relatives

Besides cattle, the true horned ruminants include buffaloes, sheep, goats, antelopes, etc. Their fossils are first found in early Miocene strata, and they seem to be both the most recently evolved and the most advanced of the even-toed ungulates. They differentiated rapidly in the Pliocene, produced species with many types of horns (Fig. 547), and are today the most numerous, varied, and successful of the ungulates. These animals are grazers, with high-crowned teeth, adapted to living in regions of plains, prairies, steppes, desert, and mountains. Bones of the lateral toes have almost completely disappeared, but small vestigial hoofs may remain. The immediate ancestors of these animals are not known, but they may have descended from either early horned ruminants or possibly some branch of the primitive cameloids.

Odd-toed Ungulates

The odd-toed ungulates, or perissodactyls, appear in the fossil record rather abruptly in the early Eocene, already possessing some features which serve to differentiate them into several groups. Though the ancestry of these animals is not known, they are generally supposed to have evolved from condylarth-like creatures of the Paleocene (Fig. 548). They are distinguished from the even-toed ungulates by details of articulation in the ankle. In early representatives, the first or inner toes were missing from all their feet, and the fifth or outer toe also had been lost from the hind feet. Metapodials were raised above the ground, and walking was accomplished on the toes, which ended in small hoofs. The tooth series was complete, and the molars were of quadrate form.

These odd-toed ungulates were evolutionarily more advanced than their even-toed contemporaries in the structure of their feet and in the character of their teeth. They evolved rapidly in the early Tertiary and were the most abundant and varied ungulates of that time. In the later Tertiary, however, they suffered a marked decline, and dominance passed to the even-toed ungulates.

Conspicuous evolutionary trends that progressed among the odd-toed ungulates include (1) development of both the running type and the strong pillar-like type of legs almost identically as in the even-toed ungulates, (2) reduction of toes to three in the forefeet and finally to one in all feet, achieved only in the horses (Fig. 552), (3) heightening of molars (Fig. 553) and growth of ridges and more complicated

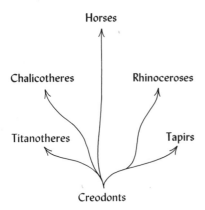

Fig. 548 Diagram showing the probable evolutionary relations of the main groups of odd-toed ungulates.

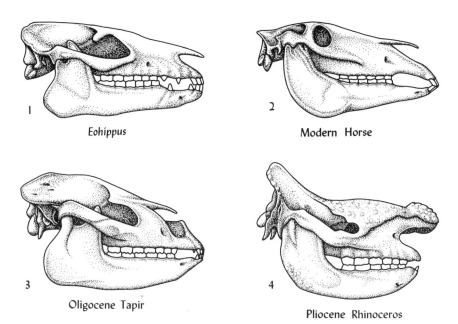

1 Eohippus

2 Modern Horse

3 Oligocene Tapir

4 Pliocene Rhinoceros

Fig. 549 Skulls of odd-toed ungulates. 1, Early Eocene four-toed horse. 2, *Equus;* this skull is ten times as long as the preceding. 3, *Protapirus.* 4, *Diceros. (After Romer,* 1966, fig. 387, p. 265.)

folding in the enamel of the crown (Figs. 506; 507), (4) molarization of premolars (Fig. 555), and (5) general increase in size. Reduction of incisors and canines did not occur to the same extent as in the even-toed ungulates.

Early Types

An early Eocene animal (Fig. 549-1) about the size and form of a small dog is generally regarded as an approximation of the prototype from which all odd-toed ungulates probably evolved. It has commonly been known as *Eohippus,* although it is possibly identical to a European beast known by another name, and is classed as the most primitive member of the horse lineage. Cusps of its teeth were low (Fig. 553-1), but ridges connecting them had begun to form (Fig. 554-3). During the Eocene Epoch evolutionary radiation produced a variety of animals more and more different from this creature. Among the differences that developed, patterns of the molar teeth are considered to be especially significant in tracing the early branching relationships. One type of pattern appearing in the upper molars is characterized by a W-shaped ridge along the outer margin, as in the horses (Fig. 557-3a,4a). The other has two transverse ridges, as in the teeth of tapirs and rhinoceroses (Fig. 557-1a,2a). Most of the lineages that did not lead on to modern animals became extinct after the early Oligocene. Two extinct groups are particularly interesting because of the bizarre creatures that evolved.

Titanotheres

The earliest titanotheres and horses of the Eocene were similar in many ways, especially in the nature of their teeth. The titanotheres rapidly evolved to produce large animals with thick legs (Fig. 550). Some stood 8 ft high at the shoulder before they became extinct after the early Oligocene. The most conspicuous features of these beasts were the stout bony horns that grew upon their noses. These began as small swellings in the late Eocene and increased greatly, to produce a variety of types. They probably were skin-covered, although there may have been tips of horn. Otherwise the titanotheres were conservative. The toes of their stubby feet were not reduced and numbered four in front and three behind. The structure of their teeth also changed very little. The upper molars became large and bore the W-shaped ridge (Fig. 557-3a), but they were low-crowned (Fig. 505a) and the enamel was not infolded. Premolars were small and only partly molarized. Incisors and first premolars were reduced and in some of the latest forms were lost. Such teeth were not efficient grinders, and these animals probably were mainly forest browsers. Perhaps the persistence of such teeth accounted for the early extinction of these animals.

Chalicotheres

The chalicotheres are believed to have evolved from moderately small, hoofed animals in the late Eocene and seem to have been closely related to the titanotheres

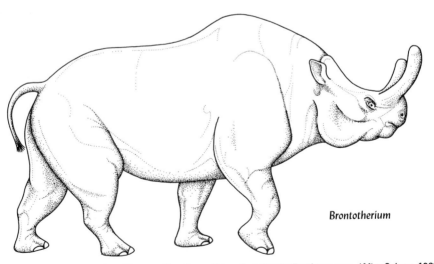

Brontotherium

Fig. 550 Restoration of an Oligocene titanothere. This animal stood taller than a man. (*After Osborn, 1929, fig. 620, p. 688.*)

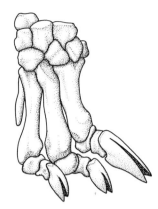

Fig. 551 Forefoot and ankle bones of a Miocene chalicothere. The metapodials are not much lengthened but stand almost vertically. (*After Romer*, 1966, *fig. 392G, p. 268.*)

Chalicothere

in their origin. Typical species from late Eocene time onward had claws (Fig. 551) instead of hoofs on toes that were soon reduced to three on the front feet. Speculation has suggested that these animals dug for roots or pulled down branches in their browsing, although the claws may have been effective as defensive weapons. The teeth were low-crowned like those of titanotheres, with the incisors reduced or lost in some species. These animals increased gradually in size until in the early Miocene they were as large as modern horses. They also resembled horses in body form and type of skull. Thereafter they changed little and persisted in diminishing numbers until the Pleistocene, when they became extinct.

Horses

The horses probably have been studied more intensively by paleontologists than any other vertebrates. They commonly are presented as the outstanding example of a fossil lineage whose evolution has been traced in detail. Perhaps these interpretations have been considerably idealized, but the evolutionary sequence is impressive and mainly satisfactory.

As previously mentioned, horse evolution is traced from a small early Eocene creature commonly known as *Eohippus*. This animal had legs with only moderately lengthened metapodials ending in four toes in front (Fig. 552-1a) and three behind. The molar teeth were quadrate, with six low cusps on the uppers (Fig. 554-3) and four on the lowers. Ridges connecting some of them had only barely begun to form. The premolars were relatively simple, and none had progressed beyond the triangular stage. In middle and late Eocene species, however, the last two premolars became square. Primitive horses of the early Oligocene grew to the size of a moderately large dog. The metapodials had lengthened, and toes of the forefeet were reduced to three (Fig. 552-2a). Well-formed ridges on the teeth formed a W above

(Fig. 554-6) and an inverted VV pattern below. All the premolars had been molarized (Fig. 555-4a). By early Miocene time pony size had been attained. These horses were browsers, with low-crowned teeth. Animals of this type continued into the Pliocene.

Other horses adapted to the grazing habit during the Miocene Epoch. Mid-Miocene species were transitional (Fig. 553-4). They grew somewhat larger, and their

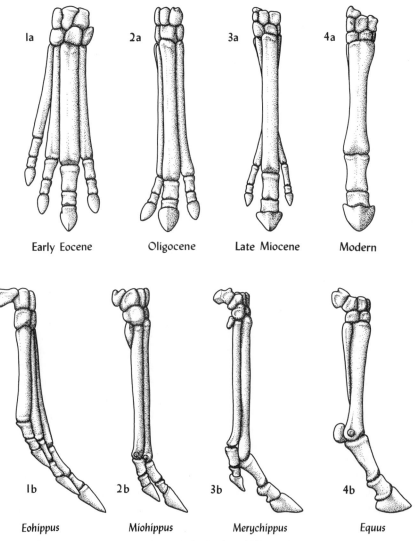

Fig. 552   Front and side views of the bones in the forefeet of horses showing progressive reduction of all but the center toe. (*After Romer*, 1966, *fig. 319, p. 422.*)

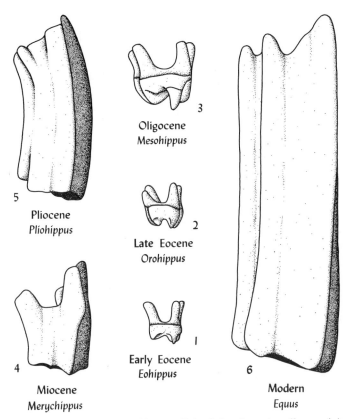

Fig. 553 Upper molar teeth of horses. (Natural size, drawn to uniform scale.)
(*After Matthew, 1926, Quart. Rev. Biol., vol. 1, fig. 27, p. 172.*)

brains increased relatively in size. Cheek teeth lengthened, the enamel patterns became more complex (Fig. 554-9), and a coating of cement covered the crowns. The side toes became smaller, and the central one was correspondingly enlarged (Fig. 552-3a). Posterior bony processes began to close the eye socket. By late Miocene time the teeth were essentially modernized. Larger horses of modern type appeared in the Pliocene. The lateral toes were gone, and the eye socket was enclosed (Fig. 549-2). Only minor changes have since occurred.

The main line of horse evolution is believed to have progressed in North America until the Pleistocene, when the unexplained extinction of these animals here occurred. Numerous collateral lineages also evolved and flourished more or less briefly during the Tertiary. An interesting one in Europe during the late Eocene and early Oligocene precociously produced beasts of relatively large size. Their teeth,

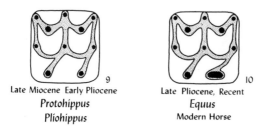

9
Late Miocene Early Pliocene
Protohippus
Pliohippus

10
Late Pliocene, Recent
Equus
Modern Horse

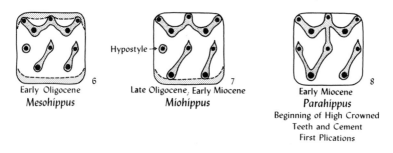

6
Early Oligocene
Mesohippus

Hypostyle →

7
Late Oligocene, Early Miocene
Miohippus

8
Early Miocene
Parahippus
Beginning of High Crowned
Teeth and Cement
First Plications

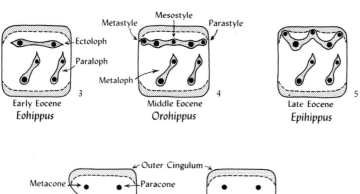

Metastyle   Mesostyle   Parastyle

←Ectoloph

Paraloph

Metaloph

3
Early Eocene
Eohippus

4
Middle Eocene
Orohippus

5
Late Eocene
Epihippus

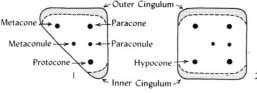

↙Outer Cingulum ↘

Metacone →   ←Paracone

Metaconule →   ←Paraconule

Protocone →   Hypocone →

1   ↖ Inner Cingulum ↗   2

Triangular Tooth   Quadrate Tooth

Fig. 554   Diagrammatic representation of morphologic stages in the evolution of upper molar teeth of horses. Below: Basic elements of the teeth, consisting of cusps and cingula, or shelves at the base of crown. The cusps are joined by ridges or lophs, new cusps, or styles, rise from the cingula on the outer margins of teeth, cingula disappear, and the crown patterns become more and more complex. High-crowned teeth, cement, and folding of enamel are advanced evolutionary characters.

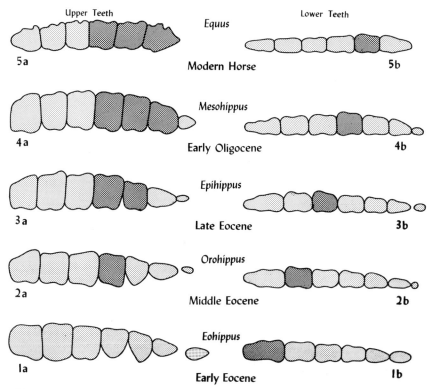

Fig. 555 Cheek teeth of horses. Darker shading in upper teeth shows progressive molarization of premolars. Darker shading in lower teeth shows that the largest one occurs at progressively more forward positions. Canines are missing from the dentition of modern horses. (*After Granger* 1908, *Bull. Am. Museum Nat. Hist.*, vol. 24, fig. 5, p. 263.)

although low-crowned, had well-developed crests and were covered with cement. Also the premolars were molarized before this character appeared in American horses. These animals were relatively short-legged, had three toes, and probably had long noses like those of tapirs.

Tapirs

Various tapir-like animals evolved from the primitive odd-toed ungulate stock in the early Eocene. The first that are considered to be true tapirs, however, were of Oligocene age, and beasts much like them are still living. The modern tapirs are very primitive with respect to their short legs with unfused bones and short feet with four toes in front and three behind. Teeth are low-crowned, of the browsing type, with simple cross ridges most prominent in their patterns (Fig. 557-1a).

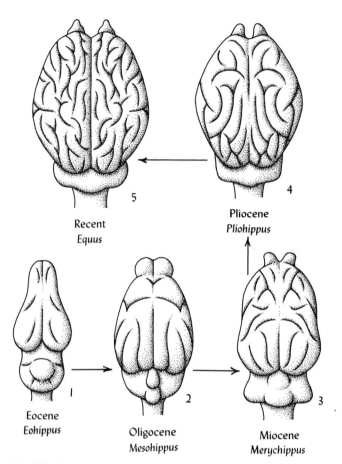

Fig. 556   Generalized sketches of the brains of horses, drawn approximately to the same scale relative to each animal's actual size. Progressive increase in development of the cerebrum is clearly shown. (See *Simpson, 1951, "Horses," fig. 26, p. 177.*)

Three premolars, however, are molarized. These animals have long flexible noses like short trunks (Fig. 549-3).

Rhinoceroses

Rhinoceroses probably evolved from tapir-like animals, which the early forms resembled. Teeth are all low-crowned, indicating browsing habits. Incisors and canines have been variously formed, and they and the anterior premolars were lost in some lineages. Cross crests of the upper molars connect with an outer ridge (Fig. 557-2a), but this lacks the W pattern seen in horses. The last molar of late Tertiary species tended to shrink in size and become simpler in structure. Rhinoceros evolution paralleled that of horses in several ways. For example, toes of the forefeet were reduced to three, and premolars were molarized. Some rhinoceroses of the late Eocene and Oligocene developed the running type of leg, but mostly these beasts had large bodies set upon short legs. One early group resembled

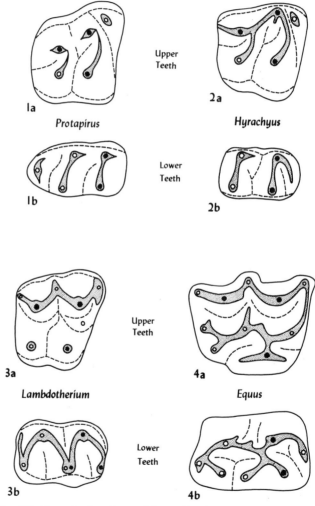

Fig. 557 Crown tooth patterns of four odd-toed ungulates. 1, Oligocene tapir. 2, Eocene rhinoceros. 3, Early Eocene beast intermediate between titanotheres and primitive horses. 4, Modern horse. (*After Osborn, 1907, "Evolution of Mammalian Molar Teeth," figs. 163A, 164, 177, pp. 177, 181; 1929, vol. 1, fig. 221, p. 268.*)

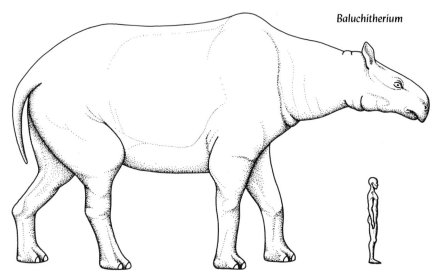

Baluchitherium

Fig. 558   Restoration of a mid-Tertiary hornless rhinoceros. This is the largest land mammal that ever lived. (*After Gregory, 1951, vol. 2, fig. 21.80, p. 849.*)

hippopotamuses in body form and probably was amphibious. Some Asiatic species of the Oligocene and Miocene grew to enormous size. One was 18 ft high at the shoulder and exceeded some of the largest dinosaurs in bulk (Fig. 558). True rhinoceroses first became prominent in the Oligocene. Small horns appeared on their noses in the Miocene and subsequently grew larger. These occur singly or in pairs arranged side by side or one before the other. Rhinoceros horns do not consist of bone but are more or less fibrous and evolved from crowded clumps of stiffened hairs (Fig. 546-4). Their presence is indicated on fossil skulls by roughened areas of attachment or support (Fig. 549-4).

Elephants and Their Relatives
The ancestry of elephant-like animals is not known. The earliest discovered representatives of this group from the late Eocene already possessed distinctive proboscidian characters. The teeth are somewhat like those of early pigs, but no other features suggest relationship to those animals. Probably evolution began with early condylarth-like creatures (Fig. 559) in some part of Africa where Paleocene and early Eocene fossils have not yet been found. A few other kinds of animals of uncertain origin, which at first sight seem very unlikely relatives but exhibit some suggestive similarities, are commonly believed to be distantly related to the elephants.

Moeritheres
The moeritheres of late Eocene and early Oligocene ages clearly stand near the base of typical proboscidian evolutionary radiation (Fig. 560-1). They were rather

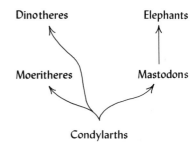

Fig. 559 Diagram showing the probable evolutionary relations of the elephants and kindred animals.

tapir-like in size and form and probably possessed long flexible noses or short trunks. Their most interesting features are the enlarged canine-like second incisors that are the forerunners of tusks. The molars, both above and below, are quadrate and bear four large but low blunt cusps, which are connected to form two transverse ridges (Fig. 562-1).

Dinotheres

No fossils have been found connecting moeritheres with the dinotheres, whose known record extends from the late Miocene to the Pliocene, but relationship is probable. The dinotheres were large animals, some of which exceeded modern elephants in size. They were all very similar and were characterized by downward and backward curving tusks extending from the lower jaws (Fig. 559a-1). Upper

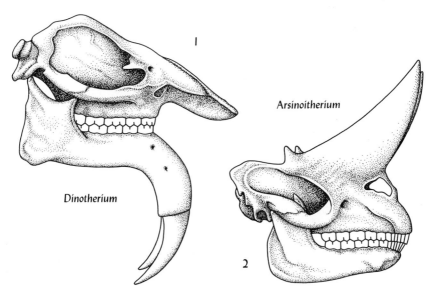

Fig. 559a Skulls of two elephant-like beasts. 1, From the Miocene. 2, From the Lower Oligocene. (*After Romer*, 1966, *figs*. 360, 363, *pp*. 249, 250.)

tusks were missing. The other teeth consisted of low-crowned grinders, five in each series, with two or three cross ridges, much like those of mastodons. These beasts seem to have had well-developed trunks.

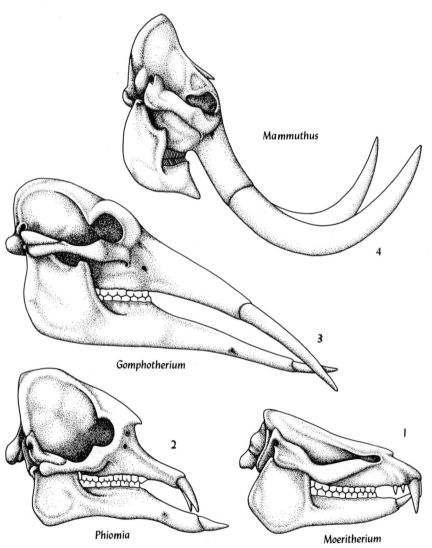

Fig. 560 Skulls showing progressive evolution of tusks in elephant-like animals. 1, Early Eocene moerithere. 2, Early Oligocene primitive mastodon. 3, Miocene mastodon. 4, Pleistocene woolly mammoth. (*After Romer, 1966, fig. 361, p. 249.*)

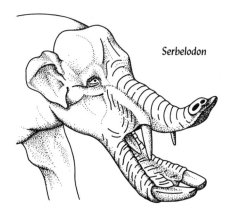

Serbelodon

Fig. 561 Restoration of the head of a late Tertiary shovel-toothed mastodon. (*After Osborn, 1936, "Proboscidea," Am. Museum Nat. Hist., fig. 406, p. 441.*)

Mastodons

The oldest known mastodon of early Oligocene age was intermediate in many ways between the moeritheres and the later better-known mastodons (Fig. 560-2). The anterior part of the skull is enlarged and consists of thickened partly spongy bone with extensive sinuses. The lower jaw is long and ends in a pair of short sharp tusks. The molar teeth have three cross ridges, and accessory cusps had begun to form between them (Fig. 562-2).

Evolutionary radiation produced a great variety of mastodons in the Miocene (Fig. 560-3) and Pliocene, and these animals continued to be abundant right up to Recent time before extinction overtook them. Most were heavy beasts with long thick legs. The feet had five stubby toes capped with thick flattened nails, and the metapodials were very short. The most conspicuous differences that developed were evident in the tusks. Generally these were present in both upper and lower jaws, but the upper ones tended to become longer and larger and the lower ones shorter and smaller until they disappeared. The large tusks lacked a coating of enamel and grew persistently. Some aberrant forms had downward-curving lower tusks like those of dinotheres, and these teeth in others were flattened and formed projecting shovel-like structures (Fig. 561).

Progressive evolutionary modifications of the cheek teeth are more interesting and led to the development of elephants. In early mastodons these teeth were low-crowned, all were functional in mature individuals, and permanent premolars replaced milk teeth in the usual manner. One conservative group with only two ridges on the molars survived until the Pleistocene. In others, however, accessory cusps combined to produce additional cross ridges (Fig. 562-3) that increased to seven or eight in the last molars. Also, permanent premolars failed to develop, and the milk teeth were pushed out of the jaws by the molars which grew up behind them. Not all the cheek teeth were in use at the same time in these animals.

The proboscidians, particularly the mastodons, demonstrate close functional

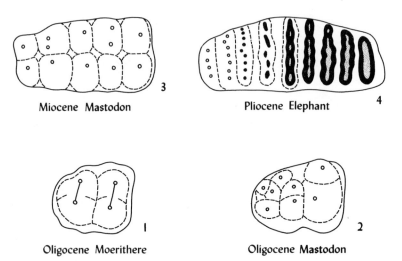

Miocene Mastodon   Pliocene Elephant

Oligocene Moerithere   Oligocene Mastodon

Fig. 562   Molar teeth of elephant-like animals showing progressive increase in the number of cusps. 1, *Moeritherium*. 2, *Palaeomastodon*. 3, *Mastodon*. 4, *Stegodon*. (After Gregory, 1951, "Evolution Emerging," vol. 2, fig. 21.34B, D, E, G, H, p. 799.)

evolutionary relationships among weight, stature, teeth, and trunks. The moeritheres and early mastodons already possessed lengthened flexible noses, as indicated by the nasal openings located far back on the upper surface of the skull. Lengthening of the lower jaws in mastodons and the general presence of forward-projecting tusk-like or shovel-like lower incisors indicate that a long agile upper lip was necessary for their feeding. Evolution among these animals produced large tall beasts with long legs but without corresponding lengthening of the neck. This disparity must have been compensated by continued lengthening of the nose and the development of functional trunks. At the same time their larger bodies required greatly increased quantities of food. Although body bulk, and therefore food requirements, increase according to a third power function, the chewing area of unmodified teeth increases only by a function of the second power. This relationship requires either that larger animals be provided with relatively increased dental areas or that their teeth function more efficiently. The latter development is evident among the more advanced mastodons and more especially in the elephants. As their teeth became more effective grinders by infolding of the enamel, the number of teeth actually was reduced.

Elephants
Elephants evolved from mastodons of the last mentioned kind. They first appeared early in the Pliocene, and by the end of that epoch typical species had developed. These beasts had long upper tusks but none below. The jaws were greatly shortened

(Fig. 560-4), and only one of the cheek teeth was fully functional at a time (Fig. 563). These teeth had lengthened, and the cross ridges increased in number. The teeth grew up in succession, and as one was worn down, the next, which had been growing meanwhile, moved forward in the jaw and took its place. The early elephants had low teeth, but higher ones rapidly evolved, and cement appeared between the ridges. The enamel became deeply folded in the high-crowned teeth, and when they were worn their upper surfaces exposed repeated laminae of cement, enamel, and dentine (Fig. 562-4). The number of ridges increased and reached a maximum of about 30 in the last molars of some species. Elephants ranged in size from giants 14 ft high at the shoulder to dwarfs no bigger than pigs. The large elephants of the Pleistocene, which became extinct only a few thousand years ago, are commonly known as mammoths (Fig. 560-4).

Conies

The modern conies and their fossil relatives constitute a group of mostly small herbivorous animals known sparingly from the early Oligocene onward. The living species superficially resemble rabbits in size, form, and habits, but their structures demonstrate that they are ungulates. Rather incongruously they are commonly referred to the same general group that includes the elephants. Their early evolutionary history is unknown, although presumably they are African products of

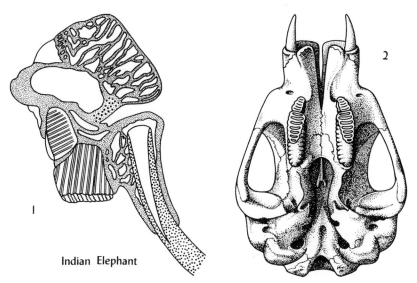

Indian Elephant

Fig. 563 Skulls of modern elephant. 1, Longitudinal off-center section showing implantation of molar teeth and tusk, and mass of spongy bone above brain cavity. 2, Ventral view of a young skull showing partly worn molars about to push out last premolars. (*After Gregory, 1951, "Evolution Emerging," vol. 2, figs. 21.31A, 21.33B1, pp. 796, 798.*)

Paleocene condylarth radiation. These animals generally possessed continuously growing rodent-like incisors, two above and four below. The lateral incisors, canines, and perhaps some of the premolars have been lost. Other premolars became molarized. Molar patterns have varied considerably, but the teeth of some species have one longitudinal ridge and two transverse ridges, as in some early rhinoceroses. The legs and feet of modern species are a curious mixture of mainly primitive but some advanced characters. The wrist complex contains a bone rarely retained in ungulates, and walking is flat-footed on a single pad. The toes, however, are reduced to four in front and three behind and bear somewhat hoof-like nails. A posterior bony bar that closes the eye socket in living species was not developed in most fossils.

Sea Cows
The sea cows, or manatees, and similar animals are fully adapted herbivorous marine mammals that never leave the water. The forelimbs have been converted to paddles without much reduction of their bones. The hind limbs and pelvis, however, are vestigial, and swimming is effected by a horizontally flattened tail. The dentition consists of a pair of short tusk-like upper incisors, not present in all species, and low-cusped cheek teeth, commonly with two cross ridges like those of primitive proboscidians. Fossil remains of animals of this kind are known from mid-Eocene and later strata. They show progressive reduction of the pelvis and hind legs and generally in the number of teeth. In one modern species, however, the cheek teeth have increased to 20 or more. These develop successively and move forward, replacing one another in the jaws, so that only five or six are functional simultaneously. The earliest fossils provide little indication of evolutionary origin; these animals are grouped with the elephants and their relations because of the dentition.

Other Fossils
A few peculiar or incompletely known extinct animals are tentatively believed to be related to the elephants and their allies. These include some creatures compared to manatees of the Oligocene and Miocene, but these were more likely marine quadrupeds somewhat resembling a hippopotamus. Another was a large rhinoceros-like beast from the Lower Oligocene, with two huge bony horns rising between its eyes (Fig. 559a-2). The pyrotheres (Fig. 559a-1), mentioned in a preceding section, were elephant-like mammals of South America. Similarities are so close that relationship with elephants certainly is suggested strongly. The difficulty of explaining how the ancestors of these creatures could have reached South America from Africa, however, is so great that a close common evolutionary origin seems doubtful.

RODENTS
Rodents are by far the most varied, abundant, and widely distributed of all mammals. With few exceptions they have been small gnawing herbivores that adapted

to almost every terrestrial environment. Fossils are known from strata as old as the late Paleocene. These already possessed greatly specialized paired chisel-like incisors enameled only on the anterior surface, but the cheek teeth were tribosphenic. The skeletons of most rodents are characteristically rather primitive. A bony bar is never present behind the eye socket, clavicles are usual in the shoulder girdle, and there is little tendency for reduction in number of toes, which all bear claws. Although the immediate ancestors of these animals are not known, they are believed to have evolved from early insectivores which may have been close to the stock that produced the primates.

Fossil rodents are relatively not well known. Their skeletons commonly being small and delicate were easily destroyed and are likely to be overlooked by collectors unless special search is made for them. Their study also has been somewhat neglected because many paleontologists prefer to devote their attention to larger, more spectacular animals that are more readily observed. Teeth have been much relied on in tracing the lineages of other mammals, but rodent teeth generally are considered to be much less useful for this purpose (Fig. 569). Modern species are so numerous that they are difficult to classify satisfactorily, and consequently they provide little aid in arriving at conclusions regarding evolutionary pathways. For all these reasons, less is known about evolutionary relations among rodents than among any other large major group of mammals.

In spite of much uncertainty with respect to details, but by applying principles derived from the study of other animals, generalizations can be reached with respect to the evolutionary succession of certain features of the rodents that are revealed by differences in their skeletal structures. These are concerned with (1) jaw musculature, (2) tooth development, and (3) limb structure. The various ways in which apparently primitive and advanced characters are combined in different species, however, indicate that much parallelism has occurred in the evolutionary development of rodents. Consequently the recognition of well-defined lineages has not been accomplished with desirable assurance.

Jaw Muscles

Closure of the jaws in mammals is effected mainly by the action of two sets of muscles. Generally stronger and more important in carnivores are muscles attached to the outer surface of the braincase within or above the temporal opening and extending to the process that rises from the lower jaw in front of its articulating area. The other muscles, known as masseters, are attached to the posterior outer surface of the lower jaw and extend upward and forward to the skull. These serve principally to move the lower jaws forward and backward and from side to side in action which is particularly important for grinding food between the cheek teeth. These muscles generally are strong and well developed in herbivores.

The masseter muscles of rodents are more than ordinarily accentuated. On each side of the jaws they are divided into inner and outer parts (Fig. 566). In the most primitive condition, both parts attach to the lower edge or inner side of

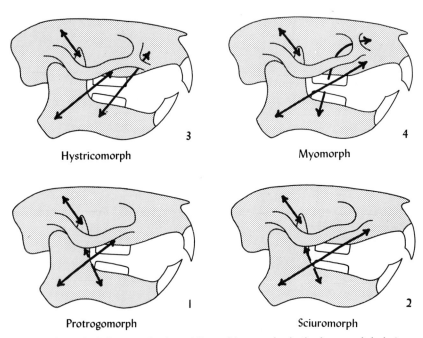

Fig. 566 Generalized diagrams showing relations of jaw muscles in the four morphologic types of rodents. (*See Romer, 1966, fig. 437, p. 305.*)

the bony bar below the temporal opening, as they do in most other mammals. Evolutionary advancement is indicated by migrations of one or both of these attachment areas. The major divisions recognized in the classification of rodents are based on the disposition of these muscles (see Table 566). Fossils can be classified similarly because the attachment areas generally are evident. This division, however, probably groups some unrelated animals whose similarities evolved independently.

Table 566 Four groups of rodents based on the location of masseter muscle attachments

Protrogomorphs: both masseter jaw muscles attached to bar of bone below temporal opening. Paleocene to Recent

Sciuromorphs: outer masseter muscles moved forward to attachment areas below or a little in advance of the eye sockets. Oligocene to Recent

Hystricomorphs: inner masseter muscles lengthened, passing through an opening in the skull bones in front of eye socket to areas of attachment. Oligocene to Recent

Myomorphs: both masseter muscles lengthened, neither remaining attached to the bar of bone below the temporal opening. Oligocene to Recent

Sciuromorphs and hystricomorphs both evolved from the more primitive protrogomorphs. Myomorphs are the most advanced and may have evolved from either sciuromophs or hystricomorphs, or from both.

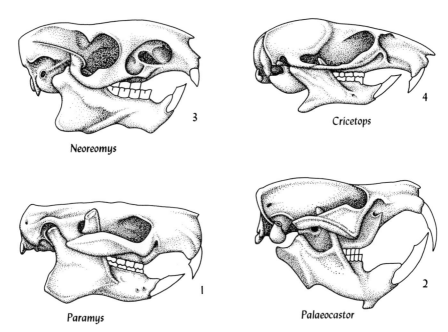

Fig. 567   Four rodent skulls. 1, A primitive Eocene protrogomorph. 2, An early Miocene beaver, a sciuromorph. 3, An early Miocene nutria, a hystricomorph. 4, A late Oligocene mouse, a myomorph. (*See Romer, 1966, fig. 438, p. 306.*)

Protrogomorphs

Protrogomorphs are rodents characterized by having both pairs of masseters attached to the bar below the temporal opening in their original position (Fig. 566-1). Fossils of this type (Fig. 567-1) occur in the Paleocene, and similar species still exist. These are believed to be the most primitive rodents, commonly squirrel-like in size and form and possibly in habits. All others probably evolved from creatures of this kind.

Sciuromorphs

In sciuromorphs the outer masseters are longer, and the upper ends have moved from the subtemporal bar to areas of attachment in depressions on the side of the skull below or slightly in advance of the eye socket (Fig. 566-2). Lengthening of these muscles probably facilitated forward movement of the lower jaws that would bring the tips of the incisors together, as is required for gnawing. Fossils of this kind are known from the Oligocene. Squirrels, beavers (Fig. 567-2), prairie dogs, and woodchucks are familiar modern examples of this group.

Hystricomorphs

Modern porcupines and guinea pigs represent the hystricomorphs, which have been an important group since at least Oligocene time. These animals probably evolved from protrogomorphs. The outer masseter muscles retain their original attachment to the subtemporal bar. The inner ones, however, have lengthened, pass through an opening in the bone in front of the eye socket, and are attached in a depression on the side of the skull (Figs. 566-3; 567-3). This muscular arrangement seems to reinforce the power of other muscles, whose principal function is closure of the jaws.

Myomorphs

Myomorphs are the most advanced rodents. Both parts of the masseter muscles are elongated as has been described, so that neither remains attached to the subtemporal bar (Figs. 566-4; 567-4). This group is represented among Oligocene fossils and includes modern rats and mice. Animals of this kind may have evolved from either sciuromorphs or hystricomorphs or from both.

Tooth Development

Dental Formula

The dentition of all Tertiary placental mammals presumably was derived from a primitive condition with 11 teeth in each jaw series, described by the formula 3,1,4,3. Reduction in these numbers has occurred in many groups of animals, and generally it can be followed through successive stages. The earliest and most primitive rodents, however, had already proceeded far along this course. No species is known with a formula exceeding 1,0,2,3 in the upper and 1,0,1,3 in the lower jaws. This, of course, indicates the loss from each series of 2 incisors, the single canine, and 2 or 3 premolars, so that the maximum number of cheek teeth is 5 above and 4 below (Fig. 569).

Two premolars above and one below are known only in protrogomorphs and sciuromorphs, and even in these animals the anterior upper ones may be small or lost. In some of the protrogomorphs the last premolar may be larger than the smaller molars. Cheek teeth in sets of four also are present in both hystricomorphs and myomorphs. This number is reduced, however, by the loss of first the lower premolar and then the upper one. Many modern rodents, particularly myomorphs, have only the three molars remaining.

Degeneration and reduction in the number of molars also has occurred in some rodent groups, particularly small burrowing forms that use their incisors for digging and chewing. Among them the last molar may be lost and the others reduced to a less-functional condition.

Rodent-like incisors evolved in a parallel or convergent manner in several other quite different groups of mammals. They seem to be characteristic of gnawing habits in every case. In the rodents these teeth are rootless and grow continuously from points far back in the bones of the skull and lower jaws. This probably was an important factor in the loss of canines and adjacent cheek teeth, because embedded parts of the incisors underlie the areas where the other teeth should be and may have interfered with their development.

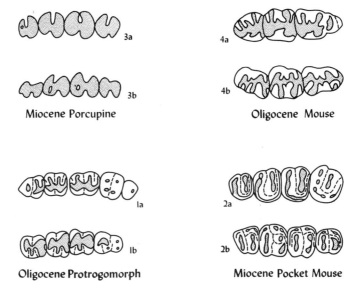

Fig. 569   Cheek teeth of rodents. 1, *Ischyromis*. 2, *Florentinamis*, a sciuromorph. 3, *Sciamys*, a hystricomorph. 4, *Eumys*, a myomorph. (*See Romer*, 1966, *fig.* 439, *p.* 307.)

Tooth Structure

The cheek teeth of rodents surely evolved from a pattern dominated by the four principal cusps present on the teeth of many other mammals. These were augmented by the development of other cusps that rose above basal shelves, external ridges, and infolds of the enamel. The cusps tended to lengthen transversely and join to form a series of roughly parallel ridges extending across the crown (Fig. 569). By late Eocene time, molars with five cross ridges had developed.

Most of the early rodents seem to have been small forest-dwelling creatures which fed on succulent vegetation, fruit, insects, and other easily chewed food. Their molar teeth, with low crowns and relatively elevated cusps, were adapted to the cutting and crushing of such readily reduced material. As the forests shrank and savannas expanded during Tertiary time, rodents moved into new habitats where their main food consisted of harsh grasses, grain, and hard-shelled seeds. Their teeth evolved in adaptation to this diet in a way paralleling that of horses when these animals made a similar environmental shift. Increasingly high crowns developed on teeth that continued to grow as the upper surfaces wore down. The enamel folds extend deeply into teeth of this kind. At the same time, the cusps decreased in prominence, producing teeth with more nearly plane upper surfaces better adapted to grinding functions.

The crown patterns of rodent teeth vary considerably even among genera that are believed to be closely related to one another, and similar evolutionary development probably proceeded independently in groups that almost certainly belong to different lineages. A general tendency is evident, however, in several rodent groups toward simplification of structure from the five-ridged type of tooth. This produced teeth that reverted to more primitive patterns, with four, three, or two ridges, and finally to a relatively simple conical tooth. The reduction of ridges in this way seems to have progressed in an order reversing that of their original appearance. This secondary simplicity is not correlated with primitiveness, as indicated by the masseter muscles.

Limb Structure
The forelegs of many rodents are capable of varied movements, and the paws commonly are used as aids in feeding. Only rarely is the first toe reduced or lost. The hind legs are much less flexible. The two bones of the lower leg may be fused at their lower ends. This seems to have occurred independently in different lineages. The toes of the hind feet of some rodents are reduced, particularly in hopping or jumping creatures. A few have very bird-like three-toed feet.

Rabbits
The rabbits and related animals resemble rodents in their prominent incisors and batteries of cheek teeth (Fig. 570). These two groups of small creatures may have evolved from the same ancestors, but they have been distinct since early Tertiary time, and transitional fossils are not known. Typical members of the rabbit group have been found in Oligocene strata, and a single late Paleocene species may belong here also. Rabbits differ from rodents in many important ways. Among the more obvious and probably significant are the following: (1) The masseter muscles are less complex and do not migrate forward on the skull. (2) A second pair of small incisors is located directly behind the larger upper pair. (3) The cheek teeth are simpler and more numerous. (4) The sides of the skull directly in front of the eye socket are peculiarly vesicular. Altogether the rabbit-like animals seem to be

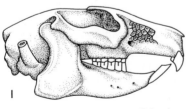

Palaeolagus

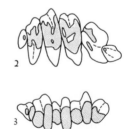

Fig. 570 Skull and cheek teeth of an Oligocene rabbit. (*After Romer, 1966, fig. 377, p. 506.*)

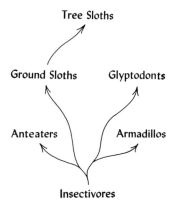

Fig. 571   Diagram showing the possible evolutionary relations of the edentates. Their derivation from insectivores is somewhat doubtful.

more primitive than rodents, except for their very strong and long hind legs, which are specialized for hopping.

The dental formulas of rabbits and their allies are 2,0,3,3 above and 1,0,2,3 below. All the teeth are rootless and grow continuously. The cheek teeth are long prisms. They bear two transverse ridges produced by the convergence of four cusps similar to those present on most quadrate teeth. No clear examples of progressive evolutionary trends have been noted among these animals.

EDENTATES

The mammals known as edentates constitute a peculiar group of aberrant creatures. Although modern representatives, the armadillos, South American anteaters, and sloths, possess curious teeth without enamel or are toothless, they present few obvious external unifying features. Closer anatomical comparisons, however, reveal a variety of significant similarities, and the fossil record provides some evidence regarding their mutual relationships (Fig. 571). For example, shoulder and hip girdles are constructed and vertebrae articulate in ways that differ from those in all other animals, and fossils show some convergence between living types.

The evolutionary origin of edentates is unknown. Some of the oldest fossils included in this group, from the late Paleocene and their younger relatives, suggest comparison with the creodonts. These animals lacked, however, some typical edentate characters. If edentates did not evolve from early creodonts, they probably differentiated directly from insectivores.

The evolutionary development of the edentates is revealed exclusively by South American fossils. If these creatures originated in some other region, they must have reached this southern continent before its isolation in the very early Tertiary. There they prospered. Edentates are not certainly known elsewhere until they migrated to North America in the Pleistocene Period. A few animals in other parts of the world resemble the edentates rather closely in some ways. Their geographic occurrence, however, makes their relations very doubtful, and the similarities probably resulted from evolutionary convergence reflecting ways of life that were much the same.

Possible Primitive Edentates

A few rare North American Paleocene to Oligocene mammals may be descendants of creatures that also were ancestral to the South American edentates. If this is so, some inferences can be drawn regarding the evolutionary origin of edentates. These northern fossils resemble armadillos in form and some structures. They lack incisors except for a pair of small ones in the lower jaws, and the cheek teeth, provided with very little enamel, are peg-like or nearly lost. The front feet were furnished with long narrow claws, and there was a tendency for the foot to be turned upon its outer side in walking. These beasts, however, had neither the peculiar vertebrae nor hip girdle of the typical edentates. Their possible relations to creodonts are suggested by large canine teeth and other rather primitive and generalized features.

Armadillos

Armadillos seem to approximate the stock from which all other South American edentates evolved. Their most obvious modern characteristic is the dorsal armor, consisting of many small bony dermal plates and thin horny epidermal scales arranged in several transverse bands that permit the animals to roll up when disturbed. Teeth are absent anteriorly in the jaws (Fig. 572-1). Cheek teeth in a

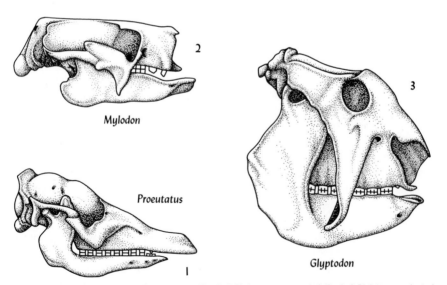

Fig. 572   Edentate skulls. 1, A Miocene armadillo. 2, A Pleistocene ground sloth. 3, A Pleistocene glyptodon. (*See Romer, 1966, fig. 424, p. 293.*)

single series generally number 7 to 10, but one modern species has 25. They are rather simple peg-like structures, totally lacking in enamel. The toes are unreduced and bear strong claws. Modern armadillos feed mainly on invertebrates and carrion. Fossils are known from the Paleocene onward. One Pleistocene species grew larger than a horse.

Glyptodons

The extinct glyptodons are known from the late Eocene to the Pleistocene, with typical forms appearing in the Pliocene. They evolved from early armadillos, which they resemble in their bony armor, but this is a solid shield something like a turtle's shell and these animals could not roll up. Glyptodons generally were larger than contemporary armadillos. Their vertebrae were much fused. These were the only edentates with a bar of bone behind the eye socket (Fig. 572-3) and without clavicles in the shoulder girdle. In both respects they were evolutionarily more advanced than armadillos. The trilobed enamel-less teeth of most glyptodons were confined to the cheek regions and numbered eight in each series. The feet generally were five-toed and provided with claws; those of the hind feet, however, had somewhat the form of hoofs. These animals probably were herbivorous.

Anteaters

The South American anteaters are known in the fossil state only from very frag-mental material of late Tertiary and Pleistocene ages that gives little information about their evolution. They probably differentiated from very primitive armadillos before these creatures developed dermal bone. If they originated later, the armor was quickly lost, because anteaters have nothing resembling these plates, and their bodies are clothed with hair. Lateral toes are reduced on the front feet, and the others, particularly the middle ones, are provided with long claws useful for digging. These are so large that the animals have acquired the peculiar habit of walking on their knuckles with the claws turned inward. The hind feet are primitive and flat, with much smaller claws. The long skull, with a very pointed snout, evidently is an adaptation to ant-eating because several other unrelated animals of similar habits have developed the same features. The lower jaws of South American anteaters are very weak, and these animals lack all teeth.

Ground Sloths

Ground sloths are known from the late Miocene to the Pleistocene but are now extinct. They seem to have evolved from early members of the anteater lineage before these animals had become highly specialized. The ground sloth's strong claws, probably originally adapted for digging, are supposed to have been used to pull down the branches of trees upon whose leaves these herbivorous creatures may have fed. They walked upon their knuckles. Sloths were more primitive than modern anteaters in their teeth, five above and four below, and the unreduced number of their toes. The first upper teeth which resemble canines were reduced in size or lost in late species. These animals, however, did not have the temporal opening closed off by a bony bar below (Fig. 572-2). Size increased rather steadily with passing time, and one Pleistocene species attained the dimensions of an elephant. Some had small rounded dermal ossicles, but none developed armor.

Tree Sloths

Although no fossil tree sloths have been discovered, these animals almost certainly evolved from early ground sloths. They are leaf eaters. Their rather slight structural differences from the other sloths are mainly adaptations to arboreal existence. These animals spend most of their lives suspended upside down from the limbs of trees which they hook with two or three long curved claws. Their teeth are similar to those of ground sloths, but their forelimbs are relatively longer.

Other Anteaters

The scaly anteaters of tropical Asia and Africa commonly have been classed as edentates, but their evolutionary relations are uncertain. These are toothless long-snouted creatures with strong digging claws. The body is covered with large over-lapping horny scales, but there are no bony dermal plates. The African aardvark is a hairy but otherwise somewhat similar beast. Adults have four or five peg-like teeth in each series. These are without enamel but are coated with cement. The feet are four-toed and furnished with nails intermediate in type between claws and hoofs. These animals probably are related to the ungulates. Fossils of both fore-going kinds have been found in Pliocene strata, and other incomplete remains that may record the existence of similar creatures have come from Tertiary beds dating back perhaps as far as the Eocene Epoch.

Adaptations to ant-eating have produced animals remarkably similar in form among the monotremes, marsupials, and placentals. Such similarities certainly are not good evidence for assumptions that phylogenetic relationships are close. Many other similarities of form are equally the products of similar ways of life.

WHALES

The whales are mammals that have adapted so completely to life in water that little remains of the structural features that must have characterized their terrestrial ancestors. Even the earliest known whales from the Middle Eocene were completely aquatic. Their forelimbs had been converted to flipper-like paddles, stiffened by toe bones in five series, but consisting of more elements than are found in other mammals. The hind legs were reduced to mere internal vestiges. The nostrils had moved some distance backward on the upper surface of the skull in a way that has marked several other groups of air-breathing vertebrates which lived in water. At the very beginning of their known fossil record, whales had evolved further in their highly specialized adaptations than modern seals.

Whales are believed to have been derived from creodonts, although no crea-tures of transitional form have been discovered. Their carnivorous ancestors per-haps inhabited freshwater somewhere on the African continent. Evolutionary radiation certainly was in progress very early in the Tertiary, and the beginnings of most major lineages seem to be represented among Eocene fossils. By Miocene time all modern groups were clearly differentiated.

Primitive Early Tertiary Whales

Skulls of the most primitive whales were elongated, much like those of fish-eating reptiles (Fig. 575-1). They had not developed the very peculiar structures and distortions that appeared in many later whales. A complete set of placental teeth was present, the incisors and canines being rather simple cones, and the cheek teeth were of primitive creodont type. These animals, unlike other whales, had relatively slim bodies, up to 70 ft in length, and perhaps swam in an undulatory snake-like manner. The fossils are mainly Eocene, but a few descendants of these creatures lived on into the early Miocene.

Advanced Toothed Whales

Most whales have had rather thick, rounded, streamlined bodies, without constricted necks. This form suggests that they probably did not evolve from the type of animals described in the last section. All modern whales, many of whose ancestors can be traced back in a general way to the late Eocene, are of this second

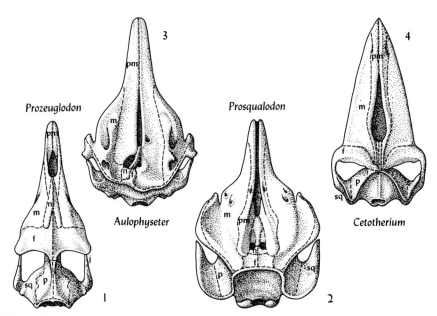

Fig. 575 Whale skulls as seen from above. 1, A primitive Eocene toothed whale. 2, An advanced Miocene toothed whale. 3, A Miocene toothed whale; notice distortion of bones around the blowhole. 4, A Miocene whalebone whale. Dashed lines outline bones; see Table 409 for identification. (*See Romer, 1966, fig. 431, p. 298.*)

type. Their bodies are not adapted to lateral undulatory swimming, and almost all their propulsive force is generated by upward and downward movements of the tail, which is provided with a pair of large horizontal flukes. Many have dorsal fins. Considerable diversification is evident among these whales. The major group consists of those with teeth.

The earliest known representatives of these toothed whales had peculiarly constructed skulls. The nostrils had moved far backward to produce a blowhole (Fig. 575-2). This had carried with it the bones that surround the nostrils, which resulted in the crowding of those bones that normally roof the skulls of mammals. The blowhole finally attained a position between or even behind the eyes, and the roofing bones eventually were greatly reduced in size or squeezed out laterally into positions where they partly covered the temporal openings. This process in several types of whales resulted in the development of unsymmetrical structures involving growth of paired elements to different sizes or rotation in position around the blowhole such as is seen in no other animals (Fig. 575-3).

The teeth were mainly conical or peg-like and not differentiated as in most other animals. In one important Oligocene group, however, the back teeth were upstanding shark-like blades. Various evolutionary trends resulted in either increase of the number of teeth, as in many porpoises one of which has about 300, or decrease, as in some of the larger whales. Reduction of the teeth involved certain positions in the jaws, or in the upper jaws more than the lower ones. Some whales retained only a few teeth, which were nonfunctional. The most peculiar example of dental modification is provided by the narwhal, which has only one incisor greatly elongated and transformed into a forwardly directed and twisted tusk.

Whalebone Whales
A much smaller group consists of whales that are entirely toothless. Primitive forms occur in the Miocene, and possible ancestors have been found of Oligocene or even late Eocene age. The skulls have their roofing bones reduced or displaced, much as in the other whales, but asymmetrical distortions do not occur (Fig. 575-4). The modern species, which probably evolved from predacious ancestors, are toothless and feed on tiny planktonic animals which are strained out of the seawater through horny laminae hanging downward in the mouth. Some of these beasts attain lengths of 100 ft. Estimates of weight extend upward to 150 tons, a far greater bulk than in any other animals that ever lived.

BIBLIOGRAPHY

G. M. Allen (1939): "Bats," Harvard University Press, Cambridge, Mass.; paperback ed., Dover Publications, Inc., New York.
This is an interesting account of bats, mostly modern ones; see pp. 100–101 and 159–187 for evolution, tooth morphology, and geologic record of these animals.

J. Buettner-Janusch (1966): "Origins of Man: Physical Anthropology," John Wiley & Sons, Inc., New York.
See chaps. 7 to 9, pp. 97–155, for an account of fossil primates. This book contains much else pertinent to primate evolution.

W. E. LeG. Clark (1958): "History of Primates," 6th ed., British Museum (Natural History), London; paperback, Phoenix Books, 2d ed., 1957, The University of Chicago Press, Chicago.
This is a popular but relatively detailed account of physical evolution leading up to man.

_____ **(1964):** "The Fossil Evidence for Human Evolution," The University of Chicago Press, Chicago.
Australopithicus, Pithecanthropus, and early *Homo sapiens* are successive evolutionary stages but are doubtfully related directly to each other. *H. neanderthalensis* probably evolved from early *H. sapiens.*

W. K. Gregory (1912): Notes on the Principles of Quadrupedal Locomotion and on the Mechanism of the Limbs in Hoofed Animals, *Ann. N.Y. Acad. Sci.,* vol. 22, pp. 267–294.
This article analyzes leg mechanisms adapted to various body forms and methods of progression.

_____ **(1920):** On the Structure and Relations of *Notharctus,* an American Eocene Primate, *Mem. Am. Museum Nat. Hist. N.S.,* vol. 3, pp. 49–243.
See pp. 233–241 for a discussion of the evolution of feet and limbs in primates.

P. Hershkovitz (1964): Evolution of Neotropical Cricetine Rodents (Muridae) with Special Reference to the Phyllotine Group, *Fieldiana, Zool.,* vol. 46.
Rodent dental evolution is discussed; see pp. 82–107.

A. B. Howell (1944): "Speed in Animals: Their Specializations for Running and Leaping," The University of Chicago Press, Chicago.
This book includes a good discussion of the form and function of limbs; see particularly chaps. 7 and 8, pp. 133–194.

R. Kellogg (1928): The History of Whales—Their Adaptation to Life in the Water, *Biol. Rev.,* vol. 3, pp. 29–76, 174–208.
This article is largely descriptive, but progressive morphologic changes are noted.

W. D. Matthew (1926): The Evolution of the Horse: A Record of Its Interpretation, *Biol. Rev.,* vol. 1, pp. 139–185.
Evolution is traced in the changing forms of teeth and limbs.

_____ **(1928):** The Evolution of Mammals in the Eocene, *Proc. Zool. Soc. London,* 1927, pp. 947–985.
The early evolutionary radiation of the mammals is sketched in a broad way.

H. F. Osborn (1898): The Extinct Rhinoceroses, *Mem. Am. Museum Nat. Hist.,* vol. 1, pp. 75–164.
Evolutionary changes in teeth and skull are described.

_____ **(1906):** Adaptive Modifications of the Limb Skeleton in Aquatic Reptiles and Mammals, *Ann. N.Y. Acad. Sci.,* vol. 46, pp. 447–476.
This article compares and attempts to explain features resulting from similar adaptations in quite different animals.

_____ **(1918):** Equidae of the Oligocene, Miocene and Pliocene of North America, Iconographic Type Revision, *Mem. Am. Museum Nat. Hist. N.S.,* vol. 2, pp. 1–331.
The introduction describes and illustrates tooth morphology and evolution.

_____ **(1929):** "The Titanotheres of Ancient Wyoming, Dakota and Nebraska," 2 vols., United States Geological Survey Monograph 55.
The reconstruction probably is complicated by the author's failure to distinguish age and sex differences from evolutionary developments.

G. E. Pilgrim (1947): The Evolution of the Buffaloes, Oxen, Sheep and Goats, *J. Linnaean Soc. London,* vol. 41, pp. 272–286.
Nature of the horns is important in indicating evolutionary relationships.

A. S. Romer (1966): "Vertebrate Paleontology," 3d ed., The University of Chicago Press, Chicago.
Nearly half of this book is devoted to Tertiary mammals.

G. G. Simpson (1945): The Principles of Classification and a Classification of Mammals, *Bull. Am. Museum Nat. Hist.,* vol. 85.
See part 3, pp. 163–272, for much information regarding phylogenetic relations.

_____ **(1950):** History of the Fauna of Latin America, *Am. Sci.,* vol. 38, pp. 361–387.
Evolution proceeded independently during most of the Tertiary when North and South America were separated.

R. A. Stirton (1941): Development of Characters in Horse Teeth and the Dental Nomenclature, *J. Mammal.,* vol. 21, pp. 434–446.
The order in which complications appear in the molar structure of horses is described.

D. M. S. Watson (1946): The Evolution of Proboscidea, *Biol. Rev.,* vol. 21, pp. 15–29.
Evolutionary changes were mainly related to increasing size and weight and greater food requirements.

A. E. Wood (1959): Eocene Radiation and Phylogeny of the Rodents, *Evolution,* vol. 13, pp. 354–361.
Parallel evolution was common, and the four groups generally recognized probably had multiple origins.

J. Z. Young (1957): "The Life of Mammals," Oxford University Press, Fair Lawn, N.J.
This is a good source for information on mammalian structural and functional biology.

TWELVE

PROBLEMATIC FOSSILS

GRAPTOLITES, CONODONTS
AND OTHERS

Most fossils are so similar to parts of modern organisms that their biologic relations are easily determined. This is not true of all, however, and there exists a considerable variety of fossils or supposed fossils whose assignment to familiar groups of organisms is uncertain. These include (1) objects or markings of doubtful organic origin, (2) markings in sediments certainly made by organisms but of such character that the nature of the organisms is variously obscure, and (3) well-preserved fossils either so different from the remains of other organisms or so lacking in diagnostic characters that they cannot be identified precisely with any commonly recognized group. With few exceptions only some of the last include specimens providing evidence of evolutionary progression.

A book like this one cannot include discussions or even mention of every kind of fossil. Probably no two persons would agree completely concerning all that should be included and what might be left out. The problematic fossils are particularly troublesome because of their great diversity and the unequal attention accorded them. Some kinds, e.g., graptolites and conodonts, are abundant, varied, and of great interest to paleontologists because they are useful for stratigraphic correlation or provide evidence for certain features of ancient environments. Consequently these have been much studied, and a great deal has been learned about their structure. They are currently identified tentatively with one or another of the recognized animal phyla. The possibility must not be overlooked, however, that such assignments are mistaken. Perhaps some of these fossils are the remains of organisms so different from other better-known ones that additional phyla, long extinct, may be needed for their reception. So much uncertainty beclouds understanding of the real nature of some fossils that speculations concerning them should not be accepted uncritically or entertained too seriously. The discovery of new evidence at any time may aid materially in their proper placement within the organic hierarchy.

Various other problematica are at present little more than curiosities, but the possibility is ever present that more careful study and better understanding may raise them to the status of useful fossils. Some, like spores and pollen and the footprints of vertebrates, may be unmistakable in their broader organic relationships but impossible to equate with other fossils of more ordinary kinds. Others are complete mysteries. The selection presented in this chapter is a miscellany of unrelated forms included for a variety of reasons, perhaps only because they are mentioned elsewhere in this book.

PRECAMBRIAN FOSSILS

For all practical purposes the fossil record of life begins with the Cambrian Period. At that time evolution had progressed so far that all the commonly recognized animal phyla probably had differentiated. Consequently the discovery of older fossils should provide interesting and important evidence regarding the earlier stages of biologic evolution.

Careful search has been made at many places for remains of life in rocks underlying those which contain fossils of the oldest familiar types. The results generally have been disappointing, although a number of different kinds of fossils has been reported. Many of these, on close examination, are now judged to be doubtfully organic, and others have come from strata of doubtful Precambrian age. None has furnished evidence of service in the solution of evolutionary problems.

The most spectacular recent discovery of possible Precambrian animal fossils was made in Australia in strata lying about 500 ft below beds containing coral-like archaeocyathids and poorly preserved early Cambrian brachiopods. These specimens commonly have been cited as Precambrian, but there is no certainty that they actually antedate Lower Cambrian faunas occurring in other areas. They include trails and burrows of worms and the impressions of soft-bodied animals in quartzitic sandstone. Most of the impressions of soft-bodied animals seem to indicate segmented worms and coelenterates similar to jellyfish and sea pens. Also present, however, are the casts of strange creatures quite different from any known in either ancient or modern faunas (Fig. 581). The main significance of these fossils and similar ones found at a few other places is that they record the occurrence of a varied lot of very ancient animals which ordinarily have left no traces.

Another interesting discovery has been made near Lake Superior in Precambrian rocks whose age has been estimated radiometrically at about two billion years. The fossils, all of microscopic size, are preserved in chert and represent what seems to be a considerable variety of algae and fungi. They demonstrate that such

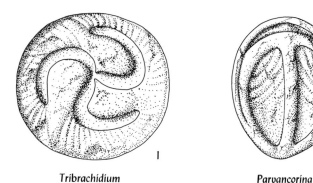

Tribrachidium   *Parvancorina*

Fig. 581   Possible late Precambrian fossils of unknown organic relations from beds in Australia below those containing an early Cambrian fauna. (× 1½.) (*After Glaessner, 1959, Records S. Australian Museum, vol.* 13*, pl.* 47*, figs.* 5, 7.)

plants probably had a very ancient origin. This is not surprising because structures believed to have been formed by blue-green algae are present in South Africa in rocks half again as old.

PLANT SPORES AND POLLEN

Plant spores and pollen grains (Figs. 79, 80), which are essentially spores containing a gametophyte, are tiny objects possessing waxy coats that are resistant to decay. They have been preserved in sediments under circumstances which resulted in the destruction of most other plant material. Their minute size and common adaptation to wind dispersal also explain their presence in strata lacking all other remains of plants. Spore-like bodies have been reported from beds as ancient as the late Precambrian, but those older than Devonian are nowhere abundant. Studies have been devoted mainly to specimens obtained from Pennsylvanian and Pleistocene deposits, the former to aid in the correlation of coal beds and the latter to reconstruct climatic fluctuations of the recent geologic past. The relative abundance of fossil spores and pollen also has been useful in locating some ancient shore lines.

Pollen ranging from the late Cretaceous to the present may be identifiable with modern families and less commonly with genera. The relations of older specimens are much more doubtful. Most older spores or pollen cannot be identified with the plants from which they came unless found within fertile fruiting structures that are related in turn to plants known from their other parts.

The contribution of most fossil spores and pollen to evolutionary knowledge is very limited. These bodies are interesting in this respect principally because some specimens, whose botanical relations are suspected or presumed, have been found in strata older than any which have yielded other identifiable fragments of these plants. This suggests that some well-known groups of plants evolved earlier than is indicated by other fossils.

Most spores and pollen grains bear one longitudinal or three radiating markings along which the outer coat splits when a spore germinates or through which the pollen tube emerges. Although these markings of spores and pollen grains resemble one another and are related to similar processes of spores formation, they are not homologous.

Both spores and pollen grains are formed in tetrads by the two successive cell divisions of meiosis. If a cell boundary membrane forms after the first division of the nucleus of the mother cell and before the second nuclear division, spores with a single marking are produced. They generally are elongated and bilateral. If the second division rapidly follows the first without the development of a separating membrane, three markings on each spore result. These spores commonly are more or less spherical, radially symmetrical, and arranged as in a tetrahedron.

The second type of spore formation seems to be evolutionarily more advanced. This is suggested by the facts that: (1) Most dicotylledonous angiosperm spores have three markings, but the more primitive kinds, such as those of the magnolias, have only one. (2) Evolution leading to the perfection of seeds involved greater changes in female than in male spores. Some plants have female megaspores with three markings and male spores or pollen grains with only one. If this conclusion regarding the differentiation of spores is correct, monocotylledonous angiosperms probably split off from the dicots very early, because their pollen grains generally have a single marking.

RECEPTACULITIDS

Receptaculitids (Fig. 583) range from the Ordovician to the Devonian, with a single specimen, probably mistakenly reported long ago, from the Carboniferous. They are widespread in their distribution and are particularly abundant in the United States in certain Middle Ordovician zones, where they may constitute excellent index fossils. Some are hollow, more or less spherical or pear-shaped bodies, but the larger ones are flat or plate-like, with slightly upturned edges. The latter, popularly known as sunflower corals, may be incomplete, with the upper parts destroyed, but this is far from certain. Some of these fossils have been identified as pine cones, calcareous algae, foraminiferans, sponges, corals, cystoids, or colonies of tunicate ascidians. Relatively modern assignments have been about equally divided between coelenterates, sponges, and an independent group of animals different from all other known fossil or modern organisms.

These fossils are composed of closely set, spirally arranged spicules consisting of a main shaft ending outwardly in a generally rhomboidal head, beneath which are four short rays directed toward the corners of the head. The inner end of the main shaft may or may not bear a somewhat similar termination. These spicules, whose forms suggest hexactinellids, seem to be pierced by central canals and

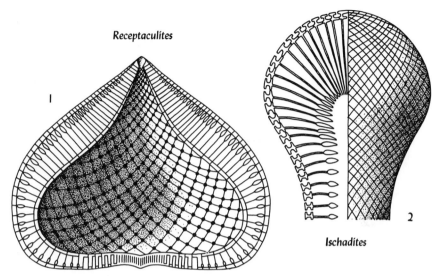

Receptaculites

I

2

Ischadites

Fig. 583   1, Section through a specimen restored as if it were a sponge-like organism. 2, Generalized representation of the structure of a specimen interpreted as a calcareous alga. (*After Billings, 1865, Paleozoic Fossils, vol. 1, fig. 353, p. 378; Kesling and Graham, 1962, J. Paleontol., vol. 36, fig. 1, p. 947.*)

provide the main reason for postulated alliance with the sponges. They are much coarser than most sponge spicules, however, and those which are now calcareous probably preserve their original composition. Some specimens are silicified, but many more, particularly those contained in dolomite, have been dissolved, leaving empty molds. The tubes formerly occupied by the spicules have been mistaken for the receptacles of polyps, and this has led to the assignment of receptaculitids to the coelenterates. Altogether, the structure of these fossils is so peculiar that they cannot be nearly related to any known animals; one of the latest expressions of opinion favors their identification as calcareous algae.

CHITINOZOANS

Chitinozoans are tiny fossils found in sediments of Ordovician to Devonian age (Fig. 584). They are the delicate horny shells of probable protozoans that inhabited the sea, as shown by their common association with other marine fossils, and many of them may have been planktonic. These little-known fossils resemble most closely the shells of some modern amoeboid and flagellated animals that live mainly in freshwater. Chemical tests indicate, however, that the modern and fossil shells are of somewhat different composition, and close relationship is not assumed. Chitinozoans have been studied by few persons, and nothing can be inferred about their

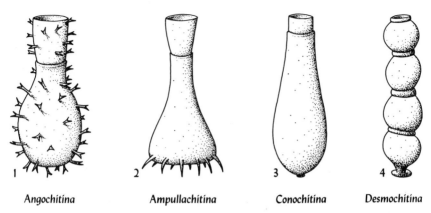

Angochitina   Ampullachitina   Conochitina   Desmochitina

Fig. 584   Chitinozoans from the Silurian and Devonian. Not drawn to uniform scale; specimens range from 0.1 to 0.5 mm in length. (*After Collinson and Schwalb, 1955, figs. 5F, 7, 9, 11C, pp. 18, 21, 28, 32, pl. 2, fig. 17.*)

evolution, but they may be useful in identifying otherwise equivocal sediments deposited in marine environments.

ARCHAEOCYATHIDS

A puzzling group of calcareous fossils known as archaeocyathids, or pleosponges, many of which superficially resemble horn corals, is locally represented by abundant specimens in Lower and Middle Cambrian limestones. A few somewhat similar objects have been reported from both older and younger strata. The animals, if they were animals, which built these skeletons lived on or attached to the sea bottom in populous communities, but contrary to some statements, they did not form true reefs. The fossils have been referred at various times and by various persons to the corals, sponges, protozoans, and calcareous algae. Their structures perhaps resemble some of the calcareous sponges most closely. The modern tendency, however, is to class them as an independent phylum different from all other known animals. In several ways this phylum, if it is to be recognized, seems to be intermediate between the sponges and coelenterates, but there is nothing to suggest that it occupies an intermediate evolutionary position.

Most of the archaeocyathids are obtusely to acutely conical or tubular in form (Fig. 585-1). They consist of delicate outer and inner walls generally concentric with each other, connected by vertical septa-like partitions or other structures, all pierced by perforations and an inner open space. The living tissue of the animals is supposed to have occupied the space between the walls, with thin layers covering the outside of the outer wall and the inside of the inner one. They are believed to have fed like sponges, i.e., by capturing food particles from water currents caused to pass through their bodies by ciliary action.

The archaeocyathids vary greatly among themselves in structural details. The pores differ in size and arrangement. The walls and partitions of some are reduced almost to spicular networks, but, in contrast, a few species have been described as being nonporous. The region between the walls may be relatively wide or narrow. It may contain variably spaced vertical septa, horizontal tabulae, irregularly inclined or curved plates, or rods or tubes. All these variables are associated in different combinations. The inner walls of some are thickened and vesicular. The inner cavity may be partly filled with vesicular calcareous material. The outer wall may be continued or overlaid by different kinds of structures that have been interpreted as (1) normal parts of the skeleton, (2) pathologic growths, and (3) different encrusting species. Some specimens are associated or intergrown in such ways that they seem to be colonial aggregates budded from a single individual (Fig. 585-5).

The interpretation of these varied structures, probably produced by animals whose nature otherwise is unknown, poses numerous problems. Many genera and species have been described and named on the basis of morphologic differences whose significance is uncertain. No careful study seems to have been made of the possible variation within species, and some evidence suggests that different names have been applied to growth stages or normal variants of the same species. Knowledge of stratigraphic ranges within the Cambrian is incomplete and does not furnish basis for the recognition of progressive morphologic series that might provide convincing evidence regarding the course of evolution. Evolutionary speculation, therefore, is dependent on what is known of the ontogeny of a few species

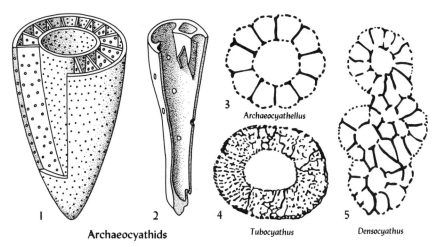

Archaeocyathids   *Archaeocyathellus*   *Tubocyathus*   *Densocyathus*

Fig. 585   1, Generalized drawing showing typical archaeocyathid structure. 2, Diagram showing origin of internal structure. 3 to 5, Cross sections. (× 3–5.) (*After Okulitch, 1955, figs. 1, 6, 8-1, -6, 13-3, pp. 2, 6, 11, 19.*)

and the assumption that increasing structural complexity is indication of evolutionary advancement.

Most archaeocyathid fossils seem to start as expanding tubes which are the beginning of the outer wall. Commonly these are empty, and such simple structure, which is duplicated by some specimens believed to be mature, suggests that this is a primitive condition. Vesicular tissue filling the interiors of some tubes may be the result of much later deposition, but this is far from certain. In ontogeny, inner structures first appear as internally directed vertical septa, which later become joined to produce the inner wall (Fig. 585-2). Some presumably mature specimens have been observed with structure transitional between these two conditions. Later evolutionary developments seem to involve progressive complication of the inner wall and the structures connecting it with the outer wall (Fig. 585-3,4), and the appearance of differences in the size, shape, and arrangement of the pores. Specimens are so variable in these respects, however, that seemingly progressive morphologic series have not been recognized. The many kinds of Lower Cambrian archaeocyathids that have been described suggest that much of the evolution in this group of animals was accomplished at an earlier time, before calcareous skeletons were produced.

GRAPTOLITES

The graptolites have been recognized for a century or more as a group of long extinct tiny colonial marine animals exhibiting several unusually clear progressive evolutionary trends. These occur well ordered in the lower Paleozoic stratigraphic column, and the graptolites are much favored as reliable zonal index fossils. They commonly are preserved as thin carbonaceous compressions, the residue of horny skeletons, in black shale that show little more than the forms of the colonies and profiles of the thecae, or receptacles, in which the individual animals lived. The thecae are arranged in linear sequences that may branch one or more times. Because of the general resemblance of these fossils to certain hydrozoans, they commonly have been classed with the coelenterates, although comparisons to the bryozoans also have been made.

The discovery of uncompressed horny specimens in limestone and chert provided material suitable for more detailed morphologic studies, because they could be sectioned or removed from their matrix by acid treatment. Specimens prepared in these ways (1) show the forms of the thecae in three dimensions, (2) show the relations of thecae to one another within the colonies, (3) reveal that some consist of regularly alternating large and small individuals, and (4) substantiate to some extent what had formerly been suspected, that the animals probably were bilaterally symmetrical. These studies have led to much better understanding of evolutionary trends and also to some radical modifications of ideas regarding the zoological affinities of the graptolites.

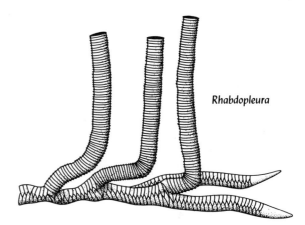

Rhabdopleura

Fig. 587 Tubes containing individual animals rise as buds from a creeping stolon. Notice that the integument of the stolon consists of half-rings like those of graptolites but the tubes do not. (*Modified after Schepotieff, 1907, Zool. Jahrb. Abt. Anat., vol. 24, pl. 22, fig. 2.*)

Relations to Hemichordates

A view that has become increasingly popular allies the graptolites with a primitive group of chordates, rather than with the coelenterates. It is encouraged principally by comparison of these fossils with living *Rhabdopleura* (Fig. 587), a colonial hemichordate. Two points of similarity have been emphasized: (1) the budding of individuals from a stolon encased in a horny sheath, and (2) the structure of the enclosing skeleton, which consists of alternating growth bands in the form of half segments meeting medially in a zigzag manner.

Neither of these similarities, however, seems to provide convincing evidence for assigning graptolites to the hemichordates. (1) *Rhabdopleura* buds from a tube which has a permanent growing tip. In contrast, graptolites budded one theca from another, and they had no similar growing tip. (2) Alternating growth bands are characteristic of graptolite thecae (Fig. 587a). In *Rhabdopleura* they mark the tube that contains the stolon but are absent in the tubes housing the individual animals

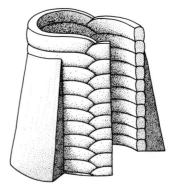

Fig. 587a Schematic drawing showing structure of the horny graptolite skeleton. The tubes are formed of alternating half-rings which may be covered by external thickening deposits. (*After Bulman, 1955, fig. 6-1, p. 22.*)

of the colony. Similarly alternating growth bands are a conspicuous feature of conularids, which may be extinct coelenterates.

Another feature that is unfavorable in the comparison of graptolites with *Rhabdopleura* and its allies is the presence in graptolites of laminated horny material with growth bands laid down on the outside of the skeleton (Fig. 587*a*). This seems to be evidence for the existence of a layer of external common colonial tissue, like that which unites the polyps in many hodrozoan and coral colonies. Nothing of this kind occurs in hemichordates. The dimorphism exhibited by one important group of graptolites (Fig. 594) might be sexual, like that of a few hemichordates, but it is equally as likely to be similar to the differentiation of feeding and protective polyps which is very characteristic of several hydroid groups. Bilateral symmetry of the graptolite thecae is not incompatible with coelenterate relationship, because most corals are bilateral in some degree.

Evolutionary Trends

Although the skeletal structure of an increasing number of graptolites is becoming known in some detail, the nature of the animals that built the skeletons probably never will be completely determined. Most graptolites are grouped in two main divisions: (1) dendroids, which are dimorphic, with large and small thecae, and (2) graptoloids, whose thecae are all of a single kind. (See Table 588.) A few others are peculiar in such ways that they are excluded from these groups. Most recognized genera and species are based on characters shown by compressed specimens which do not reveal some of the features believed to be of great importance in indicating genetic relationships. Parallel and convergent evolution surely were common among the graptolites (Fig. 589). Therefore, evolutionary lineages cannot be traced in any detail.

In spite of much uncertainty regarding the actual course of evolution, a general

---

Table 588 Principal distinguishing characters of the main morphologic groups of graptolites

---

Dendroids: dimorphic, with large and small thecae
  Dendrograptids: not suspended by thread, branching generally irregular
  Anisograptids: suspended by thread, branching becoming regular
Graptoloids: with only one kind of theca, suspended by thread, branching generally regular
  Temnograptids: four main branches forking to give off other widely spaced branches
  Goniograptids: two or four main zigzag branches giving off other closely spaced branches
  Schizograptids: two or four main branches giving off other branches laterally
  Dichograptids: eight regular branches
  Tetragraptids: four regular branches
  Didymograptids: Only two branches
  Diplograptids: Biserial, climbing up thread
  Monograptids: Uniserial, climbing up thread

This grouping is based on features seen in compressed specimens; not all the members of a group are necessarily closely related to each other.

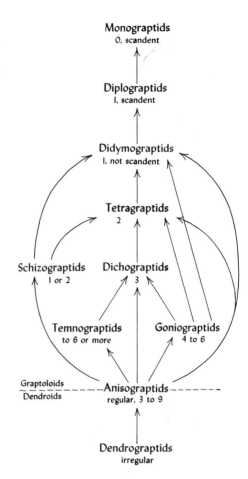

Fig. 589 Diagram showing probable evolutionary relations of the main morphologic groups of graptolites. Numbers indicated orders of branching; first order—two branches, second—four, third—eight, etc. (*Adapted in part after Bulman, 1955, fig. 49, p. 71.*)

progressive sequence is fairly clearly exhibited by the graptolites as follows: (1) Erect dendroids (Fig. 590-1,2): these primitive, conservative graptolites were attached by a thickened stalk, a root-like process, or an adhesive disk. They range from the Upper, or possibly the Middle Cambrian, to the lowermost Mississippian. (2) Pendant dendroids (Fig. 590-3): these colonies hung suspended by a hollow horny thread from seaweeds or other objects rising above the sea bottom. They are all of early Ordovician age and seem to be transitional between the benthonic dendroids and the graptoloids. (3) Attached graptoloids: many of these pendant forms are believed to have lived suspended from floating objects, probably seaweed like *Sargassum*, which accounts for their extremely wide geographic distribution. They extend from the Lower Ordovician to the Upper Silurian and probably into

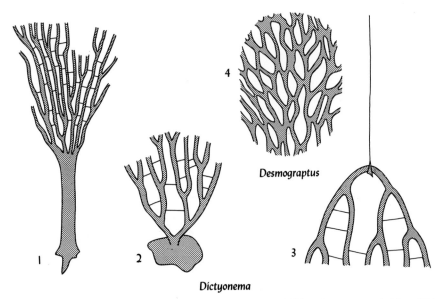

Dictyonema

Fig. 590 Dendroid graptolites. Branching colonies grew in several different ways, supported by a thickened stem (1), attached by an incrusting disk (2), and hanging by a thread (3). 4, Part of a colony of anastomosing branches. (*After Bulman, 1938, "Handbuch der Paläozoologie," O. H. Schindewolf (ed.), vol. 2D, pt. 2, fig. 9, p. 10; 1955, fig. 15-2, p. 31.*)

Table 590 List of graptolite genera selected to show progressive advancement in colony form and stratigraphic order of first appearance

9. *Monograptus:* scandent, uniserial
   Lower Silurian
8. *Diplograptus:* scandent, biserial
   Upper Ordovician
7. *Didymograptus:* pendant to reclined, branching first order
   Middle Ordovician
6. *Tetragraptus:* branching second order
   Lower Ordovician
5. *Dichograptus:* branching third order
   Lower Ordovician
4. *Loganograptus:* goniograptid, without small thecae, branching third to fifth order
   Lower Ordovician
3. *Clonograptus:* anisograptid, some with small thecae, branching regular third to ninth order
   Lower Ordovician
2. *Bryograptus:* anisograptid, with small thecae, numerous pendant irregular branches
   Lower Ordovician
1. *Dictyonema:* dendrograptid, with both large and small thecae, branches connected by crossbars, rarely
                 pendant
   Upper Cambrian

This sequence is not intended to suggest a clear evolutionary progression.

the lowermost Devonian. (4) Floating graptoloids (Fig. 591-2): structures supposed to be floating organs have been found with attached colonies of a few different kinds of Ordovician graptolites. If the interpretation is correct, these forms lived much like the modern Portuguese man-of-war, which is a coelenterate, and in this respect they were the most highly evolved colonies among the graptolites.

In addition to the general sequence outlined here, evolutionary advancement is shown by several features of the colonies and the individual thecae which compose them. The principal ones are (1) position of the bud on the first individual of a colony, (2) form of colony, (3) type of branching, (4) number of branches, (5) disposition of branches, (6) method of first branching, and (7) form of theca. Variously advanced features of these kinds occur in many different combinations. The relations are so complex that very little can be determined about direct evolutionary sequences.

Position of First Bud

Every graptolite colony began with a sexually produced individual, and all subsequent individuals grew from asexual buds. The first individual secreted a horny outer skeleton consisting of two parts. The first part seems to have been secreted as a unit, because it shows no increments of growth. An abrupt morphologic change marks the beginning of the second part, which is composed of half segments added successively on opposite sides of a tube and commonly meeting at two zigzag sutures. In what appear to be the most primitive graptolites, these segments are

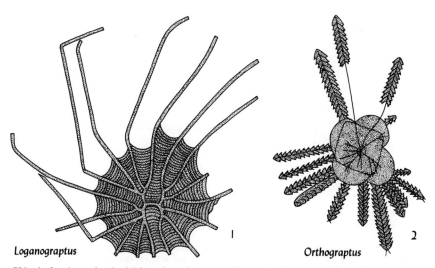

Loganograptus   *Orthograptus*

Fig. 591 1, A colony of uniserial branches whose central part is united in a disk consisting of horny material evidently secreted by communal tissue. 2, A composite colony consisting of biserial stipes that seem to have been suspended from a central structure that may have been a floating organ. (*After Bulman, 1955, fig. 52-2, p. 75; Ruedemann, 1895, 48th Ann. Rept. N.Y. State Museum, pl. 2, fig. 3, p. 52.*)

irregular and well-marked zigzag sutures are not developed. Evolutionary advancement is indicated by more and more prompt attainment of regularity, after a
beginning of this kind, until the irregular part was totally suppressed.

The bud that began the development of a colony in the most primitive graptolites broke through a resorbed pore near the base of the first part of the skeleton
that has been described. This pore and its bud appeared at successively higher
positions in more advanced graptolites, moving up into the second part of the
skeleton (Fig. 592). At last the bud formed at the margin of the growing skeleton,

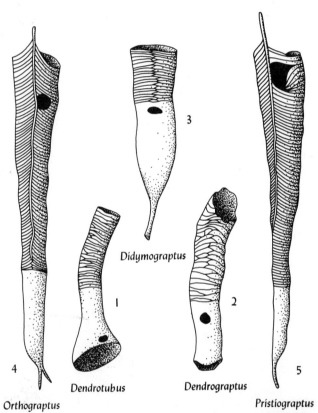

Didymograptus

Dendrotubus   *Dendrograptus*

Orthograptus   *Pristiograptus*

Fig. 592 Budding of the second graptolite individual in a colony was progressively
delayed, as shown in this sequence of drawings. Most primitively the bud
emerged through a pore resorbed in the thecal wall of the first individual near
its base (1). In more progressive forms the bud moved upward (2, 3) into the part
marked by growth lines (4) and finally appeared above the wall, where it was soon
surrounded by later growth (5). Concurrently the half segments of the first
individual's growing wall became more regular (1, 2), to produce the zigzag
suture (3) characteristic of most graptolites. (*Mainly after Kozlowski, 1966, figs. 2
to 7, pp. 493, 494.*)

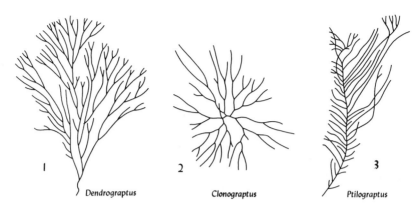

Fig. 593   Patterns of branching in graptolites. 1, A dendroid. 2, 3, Colonies intermediate in character between dendroids and graptoloids. (*After Ruedemann, 1947, pl. 18, fig. 2; pl. 33, fig. 5; pl. 44, fig. 5.*)

where resorption was not required. Here it produced a notch that was closed above by later-secreted half segments. This succession of progressive changes, beginning in dendroids or even more primitive types and continuing through various grades of graptoloids, substantiates conclusions regarding the direct evolutionary relations of the two main groups of graptolites.

Form of Colony

The forms of graptolite colonies were determined to a considerable extent by the number and disposition of their branches (Fig. 593). Colonies with many branches may have been bushy, funnel-like, or fan-like. The last two forms are most distinct in colonies whose branches are joined at intervals by crossbars (Fig. 590-1 to 3) or more rarely by lateral contact (Fig. 590-4). Junctions of these kinds provided support that was lacking in colonies with unconnected branches. The crossbars seem to have been secreted by the common colonial tissue that is believed to have overlain the main skeleton, and the anastomosing branches probably were joined by similarly deposited material. Likewise, exterior deposits thickened and strengthened the stems of some benthonic dendroids and the main branches of certain others.

The oldest known graptolites, of possible middle Cambrian age, had unconnected branches, and dendroids of this kind were common in the later Cambrian (Fig. 593). The development of crossbars and the intermittent lateral fusion of adjacent branches seem to have been progressive evolutionary developments advantageous to many-branched graptolite colonies. The only dendroids that survived the early Ordovician were provided with connecting crossbars. Dendroids of these types probably did not evolve to produce graptoloids, none of which has its branches joined in either way.

Some dendroids of Ordovician and Silurian ages consist of complexly organized branches in which elongated thecal tubes are arranged in groups or bundles. These colonies also were side developments, not in the main line of graptolite evolution.

The graptoloids probably evolved from dendroids with simple unconnected branches. A few kinds of Ordovician colonies have their diverging branches joined by a central disk-like structure secreted by the common colonial tissue (Fig. 591-1). This seems to have been an advantageous evolutionary development that aided in maintaining the spreading branches of these pendant colonies in a horizontal position.

The graptoloids, which possessed possible floating structures (Fig. 591-2), consist of several to many pendant sequences of thecae, each of which ordinarily would be considered an individual colony. Each sequence is believed to have originated by the budding of a primary individual animal in the central part of the collective colony. The primary, or founding, individuals of other graptolites probably were produced sexually, rather than by budding. In this way also the floating graptolites show evolutionary advancement beyond all others.

Types of Branching

The thecae of all graptolites budded directly from one another. In dendroids, two buds appeared simultaneously, one developing into a large and the other into a small theca (Fig. 594). The next generation of paired buds arose from the immature part of the large theca; the small theca did not bud. Repeated budding in this way produced a single continuous succession of paired thecae. Branching was accomplished by the development of large thecae from both buds of a single pair.

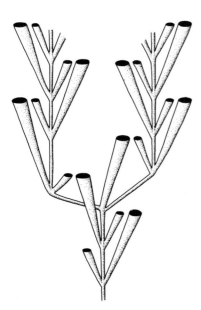

Fig. 594 Diagrammatic representation of the budding system and branching of a dendroid graptolite. Each large theca ordinarily has a smaller associate produced at the next budding.

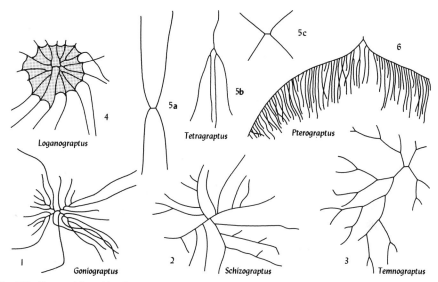

Fig. 595   Types of branching in graptoloid graptolites. (Mostly approx. × ½.) (*After Ruedemann,* 1947, *pl.* 46, *fig.* 2; *pl.* 48, *fig.* 8; *pl.* 50, *fig.* 16; *pl.* 51, *fig.* 32; *Bulman,* 1955, *figs.* 52-7B, 53, 54-1, *pp.* 75, 76, 77, 79.**)**

Graptoloids possess thecae of only one kind, corresponding to the large thecae of dendroids. Ordinarily a graptoloid theca produced a single bud. Branching resulted from the appearance of two buds growing from the same theca.

The most common type of branching involved the development of two similar sequences of thecae which diverge at almost equal angles from the projected direction of the parent sequence (Fig. 595). The branches seem to have begun their growth simultaneously and to have lengthened at approximately equal rates.

In the other type of branching, a sequence of thecae diverges laterally at a large angle from another sequence that does not change direction (Fig. 595-6). Branching of this general kind occurs in both dendroids and graptoloids whose other characters are believed to identify diverse groups not closely related to each other. Lateral branching, therefore, probably evolved independently several times by the delayed appearance of second buds.

One group of graptoloids characterized by lateral branching consists of se- quences of thecae whose forms change gradually and regularly (Fig. 596). The budding which produced a lateral branch is believed to have been delayed because (1) the bud appeared on the mature part of a theca of the parent sequence, and (2) the form of the first theca of the lateral branch is similar to that of a theca, believed to have developed at the same time, which is separated by several others of intermediate forms from the parent theca of the original sequence.

Successive thecae of these and some other graptoloids seem to change from

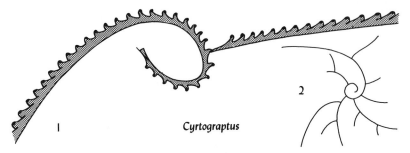

Fig. 596 1, Detail of part of a graptoloid colony whose earliest formed thecae exhibit the most advanced characters and whose later ones gradually revert to a more primitive type. The first theca of the branch resembles the ninth theca above the one from which it budded, suggesting delayed growth. (Approx. × 2.) 2, Reduced sketch of a colony with several branches. (*After Thorsteinsson*, 1955, *Geol. Mag., vol. 92, pl. 4; Bulman, 1955, fig. 69-1, p. 92.*)

advanced evolutionary types in the initial part of a sequence to later thecae of earlier evolutionary types (Fig. 596-1). These changes in thecal form have been interpreted in two quite different ways: (1) Advanced evolutionary features first appeared in the earliest thecae of a colony and probably persisted longer and longer in subsequent generations of colonies until they characterized all the thecae. (2) Regressive evolutionary tendencies appearing first in thecae produced late in the development of a colony and probably set in at earlier and earlier growth stages of descendant colonies. The details of lateral branching in other graptolites are incompletely known and may be different.

Reduction of Branches

Some typical dendroids are bushy colonies consisting of a multitude of irregularly diverging, unconnected branches. Primitively each forking produced two equal and symmetrically disposed branches (Fig. 593-1). Starting with this pattern, evolution seems to have proceeded in two directions: (1) One branch was favored over the other. It commonly was thickened and strengthened by external deposits, became a main branch, and bore a variable number of lateral branches. (2) Equal branching persisted, but all the growing branches tended to fork at the same time. Thus symmetrical colonies developed, particularly among the pendant dendroids. This type of branching probably was advantageous, because it contributed to the balance of hanging colonies. Some of the dendroids of this kind evolved to produce the graptoloids.

Pendant dendroids are known with more than 100 branches. Such numbers, however, were rapidly reduced, particularly among the graptoloids (Fig. 595), until only one forking and two branches remained. Two-branched colonies occur in the Lower Ordovician and are especially abundant in the Middle Ordovician. Finally even the single forking was suppressed.

The number of branches has been used as a primary character in the classification of graptoloids (Fig. 589). Reduction in the number of forkings, however, almost surely occurred independently in several different lineages. Therefore, classification on this basis reflects evolutionary relations in an imperfect way.

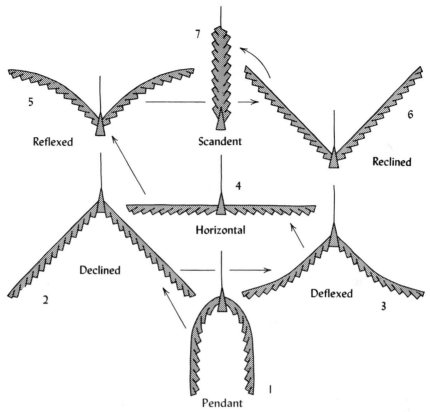

Fig. 597 Sketches illustrating the general progressive evolutionary tendency for the branches of grapto-loids to rise until they join back to back along their suspending thread, thus producing biserial colonies.

Disposition of Branches

Branches of the earliest pendant graptolites probably hung limply from their sus-pending threads. With this disposition, the individual animals of the different branches were close together, and they obtained their food from the limited volume of water that surrounded them. Advantage evidently was to be gained by lateral spreading, particularly of colonies with many branches. Evolution proceeded rapidly in this direction, and many of the pendant dendroids consisted of horizontally directed branches. Very little strengthening was required for the branches to main-tain this position, because the density of the animals was almost the same as seawater and the skeletons were very thin.

As the number of branches was reduced, less advantage accrued from spread-ing, and some of the four-branched and two-branched graptoloids may have re-verted to a limp condition. Nevertheless in others, evolution continued in the direction of rising branches (Fig. 597). This trend culminated in colonies whose branches extended upward back to back along their suspending threads (Fig. 598).

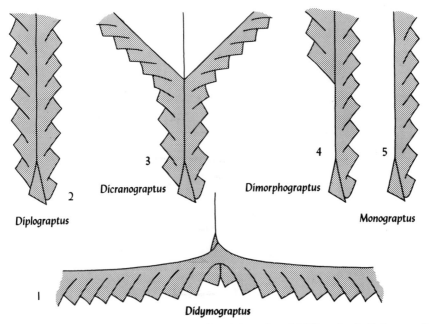

Fig. 598 Much-generalized drawings showing the relations of graptoloid branches to their suspending threads. The ascending monoserial colony (5) is the most advanced. Two of the patterns (3, 4) are particularly interesting because early growth is of an advanced type, and such colonies seem later to revert to more primitive relations.

Both four-branched and two-branched graptoloids attained this condition in the early Ordovician, and ascending two-branched, or biserial, colonies of late Ordovician age are particularly common. Most of these graptolites became extinct early in the Silurian Period. Those that survived to the late Silurian had their skeletons much reduced to a kind of spicular network.

Method of First Branching

A graptolite colony began with a single individual, which presumably developed from a fertilized egg. This animal secreted the first theca, of similar conical form in most graptoloids, which was suspended by its apex from a hollow horny thread. Further growth resulted from the successive budding of new animals and thecae from their predecessors. Branching was accomplished by the appearance and development of two buds from a single individual.

The study of uncompressed specimens has revealed the unexpected complexity of first branching in a variety of graptoloids. Double budding seems to have begun on the second theca and then was progressively delayed to later and later ones (Fig. 599). The second theca grew out from the side of the first theca which initiated the colony. In the earliest evolutionary stage it produced two buds, and one of these developed into a third theca, which crossed over to the other side of the first theca to begin the second branch. If the second theca did not produce two buds, the third one crossed over, and one of its two buds then grew out in the opposite direction, resulting in a second crossing-over. In the same way, if double budding occurred on the fourth theca, there was a third crossing-over by the fifth theca, which began the other branch.

This evolutionary series of delayed budding progressed no further in the spreading graptoloids. The relations of delayed double budding to other features are not known in many species. A single crossing-over occurs, however, in colonies with drooping branches, two in colonies with horizontally spreading branches, and three in some with rising branches. This suggests that the crossing thecae, which probably were firmly attached to each other laterally, produced stronger initial structures that aided in maintaining the gradually more and more ascending branches.

The biserial graptolites, with thecae climbing up along both sides of their suspending threads, generally began in the last-described way, i.e., they had three

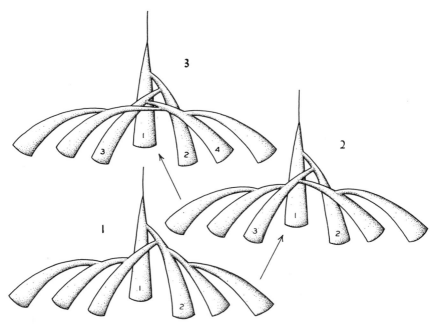

Fig. 599 Diagrammatic representations showing the details of graptoloid budding which produces two branches. 1, Most primitively, double budding occurred on the second theca; this type has been observed on pendant colonies. 2, Double budding is on the third theca; this type has been observed on horizontal colonies. 3, Double buds arise on the fourth theca; this type has been observed on reclined colonies. Further delay in double budding is known only in scandent colonies; see Fig. 600.

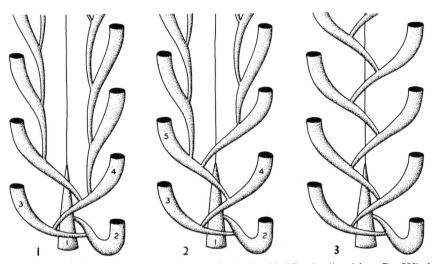

Fig. 600 Diagrammatic representations showing the details of budding (continued from Fig. 599). 1, Double budding is on the fourth theca. 2, Double budding is on the fifth theca. 3, Double budding has been completely suppressed. These differences cannot be distinguished in ordinary compressed specimens.

crossings-over. From this condition, double budding was still further delayed, until finally it seems to have been entirely suppressed in the development of some colonies (Fig. 600). All these graptolites have a similar appearance. In ordinary specimens it is not possible to distingish the parts that are potentially monoserial, but have thecae opening in opposite directions, from those that are structurally biserial, with two branches adhering back to back.

The second thecae of most biserial graptolites first are directed downward and then turn up, somewhat like the letter J. In the evolutionary progression, the second theca gradually straightened up (Fig. 601). In the monoserial colonies, with thecae climbing up on only one side of their supporting threads, the second theca rises directly from the point of its budding. The evolutionary derivation of these colonies, which first appear in Lower Silurian strata and may persist into the lowermost Devonian, from those of biserial form is something of a problem. They might have resulted simply from failure of the thecae to cross over, or there may have been suppression and loss of one or more thecae that, in ancestral colonies, crossed over. An intermediate form, first uniserial and then changing to biserial, might or might not represent an intermediate evolutionary stage.

Form of Theca

Almost all graptolite thecae are bilaterally symmetrical. The simplest and most primitive are straight tubes in their mature parts which extend outward and forward

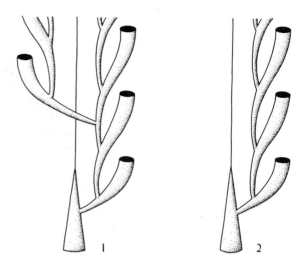

Fig. 601   Diagrammatic representations showing the details of graptoloid budding. These structures are essentially similar to those of Fig. 600, but the early thecae do not cross over. The last is a typical monoserial colony.

at an angle from the direction of the branch (Fig. 602-1). They are in contact with each other for their entire lengths. Thecae of this form are characteristic of most dendroids and of both limply hanging and spreading graptoloids. This type of theca also occurs in some biserial and a few monoserial climbing colonies.

Straight tubes rapidly evolved into S-shaped ones that, after diverging from the direction of the branch, bent forward parallel with it (Fig. 602-2,3). Gentle curvature of this kind first appeared in the early Ordovician in some of the spreading graptoloids and also in certain biserial and monoserial climbers. Sharper curvature that became almost angular was a continuation of this evolutionary trend and likewise is known in some early Ordovician species. Sigmoidal curvature may be accompanied by partial or almost complete separation of the thecae after they turn outward from the branch axis, and also by inward rather than forward direction of their apertures (Fig. 602-4,5). These evolutionary developments became noticeable in some early and middle Ordovician colonies.

The most extreme thecal modifications characterize some of the monoserial climbing graptolites. Straight tubular and S-shaped thecae are present in this group, as previously mentioned. Backwardly curved and hooked forms were new evolutionary developments (Fig. 602-6,7). The first tendency in this direction is evident in a few Ordovician biserial climbers, but its full expression was delayed until the Silurian. More than one morphologic series seems to be traceable from hooked, backwardly opening thecae to those with inwardly directed apertures, some ex-

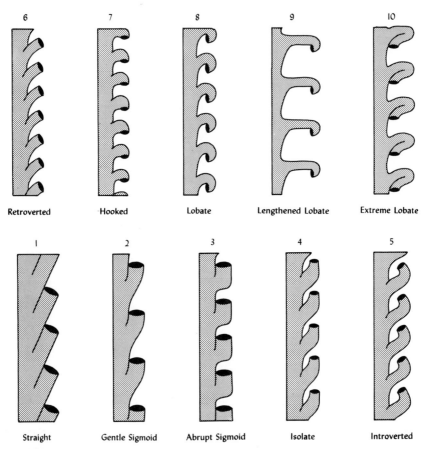

Fig. 602 Generalized drawings showing some of the differences in aperture development of mono-serial graptoloids. Beginning with straight thecae, evolution progressed in two practically opposite directions. Seemingly successive stages are illustrated in these two sequences, passing to the right. (*See Bulman,* 1955, *figs.* 30 to 33, *pp.* 46 to 50.)

tended as on lengthened stalks, and to others which turned back upon themselves (Fig. 602-8 to 10).

More and more varieties of thecae are discovered as favorably preserved, three-dimensional specimens are prepared and studied. The association of similar thecal forms with different stages in other progressive trends makes the tracing of evolutionary lineages difficult and uncertain.

Spines

No attempt seems to have been made to trace the evolutionary development of spines. These or related structures are present on many graptolites, generally rising from the most exposed parts of the colonies. Because of this, it is reasonable to conclude that they served protective functions. Probably some of them are growths secreted by the common colonial tissue. Perhaps differences in the spines and their locations may provide clues to some evolutionary sequences.

STROMATOPOROIDS

The stromatoporoids are a group of calcareous fossils consisting of layered, rounded to irregularly hemispherical masses or elongated forms that have been stated to range from the Cambrian to the Cretaceous. This is denied by some paleontologists, who class as true stromatoporoids only those specimens present in Middle Ordovician to Upper Devonian strata. The age of the reported Cambrian specimens is suspect, and those of Permian and Mesozoic ages are structurally somewhat different and are related, perhaps, to the coralline hydroids. Fossils of these kinds seem to be wholly lacking in the Mississippian and Pennsylvanian except for some specimens from strata of transitional age variously classed as latest Devonian or earliest Mississippian. This suggests that the earlier and later forms probably are not closely related to each other.

The early and middle Paleozoic stromatoporoids consist of approximately transverse, plate-like laminae commonly connected by vertical pillars (Fig. 604-2 to 5). In contrast, the late Paleozoic and Mesozoic specimens are formed of calcareous strands or rods joined to produce a more or less regular framework. Investigation of the detailed structure of these fossils requires the preparation of thin sections and microscopic study. The generally minutely porous, vesicular, or chambered nature of the specimens accounts for their assignment at one time or another to the calcareous algae, foraminiferans, sponges, coelenterates, and bryozoans. Possible intergrowths with corals, bryozoans, and worm tubes may explain part of this confusion. The Paleozoic fossils are most commonly considered to be the skeletons of some unknown kind of hydrozoan-like colonial organism. The suggestion has been made that they might have evolved from the archaeocyathids, but this does not seem probable. Paleozoic stromatoporoids are abundant fossils at some places, particularly where they contributed to the building of reefs.

Opinions differ regarding the systematic values of differences in the structure of these fossils constructed by unknown organisms. The tracing of progressive morphologic changes that might be evolutionary also is uncertain. In a general way, however, there seems to have been transition from forms consisting mainly of curved overlapping plates (Fig. 604-2) to those with more regular laminar structure. At the same time vertical pillars developed as upward growths from the transverse skeletal elements. These increased in length and continuity from layer to layer (Fig. 604-5). As the pillared structure was increasingly perfected, the intervening spaces seen in vertical sections more closely resemble tabulated tubes.

The Mesozoic specimens are less massive and conspicuous as fossils. Their evolutionary trends seem to include strengthening of the transverse elements to

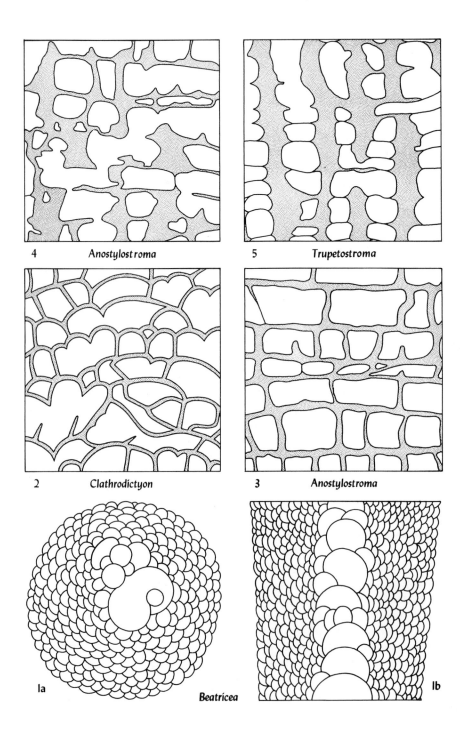

4   *Anostylostroma*

5   *Trupetostroma*

2   *Clathrodictyon*

3   *Anostylostroma*

Ia

Ib

Beatricea

Fig. 604 (See facing page.) Sections showing structures in stromatoporoids. 1, Transverse and vertical sections of a branching organism, possibly a stromatoporoid. 2 to 5, Vertical sections of typical stromatoporoids, showing in succession what seem to be progressive evolutionary morphologic stages. (*After Nicholson and Lydekker, 1887, "A Manual of Palaeontology," vol. 1, fig. 118B, C, p. 234; Parks, 1908, Univ. Toronto Stud., Geol. Ser., no. 5, pl. 7, fig. 3; 1936, no. 39, pl. 1, fig. 3, pl. 10, fig. 1, pl. 13, fig. 5.*)

produce more distinct but perhaps less-continuous laminae, and the transformation of vertical rods to open tabulated tubes. Thus these fossils also seem to show transition from dominantly horizontal to more pronounced vertical or radial structure.

CONULARIDS

Conularids are a group of peculiar, generally steeply four-sided, pyramidal fossils consisting of horny material that may be somewhat calcified. Their known range is from Middle Cambrian to Triassic, but fragmentary impressions resembling them have been reported from the Precambrian. In the past conularids have been doubtfully identified as molluscs. Their radial fourfold symmetry, however, and presumed traces of longitudinally disposed septa (Fig. 606-2) reported in a few specimens suggests comparison with coelenterates of the general scyphozoan, or jellyfish, type. The significance of these similarities certainly is doubtful.

Most of the conularids conform to a single pattern, and differences are evident mainly in minor variations of shape and ornamentation. Young individuals are supposed to have been attached by an adhesive disk at the apical end of the pyramid (Fig. 606-1). Larger ones, however, seem to have been free, with the severed apex healed or closed by a transverse partition. The shells of many conularids consist of growth increments that meet in an alternating manner along either or both the midlines of the sides and at the corners (Fig. 606-3). The shell near the larger end commonly was flexible, and the opening could be closed by infolding of the sides.

Very little can be inferred about the evolution of conularids. Their ancestry is unknown, and no clear evolutionary trends are evident, in spite of their extended range. Some of the earlier ones seem to have been rectangular rather than square in cross section.

HYOLITHIDS

Hyolithids are bilaterally symmetrical expanding calcareous shells (Fig. 607-1,2) that occur associated with marine fossils throughout the entire Paleozoic succession. They are most abundant and diverse in Cambrian strata, decline in numbers rather regularly through the Devonian, and are rare thereafter. Specimens have been reported from beds of supposed late Precambrian age. A few post-Paleozoic references probably are mistaken.

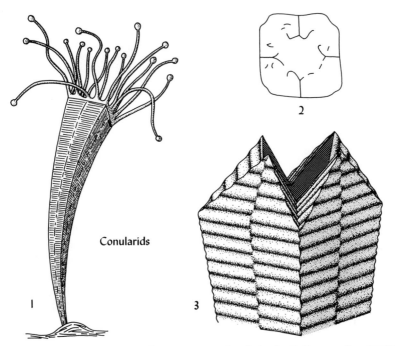

Conularids

Fig. 606 1, Imaginative restoration as a supposed coelenterate. 2, Cross section showing faint structures interpreted as septa. 3, Generalized drawing illustrating infolding of aperture. (*After Kiderlen, 1931, Neues Jahrb. Mineral., B.B., vol. 77, Abt. B, fig. 46, p. 164; Wiman, 1895, Bull. Geol. Inst. Univ. Upsala, vol. 2, pl. 5, fig. 11.*)

The zoologic relations of hyolithids are unknown, and not all fossils included in this group may be close kindred. They have been considered the shells of worms, pteropods, and cephalopods. The prevailing modern view is that they represent an otherwise unknown group of molluscs. Whatever their relations, the distinctive characters of these animals evolved in Precambrian time. The hyolithids have been little studied, and no subsequent evolutionary trends have been recognized.

TENTACULITIDS AND OTHERS
A considerable variety of mostly small, generally slender, conical shells (Fig. 607a) occurs in Paleozoic rocks, locally in great abundance. These fossils are straight or slightly curved and are distinctly annulated, smooth, or marked by longitudinal striations. Their simple forms and the lack of characters indicating what kinds of animals inhabited them are partly responsible for the little study they have received. In the past some have been identified as the remains of annelids, pteropods, scaphopods, small cephalopods, and coelenterates, or mistaken for fragments of

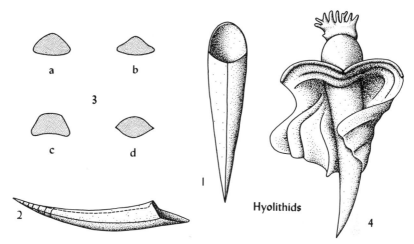

a   b

3

c   d

2

1

Hyolithids

4

Fig. 607   1, 2, Generalized drawings, top and side views, of bilaterally symmetrical calcareous shells of Paleozoic age. (Approx. × 2.) 3a to d, Cross sections showing some of the shape variations. 4, Highly imaginative restoration. (*After Fisher, 1962, fig. 64C, p. 121.*)

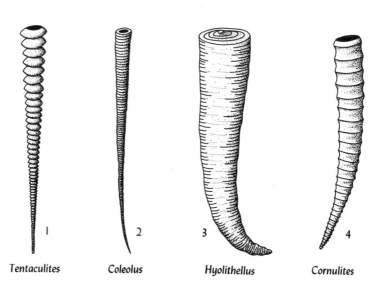

1   2   3   4

Tentaculites   *Coleolus*   *Hyolithellus*   *Cornulites*

Fig. 607a   Small horn-like or conical shells of unknown zoologic relations. 1, Calcareous shell, mid-Paleozoic. (× 5.) 2, Calcareous shell, mid-Paleozoic. (× 1½.) 3, Phosphatic shell with operculum, Cambrian. (× 4½.) 4, Calcareous shell, mid-Paleozoic. (Natural size.) (*After Fisher, 1962, figs. 55-1b, 75-1a, 76-2a, 79, pp. 112, 131, 133, 136.*)

other fossils, such as crinoid arms and the spines of echinoids or brachiopods. Modern assignments are very tentative, but relations to molluscs and annelids are most commonly suggested.

These fossils obviously represent several groups of animals not nearly related to one another. On close examination, important differences in shell structure and composition are evident, and the animals seem to have lived in quite different ways.

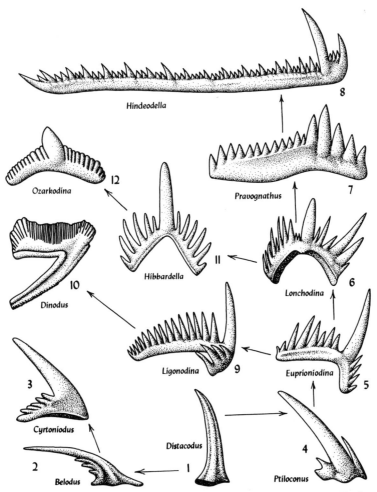

Fig. 608 Conodonts showing transitions from very simple fang-like forms to complex bars. The arrows are intended to suggest morphologic progressions, not evolutionary relations. (All much enlarged, scale not uniform.) (*After Hass, 1962, figs. 2-2, 22-1a, 23-4, 23-6a, 24-2, 27-5, 28-4, 29-9, 31-2, 34-7, 35-12, 42-2, pp. 7, 44, 45, 47, 49, 50, 51, 52, 55, 57, 66.*)

Some groups are confined exclusively to the Cambrian. Others first appear in the Ordovician, range to or into the Carboniferous, and thus have ranges long enough that evolutionary trends might be discovered. None has been reported. Some of the shells show minor ontogenetic changes in external sculpture and form which might indicate evolutionary progression, but studies leading in this direction have not been made.

CONODONTS

Conodonts are tiny tooth-like fossils occurring in a great variety of shapes (Fig. 608) and composed of calcium phosphate which generally is brownish and translucent. Their known range is from the Upper Cambrian to the Triassic, but some have been described from Upper Cretaceous strata. The last occurrence is doubtful, and all or most of the post-Paleozoic specimens may have been derived from older beds and redeposited. The nature of the animals which produced conodonts is totally unknown. These creatures probably were nektonic or pelagic, because conodonts are found in various kinds of sediment, some of which seem to record anaerobic conditions inhospitable to all bottom-living animals. Opinions assigning conodonts to fish, or some other lowly vertebrates, or worms are almost equally divided. Suggestions also have been made that they are teeth derived from the radulae of some unknown kind of mollusc, the spines of arthropods, or even structures secreted in an unexplained way by algae. The functioning of conodonts as teeth has been both asserted and denied.

Conodonts have been described as being either fibrous or lamellar in their structure. Actually both types of structure can be recognized in many of them. Fibrous structure is very conspicuous in some Ordovician forms but obscure in most others. All conodonts seem to have arisen from a basal plate (Fig. 610) which, in some "fibrous" specimens, has been believed to resemble bone. This is the principal evidence considered to favor the theory that conodonts are fish teeth. The relations of the laminae, however, demonstrate that conodonts grew on the outside by the addition of successive layers. This is not the way teeth grow, and such structure shows that conodonts must have been completely enclosed by living tissue. This conclusion is substantiated by the discovery of specimens with points repaired after being broken, indicating that they were not exposed like the cusps of teeth. Furthermore conodonts reveal no signs of wear, which should be evident on some of them if they were teeth. The preponderance of this evidence disfavors the idea that conodonts performed grasping or chewing actions in any kind of animal.

Many conodonts certainly occurred in individual animals in assemblages consisting of several different types that ordinarily are identified as different genera (Fig. 610a). Probably all were associated with others in this way. Some assemblages have been found in which five different kinds seem to be arranged naturally in bilateral pairs. This is similar to the arrangement of mouth parts in some annelids.

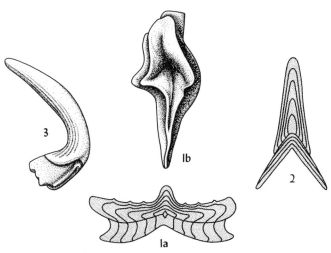

Fig. 610 Conodonts are built up of more or less cone-like laminae, and some are found attached to a laminated basal plate of differently appearing material. 1a, Transverse section through a platform type, showing the laminae in a general way. 1b, View of basal surface showing adhering basal plate. 2, Generalized section through a fang-like conodont, showing nature of lamination. 3, Side view of fang-like form with basal plate. (*After Hass, 1962, figs. 10, 11-2, 12, 48E, pp. 17, 18, 19, 90.*)

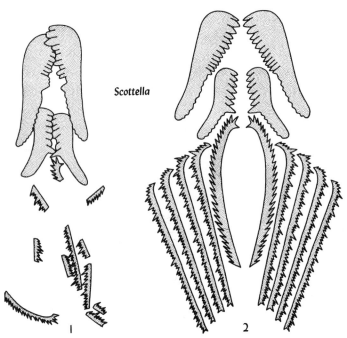

Scottella

Fig. 610a A conodont assemblage. 1, As actually observed. 2, As restored. (*After Rhodes, 1952, J. Paleontol., vol. 26, pl. 126, fig. 11, fig. 2, p. 891.*)

Conodonts, in fact, closely resemble scolecodonts, which are fairly confidently identified as the jaw parts of polychaete worms. Most scolecodonts differ from conodonts, however, by being partly siliceous and black.

Attempts to trace evolutionary changes in conodonts pose particularly difficult problems. Relatively few assemblages have been discovered. The various types of conodonts present in some of these assemblages have, as far as known, somewhat different stratigraphic ranges. Therefore the conclusion seems to be required that a particular type of conodont may have formed part of more than a single kind of assemblage. If this is so, the recognition of progressive evolution in any type of conodont is complicated by the possibility that evolution occurred independently in several different lineages of conodont-bearing animals that may not have been closely related to one another. Nevertheless, progressive evolutionary relations have been postulated for certain types of closely similar specimens (Fig. 611).

The lamellar structure of conodonts (Fig. 610-1a,2) permits the tracing of their individual ontogenies. Collections of abundant specimens may contain individuals of graduated sizes that likewise constitute ontogenetic series (Fig. 612). Evidence of both these kinds might be used in attempts to reconstruct evolutionary sequences and to infer the relations of some different types of conodonts. Few studies of this kind, however, have been made.

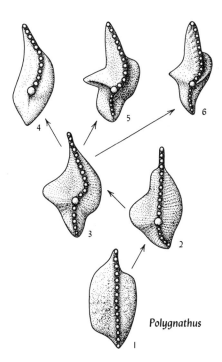

Fig. 611 Closely related Upper Devonian platform conodonts from successive stratigraphic zones, showing presumed evolutionary relations. (*After Müller*, 1962, *fig. 47, p. 88.*)

Polygnathus

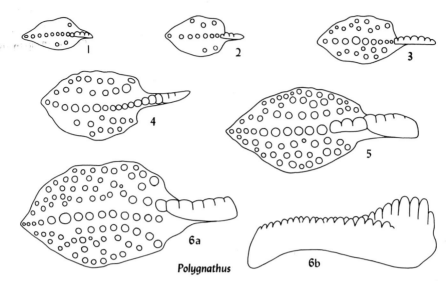

Fig. 612 Outline drawings of a series of growth stages showing progressive development of the lateral platforms and their tubercles. (× 50.) (*After Orr, 1964, Ill. State Geol. Surv., Circ. 361, fig. 4, p. 15.*)

The stratigraphic distribution of conodonts suggests that some very general evolutionary trends can be recognized. Thus simple fang-like specimens, consisting of single sharp, slender, generally curved cones (Fig. 608-1), are relatively most abundant in the Ordovician. These are superseded in importance in the Silurian by bar-like forms of various types with increasingly numerous small denticles extending in one direction or both directions from a single prominent cusp (Fig. 608-5,6 to 9,11,12). Finally the growth of lateral expansions produced platformed conodonts (Fig. 613). Some of these are present in the Ordovician, but they do not become conspicuous elements of the faunas in strata older than the Devonian. The denticles on specimens of this type seem to have been variously transformed to rows of blunt nodes or ridges or to scattered granules.

As their study has progressed, conodonts have become increasingly useful fossils for correlation. Unlike the remains of most bottom-living organisms, they are not confined to restricted sedimentary facies. They are particularly serviceable for the zonation of beds, such as some black shales, that are essentially lacking in all other kinds of fossils.

BURROWS

Burrowing animals must have been abundant at many places in the ancient seas, but evidence of their action is not commonly observed. Burrows are both easily destroyed and readily overlooked. They are not likely to be preserved and seen unless uncollapsed passages in undisturbed sediments were filled with material that contrasts in color, texture, hardness, or resistance to weathering.

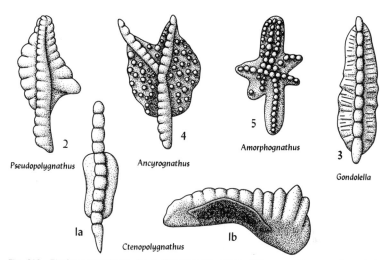

Fig. 613 Platform conodonts. 1a, b, Top and side views; platform small, suggesting primitive condition. 2, 3, Larger platform probably is an evolutionary development. 4, Branched blade standing above platform perhaps is still more advanced. 5, Nodes on platform may or may not represent blades. (*After Hass, 1962, figs. 37-3a, -4a, b, -5b, -7, -9a. p. 60.*)

Burrows provide no information directly useful in evolutionary studies. The animals that made them cannot be identified. Most of these animals probably were worms or worm-like, but some others cannot be excluded surely. Ecologic conclusions likewise are uncertain. If vertical burrows are abundant, about the only inference that can be drawn is that stable benthonic conditions persisted long enough for populations of burrowers to spread and become well established.

TRACKS AND TRAILS

Tracks and trails are markings made on sedimentary surfaces by moving animals. Mostly they are evanescent traces of living creatures that required special conditions for both their formation and their preservation. The discrimination of all markings as either organic or inorganic is not uniformly clear.

To be preserved, tracks and trails must be (1) imprinted on a relatively soft but cohesive surface, which generally rules out sand unless it is very fine and contains much clay; (2) buried promptly by some contrasting kind of sediment. More than very brief exposure is likely to result in the destruction of all markings by the erosive operation of water or air currents. This means that tracks and trails are most commonly preserved in areas of active but intermittent sedimentary deposition, whether above or below the strand line. Such areas occur mainly in very shallow water, on some tidal flats in protected situations, and on alluvial plains

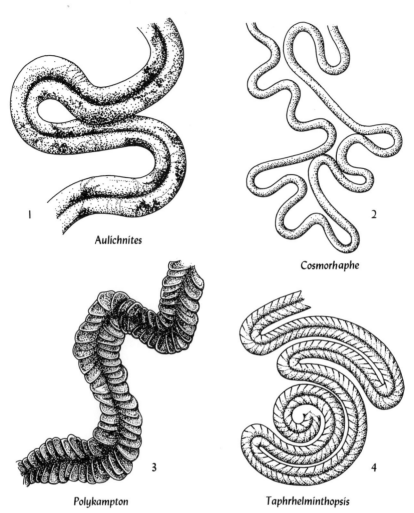

1 Aulichnites

2 Cosmorhaphe

3 Polykampton

4 Taphrhelminthopsis

Fig. 614 Invertebrate trails. 1, 2, Probably produced by worm-like animals, Pennsylvanian and Cretaceous or Tertiary. (Natural size.) 3, 4, Possibly produced by molluscs, Cretaceous and Tertiary. (Approx. × ¼.) (*After Häntzchel, 1962, figs. 110-4, 118-3, 129-8, 136-3, pp. 185, 193, 207, 217.*)

near streams. Ordinarily the burying sediment is coarser-textured than that which was imprinted, and many tracks and trails are seen as molds on the undersides of sandy layers separated by shaly partings. These markings are useful for environmental interpretations. Rarely, however, do they provide information of much interest in the consideration of evolution, because the animals that made them cannot be identified with desirable precision.

Worm Trails
Most continuous smooth trails are termed worm trails (Fig. 614-1,2). They generally are erratically curved, but some turn regularly back and forth, following parallel

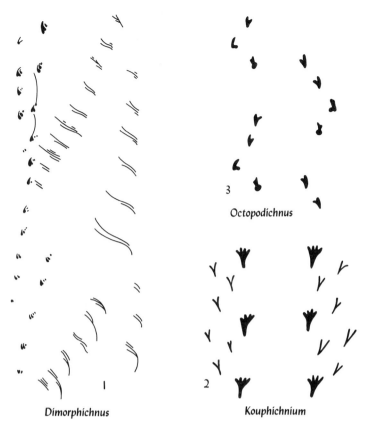

Octopodichnus

Dimorphichnus

Kouphichnium

Fig. 615 Probable arthropod tracks. 1, Probable trilobite tracks from the Lower Cambrian. 2, 3, Tracks of probable horseshoe crab-like animals from the upper Paleozoic. (All natural size.) (*After Häntzschel, 1962, figs. 119-1c, 125-1, 126-2a, pp. 194, 202, 203.*)

courses without ever crossing. The latter may be the feeding trails of molluscs. Perhaps some so-called trails are collapsed horizontal burrows made just below the sedimentary surface or at the contact between two sedimentary layers. Possibly some with a central groove or ridge are burrows of the former type (Fig. 614-1).

Molluscan Trails

Continuous curved trails with cross markings and generally divided down the middle seem to have been made by creatures that pushed themselves along by backwardly directed thrusts (Fig. 614-3,4). These are most likely to record the passage of gastropods or some similar kind of animal. Few trails supposed to have been made by pelecypods have been noted.

Arthropod Tracks

A considerable variety of tracks made by small multilegged animals has been described (Fig. 615). The only creatures capable of having made them are arthropods. Most of these tracks occur in marine strata and those of the Paleozoic, particularly its older part, probably are the marks of trilobites. Later ones were made perhaps by crustaceans and relatives of the horseshoe crabs.

Vertebrate Footprints

Vertebrate footprints (Fig. 616) are uncommon, but they have been found in strata

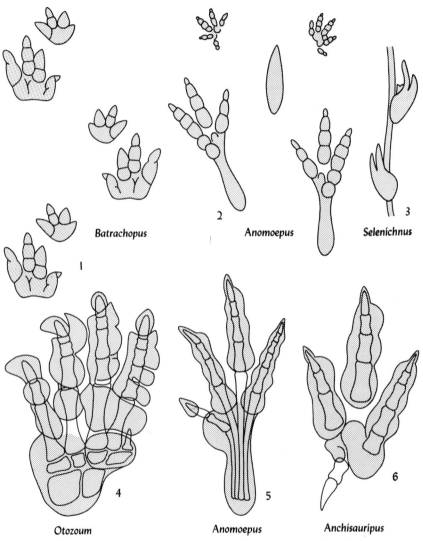

Batrachopus

Anomoepus

Selenichnus

Otozoum

Anomoepus

Anchisauripus

Fig. 616   Reptilian footprints from the Connecticut Valley Triassic. (× ½–⅛.) Probable bony structure of feet is shown in 4 to 6. (*After Lull, 1915, Conn. Geol. Surv. Bull. 24, figs. 32, 42, 59, 61, 63, 75, pp. 176, 186, 206, 209, 211, 224.*)

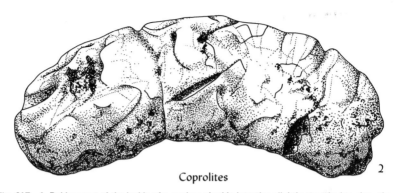

Coprolites

Fig. 617   1, Rubber cast of the inside of a modern shark's intestine slightly stretched to show the spiral form that may distinguish the coprolites of these fishes. 2, Coprolite from the Tertiary, probably of an otherwise unknown reptile. (× ¾.) (*After Zangerl and Richardson, 1963, Fieldiana, Geol. Mem., vol. 4, fig. 30, p. 145; Jepsen, 1963, Bull. Geol. Soc. Am., vol. 74, pl. 2, fig. 1A.*)

as old as the late Paleozoic. There is little doubt as to the kinds of animals that made them because they can be compared with the foot structures of contemporary beasts. This is not sufficient, however, to identify the footprints with particular animals known from their skeletal remains.

COPROLITES

Coprolites are masses of fossilized excrement. Ordinarily this name is used only for readily observed objects derived from vertebrates. The much smaller particles produced by invertebrates commonly are termed fecal pellets. These, presumably passed by sediment-ingesting organisms, are most commonly recognized in some kinds of limestone studied in thin section. Similar pellets derived from creatures which obtain food by eating ordinary mud or silt rarely retain their identity in consolidated sediment. Attempts have been made without particular success to identify various types with various kinds of animals.

Vertebrate coprolites come in many forms, few of which can be identified with the animals that produced them. Exceptions are those that show some remnant of spiral structure imparted to them by the typical coiling of a shark's intestine (Fig. 617-1). Coprolites provide no evolutionary information. They are interesting, however, because they may be evidence for the existence of larger animals, otherwise unknown, and the material on which they fed. Some contain bones, scales, or other hard parts, imperfectly recording the presence of smaller creatures.

BIBLIOGRAPHY

E. S. Barghoorn and **S. A. Tyler (1965):** Microorganisms from the Gunflint Chert, *Science,* vol. 147, pp. 563–577.
A variety of tiny Precambrian fossils, probably algae and fungi, is recorded.
O. M. B. Bulman (1955): Graptolithina, etc., in "Treatise on Invertebrate Paleontology," R. C. Moore (ed.), pt. V, Geological Society of America, Boulder, Colo., and University of Kansas Press, Lawrence, Kans.
This small volume has a good introduction devoted to the detailed morphology of graptolites.
C. Collinson and **H. Schwalb (1955):** North American Chitinozoa, *Illinois State Geol. Survey, Rept. Invest.* 186.
This article describes and illustrates a variety of tiny fossils and reviews what is known about them.
G. Erdtman (1943): "An Introduction to Pollen Analysis," Cronica Botanica, Waltham, Mass.
See pp. 43–49 for an account of the development and morphology of pollen grains.
D. W. Fisher (1962): Small Conoidal Shells of Uncertain Affinities, in "Treatise on Invertebrate Paleontology," R. C. Moore (ed.), pt. W, Miscellanea, pp. 98–143, Geological Society of America, Boulder, Colo., and University of Kansas Press, Lawrence, Kans.
See this mainly descriptive report for accounts of hyolithids, tentaculitids, and similar fossils.
J. J. Galloway (1957): Structure and Classification of the Stromatoporoidea, *Bull. Am. Paleontol.,* vol. 37, pp. 341–480.
This work is devoted mainly to the description of these puzzling fossils.
M. F. Glaessner (1961): Pre-Cambrian Fossils, *Sci. Am.,* vol. 204, no. 3, pp. 72–78.
The author describes and illustrates a fauna of puculiar late Precambrian fossils.
W. Häntzschel (1962): Trace Fossils and Problematica, in "Treatise on Invertebrate Paleontology," R. C. Moore (ed.), pt. W, Miscellanea, pp. 177–245, Geological Society of America, Boulder, Colo., and University of Kansas Press, Lawrence, Kans.
See this work for a review of many strange and doubtful fossils.
W. H. Hass (1962): Conodonts, in "Treatise on Invertebrate Paleontology," R. C. Moore (ed.), pt. W, Miscellanea, pp. 3–63, Geological Society of America, Boulder, Colo., and University of Kansas Press, Lawrence, Kans.
This article is almost exclusively descriptive and systematic.
D. Hill (1964): The Phylum Archaeocyatha, *Biol. Rev.,* vol. 39, pp. 232–258.
This article reviews the literature on archaeocyathids and shows how opinions concerning them have differed.
B. F. Howell (1962): Worms, in "Treatise on Invertebrate Paleontology," R. C. Moore (ed.), pt. W, Miscellanea, pp. 144–177, Geological Society of America, Boulder, Colo., and University of Kansas Press, Lawrence, Kans.
This is a descriptive account of fossil worms and worm-like objects and markings.
R. Kozlowski (1966): On the Structure and Relationships of Graptolites, *J. Paleontol.,* vol. 40, pp. 489–501.
The author argues in favor of his controversial theory that graptolites are hemichordates.
M. W. de Laubenfels (1955): Porifera, in "Treatise on Invertebrate Paleontology," R. C. Moore (ed.), pt. E, pp. 21–112, Geological Society of America, Boulder, Colo., and University of Kansas Press, Lawrence, Kans.
See pp. 108–110 for a brief account of receptaculitids.
M. Lecompte (1956): Stromatoporoids, in "Treatise on Invertebrate Paleontology," R. C. Moore (ed.), pt. F, Coelenterata, pp. 107–144, Geological Society of America, Boulder, Colo., and University of Kansas Press, Lawrence, Kans.
A brief discussion of evolution is presented on pp. 122–123.

M. Lindström (1964): "Conodonts," American Elsevier Publishing Company, New York.
See chaps. 6 and 7, pp. 79–107, which trace morphologic relations that may be evolutionary.

R. C. Moore and **H. J. Harrington (1956):** Conulata, in "Treatise on Invertebrate Paleontology," R. C. Moore (ed.), pt. F, Coelenterata, pp. 54–66, Geological Society of America, Boulder, Colo., and University of Kansas Press, Lawrence, Kans.
This section on conularids is entirely systematic.

K. J. Müller (1962): Taxonomy, Evolution, and Ecology of Conodonts, in "Treatise on Invertebrate Paleontology," R. C. Moore (ed.), pt. W, Miscellanea, pp. 83–91, Geological Society of America, Boulder, Colo., and University of Kansas Press, Lawrence, Kans.
The problem of homeomorphy as related to evolutionary interpretations is discussed.

G. E. Murray (1965): Indigenous Precambrian Petroleum, *Bull. Am. Assoc. Petrol. Geologists,* vol. 49, pp. 3–21.
See this article for an extensive bibliography of the literature on Precambrian fossils.

V. J. Okulitch (1955): Archaeocyatha, in "Treatise on Invertebrate Paleontology," R. C. Moore (ed.), pt. E, pp. 1–20, Geological Society of America, Boulder, Colo., and University of Kansas Press, Lawrence, Kans.
This article is almost entirely systematic. The fossils are commonly considered to represent an extinct and otherwise unknown phylum.

R. Ruedemann (1947): Graptolites of North America, *Geol. Soc. Am., Mem.* 19.
See pp. 39–46 for a discussion of graptolite phylogeny.

J. M. Schopf (1938): Spores from the Herrin (No. 6) Coal in Illinois, *Illinois State Geol. Surv., Rept. Invest.* 50.
Explanation of the different kinds of Carboniferous spores is presented on pp. 10–15.

——— **(1968):** Microflora of the Bitter Springs Formation, Late Precambrian, Central Australia, *J. Paleontol.,* vol. 42, pp. 651–688.
This article reviews previously known Precambrian microfossils and describes and interprets a varied and well-preserved, mainly algal flora about 1 billion years old.

R. P. Wodehouse (1936): Evolution of Pollen Grains, *Bot. Rev.,* vol. 2, pp. 67–84.
Relations of the basic types of spores and pollen grains are explained.

CHAPTER THIRTEEN

EVOLUTION IN PERSPECTIVE

PARADE OF LIFE
PATTERN OF EVOLUTION

The organic world as it exists today is so exceedingly complex that in its entirety it is beyond the understanding of any individual. A century ago most biologists were naturalists in the broad sense, whose interests were wide and who kept abreast of the work of their contemporaries. All this has changed. The enormous increase in detailed knowledge of plants and animals has reached a stage at which no one person can begin to follow all the research that is in progress. Most biologists now find it necessary to concentrate attention on one or another of the dozen or more well-marked but restricted fields into which biology is subdivided. As biologists have specialized, their interests and viewpoints generally have tended to narrow more and more. A narrowing of understanding also has been encouraged by the highly technical vocabularies that have grown up in various fields, so that comprehension and communication become increasingly more difficult. Thus effective barriers have been thrown up which serve to compartmentalize a great part of biologic accomplishment and thought.

Any interested observer must realize, however, that these barriers are artificial. Each biologic field has ties with several others, some strong, some weak. Nevertheless there is one aspect of biology which is outstanding because of its connection with almost every other field. This is historical biologic evolution, which contributes importantly to work and thought in other fields and draws from them for better understanding. Thus the principle of evolution is, in a very real sense, a unifying factor in biology. It gives added meaning and coherence to a great variety of studies that might otherwise drift apart more widely than they have. Moreover it is of vital interest to human beings, who themselves stand at the end of a long evolutionary sequence. In short, evolution is the single most fundamental aspect in the history of life.

THE PARADE OF LIFE

Fossils represent stages in the historical and evolutionary progression that has led up to all modern organisms. The fossil record is, in effect, a parade of ever-changing life that has passed in review for some 600 million years. Most fossils, except some in the youngest rocks, are notably different from the corresponding parts of any organisms that live today. Their order with respect to the passing of this great interval of time is the only direct evidence that points out the actual course of evolution.

In general, fossils demonstrate that a continuous transition has occurred from ancient types of organisms to more and more modern ones. This transition has involved the disappearance of older types and the appearance of newer ones to take their places. Some transitions obviously were gradual and clearly indicate the workings of evolution. Others, however, seem to have been abrupt, as shown by the sudden appearance in the fossil record of newer groups whose ancestry and relations to older ones are not yet known. This is the condition that exists with respect to some of the larger recognized taxonomic groups as well as many smaller ones. It is responsible for some of the speculation concerning the details of evolutionary mechanisms.

Precambrian Eras

The diversity and obviously advanced evolutionary status of Cambrian faunas is evidence that life must have had a very long previous history. Much the greater part of geologic time is included in the Precambrian, and much the greater part of biologic evolution must have been accomplished before the beginning of the Cambrian Period. The great scarcity of authentic Precambrian fossils, however, is difficult to explain. Because any trace of Precambrian life would be interesting and probably instructive with respect to the course of early evolution, intensive searches have been made. Structures believed to have been built by calcareous algae are abundant in the late Precambrian at some places, but otherwise the results have been almost uniformly disappointing. A considerable number of Precambrian fossils has been reported, but most of these objects or markings are either doubtfully organic or doubtfully Precambrian. The most noteworthy assemblage of impressions that probably record organisms of strange types has been discovered in the supposedly very late Precambrian of Australia.

Paleozoic Era

The Paleozoic Era, extending from the beginning of the Cambrian to the end of the Permian Period, includes more than half of the time spanned by the fossil record. The earliest fossils are exclusively marine and belong to only a few of the larger taxonomic groups. In younger rocks the diversity steadily increases until the record shows that in mid-Ordovician time marine invertebrate animals with pre-servable hard parts probably were as abundant and nearly as varied as they are in modern seas. The oldest presumed terrestrial fossils have been discovered in Silurian strata. Land plants became abundant during the Devonian Period. Vertebrates are first recognized in the rare and fragmentary fossil fishes of the Ordovician, and vertebrate creatures emerged upon the land before the beginning of the Mississippian Period. Before the end of the Paleozoic Era the earth was populated by both plants and animals much as it is today, but all the organisms that existed at that time were of archaic types.

The Paleozoic Era is particularly noteworthy, therefore, for evolution that re-sulted in colonization of the land by organisms descended from marine ancestors.

Cambrian Period

Invertebrates of the Cambrian Period seem to have been much restricted as com-pared with those of later faunas, but this probably indicates only that many types of animals lacked preservable hard parts. Trilobites are by far the most abundant and varied fossils. Metaparians characterize the Lower Cambrian and are not known in younger strata. Eodiscids are of early and mid-Cambrian ages. Agnostids range from the Lower Cambrian into the Ordovician. Most other trilobites are opistho-parians, which begin in the Lower Cambrian, but a few proparians appear in the Middle Cambrian. Brachiopods are second in importance among Cambrian fossils. Most of those present in the lowest Cambrian rocks are small inarticulates, but a few orthoids represent the articulates in strata of the latter part of the period.

Few other kinds of fossils are abundant except locally, and several important invertebrate groups have not been certainly identified. Coral-like archaeocyathids occur in the Lower and Middle Cambrian, but corals are not known except doubtfully in the Upper Cambrian. Graptolites are represented only by dendroids, which range from the middle or late Cambrian onward. Bryozoans are very rare in the Upper Cambrian. Conodonts have been found at a few places in the Upper Cambrian. Among the molluscs, primitive gastropods occur sparsely in the Lower Cambrian as well as younger strata. A few cephalopods are present in the Upper Cambrian, and there are some older doubtful forms. Pelecypods, however, are not known before the Ordovician. Fragments of echinoderms have been reported from beds as old as the early Cambrian. Most Cambrian forms seem to have been stemmed or otherwise attached species.

The most interesting Cambrian fauna is that discovered in the black Burgess Shale of the middle part of the system in western Canada. It includes a considerable variety of uniquely preserved impressions of soft-bodied creatures different from any that have been found elsewhere. Among them are arthropods which seem to be related to trilobites, chelicerates, and crustaceans (Fig. 244). This fauna provides a tantalizing glimpse of a group of animals that must have existed at other times and places but was very rarely fossilized.

Cambrian floras are practically unknown. Stromatolites probably built by marine algae are abundant locally. Some tiny plant spores and shredded woody material have been reported from the Cambrian, but their relations and significance are uncertain.

Ordovician Period

Early Ordovician faunas generally are sparse and resemble those of the late Cambrian. Great expansion is evident, however, in the preserved marine life of the mid-Ordovician.

The earliest certainly identified foraminiferans are arenaceous species from the Ordovician. Receptaculitids are particularly characteristic of strata in the midpart of this system but range onward into the Devonian. Both solitary and colonial tetracorals and tabulates became fairly common at the beginning of mid-Ordovician time. Stromatoporoids were important limestone builders. Graptolites evolved rapidly and are excellent and abundant index fossils in some of the dark Ordovician shales.

Conodonts are first abundant in some Ordovician rocks. Bryozoans are common from the Middle Ordovician onward. Massive and branching trepostomes and cyclostomes are the most characteristic kinds. Brachiopods expanded greatly in the mid-Ordovician, with long-hinged orthoids being most abundant. The earliest pelecypods appear in the Lower Ordovician, but they are less common than gastropods, which include pleurotomarids and other spired types. Remarkable expansion of the cephalopods began in the early Ordovician. Most of these fossils are orthoceroids, and all main nautiloid groups are represented in the Middle Ordovician.

Trilobites are numerous but less diversified in the Ordovician than in the Cambrian. Many of the Ordovician forms have no obvious known Cambrian ancestors. The first eurypterids appeared and ostracode history began early in this period. Echinoderms expanded greatly, and most of the main groups, except the articulate crinoids and advanced echinoids, are present in Middle Ordovician strata. Inadunate crinoids are best represented.

Scraps of bone in Lower and Middle Ordovician sandstones record the presence of ostracoderms, which are the oldest discovered vertebrates.

Almost nothing is known about Ordovician plants.

Silurian Period

The Silurian marine faunas show continuations from the Ordovician of several broad evolutionary trends. Tabulate corals reached their acme of development, and the first large coral reefs were built. The graptolites became more specialized, but only simple kinds survived into the Devonian. Nautiloid cephalopods continued to be abundant and evolved in several conspicuous ways. More of them were coiled, and species with variously restricted apertures and other peculiarities are particularly characteristic. The variety of trilobites declined, but some of the most bizarre species developed at this time. Pelecypods became somewhat more diversified, but neither they nor gastropods were particularly important elements of the known faunas.

The lace-like fenestellid bryozoans first appeared in the Silurian, but they were not abundant. All main groups of brachiopods except productoids are represented among Silurian fossils, but the inarticulates declined. The strophomenoids and pentameroids are most conspicuous. The known echinoderms are mostly crinoids, which include all main Paleozoic types. One of the most famous eurypterid faunas is of late Silurian age. It probably represents an environment that was not normally marine. The earliest scorpion that has been discovered is Silurian. It has generally been considered the first known air-breathing animal, but this interpretation is uncertain.

Vertebrates are represented by heavily armored ostracoderms, which seem to have reached their climax in the late Silurian or early Devonian. Other Silurian fishes were placoderms, which are believed to have possessed lungs.

Land plants are reported from the Upper Silurian of Australia, but their age is somewhat doubtful; they may be early Devonian. They are psilophytes, with primitive vascular stems and without leaves. A few associated fossils may be lycopods.

Devonian Period

Among all Paleozoic marine faunas, those of the Devonian are outstanding for their diversity and richness. The brachiopods and tetracorals both attained their greatest development during this period. Pelecypods and gastropods are locally more abundant fossils and more diverse than in any older rocks. Nautiloid cephalopods declined in importance, and some groups did not survive into the Mississippian, but ammonoids in the form of goniatites made their first appearance. Trilobites were not rare, but their variety was much restricted.

Foraminiferan fossils of the Devonian are mostly arenaceous but include the earliest calcareous and chambered specimens. Trepostome bryozoans declined in mid-Devonian time, and the cryptostomes rose rapidly to dominance. Small tentaculitids are very abundant in some Upper Devonian strata, but they became extinct at the end of this period. Dendroid graptolites continued through the Devonian but are uncommon and are not known after the earliest Mississippian. The crinoids were mostly inadunates and monocyclic camerates. The last cystoids lived during the Devonian Period.

Devonian eurypterids probably were restricted to freshwater. The earliest insects, which were wingless, are from Middle Devonian strata, and the oldest known spiders also are of Devonian age.

Vertebrates seem to have evolved relatively rapidly during the Devonian Period, and before its end all major groups of fishes were in existence. Ostracoderms and placoderms are well represented among Devonian fossils, but only a few of the latter survived the ending of this period. Lungfishes, crossopterygians, and other bony fishes all occur in Lower Devonian strata, and the first known primitive sharks are of late Devonian age. The Devonian sharks were all marine, but many of the bony fishes seem to have inhabited freshwater. The earliest discovered amphibians have been found in the uppermost Devonian or lowermost Mississippian strata of Green-land.

Land plants also evolved rapidly. Small psilophytes, which probably grew like rushes, dominated the older Devonian floras. These were followed by ferns, sphen-opsids, and lycopods, and by the end of the period cordaites had appeared. The mid-Devonian plants included few trees, but forests developed in the late Devonian, foreshadowing the coal swamp floras of the Carboniferous.

Mississippian Period

Mississippian, or Lower Carboniferous, fossil faunas are considerably less diverse than those of the Devonian. They are best known for the great profusion of crinoids in the Middle Mississippian. Camerates reached their climax at this time and then almost disappeared. Blastoids, which range upward from the Silurian, also achieved their greatest diversity in the Middle Mississippian but are numerically most abun-dant in the Upper Mississippian of some regions. A further noteworthy feature is the occurrence of calcareous foraminiferans in such abundance that they were important in the formation of Middle Mississippian limestones in some areas. Cryptostome bryozoans, represented principally by fenestellids, are common and characteristic fossils.

Corals and brachiopods declined considerably after the Devonian. Most of the Mississippian corals are solitary horn-like forms. Spiriferoids are conspicuous among the brachiopods, and productoids, which appeared very sparingly in the Devonian, are especially characteristic of the Middle and Upper Mississippian and younger Paleozoic strata. Pelecypods and gastropods are relatively unimportant among Mississippian fossils. Nautiloid cephalopods were less abundant and diverse and generally smaller than in the Devonian. Goniatites, however, are represented in greater numbers, and the first ceratites appeared. Trilobites are uncommon fossils, and most are rather generalized opisthoparians.

The vertebrate record in the Mississippian is meager. Sharks of somewhat modern types occur in the earliest strata. Very few terrestrial vertebrates have been discovered. All that are known were amphibians.

The Mississippian land flora was cosmopolitan and gradually became more diversified. Many kinds of plants are present among the fossils, and many species, including fern-like forms, were arborescent. The lycopods, particularly, grew to be large trees.

Pennsylvanian Period

Both the faunas and floras of the Pennsylvanian Period, or Upper Carboniferous, are more diverse than those of the Mississippian, probably because a greater variety of environments have left fossil records. Many kinds of foraminiferans occur, of which the fusulinids are most important. They evolved relatively rapidly and are useful index fossils. Brachiopods fell off further in variety, but this was compensated by a rise in the numbers of gastropods and pelecypods. Nautiloid cephalopods are somewhat more abundant than in the Mississippian. Goniatites and ceratites also are more diverse as fossils, and forms transitional to ammonites appear in the Upper Pennsylvanian.

Fenestellids continue to be common among the fossil bryozoans. Pennsylvanian crinoids are mostly dicyclic inadunates. Camerates are very rare, and blastoids are known only from strata near the base of the system.

Both brackish and freshwater Pennsylvanian faunas have been preserved. The fossils are mostly small pelecypods, gastropods, ostracodes, and branchiopod crustaceans. A considerable number of winged insects has been discovered in the Pennsylvanian. Among the most common fossils are roaches and dragonfly-like forms. The oldest known myriapods are of Pennsylvanian age.

The number of wholly terrestrial Pennsylvanian vertebrates known is small. Reptiles are represented, and their evolution probably began during the Mississippian Period.

Fossil land floras of the Pennsylvanian are relatively rich in species and include many kinds of trees. Seed ferns may have been the dominant plants at this time, but lycopods also were abundant. The earliest record of conifers appears in the Middle Pennsylvanian and some of the oldest identified bryophytes date from this period.

Permian Period

Permian faunas and floras obviously are direct continuations of those of the Pennsylvanian Period. Fusulinid evolution progressed regularly to more complex forms but ended in extinction before the end of the Permian Period. Tetracorals declined and did not survive into the Triassic. The first supposed octocorals have been reported from the Permian. Trepostome and cryptostome bryozoans declined and became extinct. Brachiopods were generally similar to those of the Pennsylvanian, but several highly specialized forms developed as members of a reef fauna. Orthoids and productoids did not survive the Permian.

Pelecypods and gastropods of the Permian resemble those of the Pennsylvanian. The nautiloids are mostly coiled. Ammonoids continued to evolve, and the first true ammonites appeared. Trilobites and eurypterids died out before the end of the Permian Period. Echinoderms are not common fossils. A unique fauna, known only in the East Indies, however, includes the last known camerate crinoids and blastoids.

Insects seem to have evolved rapidly late in Paleozoic time. The first representatives of several important modern groups have been collected from Permian strata.

The last of the placoderms and most of the Paleozoic sharks died out before the end of the Permian Period. The first well-known faunas of land vertebrates are of Permian age. Amphibians are abundantly represented among the fossils but were progressively outnumbered by the reptiles. Primitive mammal-like reptiles made their appearance in the Permian.

Changing climate in the Permian Period resulted in important floral transformations. Lycopods, seed ferns, sphenopsids, and cordaites all declined. Ferns were more successful in meeting the new conditions, and conifers likewise became relatively more important. The first ginkgos and cycads appeared near the end of the Permian Period. A new flora identified by the presence of *Glossopteris* arose in the southern hemisphere and spread northward.

Mesozoic Era

The Mesozoic Era represents about three-quarters of all post-Paleozoic time. It witnessed profound changes that affected all living things. One or more of the great Paleozoic groups of plants, coelenterates, bryozoans, brachiopods, arthropods, and echinoderms became extinct and were replaced by new forms, which evolved rapidly and achieved prominence in the Mesozoic Era. In contrast, the foraminiferans, molluscs, and vertebrates continued along evolutionary paths which had been plainly foreshadowed during late Paleozoic time.

Mesozoic life was not simply a sequel to that of the Paleozoic. It is particularly noteworthy for the rise to dominance of molluscs in the sea and reptiles on the land and for the modernization of terrestrial floras.

Triassic Period

Fossil invertebrate faunas of the Triassic Period are dominated by the molluscs. Brachiopods are not abundant and are mostly holdovers from the late Paleozoic. They seem to have been replaced progressively by pelecypods, which include several new types such as the oysters. Several of the older kinds of gastropods were reduced in numbers or had disappeared, and new ones like the naticoids took their places. The nautiloid cephalopods fared similarly. Many of the ammonoids had ceratitic sutures, and numerous species of narrow form developed with conspicuous ribs or nodes.

Fossil foraminiferans of the Triassic are not abundant. Hexacorals appeared in Middle Triassic strata, and the first authentic octocorals are recognized, although they may have had an earlier origin. The last conodonts may be of Triassic age. Few echinoderms of Paleozoic type continued into the Triassic. Cidaroid echinoids, which are rare fossils from the Devonian onward, became important in late Triassic time. Almost all fossil crinoids are articulates.

Several additional modern kinds of insects appeared in the Triassic. Among them were the true flies and dragonflies.

Terrestrial vertebrates include holdovers from the Permian, most of which soon became extinct, and several entirely new types. Few nonaquatic amphibians remained. The thecodont reptiles were exclusively Triassic, but they were the stock from which several more specialized groups evolved. Among those which appeared during the Triassic Period were the dinosaurs and crocodilians. Other reptiles that first appeared in the Triassic Period were ichthyosaurs, plesiosaurs, turtles, and lizards. Most important of all, the first mammals evolved in very late Triassic time from the therapsids, which had their origin in the Middle Permian.

Early Triassic floras are not well known. Ferns and cycadophytes were dominant in the late Triassic, and ginkgos were well established. Cordaites had disappeared, lycopods and sphenopsids were limited in variety, and conifers and seed ferns were only a little more common. Primitive angiosperms may have originated in the Triassic Period, although this is generally doubted.

Jurassic Period
Brachiopods, mainly terebratuloids and rhynchonelloids, were somewhat more abundant in the Jurassic Period than in the Triassic. Pelecypods, however, continued to expand, and several new types, including the rudistids, made their first appearance. Gastropods were much like those of the Triassic. Most Jurassic ammonoids have ammonitic sutures and seem to have developed from a single group of Triassic ancestors. Belemnites, first known from the Mississippian, were abundant locally, and fossil sepioids and squids have been discovered. Foraminiferans are well represented among the fossils, and several other kinds of Jurassic protozoans have been recognized. The most abundant bryozoans are cyclostomes. Regular echinoids with compound ambulacral plates first appeared in the Jurassic.

Fishes began to take on more familiar forms in the Jurassic Period. Sharks evolved along modern lines, and teleosts originated. Few land vertebrates are known except from the late Jurassic. By that time the saurischian dinosaurs had produced their largest species. Plesiosaurs increased in size, and pterosaurs, including pterodactyls, were present. Mammals were small and are known mainly from teeth of archaic kinds. The first birds occur as fossils. They were very reptile-like, but impressions of their feathers have been preserved.

The first diatoms have been found in Jurassic marine sediments. The land floras were cosmopolitan and rather similar to those of the late Triassic. Seed ferns became extinct before the end of the period, and ferns declined in importance. The cycadophytes and ginkgos reached their climax in the late Jurassic. Angiosperm pollen has been reported.

Cretaceous Period
Cretaceous faunas include a large variety of pelecypods, many of modern types. Rudistids made small reefs but became extinct before the end of the period. Advancement among gastropods is shown by the appearance of the first shells with well-developed siphonal canals. Ammonites achieved their greatest diversity in the Cretaceous. Numerous aberrant forms appeared, and the sutures of some late Cretaceous species reverted to simpler patterns. Belemnites locally were abundant and continued into the early Tertiary. The oldest known octopods are of Cretaceous age. Irregular echinoids, which were rare in the Jurassic, expanded considerably in numbers and variety.

Foraminiferans were abundant and varied in the Cretaceous. Planktonic forms became important in the latter part of the period, and they are associated in the fossil faunas with many other modern types. Corals flourished and built reefs. Hydrocoralines, reported as early as the Triassic, certainly were present. Cyclostome bryozoans continued to be abundant, but cheilostomes rose to dominance. Brachiopods became progressively less important and included only forms of modern type.

A great expansion took place among the bony fishes in late Cretaceous time, and many modern kinds developed. The first salamanders appeared among amphibians. Both groups of dinosaurs prospered, and some of the ornithischians, particularly, were huge beasts. The late Cretaceous saw the culmination of the Mesozoic reptilian faunas, and maximum sizes were attained in several other groups. Snakes appeared. Near the end of the Cretaceous Period, however, most of the reptiles became extinct.

The fossil record of Cretaceous birds is poor, but by the end of the period these creatures were essentially modernized. Cretaceous mammals were small and constituted a very minor part of the land faunas. Primitive modern types representing both marsupials and placentals had evolved before the end of the period.

Cretaceous floras exhibit the transition to modern plants. Angiosperms occur in the Lower Cretaceous in association with cycadophytes and ferns. Cycadeoids did not survive the mid-Cretaceous. Upper Cretaceous floras are very modern in their composition. The fossils represent mostly flowering plants, with abundant deciduous trees, and many are identified with living genera. Dicotyledons predominate. Palms were the most abundant monocotyledons of the late Cretaceous.

Cenozoic Era

With the extinction of large reptiles on the land and ammonites in the sea at the end of the Cretaceous Period, Cenozoic life acquired a distinctly modern aspect. In fact, plants and invertebrate animals have evolved relatively little since that time. Most of the families and many of the genera now living existed at the beginning of the Tertiary Period.

The situation with respect to terrestrial vertebrates, however, was vastly different. At the end of the Mesozoic Era mammals were small, primitive, and relatively rare. Thereafter they rapidly evolved in many ways, taking advantage of the habitats left vacant by the vanished reptiles. Changes in the life history of the Cenozoic, therefore, resulted largely from mammalian evolution.

Paleocene Epoch

Both marsupial and placental mammals are represented in late Cretaceous faunas. The marsupials soon were isolated in Australia and South America, where they seem to have evolved entirely independently. In the northern hemisphere these more primitive mammals were promptly displaced by the placentals, whose primary evolutionary deployment was the main event of the Paleocene Epoch at the beginning of the Tertiary.

At least six important groups of placental mammals are known from fossils of Paleocene age, but all of them were primitive or of archaic types. Insectivores constituted the central stock from which all the others are believed to have evolved. Before the end of the Paleocene carnivores had differentiated into three types, one of which was ancestral to modern forms. Primates were represented by tree shrews, lemuroids, and tarsiers. The condylarths later developed into more specialized ungulates. Both rodents and edentates first appeared in the late Paleocene. Rodents were all of the primitive protrogomorph variety. Edentates, represented by armadillo-like creatures, seem to have originated and achieved most of their subsequent evolution in South America.

Eocene Epoch

Abundant placental faunas obtained from Eocene deposits demonstrate that evolution continued at a relatively rapid rate. Some of the oldest bats were of Eocene age. Carnivores of modern types are well represented among the fossils. Some were dog-like, and others were true cats, mostly with long sabre-like canines. Jaws resembling those of the great apes occur in the Upper Eocene. The last condylarths were present, but more advanced ungulates show remarkable differentiation. The odd-toed group was represented by very small and primitive horses, titanotheres, chalicotheres, and tapir-like and rhinoceros-like beasts. The even-toed group included early pigs, camels, and relatives of the deer. Proboscidians were represented by moeritheres, and sea cows appeared in the mid-Eocene. Glyptodons had evolved among the edentates. The first whales are known from Middle Eocene strata, and advanced toothed whales occur in later sediments of this epoch.

Oligocene Epoch

By Oligocene time most of the primitive carnivores had disappeared, and their places were taken by more familiar forms, including bear-like dogs. The earliest known old world monkeys were of Oligocene age, but they probably originated earlier. Giant pigs and oreodonts appeared among the even-toed ungulates, some of which were horned. Titanotheres died out in the Oligocene. Horses grew to the size of large dogs. True tapirs and rhinoceroses were present, and some of the latter grew to gigantic size. All three kinds of rodents with advanced jaw musculature had evolved, and rabbit-like animals occurred. The first ground sloths appeared in South America.

Miocene Epoch

Viverrid and mustelid carnivores were well differentiated in the Miocene. Raccoons, true bears, hyenas, and seals were all represented in the faunas of this epoch. Deer and cattle-like beasts were present. Horses grew to pony size and some had high-crowned teeth. Late in the Miocene these animals were essentially modernized and had adapted to a grassland habitat. Proboscidian dinotheres and mastodons both appeared in the late Miocene. This epoch saw the last of primitive toothed whales and the first occurrence of whalebone whales.

Pliocene Epoch

The first modern cats with short canine teeth were present in Pliocene faunas. Hippopotamuses appeared, although possible ancestors are known from older strata. Cattle-like animals differentiated to produce numerous kinds of antelopes and other horned creatures. The earliest elephants were of Pliocene age. Anteaters had evolved by the late Tertiary in South America.

Pleistocene Period
In terms of geologic time, the Pleistocene Period was relatively short, too brief in fact to have witnessed much evolutionary change. It is noteworthy in a biologic way, however, for the local or total extinction of a considerable number of terrestrial animals, which probably resulted in part at least from glaciation and climatic change. Horses and camels disappeared from North America, which long had been their home, although they continued to exist in other lands. Such creatures as mammoths, saber-toothed cats, and ground sloths among others became extinct. A few South American animals migrated to North America for the first time. The opossum and armadillo are still extending their northern ranges.

More important and far reaching in its results, however, were the advent and expansion of the human race. The most ancient fossils of human type have been found in deposits judged to be early Pleistocene. In the last interglacial age human beings were abundant in Europe and probably also in Africa and Asia. In postglacial or Recent time modern man has spread to all habitable parts of the world. His activities have upset the previous balance of nature in many ways and at a rapidly accelerating rate.

THE PATTERN OF EVOLUTION
A review of the geologic succession of faunas and floras, even as brief and generalized a one as the foregoing, clearly shows the profound changes that have occurred in the earth's inhabitants with passing time. A person does not need to have an intimate acquaintance with fossils to recognize these changes. After the general aspects of the successive faunas and floras have been pointed out, the relative ages of many fossil associations can be approximated by an observant person even though he may be unable to identify a single species.

Evolutionary progression is plainly indicated by some groups of fossils in almost every phylum. Many of the invertebrates and plants, however, are less satisfactory than the vertebrates for such a demonstration. Vertebrate fossils have the advantage of providing many morphologic features that can be compared and that reveal a great deal about the lives and habits of these ancient animals. Conclusions regarding the patterns of their evolution substantiate and strengthen similar conclusions less securely based on the studies of invertebrates and plants.

Evolutionary Lineages
Although fossils seem to demonstrate numerous general but unmistakable evolutionary trends, very rarely can any of these be traced in detail through a long interval of time. Many hypothetical lineages or family trees have been built up indicating complicated ancestor-descendant relationships of genera (Fig. 633). Few of these

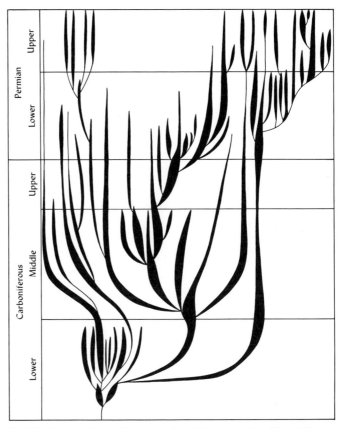

Fig. 633   Diagram showing the presumed evolutionary radiation of foraminiferans, mainly fusulinids, from Lower Carboniferous *Endothyra* in the U.S.S.R. Each swelling of the lines represents a genus. (*After Reitlinger, 1962, 4th Cong. avanc. études strat. géol. carbon., C.R., vol. 3, fig. 1, p. 592.*)

can be accepted without considerable reservations. Conclusions of this kind ordinarily depend upon which of several morphologic features are considered to be most significant in indicating close relationships, and disagreements among experts are not uncommon. Much of this uncertainty is a direct result of the incompleteness of the fossil record.

Estimates of the number of species that may have inhabited the earth before the present epoch have varied between such wide extremes that obviously they have little value. Actually the number of species does not greatly matter. More interesting and pertinent are estimates of the number of ancient evolutionary lineages that have survived the vicissitudes of time and are represented by the organisms of the modern world. All estimates of this kind are, of course, speculative, and few of them are more than guesses. Probably estimates for the vertebrates can be made with the most assurance.

The actual ancient species that were ancestral to modern organisms are very rarely or never known. The very small number that is believed to have existed at some particular time in the far distant past, compared to the enormously abundant and diverse life of the modern world, points up one of the principal accomplishments of evolution, i.e., the enormous differentiation that has occurred among living things. Another factor related to evolution that is not so obvious or generally realized is that only a minute fraction of the fossil species stood in the direct ancestral lines of modern organisms. In other words, most fossil species belong to lineages that became extinct. For example, all modern terrestrial vertebrates probably evolved from no more than 8 or 10 lineages that were in existence late in the Permian Period. The faunas of that time are far from completely known, but the thousand or more species of vertebrates that have been discovered undoubtedly belonged to

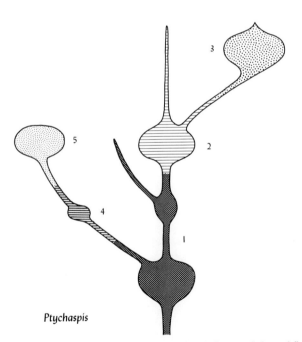

Ptychaspis

Fig. 634 Diagram showing the presumed evolutionary relations of five closely related Upper Cambrian trilobite species. (*After Grant, 1962, J. Paleontol., vol. 36, fig. 10, p. 983.*)

Table 635 Approximate number of known modern species in various groups of organisms and estimates of the number of ancestral species at certain times in the geologic past

	KNOWN MODERN SPECIES	ANCESTRAL SPECIES	
		NUMBER	AGE
All bryozoans	3,000	3	Cambrian
All brachiopods	250	4	Late Paleozoic
All molluscs	120,000	5	Late Precambrian
All arthropods	875,000	3	Late Precambrian
All insects	850,000	1	Devonian
All echinoderms	5,000	5	Early Cambrian
All vertebrates	40,000	1	Late Precambrian chordate
		5	Early Devonian fishes
All fishes	20,000	8–9	Permian
All amphibians	1,000	2–3	Permian
All reptiles	4,750	4	Permian
All birds	7,500	1	Triassic
All mammals	3,500	1–2	Permian
All thalophytes	350,000	4	Early Devonian
All angiosperms	200,000	1	Permian

several hundred lineages. The great majority of these sooner or later died out, leaving no descendants. They failed because they did not adapt to the physical or biologic elements of their changing environments so successfully as the few lineages that managed to persist.

The Tree of Life

These two essential aspects of a single continuing process have characterized the pattern of all life since its first appearance upon the earth. Part of this pattern is effectively represented by the metaphorical tree of life (Fig. 636). Such a visual device commonly is employed to show the presumed evolutionary relationships of organisms as their ancestors are traced backward through past time. Of course it can be looked at in the opposite way as a representation of the progressive diversification or branching that resulted mainly from adaptational adjustments to an ever-changing succession of environments. Generally, however, the tree model neglects or obscures the fact that most of its branches sooner or later were pruned away by natural selection (Fig. 637).

Such diagrams cannot represent the pattern of evolution adequately. Relations ordinarily are not so simple as the successive divisions of the branches seem to indicate. The ancestry of most organisms is, in its details, a tangled skein of crossing lineages. This is true both for creatures descended simply from two parents, four grandparents, etc., all belonging to the same species, but more particularly for those whose antecedents are complicated by hybridization. Many plant species are believed to have originated as hybrids (Fig. 638). Similar patterns may have been common in the ancestry of some invertebrate animals whose genetics are very little understood.

The pattern of evolution may be likened to a chance trial-and-error process. In-

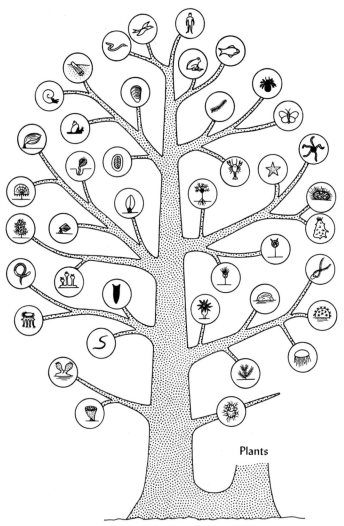

Plants

Fig. 636 Diagrammatic and generalized tree of life. This is the traditional kind of diagram that is intended to show the relative evolutionary positions of the animal phyla and their principal divisions in a hierarchial manner. The branching of all phyla from a single stem is unrealistic. (*After Twenhofel and Shrock, 1935, "Invertebrate Paleontology," 1st ed., fig. 2, p. 6.*)

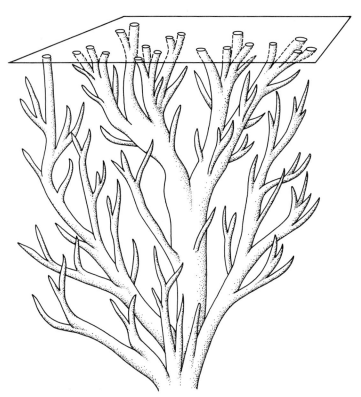

Fig. 637 Diagrammatic representation of a small portion of the tree of life. The vertical dimension is time. At the top the tree intersects and is terminated by a plane which is the present. Branching evolution and the extinction of many lineages is well shown. This diagram also is unrealistic because each successful new branch should start narrow and then broaden in a way not seen in any actual tree. (*After Moore, Lalicker, and Fischer, 1952, "Invertebrate Fossils," fig. 1-6, p. 15.*)

numerable unconscious trials were made by organisms branching out in a multitude of ways. Presumably every trial that produced new species was successful in a particular environment and for a certain length of time. Each evolutionary accomplishment, however, limited to some degree the possibility of subsequent evolutionary change. Most trials, including the temporarily successful ones, ended eventually in failure. Somewhere in most lineages an evolutionary happening occurred which, as viewed in backsight, was an error if judged on the basis of long-range consequences.

Teleology

The advantages gained by species in their progressive evolution are assessed in terms of their utility in particular situations. This is consistent with an explanation for evolution guided or controlled by natural selection. The human mind, however, slips easily back and forth between the concepts of utility and purpose. The differ-

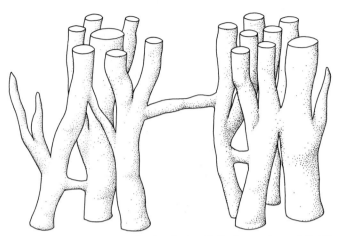

Fig. 638   Schematic diagram showing the hypothetically complex evolutionary relationships of the numerous species of two genera of modern plants. Somewhat similar relationships may be characteristic of some animals. (*Compare Clausen, 1951, "Stages in the Evolution of Plant Species," fig. 76, p. 179.*)

ence is subtle, but in some degree purpose implies predetermination. Strictly expressed, purpose is the motivation that directs action toward some desired but unattained goal. Likening evolution to a trial-and-error process is dangerous because it may induce the mistaken idea of directed action or purpose that was variously successful.

Teleology, or the doctrine of purposiveness in nature, has played an important part in the development of evolutionary theory. It is an inherent element of most Lamarckism. Paleontologists who distinguish long-continued seemingly orthogenetic trends have been particularly receptive to this idea in one or another form. Their concepts of evolution seem to have been shaped by efforts to trace lineages backward. When viewed in this way, the succession of ancestral organisms is not complicated by branching evolution because each stage leads to one and only one antecedent if this can be recognized. Branching, however, has been characteristic of almost all evolution. At each branching two lineages were initiated, and choice must be made if one branch rather than the other is to be considered the more direct continuation of the parent lineage. When viewed in this way evolution has a different aspect. Orthogenetic trends lose much of their distinctness, and evolution is seen in a perspective that is more realistic.

From a practical or materialistic standpoint, the idea of purposiveness in evolution smacks of mysticism and is nonsensical. No real evidence can be presented in its favor. Anyone who believes in the orderliness of all nature, and in the invariable necessity for a rational cause to precede every effect, must conclude that evolution has been purposeless and unpredictable in its details. The conclusion also is inescapable that all evolution has been opportunistic, in the sense that it has followed whatever path or paths happened to be open to it. Its immediate course at any time was determined by (1) the then-current nature of environments and their distributions, and (2) the qualities of contemporary organisms with respect to both their possession of chance preadaptive traits and their potentialities for further advantageous adaptations.

Accomplishments of Evolution

The succession of faunas and floras preserved in vertical sequences of stratified rocks shows that life has changed in many ways during the course of some 600 million years. When this record of evolution, persisting for so long, is regarded in its entirety, two questions rather naturally arise. These are: (1) What are the major accomplishments of evolution after such an extended interval of time? (2) Is the overall tempo of evolution appreciably different now from what it was in the distant past? Neither of these questions can be answered explicitly in ways that are wholly satisfactory, but it is interesting to consider them.

The two principal accomplishments of evolution have been (1) the appearance and gradual perfection of more and more complex organisms increasingly well adapted to specialized environments, and (2) a progressive increase in the number and variety of different organisms. Both certainly are true if all plants and animals evolved from a very primitive form of life that appeared as a natural development in a previously inanimate world. They also are true, but somewhat less obviously so, if consideration of evolution is restricted to what is revealed by the preserved fossil record, beginning in the early Cambrian.

Increase in Complexity

Many groups of invertebrate fossils demonstrate progression from ancient to modern types. Most of the earliest fossils, however, seem to represent well-known phyla just as typically as any modern species. This means that the increasing complexity which led to the differential development of these phyla was mainly if not entirely completed in Precambrian time. Most subsequent morphologic changes of the hard parts probably reflect specializations and improved adjustments to environmental conditions rather than noteworthy increase in complexity of organization.

Structural complexity in the hard parts of certain groups of invertebrate fossils did increase, as, for example, in the sutures of ammonites. In contrast, other groups show progressive structural simplification, as is seen in the plating of several crinoid lineages. Changes of these kinds cannot be relied upon as evidence either for or against greater organic or functional complexity. Probably, however, complexity did slowly increase in many groups of invertebrate animals after the Precambrian. If so, it was concerned mainly with physiologic processes and behavior which ordinarily are not reflected in the structure of hard parts.

Better evidence of increasing complexity is provided by terrestrial organisms.

These originated and evolved much later than the marine invertebrates, and, therefore, their history is more fully known. The vertebrates exhibit many excellent examples of specialization in their skeletal structures. The musculature of land animals grew more complex than that of their fish ancestors, because of the change from undulatory swimming to walking and running upon legs. This and other adaptations to land life, as well as intensive competition and active predator-prey relations, were accompanied by improved sense reception and a more efficient nervous control system. The last, particularly, is indicated by increase in both size and complexity of the brain, which is known from the study of fossil skulls.

Physiologic complexity certainly increased among the terrestrial vertebrates, although this generally is not apparent from any changes in skeletal structure. For example, reproductive processes became more intricate, and the sexual and related organs were modified from amphibian to mammalian type. The development of warm-bloodedness also involved noteworthy increase in physiologic complexity, to which the greater structural complexity of the heart probably was related.

Land plants, in contrast to the vertebrates, changed very little in fundamental physiology, except for reproductive details, but advanced in structural complexity. Photosynthesis, which had been perfected in the far-distant past, is similar in all green plants. The efficiency of this process was promoted, however, by the evolution of leaves and perhaps by the development of more complex leaf venation. The circulation of fluids also was improved through the development of better conducting cells, both vertical and radial, within the woody stem. The most noteworthy advance, however, was made in reproduction and the transition from spores to seeds. This involved increase in the complication of sexual organs and related structures and the almost complete suppression of alternating sexual and asexual generations.

The insects are another important group of terrestrial organisms whose evolution began in mid-Paleozoic time. Their known history, unfortunately, is disconnected because of the very sporadic occurrence of their fossils, and most details of their evolution are unclear. Nevertheless they are remarkable for the development of wings in a way very different from that of the pterosaurs, birds, and bats. The insect wing was a new structure, and it introduced greatly increased complications of the external skeleton, muscles, and trachiation. The vertebrate wing, in contrast, is simply the modification of a previously existing structure that required the introduction of no new parts. The metamorphosis of insects, which is related to abrupt changes in environmental adaptations and ways of life, differs from that of all other animals. It also involved profound changes in structure and physiology.

Increase in Diversity

Branching, or radiating, evolution has steadily tended to increase the variety of organisms that lived contemporaneously. This generally has been balanced by the extinction of older lineages. If extinction had not been very common, the variety of organisms in the modern world would be infinitely more bewildering than it actually is. This balance between evolutionary radiation and extinction seems to have fluctuated considerably mainly because: (1) Radiation gained on extinction when organisms adapted to a greater diversity of environments, either those that had not previously been exploited, or new ones created by changed physical conditions or by interactions with other organisms. (2) Extinction gained over radiation when relatively abrupt changes, mainly physical, produced conditions inimicable to many then currently existing organisms.

Some of the more conspicuous fluctuations in the relative predominance of evolutionary radiation or extinction as indicated by the fossil records of important groups of organisms are shown in Table 641. Similar data might be presented for numerous more restricted groups. Data of this kind may be true and accurate, but they do not provide a uniformly reliable basis for evolutionary conclusions. This is because the fossil record is very incomplete and its degree of incompleteness un-

Table 641   Distribution in time of the main episodes of relative radiation and extinction in the evolutionary history of selected groups of organisms as indicated by the number of known genera in the fossil record

	RADIATION	EXTINCTION
Marine invertebrates as a whole	Early Paleozoic	Permian
	Jurassic	Late Cretaceous
Foraminiferans	Silurian	Triassic
	Mississippian	
	Jurassic-Cretaceous	
Graptolites	Ordovician	Silurian
Brachiopods	Ordovician	Mississippian-Pennsylvanian
	Jurassic	Cretaceous
Nautiloids	Ordovician	Silurian
		Mississippian
Ammonoids	Devonian	Early Mississippian
	Early Triassic	Late Triassic
	Early Jurassic	Late Jurassic
	Mid-Cretaceous	Late Cretaceous
Trilobites	Cambrian	Silurian
	Devonian	Mississippian
Crinoids	Ordovician	Late Mississippian
	Early Mississippian	Late Permian
Fishes	Devonian	Pennsylvanian
	Cretaceous	
Land vertebrates as a whole	Permian	Late Cretaceous
	Jurassic	Pleistocene
	Paleocene-Eocene	
Amphibians	Pennsylvanian	Permian-Triassic
Reptiles	Permian	Late Triassic
	Jurassic	Late Cretaceous
Mammals	Paleocene-Eocene	Pleistocene
Insects	Late Paleozoic	
	Late Cretaceous	
Land plants	Devonian	Permian
	Pennsylvanian	
	Late Cretaceous	

doubtedly varies greatly for different groups of organisms and for different intervals of geologic time. For example, a primary uncertainty concerns the meaning of the increasing variety of fossils known from strata ranging from the early Cambrian to the mid-Ordovician. Perhaps this was a time when radiation did predominate in most groups of marine invertebrates. Another conclusion, however, which seems more probable, is that the organisms known as fossils were preceded by soft-bodied ancestors that left no traces and that important radiation in these groups occurred mainly during some earlier interval of time. Obviously allowances must be made for uncertainties of this and other kinds in any attempt to assess the accomplishments of evolution.

Rhythms in Evolution

In spite of numerous uncertainties, the fossil record almost surely indicates that evolution has not pursued a steady course with respect to either the increasing complexity of organisms or the increase in their diversity by radiation. The primary differentiation of most groups of animals or plants probably proceeded relatively rapidly in small populations during the early stages of their branching evolution. At this time the increasing physiologic or morphologic complexity of a group in many instances seems to have been related to a breakthrough into a new way of life different from the old in one or more respects. Thereafter relative stability ordinarily was achieved. Further evolution seems to have proceeded more slowly and produced less-spectacular results. Progression in this phase generally was toward closer adaptation to more restrictive environments. The whole process can be likened to (1) a brief period of pioneering in new territory requiring major change, and (2) a longer period devoted to consolidation of the gains that had been made requiring only minor changes.

Simplification in some respects may have marked the second stage, particularly in parasitic organisms. Such a reversal, however, is not at all related to the concept of racial aging, which has been postulated mainly on the basis of relative measures of evolutionary radiation rather than the degree of continuing morphologic or physiologic modification.

This pattern repeated as a sort of rhythm is evident within the records of many large groups or organisms. In a wider view encompassing all contemporaneous life, the alternating dominance of first radiating evolution and then extinction suggests a greater rhythm, although one that was irregular and obscure. Parts of the apparent fluctuations probably reflect only variable imperfections of the fossil record. Others, however, seem to indicate widespread changes in physical environments or biologic relations or in opportunities for the expansion of life in environments not previously exploited.

Evolutionary Radiation

Evolution presumably has resulted in an increasing diversity of life, and more different kinds of organisms may be living now than at any time in the past history of the earth. The fossil record, unfortunately because of its incompleteness, is not adequate to indicate (1) the rate at which diversity possibly increased, or (2) the degree to which the rate was steady, uneven, or fluctuating between alternate extremes. Assumptions based on very little evidence range from opinions that diversity has not increased notably since early in Paleozoic time to the idea that the number of species doubled on the average in each interval of from 60 to 100 million years.

This problem cannot be attacked except in a theoretical and intuitive way. The kinds of evidence that may be pertinent in its consideration include: (1) Marine invertebrate fossil faunas of the mid-Ordovician are not exceeded in diversity or abundance by the fossil faunas of any later time. (2) Fishes first recorded in the Ordovician are not known to have become important elements of marine faunas before the Devonian Period. (3) Terrestrial plants and animals are unknown before the late Silurian, and land life seems to have been sparse before the late Devonian. (4) As far back as the record can be judged with any reasonable certainty, the earth's climate generally has been milder than at present and less marked by north-south temperature zonation.

One conclusion of great importance can be drawn with reasonable assurance. This is that the evolutionary diversification of life progressed in two principal pulses when radiation consistently and for considerable periods of time exceeded extinction. These corresponded to the times when first the sea and then the land was populated.

Very little can be guessed about the evolutionary history of marine plants. They certainly had a very ancient origin. The presence of all main animal phyla in the early Paleozoic seems to be good evidence that their diversification also dates back well into the Precambrian. Evolutionary radiation in the sea in times antedating the beginning of the fossil record probably was slow at first when the principal influences were differences in physical marine environments. Subsequently it may have speeded up as the interrelations of increasingly diverse organisms created new, more numerous, and more restricted environmental opportunities. This evolutionary phase probably came slowly to an end when invertebrate diversification approached that which is evident in mid-Ordovician faunas. Thereafter additional very slow diversification may have continued to the present time, interrupted temporarily by episodes when environmental conditions changed and became unfavorable for some important groups of the then-existing plants and animals. The appearance of abundant fishes in the Devonian Period may indicate a secondary and minor increase in the variety of marine animals.

It is not known when the earliest terrestrial plants evolved from marine algae. Primitive vascular plants had become established on the land by late Silurian or early Devonian time, however, and subsequent evolution and diversification were relatively rapid. Before the end of the latter period the moist lowlands at least were well vegetated. The first animals to achieve adaptation to terrestrial conditions were arthropods. Among them the creatures that became ancestral to the insects were in the long run especially important. Probably the presence of abundant plants was prerequisite to insect evolution.

The first land vertebrates were amphibians whose reproduction bound them to the water. The earliest reptiles also found most of their food in streams and ponds and did not wander far from water. Therefore, a typical wide-ranging vertebrate land fauna probably did not appear until herbivorous reptiles had evolved. This is believed to have occurred during the Permian Period. By mid-Mesozoic time land life certainly had diversified greatly, and terrestrial floras and faunas probably approached those of the present day in variety and interrelations. Most of the plants and animals, however, were not of modern types.

The second great pulse in evolutionary radiation and diversification which accompanied the spreading of life across the land began, therefore, at about the beginning of the Devonian Period. The number of different land organisms probably increased rather steadily until at least late Triassic time. Some fluctuations undoubtedly occurred, however, in individual groups of terrestrial organisms, and these continued into the Cretaceous, but they do not seem to have affected the general evolutionary trend importantly.

A secondary phase of moderately increasing diversity, both in the sea and on the land, may have come in the latter half of the Cretaceous Period. This was the time when angiosperms, modern bony fishes, and pelagic foraminiferans first became abundant. Subsequent evolutionary advancement and radiation of the mammals in the early Tertiary may have more than balanced out extinctions among the reptiles and cephalopods in the very late Cretaceous.

Extinction

Major episodes of extinction seem to have been somewhat fewer and less randomly distributed in time. Many extinctions were concentrated at or near the transition from the Paleozoic to the Mesozoic Era, when the sedimentary rock record indicates that physical conditions related to topography and climate were undergoing more than ordinarily important changes. Although biologic influences cannot be excluded, these probably were secondary in importance to the altered physical environments, beginning as early as the late Mississippian and extending into the Triassic Period, which affected many forms of life.

Causes for the extermination of ammonites and most reptiles at the end of the Cretaceous Period are more doubtful, but they probably were related mainly to the alteration of physical conditions in some unknown way. Finally after the mid-Tertiary, the previously mild climates slowly but steadily deteriorated, and relatively extreme climatic cycles marked the Pleistocene. The overall evolutionary results of these conditions are difficult to assess. Mammals seem to have been adversely affected in the Pleistocene, but diversification among some herbaceous plants, insects, and marine invertebrates may have been mildly stimulated.

Probably more different kinds of organisms are alive in the world today than at any time in the past history of the earth.

PRESENT STATUS OF EVOLUTION

Small reason exists for the conclusion that evolution is not occurring now or that it will not continue as an essentially unchanged process as long as life persists. Nevertheless, the opposite conclusion has been argued by a few paleontologists and biologists, who believe that evolution is slowing down and may eventually cease. Their conclusion is based on an interpretation of counts of new species and the numbers of other taxonomic groups that have appeared during successive intervals of past time.

Counts of the known new appearances of species, genera, etc., in related groups of fossils show maxima at one or more points in geologic time. If the fossil group became extinct or evolved into descendants that are differently classified, the counts tail off to zero. On the other hand, if the group still lives, the number of species and genera is likely to show a sharp rise to a modern maximum. This undoubtedly reflects the fact that all modern representatives of the group can be known whereas the fossil record certainly is incomplete. Such modern increase, therefore, does not accurately reflect correspondingly accelerated evolutionary radiation of these lower taxonomic groups in very recent time.

In contrast to the rising number of species and genera in most groups of modern organisms, counts of families and higher taxonomic categories generally show progressive decrease from a maximum at some past point in time (Fig. 645).

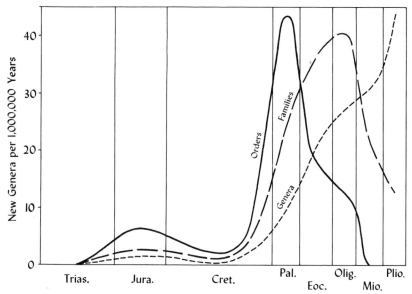

Fig. 645 Graph showing the appearance of new orders, families, and genera of mammals scaled against the passage of Mesozoic and Tertiary time. (Vertical scale: Families, $\times \frac{1}{10}$; Orders, $\times \frac{1}{50}$.) (*Data from Simpson, 1952, J. Paleontol., vol. 26, fig. 3, p. 362.*)

These distributions exhibit a pattern very similar to that of other earlier groups as these approached extinction. The conclusion, however, that this indicates slowing down of evolution in its major operation is entirely unjustified. If projection could be made into the distant future, evolution would be seen to have produced divergent stocks whose characters would warrant their recognition as new taxonomic groups of family or higher rank. The ancestors of these groups would be traced backwards. Their origins would be established as occurring among fossils which had not diverged from their colateral relatives sufficiently to be clearly differentiated taxonomically without knowledge of subsequent evolution. The origin of no taxonomic group can be recognized before that group has clearly diverged from others and has attained distinctly separate identity.

BIBLIOGRAPHY

J. Brough (1958): Time and Evolution, in "Studies on Fossil Vertebrates," T. S. Westoll (ed.), pp. 16–38, The Athlone Press, University of London.
The author concludes that early evolution was very rapid and directional and that natural selection became important only later when evolution was much slower.

J. W. Durham (1967): The Incompleteness of Our Knowledge of the Fossil Record, *J. Paleontol.*, vol. 41, pp. 559–565.
Perhaps only 1 percent of all preservable species of the past are now known as fossils.

M. Kay and **E. H. Colbert (1965):** "Stratigraphy and Life History," John Wiley & Sons, Inc., New York.
The succession of life through geologic time is presented without much explanation but more completely than in most historical geology texts.

N. D. Newell (1962): Paleontological Gaps and Geochronology, *J. Paleontol.*, vol. 36, pp. 592–610.
The problem of the mass extinction of fossil organisms is considered.

D. Nicol, G. A., Desborough, and **J. R. Solliday (1959):** Paleontologic Record of the Primary Differentiation in Some Major Invertebrate Groups, *J. Washington Acad. Sci.*, vol. 49, pp. 351–366.
Major divergence in evolution was most important during the early history of a group and was followed by evolution that affected only details of the early pattern.

K. P. Oakley and **H. M. Muir-Wood (1964):** "The Succession of Life through Geologic Time," 6th ed., British Museum (Natural History), London.
This rather superficial and quite conventional account was intended to supplement museum displays.

G. G. Simpson (1952): Periodicity in Vertebrate Evolution, *J. Paleontol.*, vol. 26, pp. 359–385.
The rise and fall of several groups of vertebrates was not related to the boundaries of geologic periods.

C. Teichert (1956): How Many Fossil Species? *J. Paleontol.*, vol. 30, pp. 967–969.
The bibliography lists references to several estimates of the number of species that have inhabited the earth before the modern era.

CHAPTER FOURTEEN

HUMAN EVOLUTION AND THE FUTURE

NONGENETIC EVOLUTION
FUTURE OF EVOLUTION AND THE HUMAN RACE

Any objective account of the course of biologic evolution is retrospective and attempts an understanding and explanation of what has happened in the past. Many details are obscure and much remains to be discovered, but the main pathways of evolution leading to the modern organic world seem to be reasonably clear. The present may appear to be a climax to all that has gone before, but that is only because we happen to be living now. Time surely will go on, and the same principles that have governed natural processes in the past will continue to operate just as they always have.

After reviewing the past one cannot help wondering what the future will bring forth. Man in his egocentric way is interested in himself above all else. Does knowledge of the past provide basis for predicting human prospects?

NONGENETIC HUMAN EVOLUTION

The idea of human cultural evolution progressing through stone, bronze, and iron ages antedates the concept of biologic or organic evolution directed by natural selection. The advantages gained by men possessing more advanced and more efficient implements are unmistakable. It probably was inevitable that enthusiasm engendered by the later theory of biologic evolution soon resulted in the appropriation of its principles by anthropologists and their application in interpretations of human cultural history and prehistory. Subsequently the realization that biologic and cultural evolution, in spite of their analogies, are fundamentally different in many ways seemed to necessitate a change in viewpoint, and close correlation of these types of evolution fell into disrepute. Actually, however, no sharp division can be made in the study of early human development between (1) the physical evolution of modern man from his immediate ancestors, and (2) the history of human culture. Both are concerned with a series of stages that succeeded one another in time and are considered to have progressed in the sense that they have culminated in the physical being and the social animal that is modern man. Viewpoints, therefore, have changed again in more recent time and most anthropologists now recognize that human biologic and cultural evolution have been closely linked and each probably is a significant aspect of the other.

Human Attributes

Man is unique in the animal world by virtue of his ability (1) to communicate by symbols, as exemplified by both spoken and written language, (2) to reason concretely, as in drawing logical inferences from observations, and abstractly, as in developing sets of values, (3) to imagine situations not existing at the moment and creating means for meeting them, (4) to think self-consciously, and (5) to appreciate aesthetically. These characteristics are not shared by any other animals, except to a very imperfect and primitive degree. They have a genetic basis; i.e., they evolved in the normal biologic way. It is easy to understand how possession of such abilities was advantageous and how natural selection led to gradual improvement that was passed on by inheritance.

Cultural Evolution

Culture, on the other hand, is nongenetic. It is something that must be learned by every individual in a society. It is passed on from one generation to the next but this is done by tradition, example, and imitation rather than directly by genetic inheritance. Cultural evolution, therefore, is Lamarckian in the sense that it is advanced by the transmission of acquired characters.

Culture certainly has had important adaptive values, but it confers advantages primarily on groups of men, rather than on individuals. As in biologic evolution, cultures may diverge from a common origin and then follow independent courses. Parallel and convergent evolution probably have been common. Cultures, however, do not diversify indefinitely like evolving organisms. They can be isolated from each other only by space or time. When different cultures come in contact, they hybridize or one assimilates features of the other before the second is suppressed. A kind of natural selection commonly has operated, but the superior or more advanced culture has not prevailed in every instance.

Culture is the synthesis of man's conditions with respect to social status, ideas, beliefs, customs, ethics, institutions, technology, and way of life. Culture may be classified in ascending scales and graded in several different practical or ideologic ways, all of which seem to be significant. Opinions regarding relative significance, however, are likely to vary greatly, and the status values of some grades certainly are controversial. Among the scales are: (1) Economic; searchers and gatherers of food, hunters, herdsmen, farmers, artisans, traders, and industrialists; or communism in many small societies, private property, capitalism, and socialism. (2) Social or political; family, tribal, feudal, monarchical, and republican; or nomadic, semipermanent settlement, village, town, city, state, empire, and world unity. (3) Technologic; stone, bronze, iron, steam, electric, and atomic. (4) Theologic; nature worship, polytheism, monotheism, agnosticism, and atheism. Cultural evolution has been orderly, but not all these stages are clear-cut, and there are many irregular overlaps between the stages in the same and in different sequences. Moreover, some of the camparable aspects of equally advanced cultures have differed greatly.

In another way, culture may be considered to have evolved by a series of so-called revolutions, each of which changed men's lives in many ways. Those generally recognized as having affected Western civilization were: (1) Agricultural; Near East, about 7000 B.C.; food produced in excess of immediate requirements. (2) Urbanization; eastern Mediterranean, about 3000 B.C.; beginning of writing, metallurgy, law, and organized politics. (3) Theologic; Near East and southern Europe, about 600 A.D.; rise of religious philosophy, organization, institutions, and social control. (4) Industrial; western Europe, about 1700 A.D.; increasingly rapid growth of invention, science, technology, manufacturing, and wealth. Cultural advancement in many other parts of the world, however, has lagged far behind these dates. Most of it everywhere has involved increase in the complexity of social organization, in the stratification of society by craft, class, or economic status, in increasing collective political, religious, and military power, and in some loss of personal freedom. On the other hand, it has in general contributed to the greater security, comfort, and satisfaction of individuals.

Prerequisites of Culture

The origin and evolution of human culture required several prerequisites, such as: (1) Gregariousness, because culture could have had beginnings only among animals that instinctively associated. (2) The capacity to communicate more elaborately than by the calls of animals, which are little more than simple signals. (3) Hands capable of manipulating objects. This last characteristic was promoted by bipedalism, which freed forefeet from their original function. These qualities appeared in the ancestors of the human race as the results of genetic mutations. Because they were advantageous adaptations to the environment of prehuman beings, natural selection favored the individuals so endowed. These qualities, therefore, accumulated in populations and slowly were accentuated.

The communication of experience and later of ideas, even in a very primitive way, was advantageous to groups of individuals, and those groups which communicated more effectively were favored by natural selection. Learning from others, originating in this way, was so advantageous that it greatly enhanced the adaptive values of both mental and physical mutations, which increased the ability to communicate. Thus a mutual and circular relationship certainly existed between the genetic evolution of heritable abilities and the progressive development or evolution of acquired behavior, in which each was stimulated importantly by the other.

Man has achieved supreme dominance in the organic world by the superimposition of cultural evolution on biologic evolution. For a very long period of time these went hand in hand, and both were adaptations to natural environments. Human biologic evolution has been slow, comparable to that of many other mammals. Cultural evolution, however, has progressively accelerated. Since before the dawn of recorded human history, cultural evolution has increasingly involved adaptation to social rather than physical environmental factors. In the present century man has largely overcome most elements of his physical environment, and now the social environment is by far the most important.

Suppression of Natural Selection

Progressive decline in the necessity for continued adaptation to the physical environment has profoundly altered the effectiveness of natural selection in human evolution. Superior physical hardiness, intelligence, and skill are no longer such decisive survival factors. Competition among individuals still persists, but it is largely economic. Relative failure is no longer punished by the severe penalties that were natural consequences in the earlier stages of human evolution. Disadvantageous physical and mental mutations undoubtedly still appear in the human population with undiminished frequency, but the unfortunate individuals who suffer from them are not so regularly eliminated. In comparison with the past, natural selection is now practically at a standstill so far as the adaptations of many human beings to the physical environment are concerned.

Old standards of genetic superiority certainly are not applicable to modern human society. Purely physical attributes are not nearly so essential as they once were. Millions of people, contributing constructively to society, certainly are alive today, preserved by the amazing advances in public health, medicine, and surgery, who would not have survived to maturity in former times. Intelligence and good judgment, however, have not diminished in importance. They actually are more essential now than ever for the welfare of the human race in meeting the constantly more complex social problems arising from a rapidly increasing population and an advancing technology.

FUTURE OF EVOLUTION AND THE HUMAN RACE

There is every reason to believe that evolution of most organisms except man will continue in the future by genetic modification in ways unaltered from the past. Natural selection will continue to act on species of plants and animals, which will adapt to changes in environment or decline and eventually become extinct. Evolutionary radiation will occur, and new species will appear and increasingly differentiate.

Human Influence

An entirely new aspect of evolution, however, has been introduced by the activities of man. These influences are principally of four kinds: (1) Environments of extensive regions have been profoundly modified by agriculture, deforestation, irrigation, and urbanization. Perhaps in the future climate will be controlled to some extent. (2) Environments also have been modified more locally and in less-drastic ways. Some have been improved for favored organisms, as by fertilization and cultivation and the elimination of predatory enemies. Others adverse to unfavored organisms have been created by the use of pesticides, the introduction of parasites, or by other biologic means. (3) Many plants and animals have been introduced, either purposely or accidentally, into new areas far beyond their former ranges. Some introductions have upset previous natural biologic relations in catastrophic ways. (4) The genetics of organisms have been manipulated by selective breeding and hybridization to produce a great variety of domesticated plants and animals. Most of these could not survive except in the artificial environments that man provides for them. In the future, man's influence, both direct and indirect, on the evolution of other organisms is almost certain to increase progressively.

Human Evolution

The prospects for future human evolution are fundamentally different from those of other organisms in two respects: (1) Knowledge of genetics provides the human race with the possibility of consciously guiding its own evolution in preferred directions, just as domestic plants and animals are directed by selective breeding. Whether available knowledge ever will be utilized in this way to any significant degree, however, is an open question. (2) Natural selection no longer operates in the normal way with respect to human beings. This is extremely important in any consideration of the long-range results that may accrue.

Protection of Disadvantaged

Modern society is increasingly solicitous in protecting and preserving its weaker and less well-endowed members. No matter how ill equipped, they are supported, and generally, if there is opportunity, these people reproduce at greater rates than more fortunate, better-educated, and more successful individuals. Many people of this class simply have lacked advantages available to others, but included here are all those who do not have the ability to advance themselves socially or economically because of genetic mental or physical deficiencies. Formerly many persons in the latter group performed unskilled or menial labor and were useful to society. As technology advances, however, work of this kind is performed increasingly by machines and employment opportunities for the unskilled steadily decline. At the present time a class of permanently unemployed and unemployables seems to be increasing. Perhaps a new kind of leisure class is developing consisting of the least able members of society. Already such a trend is imposing a severe handicap on advanced societies. This is largely a current and perhaps a short-range economic problem. It is not yet important as far as evolution is concerned but may become extremely important in the future. In a somewhat similar way, but on a larger scale, underdeveloped nations, unless they can industrialize quickly, pose a serious threat to the future economic progress of all mankind.

The Population Explosion

Public health measures during a period of less than 100 years have reduced human death rates remarkably, especially among children. A much greater proportion of those now born lives to maturity and reproduces. Because birthrates have not been reduced correspondingly, the human population of the world is increasing very rapidly, in what is aptly termed a population explosion (Fig. 654). This is particularly true in the areas of the world that are less advanced economically. Experience has shown that birthrates are likely to drop sharply as a society becomes more highly industrialized and more prosperous. This drop has occurred mostly, however, among the better-educated and more gifted portion of the population. In contrast, birthrates among the ignorant and less well-to-do fraction have not been reduced to the same extent. This means that in every generation a larger proportion of the population is being derived from less-favored parents, who include a dispropor-tionally large number of persons afflicted by hereditary defects or otherwise having less desirable genetic endowments. Over an extended period of time the continua-

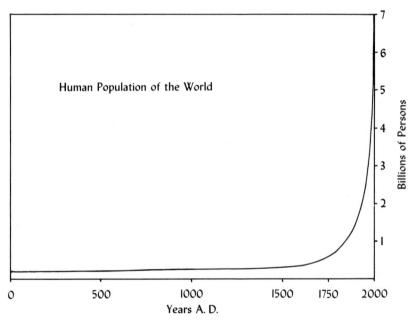

Fig. 654 Graph showing estimated increase of the earth's human population during the last 2,000 years. (After Dorn, 1962, *Science, vol. 135, fig. 1, p. 284.*)

tion of this trend can result only in steady deterioration of the average genetic equipment of the human race.

The human population undoubtedly will continue to increase, but obviously this cannot keep on indefinitely. A limit must be reached sometime, and what this limit is and how it is attained will determine to a very large extent the long-range future of the human race. The seriousness of the population problem cannot be over-stated. Man's economic well-being and his evolutionary future hinge upon its solution.

Without controls, the population of any organism, including man, will increase at a constantly accelerating rate and, if continued, this rate would soon become fantastic. Natural selection checks such increase and maintains the populations of all organisms in a dynamically fluctuating equilibrium. Man has at least tempo-rarily escaped from this natural system of control by creating his own environment in ways impossible for any other organism. Either he must manage to restrict his population increase or natural selection will again assert itself in the ordinary way, much to the detriment of the human species.

Limitations on Population

The density of any population is limited ultimately by the available food supply, although other environmental factors are equally important for most organisms. Man's control of his environment, however, is so effective that only food supply provides a direct limit for his population growth. At the present time there is no appreciable excess food supply for the 3 billion or so persons now living in the world. Some regions produce surpluses, but in many others people do not have enough to eat to maintain a healthful existence. If the population of a predominantly agricul-tural society increases faster than its food supply, existence becomes progressively more precarious, and starvation may result. This is the situation in several parts of the world today.

Thoughtful estimates indicate that without too drastic changes the earth is capable of producing sufficient food to provide several times the number of human beings in the present population with an adequate and satisfying diet. Such an increase in food production, however, would require great advances in science and technology to develop more efficient food sources, to utilize much land not now suitable for agriculture, to cultivate the land more intensively, to process many kinds of food materials, and to distribute food effectively. All this would require the services of relatively more persons in occupations subsidiary to agriculture, relatively more equipment and machines, relatively more heavy industry to produce them, and relatively more exploitation of mineral and energy resources. In other words, it would require relatively much more energy and capital. Society would become more complex and even more interdependent than it is today. The problem facing mankind, therefore, is very intricate and involves the closely interrelated factors of increase in (1) population, (2) food supply, (3) science, technology, and industry, and (4) natural resources.

Possible Accomplishments

The limits of human ingenuity in the development of science, technology, and industry are incalculable. Progress in these respects has been so rapid and spectacular in the present century, and so many things formerly considered to be beyond attainment have been accomplished, that almost nothing can be stated certainly to be impossible. If circumstances are favorable, means are available, and man is determined, he can learn to accomplish almost anything. He surely will be able for an indefinitely long period of future time to extract useful substances from the earth, the sea, and air in greater quantities and from leaner sources than he does at present, and he surely will be able to cultivate more land more intensively than he does now. Perhaps harvesting the sea will add importantly to the future food supply. Inevitably, however, the law of diminishing returns will apply to man's activities, and eventually a point will be reached beyond which greater effort will fail to produce satisfactory results. At that point agricultural and industrial expansion must slow and cease. For all practical purposes, the limits of human accomplishment will have been reached.

A high plateau of human satisfaction could be maintained almost indefinitely at that point. This is a possibility, however, only if the human population is stabilized and means are devised to prevent its genetic deterioration. Efforts at population control to be effective cannot be long delayed. Unfortunately powerful influences are active in the world today that seriously restrict positive action directed toward this end. International tensions siphon off for armaments and other unproductive uses the means and resources that might be devoted to the betterment of backward regions for rapidly increasing human needs. Political instability and unsound economic practices in many nations also delay the technologic progress that is so much desired.

Possibility of Failure

The human race may miss what seems to be its only chance to escape eventual disaster. If population increase is not checked in time, the high plateau of accomplishment may be only a brief episode to be looked back upon perhaps as the golden age of human attainment. Mounting population pressure will be followed by increasingly widespread want. Hunger will return as an ever-present irritant. The instinct for self-preservation will range class against class and region against region, with the more fortunate attempting to hold whatever advantages they have. Civil strife probably will grow, and if worse comes to worst the human race is likely to begin a long descending spiral, reproducing in reverse the history of its rise from barbarism. The human population will decline as food supplies shrink. Eventually science and technology may be forgotten and man may revert to the status of an intelligent savage, dependent wholly on his natural environment for survival. Natural selection again will direct his evolution in older adaptive ways. No matter what he may become, however, the rebirth of an advanced culture comparable to the present one will be impossible, because the necessary easily obtained natural resources will have been thoroughly depleted.

Hope for the Future

If all goes well, however, and man uses his wisdom and abilities in constructive ways, the human race has a long future to look forward to, a future safe from want. Evolution of the human species will progress slowly in ways that cannot be foretold, but men of the distant future will not be inferior to what they are today. Man can direct his future if he has the wisdom and the will.

BIBLIOGRAPHY

H. J. Barnett and **C. Morse (1963):** "Scarcity and Growth: The Economics of Natural Resource Availability," The Johns Hopkins University Press, Baltimore.
The authors argue that technologic advances will compensate for increasing natural resource scarcity.

G. Beadle and **M. Beadle (1966):** "The Language of Life," Doubleday & Company, Inc., Garden City, N.Y.
See chap. 20, pp. 219–238, for a geneticist's assessment of the future of the human race.

D. J. Bogue (1967): The Demographic Moment of Truth, *Chicago Today,* vol. 4, no. 1, pp. 37–41.
Recent developments in the control of human population are optimistically reviewed.

J. Bonar (1924): Malthus and His Work, George Allen & Unwin, Ltd., London.
This is an interesting account of Malthus, who first called attention to problems of resources and an increasing human population.

H. Brown, J. Bonner, and **J. Weir (1957):** "The Next Hundred Years," Viking Press, Inc., New York.
The earth possesses ample material resources to provide adequately for a much larger population than exists today for a very long period of time.

C. G. Darwin (1958): "The Problem of World Population," Rede Lecture, Cambridge University Press, New York.
Rapid expansion of the human population necessitates a pessimistic appraisal of man's future.

T. Dobzhansky (1962): "Mankind Evolving," Yale University Press, New Haven, Conn.
This book presents a broad consideration of human evolution, both biologic and cultural. Chapters 11 and 12, pp. 287–348, are especially interesting.

J. B. S. Haldane (1947): Human Evolution: Past and Future, *Atlantic Monthly,* vol. 179, no. 3, pp. 45–51; reprinted, **1949,** in "Genetics, Paleontology, and Evolution," G. L. Jepsen, E. Mayr, and G. G. Simpson (eds.), Princeton University Press, Princeton, N.J., pp. 405–418.
The prospects for the genetic future of the human race are assessed.

B. S. Kraus (1964): "The Basis of Human Evolution," Harper & Row, Publishers, Incorporated, New York.
See chaps. 8 to 11, pp. 267–360, for a presumed history of cultural development and assessment of possible human deterioration.

R. L. Meier (1956): "Science and Economic Development: New Patterns of Living," The Technology Press of the Massachusetts Institute of Technology, Cambridge, Mass., and John Wiley & Sons, Inc., New York.
The earth probably can support 50 billion people at a comfortable standard of living, but present knowledge is inadequate to attain such a level for all.

M. F. A. Montagu (1962): "Culture and the Evolution of Man," Oxford University Press, Fair Lawn, N.J.
This is a collection of 19 articles, mostly reprinted from their original sources, emphasizing the part that culture has played in human evolution.

M. Roberts (1951): "The Estate of Man," Faber & Faber, Ltd., London.
This well-reasoned little book presents a depressing picture of increasing population, decreasing natural resources, and declining human intelligence.

K. Sax (1956): "The Population Explosion," Foreign Policy Association, Headline Series, no. 120.
Birth control is essential for future human well-being; even with birth control depletion of mineral resources will raise grave problems.

L. D. Stamp (1960): "Our Developing World," Faber & Faber, Ltd., London.

The optimistic estimate is made that a world population four to five times the present one can be fed by expanding agricultural land and improving agricultural techniques.

S. Tax (ed.) **(1960):** "Evolution after Darwin," vol. 2, "The Evolution of Man," The University of Chicago Press, Chicago.

This symposium of 22 papers includes several which deal with cultural evolution.

W. L. Thomas, Jr. (ed.) **(1956):** "Man's Role in Changing the Face of the Earth," The University of Chicago Press, Chicago.

This symposium of more than 75 papers, discussions, and summaries traces the history of man's activities, their results, and the possible future effects with respect to the welfare of the human race.

R. Thomlinson (1965): "Population Dynamics: Causes and Consequences of World Demographic Change," Random House, Inc., New York.

This is a broad exposition of population facts and theories.

W. S. Thompson (1965): "Population Problems," 5th ed., McGraw-Hill Book Company, New York.

The author denies that future genetic degeneration of the human race is probable; see particularly chap. 21, pp. 417–445.

W. Vogt (1960): "People! Challenge to Survival," William Sloane Associates, New York.

If radical changes are not soon made, the human race is headed toward disaster.

L. A. White (1959): "The Evolution of Culture," McGraw-Hill Book Company, New York.

Human culture, founded on the ability to communicate, has surged forward whenever a new source of energy has been exploited.

INDEX

INDEX

Italics indicate references to figures and tables. Asterisks identify explanatory references or definitions of terms. Boldface references are those which include the principal considerations of groups of organisms or their features.

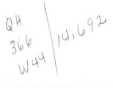